Osseous Projectile Weaponry

Vertebrate Paleobiology and Paleoanthropology Series

Edited by

Eric Delson
Vertebrate Paleontology, American Museum of Natural History
New York, NY 10024,USA
delson@amnh.org

Eric J. Sargis
Anthropology, Yale University
New Haven, CT 06520,USA
eric.sargis@yale.edu

Focal topics for volumes in the series will include systematic paleontology of all vertebrates (from agnathans to humans), phylogeny reconstruction, functional morphology, Paleolithic archaeology, taphonomy, geochronology, historical biogeography, and biostratigraphy. Other fields (e.g., paleoclimatology, paleoecology, ancient DNA, total organismal community structure) may be considered if the volume theme emphasizes paleobiology (or archaeology). Fields such as modeling of physical processes, genetic methodology, nonvertebrates or neontology are out of our scope.

Volumes in the series may either be monographic treatments (including unpublished but fully revised dissertations) or edited collections, especially those focusing on problem-oriented issues, with multidisciplinary coverage where possible.

More information about this series at http://www.springer.com/series/6978

Osseous Projectile Weaponry

Towards an Understanding of Pleistocene Cultural Variability

Edited by

Michelle C. Langley

Institute of Archaeology, University of Oxford, Oxford, UK

Archaeology and Natural History, College of Asia and the Pacific, Australian National University, Canberra, ACT, Australia

Springer

Editor
Michelle C. Langley
Institute of Archaeology
University of Oxford
Oxford, UK

Archaeology and Natural History
College of Asia and the Pacific
Australian National University
Canberra, ACT, Australia

ISSN 1877-9077 ISSN 1877-9085 (electronic)
Vertebrate Paleobiology and Paleoanthropology
ISBN 978-94-024-0897-3 ISBN 978-94-024-0899-7 (eBook)
DOI 10.1007/978-94-024-0899-7

Library of Congress Control Number: 2016950859

Cover illustration: View from inside Vindija Cave, Croatia (Photograph: I. Karavanić), with overlay of Magdalenian antler projectile technology (single bevel based point, baguette demi-ronde, and double bevel based point, all from Isturitz, France), and tracings of Magdalenian parietal images of projectile technology

Printed on acid-free paper

This Springer imprint is published by Springer Nature
The registered company is Springer Science+Business Media B.V.
The registered company address is: Van Godewijckstraat 30, 3311 GX Dordrecht, The Netherlands

Foreword: Invention, Innovation, and Creative Imagination

Originally: An Invention

Hunted prey can be grouped into either aggressive and fierce or timid and cautious animals. To avoid dangerous hand-to-hand combat (with the former) or to catch the animals before they vanish from sight (the latter), hunters have created weapons able to mortally wound animals from a safe distance. This concept of risk reduction, developed alongside increasing efficiency of projectile weaponry in various countries around the world during prehistoric times, constitutes what A. Leroi-Gourhan termed a "tendance" ("general trend") (Leroi-Gourhan 1971). The contributions published in this book demonstrate that these projectile hunting inventions are indeed a worldwide phenomenon.

The primeval invention must have been the spear or javelin made from a single piece of wood, the idea for which must have "sprung" into the hunters' mind since no reference existed concerning a previous device or technique. As noted by Voltaire, "the genius of invention opens a way to where nobody walked before" (as quoted in Héritier [2001]). This first projectile weapon was likely utilized throughout the world; however, only exceptional preservation conditions have permitted the survival to the present day of a very small number of examples. The eight wooden spears from Schöningen (Germany), aged 300 ka, are such an example and are among the oldest identified artifacts of this kind (Conard 2005:302).

Later: Multiple Innovations

"Whereas an invention is the discovery or the recognition of a new process, object, device, or technique, an innovation is the adoption, application, or utilization of a newly founded process, object, device" (Knecht 1991:116). Later, hunters contrived new weapons which included a point tied to a wooden shaft and manufactured in a hard material such as bone, ivory, stone, or in some tropical environments, hardwood. The purpose of this technological alteration was to make the extremity of the weapon more solid and acute, causing the weapon to penetrate deeper and result in a significantly more lethal injury to the prey. This change can be considered as an improvement or an "innovation."

Overall the morphology, volume, and weight of the points fastened onto shaft extremities depended on several variables: the environmental milieu, the system of propulsion utilized, and the technical traditions practiced. In particular, environmental resources determine the raw materials able to be selected for use in weapon construction. Wood was probably the only material used for the fabrication of shafts, since only tree branches split along the grain can yield the required long straight rods. More recent evidence (i.e., Mesolithic, Neolithic) suggests that pine or yew wood may have been used during the European Paleolithic, but other vegetal species would have been selected according to climates and landscapes. The use of driftwood was also certainly not neglected, as it has been shown that this resource was utilized in Arctic countries (where wood is scarce) for millennia. For instance, in several remarkably preserved archaeological Thule sites, elements of drift conifers (*Picea* sp.) were selected for making weapon shafts along with other objects (Alix 2007). In other geographical areas, ethnographic records provide additional information regarding the use of wood for projectile weapon components. For example, in the archipelagoes of Austral America, the Alakaluf still used branches of *canelo* (*Drimys winteri*) for the shaft of their harpoons at the beginning of last century (Emperaire 2003:329). In the relatively stable environment of this latter area, we might assume that their ancestors did the same 6000 years ago.

The hard animal materials commonly adapted to the manufacture of projectile points are ivory, antler, and bone. Their natural elongate shapes and thick cross sections permit blanks suited to point morphology. Shell elements, which can be very breakable, seem less propitious. Apart from the raw materials supplied through hunting, it has been found that ivory tusks were also recovered from carcasses of animals which died naturally, while antlers that were shed seasonally by the animals were collected from the landscape. The choice of point raw material depended on the animal resources available in each region, with mammoth ivory and reindeer antler constituting choice materials for the European Paleolithic, in addition to antler from *Cervus elaphus* (red deer) in Mediterranean regions where reindeer (*Rangifer tarandus*) did not venture.

The specific origin of the bones utilized in projectile point manufacture in Eurasia is less well documented (apart from mammoth ribs, which are easily recognizable), owing to the complete modification of the point blanks while in manufacture, but in a few cases horse or reindeer bones have been able to be identified. Recently, however, a few projectile points and foreshafts made of whale bone were identified in the French Pyrenees and central German Rhineland, confirming a Paleolithic relationship with the ocean shore (Pétillon 2009; Langley and Street 2013). Ethnographically, whale bones were preferred for manufacturing barbed points and harpoons in the archipelagoes of South America, though some of these items were made from huemul (*Hippocamelus bisulcus*; a small cervid) antler. Apart from possible points in mastodon (*Mammut americanum*) ivory in North America, or of walrus (*Odobenus rosmarus*) and narwhal (*Monodon monoceros*) in Arctic areas, bone and antler are generally the most frequent materials used for projectile points both ethnographically and prehistorically.

The volume and the weight of spearheads should normally be related to the propulsion system, either directly propelled by the strength of the muscles (hand-thrown) or by a range of implements intended for increasing efficiency, such as spearthrowers, bows, and blowpipes ("sarbacanes"). These various devices, which themselves constitute other significant inventions, are relatively common around the globe and together represent another great "tendance" for hunter-gatherer societies. Though very few prehistoric artifacts have been preserved, ethnographic specimens show a precise correlation between the propulsion system, the weight of the head point, the length and diameter of the shaft, and the center of gravity. Since prehistoric osseous projectile points have not preserved their original wooden shafts, numerous experiments have been conducted in order to identify correlations between the morphometric criteria of flint or osseous points and their corresponding propulsion system. In spite of the numerous trials, it appears that the diagnostic correlations remain blurred (e.g., Van Buren 1974; Rozoy 1992; Cattelain 1997; Pétillon 2009).

The existence of Paleolithic spearthrower heads manufactured from reindeer antler confirms the use of this implement from at least the Solutrean until the Magdalenian in Europe; however, their scarcity suggests that many other specimens made from wood may have also existed, and their use was perhaps more extended in time. Moreover, the morphometric standardization of many Gravettian antler points, extracted from blanks by the groove and splinter technique, could be consistent with an early use of the bow (Goutas 2016). And, given the morphometric variability of point types during the Paleolithic, most specialists assume that the two propulsion systems may have coexisted or have been reinvented several times during this period. As far as the very small osseous points are concerned, if they were not inserted in composite weapons, they could only be delivered by a vegetal blowpipe, as is the case in Southeast Asian or Amazonian forests. Blowpipes made of bird bones may have been used in other countries.

Creative Imagination

The technical response of hunter-gatherers to the need to kill animals for subsistence is at the same time universal in its main features and diversified in its execution. The technical determinism—Leroi-Gourhan's "tendance"—was not only confronted with the environmental or "external milieu" (availability of raw materials) but also with the "technical inner milieu" in which previously existing mental traditions influenced the innovation processes. The conception of new weapons certainly included preexisting elements specific to each considered community (Leroi-Gourhan 1973). Thus, the evolution of the European Paleolithic osseous projectile points constitutes an excellent field of investigation given the climatic changes, the relative stability of the animal stock, and the well-known succession of techno-complexes along a long span of time. In parallel to the continuing existence of the long and robust ivory spears which were probably hand-delivered, a wide range of points with a great variety of technological designs appear. The adaptations and improvements, with a possible number of mistaken attempts, have concerned various hafting systems, the insertion of flint bladelets into osseous shafts, the carving of barbs on the shaft itself, the probable coupling of carved barbed and flint inserts into the same point (Julien 1999), the innovation of mobile heads for uniserial and biserial harpoons, and, of course, the invention of throwing implements. In the course of these 30 kyr, everything was probably experimented by European hunter-gatherers except the socket toggle harpoon heads which are found later in the North Pacific regions. In other areas, simple forms predominated, but remarkable technical convergences can be seen between barbed points and the harpoons of the Late Paleolithic, Azilian, and Mesolithic and those of the Pacific Coast of America down to the southernmost extremity of the continent (Christensen et al. 2016). Hypotheses for these patterns are still debated: was it a diffusion process or a simple technical convergence? The very early invention of the Katanga barbed points (~90–60 kyr) in Africa (Backwell and d'Errico 2016) proves that human genius always created specific solutions to predation problems. But if this barbed device was afterward forgotten during tens of millennia, it is probably because the following human groups did not find any interest in adopting it. "It is not enough for an innovation to be possible for getting materialized […], it must above all be thinkable, i.e., accepted in the mind of the people to whom it is proposed…" (Héritier 2001:7).

CNRS – ArScan Équipe Ethnologie Préhistorique
Nanterre, Paris, France
michele.julien@mae.u-paris10.fr

Michèle Julien

References

Alix, C. (2007). Ethnoarchéologie de la production des objets en bois dans l'Arctique nord-américain. In S. Beyries & V. Vaté (Eds.), *Les civilisations du Renne d'hier et d'aujourd'hui. Approches ethnohistoriques, archéologiques et anthropologiques, XVIIe rencontres internationales d'archéologie et d'histoire d'Antibes* (pp. 377–392). Antibes, France: Editions ADPCA.

Backwell, L. R., & d'Errico, F. (2016). Osseous projectile weaponry from early to late Middle Stone Age Africa. In M. C. Langley (Ed.), *Osseous projectile weaponry: Towards an understanding of Pleistocene cultural variability* (pp. 15–29). Dordrecht: Springer.

Cattelain, P. (1997). Hunting during the upper Paleolithic: Bow, spearthrower, or both? In H. Knecht (Ed.), *Projectile technology*, (pp. 213–240). New York: Plenum Press.

Christensen, M., Legoupil, D., & Pétillon, J.-M. (2016). Hunter-gatherers of the old and new world: Morphological and functional comparisons of osseous projectile points. In M. C. Langley (Ed.), *Osseous projectile weaponry: Towards an understanding of Pleistocene cultural variability* (pp. 237–252). Dordrecht: Springer.

Conard, N. (2005). An overview of the patterns of behavioural change in Africa and Eurasia during the Middle and Late Pleistocene. In F. d'Errico & L. Backwell (Eds.), *From tools to symbols. From early hominids to modern humans*, (pp. 294–325). Johannesburg, South Africa: Witwatersrand University Press.

Emperaire, J. (2003). *Les Nomades de la Mer.* Paris: Gallimard.

Goutas, N. (2016). Gravettian projectile points: Considerations about the evolution of osseous hunting weapons in France. In M. C. Langley (Ed.), *Osseous projectile weaponry: Towards an understanding of Pleistocene cultural variability* (pp. 89–107). Dordrecht: Springer.Héritier, F. (2001). Innovation, invention, découverte. Discours introductif au XIIe Festival International de Géographie. *La Géographie, 1503*, 5–22.

Julien, M. (1999). Une tendance créatrice au Magdalénien: à propos de stries d'adhérence sur quelques harpons. In M. Julien, A. Averbouh, D. Ramseyer, C. Bellier, D. Buisson, P. Cattelain, et al. (Eds.), *Préhistoire d'os. Recueil d'études sur l'industrie osseuse préhistorique offert à H. Camps-Fabrer* (pp. 133–142). Aix-en-Provence, France: Publications de l'Université de Provence.

Knecht, H. (1991). The role of innovation in changing Early Upper Palaeolithic organic projectile technologies. *Techniques et Culture, 17–18*, 115–144.

Langley, M. C., & Street, M. (2013). Long range inland-coastal networks during the late Magdalenian: Evidence for individual acquisition of marine resources at Andernach-Martinsberg, German Central Rhineland. *Journal of Human Evolution, 64*, 457–465.

Leroi-Gourhan, A. (1971). *L'Homme et la Matière*. Paris: Editions Albin Michel.

Leroi-Gourhan, A. (1973). *Milieu et Techniques*. Paris: Editions Albin Michel.

Pétillon, J. -M. (2009). Bilan d'une réalisation: tir expérimental d'armatures de sagaies composites: Premiers résultats. Résumés des exposés au séminaire du 18/03/09. In B. Valentin (Ed.), *Paléolithique Final et Mésolithique dans le Bassin Parisien et ses Marges, habitats, sociétés et environnements* (pp. 23–66). Projet Collectif de Recherche, Rapport d'Activités pour 2009. UMR 7041/SRA d'Ile-de-France, Nanterre/Saint-Denis. http://hdl.handle.net/2332/1610. Accessed October 2013.

Rozoy, J. -G. (1992). Le propulseur et l'arc chez les chasseurs préhistoriques. Techniques et démographie comparée. *Paléo, 4*, 175–192.

Van Buren, G. E. (1974). *Arrowheads and Projectile Points with a classification guide for Lithic Artifacts*. Garden Grove, CA: Arrowhead Publishing Company.

Preface

As the reader will discover, weapon components constructed from hard animal materials (bone, antler, ivory, horn, and shell) were ubiquitous in ancient hunter-gatherer-fisher communities across the globe. Furthermore, these tools were not only used to obtain food and raw materials essential to everyday life but often also played a role in the social landscape of the society which produced them. Additionally, projectile points (including those manufactured from antler, bone, and ivory) have long served as indicators for identifying archaeological cultures, and so, these artifacts have come to play a central role in archaeological research.

While a tremendous volume of literature concerning Pleistocene projectile technology has been generated over the course of the past century, at the beginning of 2011, a synthesis of those components manufactured from osseous materials was lacking. This situation existed in sharp contrast to their lithic counterparts, of which a great many reviews and syntheses were available to the intrepid PhD student beginning her study of Upper Paleolithic weaponry. Being new to this area of archaeological research, the lack of a good "first port of call" made grounding oneself in the idioms and idiosyncrasies of "bone" projectile point traditions an exhausting experience. On thinking that a volume which provided an outline of the various hard animal material weaponry thus far recovered from the Pleistocene archaeological record would be very useful for both students and advanced researchers alike, and with the encouragement of colleagues, the idea for this volume was born.

The importance and relevance of this volume was reinforced during its development, which coincided with a period in modern human origins research involving intense discussion concerning the viability of the search for "behavioral modernity" and a proposed redirection in research focus to the study of early "cultural variability." With archaeologists now turning their focus onto exploring the range of adaptations practiced at various times in the deep past, discussions rarely mentioned the vast wealth of osseous weapons technology in any detail despite their enormous potential to contribute to a great many aspects of the issues being debated. A glimpse at the substantial datasets presented and range of archaeological issues discussed in this volume demonstrates this point entirely.

The temporal span covered in the volume is as expansive as the spatial territory covered. Osseous projectile technology from the five inhabited continents is presented herein, providing the reader with a comprehensive outline of the state of the art of Pleistocene osseous projectile weapons technology from the archaeological perspective.

The book is divided into five parts. The first part offers an introduction to osseous projectile weaponry, via, first, a discussion of the importance of this technology for our understanding of past cultural variability and, second, a brief overview of Late Pleistocene osseous projectile weaponry. Analyses of archaeological projectile technologies form the bases of the chapters which follow. These chapters make up the remaining four parts of the volume, organized by geographical region: Africa, Europe, Southeast Asia and Australia, and the Americas.

While each chapter stands on its own, this volume has been organized to provide a basic background to each of the archaeological osseous techno-complexes researchers will encounter in their study of Late Pleistocene cultural variability: moving from the earliest examples in the archaeological record (in Africa) to modern human population movement across Europe and Asia and their eventual appearance in North America. Understandings of the studied weaponry are provided from various perspectives, including everything from identification of micro-traces to ethnographic analogy. It is hoped that the data presented, the methods used, and the conclusions drawn by these studies will enable nonspecialist researchers to initiate the integration of osseous technologies into the many vibrant aspects of early hunter-gatherer-fisher life currently being debated in the literature.

It has been a pleasure working with all of the people who have contributed to this volume. Special thanks go to each of the contributors for sharing their research and, most of all, for their enthusiasm for this endeavor. I am also grateful to the many colleagues who reviewed each chapter manuscript. Thanks also go to the series editors, Eric Delson and Eric Sargis, for their advice, patience, and availability throughout the generation of this volume.

Michelle C. Langley
Canberra, ACT, Australia

Contents

1 **Late Pleistocene Osseous Projectile Technology and Cultural Variability** 1
Michelle C. Langley

Part I Africa

2 **Osseous Projectile Weaponry from Early to Late Middle Stone Age Africa** 15
Lucinda Backwell and Francesco d'Errico

3 **Bone Point Functional Diversity: A Cautionary Tale from Southern Africa** 31
Justin Bradfield

Part II Europe

4 **Early Upper Paleolithic Osseous Points from Croatia** 43
Ivor Karavanić

5 **Spanish Aurignacian Projectile Points: An Example of the First European Paleolithic Hunting Weapons in Osseous Materials** 55
José-Miguel Tejero

6 **Projectile Weaponry from the Aurignacian to the Gravettian of the Swabian Jura (Southwest Germany): Raw Materials, Manufacturing and Typology** 71
Sibylle Wolf, Susanne C. Münzel, Krista Dotzel, Martina M. Barth, and Nicholas J. Conard

7 **Gravettian Projectile Points: Considerations About the Evolution of Osseous Hunting Weapons in France** 89
Nejma Goutas

8 **Upper Paleolithic Bone and Antler Projectiles in the Spanish Mediterranean Region: The Magdalenian Period** 109
Valentín Villaverde, J. Emili Aura Tortosa, María Borao, and Didac Roman

9 **A Review of the Osseous Projectile Points from the Upper Paleolithic of Portugal** 131
Marina Almeida Évora

10 Diversity and Evolution of Osseous Hunting Equipment During the Magdalenian (21,000–14,000 cal BP) 143
Michelle C. Langley, Jean-Marc Pétillon, and Marianne Christensen

11 Osseous Projectile Points in the Magdalenian: 'True' Points Versus Pointed Waste-Products 161
François-Xavier Chauvière

Part III Southeast Asia & Australia

12 Bone Technology from Late Pleistocene Caves and Rockshelters of Sri Lanka 173
Nimal Perera, Patrick Roberts, and Michael Petraglia

13 The Walandawe Tradition from Southeast Sulawesi and Osseous Artifact Traditions in Island Southeast Asia 189
Ken Aplin, Sue O'Connor, David Bulbeck, Philip J. Piper, Ben Marwick, Emma St Pierre, and Fadhila Aziz

14 Bone Projectile Points in Prehistoric Australia: Evidence from Archaeologically Recovered Implements, Ethnography, and Rock Art 209
Harry Allen, Michelle C. Langley, and Paul S. C. Taçon

Part IV The Americas

15 A Review of Late Pleistocene North American Bone and Ivory Tools 221
Michael J. O'Brien, R. Lee Lyman, Briggs Buchanan, and Mark Collard

16 Hunter-Gatherers of the Old and New Worlds: Morphological and Functional Comparisons of Osseous Projectile Points 237
Marianne Christensen, Dominique Legoupil, and Jean-Marc Pétillon

Index 253

Contributors

Harry Allen Department of Anthropology, University of Auckland, Auckland, New Zealand

Ken Aplin Department of Archaeology and Natural History, College of Asia and the Pacific, The Australian National University, Canberra, ACT, Australia

J. Emili Aura Departament de Prehistòria i Arqueologia, Universitat de València, València, Spain

Fadhila Aziz Department of Prehistory, National Center for Archaeological Research and Development, Jakarta, Indonesia

Lucinda Backwell Evolutionary Studies Institute and DST/NRF Centre of Excellence in Palaeosciences and School of Geosciences, University of the Witwatersrand, Johannesburg, South Africa

Martina M. Barth Kultusministerium des Landes Sachsen-Anhalt, Magdeburg, Germany

María Borao Departament de Prehistòria i Arqueologia, Universitat de València, València, Spain

Justin Bradfield Evolutionary Studies Institute, University of the Witwatersrand, Johannesburg, South Africa

Briggs Buchanan Department of Anthropology, University of Tulsa, Tulsa, OK, USA

David Bulbeck Department of Archaeology and Natural History, College of Asia and the Pacific, The Australian National University, Canberra, ACT, Australia

François-Xavier Chauvière Office du Patrimoine et de l'Archéologie de Neuchâtel, Section Archéologie, Laténium, Hauterive, Switzerland

Marianne Christensen Institut d'Art et d'Archéologie, Université de Paris 1 Panthéon-Sorbonne, UMR 7041 – ArScAn, Équipe Ethnologie Préhistorique, Paris, France

Mark Collard Human Evolutionary Studies Program and Department of Archaeology, Simon Fraser University, Burnaby, BC, Canada

Nicholas J. Conard Institute of Pre- & Protohistory and Medieval Archaeology, Early Prehistory and Quaternary Ecology, University of Tübingen, Tübingen, Germany

Senckenberg Center for Human Evolution and Palaeoenvironment, Tübingen, Germany

Francesco d'Errico CNRS UMR 5199, PACEA, Université de Bordeaux, Pessac Cedex, France

Evolutionary Studies Institute and DST/NRF Centre of Excellence in Palaeosciences and School of Geosciences, University of the Witwatersrand, Johannesburg, South Africa

Krista Dotzel Institute of Pre- & Protohistory and Medieval Archaeology, Early Prehistory and Quaternary Ecology, University of Tübingen, Tübingen, Germany

Marina Almeida Évora ICArEHB – Interdisciplinary Center for Archaeology and Evolution of Human Behaviour, FCHS da Universidade do Algarve, Faro, Portugal

Nejma Goutas CNRS, ArScan, UMR-7041, Équipe Ethnologie Préhistorique, Nanterre, Cedex, France

MAE – Maison René Ginouvès – Archéologie and Ethnologie, Nanterre, Cedex, France

Ivor Karavanić Faculty of Humanities and Social Sciences, Department of Archaeology, University of Zagreb, Zagreb, Croatia

Michelle C. Langley Institute of Archaeology, University of Oxford, Oxford, UK

Archaeology and Natural History, College of Asia and the Pacific, Australian National University, Canberra, ACT, Australia

Dominique Legoupil CNRS – UMR 7041 – ArScAn, Équipe Ethnologie Préhistorique, Maison de l'Archéologie et de l'Ethnologie, Nanterre, France

R. Lee Lyman Department of Anthropology, University of Missouri, Columbia, MO, USA

Ben Marwick Department of Anthropology, University of Washington, Seattle, WA, USA

Susanne C. Münzel Institute for Archaeological Sciences, University of Tübingen, Tübingen, Archaeozoology, Germany

Michael J. O'Brien Department of Anthropology, University of Missouri, Columbia, MO, USA

Sue O'Connor Department of Archaeology and Natural History, College of Asia and the Pacific, The Australian National University, Canberra, ACT, Australia

Nimal Perera Postgraduate Institute of Archaeology, University of Kelaniya, Colombo, Sri Lanka

Jean-Marc Pétillon CNRS – UMR 5608, Laboratoire TRACES, Université Toulouse Jean-Jaurès, Toulouse, France

Michael Petraglia Research Laboratory for Archaeology and the History of Art, School of Archaeology, University of Oxford, Oxford, UK

Emma St Pierre School of Social Science, The University of Queensland, Brisbane, QLD, Australia

Philip J. Piper School of Archaeology and Anthropology, Australian National University, Canberra, Australia

Patrick Roberts Research Laboratory for Archaeology and the History of Art, School of Archaeology, University of Oxford, Oxford, UK

Didac Roman SERP (Seminari d'Estudis i Recerques Prehistòriques). Secció de Prehistòria i Arqueologia, Universitat de Barcelona, València, Spain

Université de Toulouse-Le Mirail, TRACES-UMR 5608, Toulouse, France

Paul S.C. Taçon PERAHU, School of Humanities, Languages and Social Science, Griffith University, Gold Coast, QLD, Australia

José-Miguel Tejero CNRS, ArScAn, Équipe Ethnologie Préhistorique, Paris, France

SERP Barcelona University, Barcelona, Spain

Valentín Villaverde Departament de Prehistòria i Arqueologia, Universitat de València, València, Spain

Sibylle Wolf Institute of Pre- & Protohistory and Medieval Archaeology, Early Prehistory and Quaternary Ecology, University of Tübingen, Tübingen, Germany

Senckenberg Center for Human Evolution and Palaeoenvironment, Tübingen, Germany

Chapter 1
Late Pleistocene Osseous Projectile Technology and Cultural Variability

Michelle C. Langley

Abstract Modern human evolution and the development of cultural complexity and variability during the Pleistocene have long been central issues in archaeology. This chapter situates the study of osseous projectile weaponry in this wider context of archaeological research, before outlining the challenges that this field currently faces. A brief overview of the evidence for Pleistocene osseous projectile weaponry is then presented in order to demonstrate the temporal and spatial breadth of these material culture items, as well as their ability to contribute to wider anthropological debates about human uniqueness and cultural variability.

Keywords Hunting • Fighting • Spear • Harpoon • Fishhook • Bow-and-arrow • Spearthrower • Boomerang

Introduction

By the end of the Pleistocene (10,000 radiocarbon years BP or 11,700 cal BP), modern humans (*Homo sapiens sapiens*) had already spent tens of thousands of years experimenting with and perfecting an astounding range of projectile weaponry manufactured from hard animal materials (bone, antler, ivory, horn, shell). Projectile weaponry, being launched weapons used in both hunting and warfare, are a technical solution ensuring the capture of vital nutritional and raw material resources used in various aspects of hunter-gatherer life. Projectile point tips, foreshafts, fishhooks, boomerangs, spearthrowers, and bow components, not to mention the various tools used in the manufacture of these weapons (such as spear-straighteners) were all fashioned out of the most durable of organic materials available. The selection of these raw materials for making (arguably) the most important tools for day-to-day life, was the result of deliberate choices made by numerous temporally and spatially dispersed communities. These choices reflect an understanding of the physical properties of osseous materials that render them supremely suitable for use as projectile weaponry.

Despite these factors, however, weaponry made from osseous (bone, antler, ivory) materials have consistently received less attention in the archaeological literature than other artefact classes, particularly their lithic counterparts. This situation is exemplified in the recently published *Oxford Handbook of the Archaeology and Anthropology of Hunter-Gatherers* (Cummings et al. 2014) where both lithic and ceramic (among other) technologies comprise a chapter each, while mentions of bone and antler technology (not limited to weapons technology) mentioned through the entire volume could fit on one to two pages. Given that a survey of ethnographic hunter-gatherer societies found that 42.37 % employed projectile points manufactured from bone and/or antler (Waguespak et al. 2009), this appears to be a tremendous oversight.

While lithic technology is less susceptible to preservation biases and taphonomic filters than organic-based evidence, focus on this class of material culture alone results in a perspective that is too narrow a sub-set on which to construct robust frameworks of Pleistocene lifeways and cultural variability. Furthermore, as the overwhelming majority of archaeological remains directly associated with hunting are constructed from bone, antler, ivory, and shell elements, integration of this organic dataset is critical if we are to move towards more holistic understandings of technology, economy, and society during the Pleistocene epoch.

This volume aims to contribute to this endeavor through providing the academic community with a summary of the osseous projectile weaponry thus far recovered from Pleistocene contexts in Africa, Europe, Southeast Asia, Australia, and the Americas. While the inevitable constraints

M.C. Langley (✉)
Institute of Archaeology, University of Oxford, Oxford, UK

Archaeology and Natural History, College of Asia and the Pacific, Australian National University, Canberra, ACT, Australia
e-mail: michelle.langley@anu.edu.au

Michelle C. Langley (ed.), *Osseous Projectile Weaponry*, Vertebrate Paleobiology and Paleoanthropology,
DOI 10.1007/978-94-024-0899-7_1

of an edited volume prohibit a discussion of every assemblage or site from which these artefacts have been recovered, the chapters herein do reflect the wealth of information held and the potential of this enigmatic class of material culture. As the reader will see, these organic artefacts, so frequently overlooked, have a remarkable ability to contribute to current debates and narratives on the development of cultural variability in early human communities.

Archaeological Perspectives on Cultural Variability

Humans differ from our nearest primate relatives in our capacity for behavioral and cultural variability (Potts 1998). For almost every activity we undertake, people have found more than one way to achieve the same or similar outcome, resulting in the diverse cultural traditions enacted both in the past and the present. We do not know the antiquity of our species' current degree of behavioral and cultural variability, however, as Shea (2011:2) points out, there are only three possibilities: "(1) it evolved after our species' origin and is a characteristic of only some *H. sapiens*, (2) it evolved at the same time our species split from ancestral hominins and is a species-specific characteristic, and (3) it evolved before *H. sapiens*' origin and is a characteristic shared by more than one hominin species". Without entering into this hotly contested debate, it can be said that by 40,000–50,000 years ago a vast range of well developed, complex, and extremely diverse cultures are in evidence throughout Africa, Europe, Asia, and Australasia.

While exploring the extent of Pleistocene cultural variability is a topic which sees increasing attention from archaeologists, the understanding that cultural variability has an antiquity in the tens of thousands of years is a relatively new standpoint for the discipline. It was not until the nineteenth century, that anthropologists and archaeologists, prompted by increasing knowledge of hunter-gatherer (and other 'exotic' or even 'savage') peoples and an increasing appreciation for the antiquity of humankind, began to focus on pinpointing the developments which led to the observed contemporary diversity in cultural practices. Indeed, it was only after this time that it was recognized that most of human (pre)history was spent as hunter-gatherer-fishers—the technology of which we are concerned with in this volume. Consequently, the decades following this early phase of anthropological thought saw many of the founding researchers for both modern anthropology and archaeology undertaking extensive fieldwork in various parts of the world in order to better understand contemporary hunter-gatherer societies and how they might inform us about past ways of life (e.g., Kroeber 1925; Radcliffe-Brown 1931; Lowie 1935; Boas 1966; Lee and Devore 1968; Lévi-Strauss 1969; Binford 1978; Hayden 1979; Gould 1980).

While archaeologists now caution the overuse of ethnographic analogies for understanding Pleistocene lifeways (e.g., Wobst 1978; Hiscock 2008), it has been argued that ethnographic studies provide the only glimpse of technologies (in particular) with which most researchers have no first-hand experience (Dibble 1995). As O'Connell (1995) puts it, archaeologists have only two sources of information from which to explore past human behavior; (1) the artefacts and patterns within the archaeological record itself, and (2) knowledge of present day human behavior and the consequences of these behaviors for the materials they use and discard. Focusing on only the former allows for little more than mere description of artefacts along with their spatial and temporal distribution. Integrating the latter provides insights into the processes which led to the appearance of these patterns in the archaeological record, allowing the researcher to build stories around their data.

Today, determining how, when, and why ancient communities developed their regionally and/or temporally specific behavioral or technical solutions remains central to Pleistocene hunter-gatherer archaeology, and osseous technologies have been particularly valuable in tracking changes through time and across space. For example, they have a demonstrated ability to identify regional patterns (e.g., Julien 1982; Weniger 1992; Vanhaeren and d'Errico 2005; Pétillon 2013), characterize technical systems (e.g., Peyrony 1909; Cheynier 1958; Barandiaran 1967; Delporte et al. 1988), and infer social and cultural adaptations and practices (e.g., Conkey 1980; Bahn 1982; Jochim 1987).

Osseous hunting weaponry, in particular, is an excellent class of material culture for exploring cultural variability as not only do these artefacts have "widespread social, economic, political and symbolic import" (Wiessner 1983:272), they are highly visible to persons who are both intimately familiar with the individual carrying the implement as well as those encountered on the landscape during the course of subsistence activities (Tostevin 2007). This visibility makes them prime candidates for use in transmitting social messages to those in 'the middle distance' (Wobst 1977), and thus, are not always simple food getting implements. Additionally, while these tools are carried around the landscape by the *individual*, the object's mode of production and morphology was guided by its *community* of practice (Lave and Wenger 1991; Dobres and Hoffman 1994). Consequently, its form and use should conform to cultural ideals regarding manufacturing techniques, morphology, use, and discard. As communities generally inhabit and exploit a certain territory (of varying sizes depending on their residential mobility strategy, resource richness, etc), the 'style' of weapon carried by an individual can become indicative of that region and the people who occupy it. Because of the production and use of these material culture items within cultural constraints, and because of the tendency of weapons to be utilized in a certain way and in a certain (at least semi-) confined spatial area, hunting equipment has been utilized as a central element for investigating cultural variability in the deep past.

These factors are but some of the reasons that osseous projectile weaponry deserves more space in discussions on cultural variability in the deep past, and why the papers drawn together in this volume are so valuable for future research into this area.

Issues and Challenges in Osseous Projectile Weaponry Research

As most readers are aware, there is a mountain of literature regarding the place of lithic technologies in investigating human behavioral evolution, dispersal and interaction of Pleistocene human groups, and temporal changes in subsistence, technological, and social practices. While this inorganic evidence has a significant place in our exploration of these contexts, osseous technologies are no less deserving. Like lithic techno-complexes, osseous assemblages are found in the earliest Modern Human and Neanderthal sites, in a wide range of environmental contexts, were used to exploit a vast array of fauna, and required specialist knowledge of the properties inherent to the raw materials used in their manufacture. These assemblages also share a number of the same issues.

For example, as is the case for Australia (Langley et al. 2011), vast areas of Africa remain uninvestigated archaeologically (Yellen 1998), this latter continent also having a small number of post-200 ka sites properly excavated or dated using recent standards (Backwell and d'Errico 2016). Europe remains the most thoroughly explored, though issues of excavation and dating quality are present here also. For example, Évora (2016) comments on the differential attention archaeologists have given the Estremadura region of Portugal, to the detriment of Algarve. Consequently, we cannot be sure that identified spatial patterns represent past realities in osseous weaponry distribution and use, just as is the case for identified lithic patterning.

Another shared issue between lithic and osseous projectile weaponry is the use of a plethora of terms when describing and discussing the technology. Spears that are thrown by hand are commonly referred to as 'spears', 'darts' or '*sagaies*', with these latter two terms also used to describe lighter projectiles which were propelled with a spearthrower (commonly termed an '*atlatl*' in America and a '*woomera*' in Australia). A 'dart' can describe a weapon launched with a spearthrower, or a small, lightweight point propelled with the use of a blowgun. Perhaps the most debated term in Pleistocene osseous weaponry research in the last few years, however, is the use of the term 'harpoon' ('*harpon*'). Having been widely used to describe the uni- and bilaterally barbed antler points of the European Magdalenian (e.g., Lartet 1861; Julien 1982), whether these weapon tips functioned as a true mobile harpoon head—that is, as "a hunting weapon, thrust or thrown, whose tip is mobile and linked by a line to the shaft, to another object or to the user" (Weniger 1995:20)—continues to be debated (e.g., Lartet and Christy 1875; Piette 1895; Girod and Massénat 1900; Julien 1982; Weniger 2000; Pétillon 2008; Langley 2015). The uncertainty surrounding the true function/s of these distinctive European Paleolithic implements has resulted in researchers recently divorcing these artefacts from their traditional French name of '*harpon*' with all its functional connotations, in favor of the much less presumptuous term, 'barbed point' (Weniger 1995; Pétillon 2008; Langlais et al. 2012).

Another issue that osseous weapons share with their lithic counterparts is the difficulty in identifying the method by which they were launched towards their target. Comparison with ethnographic weapons from various cultural contexts have shown there is significant cross over in point size between those shot with a bow and those shot with a spearthrower, and then those shot with a spearthrower and those thrown by hand (Cattelain 1997). Consequently, direct evidence for launch type is often the only way to positively determine how the weapons were launched. Launch systems, however, are far rarer in the archaeological record than the points they were projecting, limiting data sources for both lithic and osseous point specialists alike.

Thus, while osseous projectile weaponry is more susceptible to taphonomic processes and differential preservation than lithic versions, they are nevertheless, recovered from the Pleistocene archaeological record in numbers great enough to warrant their systematic integration into narratives of Pleistocene cultural variability. The next section provides a brief outline of osseous projectile weaponry thus far recovered, in order to orientate the reader for the chapters that follow.

An Overview of Late Pleistocene Osseous Projectile Weaponry

While bone tools are reported for both South and East African early hominid sites dated to between 1.8 and 1 million years ago (Backwell and d'Errico 2005), and both bone and antler were similarly used as raw materials for implements in pre-400 ka contexts in Europe (Villa and d'Errico 2001), the first examples of osseous projectile weapons do not appear until significantly later in the archaeological record.

Within Africa, a bone artefact from El Mnasra (Morocco) which is interpreted as a broken spear point and is dated by optically stimulated luminescence (OSL) to between c. 107,000 and 106,000 years BP (Jacobs et al. 2012; El Hajraoui and Debénath 2012), constitutes the oldest probable osseous projectile identified. As Backwell and d'Errico (2016) note, if the function and age of this (and associated) artefacts are confirmed by new discoveries, the Aterian (a techno-complex of North Africa) may represent a cultural adaptation in which osseous points for projectile use were systematically created for the first time (Campmas et al. 2015). Dating to some 10,000–20,000 years later, the barbed and unbarbed bone points of Katanda (Democratic Republic of the Congo) represent the oldest known hafted points within Africa. The layer from which these artefacts originate has been attributed an age of c. 90,000 years BP (Brooks et al. 1995; Yellen et al. 1995; Yellen 1998; Brooks et al. 2006). Although considered by some as possibly younger (Ambrose 1998; Klein 1999, 2000, 2008), more recent dating of the site confirms an age in excess of 60,000–70,000 years, and certainly no younger than 50,000 years BP (Feathers and Migliorini 2001).

After this precocious appearance of the Katanda barbed technology, barbed points do not reappear in sub-Saharan Africa until c. 20,000 years BP at Ishango (also in the Democratic Republic of the Congo) (de Heinzelin 1962; Brooks and Smith 1987), before becoming a common occurrence in Later Stone Age (LSA) sites throughout the region. These LSA deposits, usually associate barbed points with conditions of increased precipitation, higher water levels, fossil shorelines, aquatic resource exploitation, and commonly date to between 10,000 and 4000 years BP (e.g., Monod and Mauny 1957; Wendt 1966; Robbins 1974; Barthelme 1977; Phillipson 1977; Petit-Marie et al. 1983; Barthelme 1985; Yellen 1998). The southernmost African barbed bone points have been recovered from White Paintings Shelter (Botswana), with the oldest example at this site perhaps dating to the Late Pleistocene, and the youngest overlying a charcoal date of 2260 years BP (Robbins et al. 1994).

Interestingly, barbed bone points have not been recovered from either Middle Stone Age (MSA) or LSA deposits in Zimbabwe or South Africa, though a single fragmentary unilaterally barbed bone point associated with an Iberomaurusien industry (bracketed by radiocarbon dates of 10,800 years BP and 12,070 years BP) has been found at Taforalt (Morocco) (Camps 1974). Given the similarities of this particular artefact to those made and utilized by contemporaneous European peoples across the Straits of Gibraltar (see Villaverde et al. 2016), some researchers have speculated on the possibility that this particular artefact was inspired by the neighboring Magdalenian technology (e.g., Otte 1997; Straus 2001).

On the European continent, barbed points were manufactured from both terrestrial and marine bone, red deer and reindeer antler, along with mammoth ivory. These points are found within the well-studied Magdalenian techno-complex, with between 1500 and 2000 examples thus far recovered (Langley et al. 2016). The earliest well-dated (including direct dates on objects) examples of barbed points on this continent come from Fontalès, d'Espalunge, Isturitz, and Tito Bustillo—all of which date to around 16,000–15,500 cal BP (Tisnerat-Laborde et al. 1997; Szmidt et al. 2009; Pétillon et al. 2015), and continue to the end of the Late Magdalenian (c. 12,200 years cal BP) before being replaced by Azilian (an Epipaleolithic culture) barbed points in these same regions. While barbed points are found throughout the Magdalenian territory, regional differences in morphology have been noted (e.g., Julien 1982; Weniger 1992). In general, bilaterally barbed points dominate assemblages in the Périgord and Quercy regions, whereas unilaterally barbed points have a statistical and numerical dominance in areas both to the north and south of this area. Other noted regional differences in Magdalenian barbed point technology include the restriction of perforated bases to the northern Spanish coast (commonly known as the 'Cantabrian type'; Weniger 1987), and unilaterally barbed points to the Mediterranean coast of Spain (Villaverde and Roman 2005–2006) (Villaverde et al. 2016).

Barbed bone points are rare in the Natufian of the Levant, with the largest sample (consisting of only seven artefacts) recovered from Kebara Cave (Israel) and dated to around 11,000 years BP (Yellen 1998). As noted by Henry (1989:197–202), bone artefacts (including barbed points, bi-points, fishhooks, and gouges) become common in the Natufian (see Campana 1991 and Stordeur 1991 for an overview), perhaps because they represent higher investments in production than wooden equivalents (also see Hayden 2004). Here in the Levant, barbed points are traditionally associated with fishing, though Henry (1989) suggests they were more likely utilized in hunting terrestrial prey.

Stingray spines (probably the estuarine species *Pastinaachus sephen*/cowtail stringray) which display natural barbs are known to have been utilized to tip projectiles during the Terminal Pleistocene/early Holocene (stratigraphically associated to 8982–8645 cal BP [OxA-18358] and slightly before 9030–8650 cal BP [OxA-11864]) of Southeast Asia thanks to finds in the Niah Caves (Sarawak) (Rabett 2005; Barton et al. 2009; Rabett et al. 2013). Bone points (unbarbed and including bipoints) were also recovered from these same contexts, with several examples retaining evidence for hafting in the form of mastics and fibres (Barton et al. 2009). Historically, stingray spine armatures were employed as both spears and arrows and were used in hunting, fishing, and warfare in Southeast Asia (Barton et al. 2009), along with Australia (Davidson 1934; Allen et al. 2016), though examples dating to the Pleistocene are yet to be identified for this latter region.

While barbed projectile points are a fascinating aspect of osseous projectile weaponry, it is their unbarbed counterparts which demonstrate both greater antiquity and spatial distribution. In Africa, the aforementioned early examples from El Mnasara and Katanda are joined by small assemblages from the Still Bay layers of Blombos Cave dated to 75,000 years BP (d'Errico and Henshilwood 2007; Henshilwood et al. 2009), as well as the pre-Still Bay layers at Sibudu Cave dated to c. 72,000 years BP (Jacobs et al. 2008a, b; d'Errico et al. 2012; Backwell and d'Errico 2016). The morphological variability in the bone points from Blombos Cave, along with the morphology of the three most complete specimens and a proximally broken specimen, suggest they were more likely the tips of spears than arrows, with the later technology (bow-and-arrow) suggested to have been present from around 60,000 years BP in South Africa (Backwell et al. 2008; Lombard and Phillipson 2010; Villa et al. 2010; Lombard 2011). As will be outlined below, African bow-and-arrow technology was likely established during the MSA (Howiesons Poort), with current evidence suggesting that poison-tipped (bone) arrows were one of the key innovations of the LSA (Backwell et al. 2008). Previously thought to have been employed only after about 8000 years BP (Deacon 1976; Opperman 1987; Mitchell 2002), the use of floral-based poisons is now demonstrated to have an antiquity of at least 24,000 years BP owing to the recovery of a wooden poison applicator from Border Cave, South Africa (d'Errico et al. 2012).

Unbarbed points are well-represented throughout Western and Central Europe, becoming more numerous and diverse in design throughout the European Upper Paleolithic, prompting Straus (1993:83) to dub these artefacts as "the most dynamic component of the Upper Paleolithic technologies". Aurignacian (starting around 43,000 cal. BP the chronology of the Aurignacian is highly contested: see Zilhão and d'Errico 1999; Higham et al. 2012) hunting kits contained split-based points and simple-based points (Karavanić 2016; Tejero 2016). Gravettian (c. 29,000–22,000 years BP) assemblages comprise simple-based points, bevelled-based points, Isturitz points, simple-based points with mesial incisions (Goutas 2016), and Magdalenian (20,700–14,000 cal BP) kits exhibit a huge range of osseous and composite osseous-lithic types (Chauviére 2016; Langley et al. 2016; Villaverde et al. 2016).

Unbarbed points of various sizes and proximal (hafting) morphologies are found in South Asia (Perera et al. 2016), Island Southeast Asia (Aplin et al. 2016), Australia (Allen et al. 2016), and North America (O'Brien et al. 2016). The attributes of each weapon tip being dictated according to the raw materials available for point manufacture, the target fauna, the environmental landscape in which the hunting occurred, and the community of practice in which the weapon was produced.

Both barbed and unbarbed points were attached to foreshafts and sometimes link-shafts, with the former being a piece of hard wood, bone, antler, ivory etc. to which the projectile point is attached at one end and the main shaft (usually made from vegetal material) to the other. This element acts to add weight to the projectile, and serves to save the (usually) wooden shaft from breakage in use. The term 'link-shaft' is sometimes used to refer to the foreshaft (e.g., Sackett 1985), though they are also a distinct element often described from African contexts. Here, they are usually made from bone, are thick (compared with the point), symmetrical, and bipointed (Deacon 1984; Inskeep 1987). In Africa, where link-shafts are utilized, the foreshaft is generally made of reed or wood (Deacon 1984). Interestingly, some authors argue that they are "good indicators for the use of the bow-and-arrow" when found in these archaeological contexts (Inskeep 1987:165).

In Europe, perhaps the most interesting recent discovery concerning these mid-shaft elements is the identification of numerous foreshafts (and points) made from cetacean bone (probably whale) dating to the Middle to Late Magdalenian (c. 17,500–15,000 years cal BP) (Langley and Street 2013; Pétillon 2013; Langley et al. 2016). Analysis of the spatial distribution of these artefacts allowed Pétillon (2013) to conclude that the bone was exclusively of Atlantic origin, with objects made from this material being transported along the Pyrenees up to the central part of the range some 350 km away. Similarly, the discovery of a sole foreshaft in cetacean bone from the Late Magdalenian open air site of Andernach-Martinsberg (Central German Rhineland) indicated that these artefacts were sometimes transported over significantly longer distances—in this case, more than 1000 km (Langley and Street 2013).

Through comparison with ethnographic datasets (Christensen et al. 2016), it has been determined that the various Pleistocene projectiles were launched either by hand, with a spearthrower, or with a bow. The choice of which launch system was employed rested on a variety of factors including: the target prey, the environment (open, forested, etc.) in which hunting was taking place, the required success rate, the proximity to the target needed, and the raw materials available for launch aid manufacture (among other factors).

The spearthrower is made up of a rod or plank with a hook or gutter, with or without a spur, on or in which is inserted a fletched or inflected projectile (spear, harpoon, long arrow) for use in hunting, fishing, or warfare. The use of this tool increases the initial velocity of the projectile thrown, and thus, increases the force at impact and its efficiency in inflicting a mortal wound to the target. It also significantly increases the range over which the hunter can effectively hurl the weapon. The earliest archaeological evidence for a known spearthrower is a Solutrean antler spearthrower hook from Combe Saunière dated to 17,470 years BP (Geneste and Plisson 1986; Cattelain 1989), with components found from this period through to the Magdalenian (Cattelain 1988, 1989; Bellier and Cattelain 1990; Stodiek 1993; Cattelain 1997). Only the distal extremities of the spearthrower, made of antler, bone, or ivory, survived to recovery, however, the handles presumably were made of wood, and attached via the use of ligatures and possibly also adhesives (Cattelain 1997).

The bow, like the spearthrower, acts to increase the accuracy and force by which a projectile point is launched. It is essentially a spring made from two flexible, elastic limbs held under tension by a string, which when the latter is pulled back, allows energy to accumulate in the bow. This energy is transmitted directly to the arrow when it is launched forward (Hamilton 1982; Cattelain 1997). Bows show higher variability in form than spearthrowers, and can be made from wood, horn, antler, sinew, and vegetal fibers among other organic materials. Determining when the bow first appeared in a hunting toolkit is extremely difficult owing to the simple fact that most are made from wood which only preserve over long time periods in very particular depositional environments. The earliest known fragments of bows (along with arrows) were found in a peat bog at Stellmoor, an Ahrensburgian site dated to the final Paleolithic in Europe (c. 11,000 years BP) (Beckhoff 1968). These examples were made from pine heartwood. At Holmegaard (Denmark), several bows made from elm dated to around 8000 years BP have similarly been recovered (Beckhoff 1968; Cattelain 1997). Recent discoveries in South Africa have suggested that bow-and-arrow technology was utilized from around 60,000 years BP, as mentioned above, with arrows being tipped with both bone and stone points (Backwell et al. 2008; Lombard and Phillipson 2010; Villa et al. 2010; Lombard 2011).

Identifying the oldest evidence for the bow in North America remains highly contested, however, most researchers agree on several key facts. First, researchers currently hypothesize that the bow was an intrusive element diffused from Northeast Asia, as no one has yet been able to demonstrate its independent invention in America (Nassaney and Pyle 1999). Secondly, it is generally agreed that the spearthrower (often termed an '*atlatl*' in this region) preceded the use of the bow by several thousand years (e.g., Kellar 1955; Fagan 1995; cf. Amick 1994), and while it is widely accepted that the bow was in use around 2300 years BP, it has been suggested that the bow and arrow may have appeared at around or even before 4000 years BP in Eastern and Central United States (Ames et al. 2010), before spreading further south by around 2500–3000 years BP (Yohe 1998). The spearthrower remained in use for several centuries before complete replacement by the bow c. 1000 years BP (Bettinger and Eerkins 1999; Nassaney and Pyle 1999; Chatters 2004; Ames et al. 2010).

Further south, the bow is thought to be present after 1400 BP (as based on lithic point size and morphology: Piana and Orquera 2009), with a significant period of overlap between spearthrower and bow use suggested (Charlin and González-José 2012). Interestingly, while the bow and arrow were likely utilized in Pleistocene Southeast Asia (though its exact antiquity here is unknown; Piper and Rabett 2009), it never saw use on the Australian continent. An organic weapon that has become iconic for Australia in recent times, the boomerang (also known as a '*Karli*'), was in fact used on virtually all inhabited continents in the past (Ferguson 1843; Peter 1986; Jones 2004). Boomerangs are a thrown weapon, typically constructed from wood in a flat aero foil design which results in the object spinning about an axis perpendicular to the direction of its flights. These items are best known for the returning variety, which when thrown, return to the person who threw the weapon via a wide arc. These weapons were traditionally used to hunt smaller prey such as birds and small mammals, though are also known to have seen use in warfare (Taçon and Chippendale 1994; Jones 2004).

Of particular interest to our discussion of osseous projectile weapons, is a near complete boomerang manufactured from mammoth ivory recovered from Gravettian (c. 23,000 years BP) levels at the Polish site of Oblazowa Rock (Valde-Nowak et al. 1987). Another, smaller fragment, this time in mammoth bone was found in Upper Paleolithic contexts at Stillfried (Austria), though its exact cultural affiliation remains unknown (Kriegler 1962).

The final type of projectile weapon to be considered here are fishhooks, which may be classified as such because they are *thrown* into rivers or oceans in order to retrieve fauna. Fishhooks have been manufactured from a great range of organic raw materials, though those made from shell and bone have survived from the widest variety of depositional contexts and returned the oldest dates thus far.

Currently, the oldest fishhook in the world was recovered from Jerimalai in Timor-Leste and is dated to between c. 23,000 and 16,000 cal BP (O'Connor et al. 2011). This artefact, along with another example from this same site, and a third from the nearby site of Lene Hara are dated to c. 11,000 cal BP. All are manufactured from *Trochus* shell. These examples are single-piece baited hooks and do not seem suitable for pelagic fishing, though bone points from these same sites may have formed parts of composite fishhooks which could have been used for the fishing of *Scombridae* which is represented at levels dated to around 42,000 cal BP (O'Connor et al. 2011).

The possibility that at least some of the bone bipoints recovered from various Paleolithic contexts functioned as parts of composite fishing technology (including both gorges and multi-component hooks) is an issue which comes up in several of the chapters presented herein (Allen et al. 2016; Évora 2016; Goutas 2016; Perera et al. 2016). If some of these artefacts were indeed used as fishhook components, the spatial and temporal understanding of this approach to food getting would be considerably expanded. Clearly further experimentation and microscopy is required in investigating this aspect of early food gathering.

Known bone fishhooks are present in the Southwest Sahara (Africa) from around 6500–3700 years BP (Petit-Marie et al.

1983), while in Europe, fishhooks are known from Final Paleolithic contexts, including those at the sites of Klein Lieskow, Germany (Pasda 2001; Street et al. 2002); Grotte du Bois-Ragot and Pont d'Ambon, France (Chollet et al. 1980; Cleyet-Merle 1990); and Gratkorn, Austria (Pittioni 1954). Recently, however, a fishhook made from mammoth ivory was recovered from Federmessergroupen contexts dated to around 12,300 cal. BP at Wustermark 22, Germany, and now constitutes the oldest example in Europe (Gramsch et al. 2013). While fishhooks have not been found in earlier Paleolithic contexts, we know that fishing played an important role in Late Magdalenian life (at least) owing to significant quantities of salmon (and other) fish bones which were recovered from numerous sites (Cziesla 2004). Importantly, evidence for other examples of Late Pleistocene fishing often imply the use of nets rather than hooks, either owing to the recovery of net sinkers or the size the fish constituting the faunal assemblages (e.g., Ohalo II, Israel: Nadel and Zaidner 2002; Lake Tandou, Australia: Balme 1983, 1995).

While this overview of osseous projectile weaponry is far from exhaustive, it does demonstrate the temporal and spatial distribution of these diverse and often ingenious technologies, and consequently, their ability to significantly contribute to future dialogues regarding Late Pleistocene cultural variability.

Conclusion

Through outlining the osseous projectile technologies which were in use prior to 10,000 years BP, we are able to fill a gap in the literature, and ultimately, contribute to wider anthropological debates about human uniqueness and cultural variability. Given that around 90 % of Modern Human prehistory occurred during the Pleistocene period, we cannot restrict ourselves to ethnographies of modern day hunter-gatherer peoples, nor those who lived in more recent prehistoric eras if we hope to encompass the breadth and depth of human behavioral and cultural variability. We must consider that past peoples developed technologies to act within subsistence and social strategies which may not be consonant with those present within the past few hundred or even thousand years.

As plastic media, detailed analysis of bone, antler, and ivory objects may provide the kinds of data required for tracking micro-scalar changes in technological strategies. Furthermore, data drawn from weaponry made from hard animal materials provide insights into more than just the technological realm owing to its ability to supply details on the time/s of year that the utilized raw material was collected and possible associations with subsistence strategies. Such data complements lithic (and other) datasets and allows researchers to draw more informed conclusions on technological organization, social interaction, and group movements during this most ancient period of human life. That this information has remained on the peripheries of Pleistocene archaeological research (with notable exceptions), is perhaps owing to a general lack of understanding of the potential these materials have within the wider field of research.

Thus, this volume provides an overview of the current state of osseous projectile weaponry research. It is hoped that the chapters that follow will provide readers with an insight into the richness and diversity of this technology, and inspire the wider integration of these collections into debates and narratives on Pleistocene cultural variability.

Acknowledgements My thanks to Mirani Litster, Sean Ulm, and Eric Delson for reading drafts of this manuscript.

References

Allen, H., Langley, M. C., & Taçon, P. S. C. (2016). Bone projectile points in prehistoric Australia: Evidence from implements, ethnography, and rock art. In M. C. Langley (Ed.), *Osseous projectile weaponry: Towards an understanding of Pleistocene cultural variability* (pp. 209–218). Dordrecht: Springer.

Ambrose, S. H. (1998). Chronology of the Later Stone Age and food production in East Africa. *Journal of Archaeological Science, 25*, 337–392.

Ames, K. M., Fuld, K. A., & Davis, S. (2010). Dart and arrow points on the Columbia Plateau of western North America. *American Antiquity, 75*, 287–325.

Amick, D. S. (1994). Technological organization and the structure of inference in lithic analysis: An examination of Folsom hunting behavior in the American southwest. In P. J. Carr (Ed.), *The organization of North American prehistoric chipped stone tool technologies* (pp. 9–34). Ann Arbor, MI: International Monographs in Prehistory.

Aplin, K., O'Connor, S., Bulbeck, D., Piper, P., Marwick, B., St Pierre, E., et al. (2016). The Walandawe Tradition from Southeast Sulawesi and Osseous Artifact Traditions in Island Southeast Asia. In M. C. Langley (Ed.), *Osseous projectile weaponry: Towards an understanding of Pleistocene cultural variability* (pp. 189–208). Dordrecht: Springer.

Backwell, L., d'Errico, F., & Wadley, L. (2008). Middle Stone Age bone tools from the Howiesons Poort layers, Sibudu Cave, South Africa. *Journal of Archaeological Science, 35*, 1566–1580.

Backwell, L. R., & d'Errico, F. (2005). The origin of bone tool technology and the identification of early hominid cultural traditions. In F. d'Errico & L. R. Backwell (Eds.), *From tools to symbols. From early hominids to modern humans* (pp. 238–275). Johannesburg, South Africa: Wits University Press.

Backwell, L. R., & d'Errico, F. (2016). Osseous projectile weaponry from early to late Middle Stone Age Africa. In M. C. Langley (Ed.), *Osseous projectile weaponry: Towards an understanding of Pleistocene cultural variability* (pp. 15–29). Dordrecht: Springer.

Balme, J. (1983). Prehistoric fishing in the lower Darling, western New South Wales. In J. Clutton-Brock & C. Grigson (Eds.), *Animals and archaeology: Shell middens, fishes and birds* (pp. 19–32). BAR International Series 183. Oxford, England: British Archaeological Reports.

Balme, J. (1995). 30,000 years of fishery in western New South Wales. *Archaeology in Oceania, 30*(1), 1–21.

Barandiaràn, I. (1967). *El Paleomesolithico del Pirineo Occidental: Bases para una sistematización del Mateiral Óseo Paleolitico*. Zaragoza: Seminario de Prehistoria y Protohistoria.

Barthelme, J. W. (1977). Holocene sites northeast of Lake Turkana: A preliminary report. *Azania, 12*, 33–41.

Barthelme, J. W. (1985). *Fisher-hunters and Neolithic pastoralists in East Turkana, Kenya*. Oxford, England: British Archaeological Reports International Series 254.

Barton, H., Piper, P. J., Rabett, R., & Reeds, I. (2009). Composite hunting technologies from the Terminal Pleistocene and early Holocene, Niah Cave, Borneo. *Journal of Archaeological Science, 36*, 1708–1714.

Beckhoff, K. (1968). Eignung und Verwendung einheimischer Holzarten für prähistorische Pfeilbogen. *Die Kunde NF, 19*, 85–101.

Bellier, C., & Cattelain, P. (1990). *La chasse dans la prehistoire du Paléolithique au Néolithique en Europe*. Terriens, Belgium: CEDARC.

Bettinger, R. L., & Eerkins, J. (1999). Point typologies, cultural transmission, and the spread of bow-and-arrow technology in the prehistoric Great Basin. *American Antiquity, 64*, 231–242.

Binford, L. (1978). *Nunamiut ethnoarchaeology*. New York: Academic Press.

Boas, F. (1966). *Kwakiutl ethnography*. Chicago: University of Chicago Press.

Brooks, A. S., Helgren, D. M., Cramer, J. S., Franklin, A., Hornyak, W., Keating, J. M., et al. (1995). Dating and context of three Middle Stone Age sites with bone points in the Upper Semliki Valley, Zaire. *Science, 268*, 548–553.

Brooks, A. S., Nevell, L., Yellen, J. E., & Hartman, G. (2006). Projectile technologies of the MSA: Implications for modern human origins. In E. Hovers & S. L. Kuhn (Eds.), *Transitions before the transition: Evolution and stability in the Middle Paleolithic and Middle Stone Age* (pp. 233–255). New York: Springer.

Brooks, A. S., & Smith, C. C. (1987). Ishango revisited: New age determinations and cultural interpretations. *African Archaeological Review, 5*, 65–78.

Campana, D. (1991). Bone implements from Hayonim Cave: Some relevant issues. In O. Bar-Yosef & F. R. Valla (Eds.), *The Natufian culture in the Levant* (pp. 459–466). Ann Arbor, MI: International Monographs in Prehistory.

Campmas, E., Michel, P., Costamagno, S., Amani, F., Stoetzel, E., Nespoulet, R., et al. (2015). Were Upper Pleistocene human/non-human predator occupations at the Témara caves (El Harhoura 2 and El Mnasra, Morocco) influenced by climate change? *Journal of Human Evolution, 78*, 122–143.

Camps, G. (1974). *Les civilisations préhistoriques de l'Afrique du Nord et du Sahara*. Paris: Doin.

Cattelain, P. (1988). *Fiches typologiques de l'industrie de l'os préhistorique. Cahier II: Propulseurs*. Aix-en-Provence, France: Université de Provence.

Cattelain, P. (1989). Un crochet de propulseur Solutréen de la Grotte de Combe-Saunière I (Dordogne). *Bulletin de la Société Préhistorique Française, 86*, 213–216.

Cattelain, P. (1997). Hunting during the Upper Paleolithic: Bow, spear-thrower, or both? In H. Knecht (Ed.), *Projectile technology* (pp. 213–240). New York: Plenum Press.

Charlin, J., & González-José, R. (2012). Size and shape variation in Late Holocene projectile points of southern Patagonia: A geometric morphometric study. *American Antiquity, 77*, 221–242.

Chatters, J. C. (2004). Safety in numbers: The influence of the bow and arrow on village formation on the Columbia Plateau. In W. C. Prints & I. Juijt (Eds.), *Complex hunter-gatherers: Evolution and organization of prehistoric communities on the Plateau of Northwestern North America* (pp. 67–83). Salt Lake City, UT: University of Utah Press.

Chauviére, F.-X. (2016). Osseous projectile points in the Magdalenian: 'True' points versus pointed waste-products. In M. C. Langley (Ed.), *Osseous projectile weaponry: Towards an understanding of Pleistocene cultural variability* (pp. 161–169). Dordrecht: Springer.

Cheynier, A. (1958). Impromptu sur la séquence des pointes du Paléolithique Supérieur. *Bulletin de la Société Préhistorique Française, 55*, 190–205.

Chollet, A., Boutin, G., & Célérier, G. (1980). Crochets en bois de cerf de l'azilien du Sud- Ouest de la France. *Bulletin de la Société Préhistorique Française, 77*, 11–16.

Christensen, M., Legoupil, D., & Pétillon, J.-M. (2016). Hunter-gatherers of the old and new world: Morphological and functional comparisons of osseous projectile points. In M. C. Langley (Ed.), *Osseous projectile weaponry: Towards an understanding of Pleistocene cultural variability* (pp. 237–252). Dordrecht: Springer.

Cleyet-Merle, J.-J. (1990). *La préhistoire de la pêche*. Paris: Éditions Errance.

Cummings, V., Jourdan, P., & Zvelebil, M. (2014). *The Oxford handbook of the archaeology and anthropology of hunter-gatherers*. Oxford, England: Oxford University Press.

Cziesla, E. (2004). Late Upper Paleolithic and Mesolithic cultural continuity - or: bone and antler objects from the Havelland. In T. Terberger & B. V. Eriksen (Eds.), *Hunters in a changing world. Environment and archaeology of the Pleistocene - Holocene transition (ca. 11000-9000 B. C.) in Northern Central Europe* (pp. 165–182). Rahden, Germany: Marie Leidorf Publisher.

Davidson, D. S. (1934). Australian spear-traits and their derivations. *Journal of the Polynesian Society, 43*(41–72), 142–162.

Deacon, H. J. (1976). *Where hunters gathered: A study of stone age people in the Eastern Cape*. Claremont: South African Archaeological Society.

Deacon, J. (1984). *The Later Stone Age of southernmost Africa*. Oxford: Archaeopress.

Delporte, H., Hahn, J., Mons, G., Pinçon, G., & de Sonneville-Bordes, D. (1988). *Fiches typologiques de l'industrie osseuse préhistorique. Cahier I. Sagaies*. Aix-en-Provence: Université de Provence.

d'Errico, F., & Henshilwood, C. (2007). Additional evidence for bone technology in the southern African Middle Stone Age. *Journal of Human Evolution, 52*, 142–163.

d'Errico, F., Backwell, L. R., & Wadley, L. (2012). Identifying regional variability in Middle Stone Age bone technology. The case of Sibudu Cave. *Journal of Archaeological Science, 39*, 2479–2495.

de Heinzelin, J. (1962). Ishango. *Scientific American, 26*, 105–116.

Dibble, H. L. (1995). Middle Paleolithic scraper reduction: Background, clarification, and review of the evidence to date. *Journal of Archaeological Method and Theory, 2*(4), 299–368.

Dobres, M.-A., & Hoffman, C. R. (1994). Social agency and the dynamics of prehistoric technology. *Journal of Archaeological Method and Theory, 1*, 211–258.

Évora, M. A. (2016). A review of the osseous projectile points from the Upper Paleolithic of Portugal. In M. C. Langley (Ed.), *Osseous projectile weaponry: Towards an understanding of Pleistocene cultural variability* (pp. 131–141). Dordrecht: Springer.

El Hajraoui, A. M., & Debénath, A. (2012). El Mnasra - Chapter XXIV. L'industrie osseuse. Préhistoire de la région de Rabat-Témara. In A. M. El Hajraoui, R. Nespoulet, A. Debénath, & H. Dibble (Eds.), *Villes et sites d'Archaeologie Marocaine, Royaume du Maroc* (pp. 179–188). Rabat: Ministère de la culture et Institut National des Sciences de l'Archéologie et du Partrimoine.

Fagan, B. (1995). *Ancient North America: The archaeology of a continent*. London: Thames and Hudson.

Feathers, J. K., & Migliorini, E. (2001). Luminescence dating at Katanda - A reassessment. *Quaternary Science Reviews, 20*, 961–966.

Ferguson, S. (1843). On the antiquity of the Kiliee, or Boomerang. *The Transactions of the Royal Irish Academy, 19*, 22–47.

Geneste, J. M., & Plisson, H. (1986). Le Solutréen de la Grotte de Combe Saunière I (Dordogne). *Gallia Préhistoire, 29*, 9–27.

Girod, P., & Massénat, E. (1900). *Les Stations de l'Âge du Renne dans les Vallées de la Vézère et de la Corrèze*. Paris: Libraire J.B. Balillière et fils.

Gould, R. (1980). *Living archaeology*. Cambridge, England: Cambridge University Press.

Goutas, N. (2016). Gravettian projectile points: Considerations about the evolution of osseous hunting weapons in France. In M. C. Langley (Ed.), *Osseous projectile weaponry: Towards an understanding of Pleistocene cultural variability* (pp. 89–107). Dordrecht: Springer.

Gramsch, B., Beran, J., Hanik, S., & Sommer, R. S. (2013). A Paleolithic fishhook made of ivory and the earliest fishhook tradition in Europe. *Journal of Archaeological Science, 40*, 2458–2463.

Hamilton, T. M. (1982). *Native American bows* (p. 5). Columbia, SC: Missouri Archaeological Society Special Publication No.

Hayden, B. (1979). *Paleolithic reflections: Lithic technology and ethnographic excavation among the Australian Aborigines*. New Jersey, NJ: Humanities Press.

Hayden, B. (2004). Sociopolitical organization in the Natufian. In C. Delage (Ed.), *The last hunter-gatherer societies in the Near East* (pp. 263–308). Oxford, England: British Archaeological Reports International Series.

Henry, D. O. (1989). *From foraging to agriculture. The Levant at the end of the Ice Age*. Philadelphia: University of Pennsylvania Press.

Higham, T., Basell, L., Jacobi, R., Wood, R., Bronk Ramsey, C., & Conard, N. J. (2012). Testing models for the beginnings of the Aurignacian and the advent of figurative art and music: The radiocarbon chronology of Geißenklösterle. *Journal of Human Evolution, 6*, 664–676.

Hiscock, P. (2008). *Archaeology of ancient Australia*. London: Routledge.

Inskeep, R. R. (1987). *Nelson Bay Cave, Cape Province, South Africa: The holocene levels*. Oxford: Archaeopress.

Jacobs, Z., Roberts, R. G., Galbraith, R. F., Deacon, H. J., Grün, R., Mackay, A., et al. (2008a). Ages for the Middle Stone Age of southern Africa: Implications for human behavior and dispersal. *Science, 322*, 733–735.

Jacobs, Z., Wintle, A. G., Duller, G. A. T., Roberts, R. G., & Wadley, L. (2008b). New ages for the post-Howiesons Poort, late and final Middle Stone Age at Sibudu Cave, South Africa. *Journal of Archaeological Science, 35*, 1790–1807.

Jacobs, Z., Roberts, R. G., Nespoulet, R., El Hajraoui, M. A., & Debénath, A. (2012). Single-grain OSL chronologies for Middle Palaeolithic deposits at El Mnasra and El Harhoura 2, Morocco: Implications for Late Pleistocene human-environment interactions along the Atlantic coast of northwest Africa. *Journal of Human Evolution, 62*, 377–394.

Jochim, M. A. (1987). Late Pleistocene refugia in Europe. In O. Soffer (Ed.), *The Pleistocene Old World: Regional perspectives* (pp. 317–332). New York: Plenum Press.

Jones, P. (2004). *Boomerang: Behind an Australian icon*. Kent Town, Australia: Wakefield Press.

Julien, M. (1982). *Les harpons Magdaléniens. Supplement a Gallia Préhistoire 17*. Paris: Editions du Centre National de la Recherche Scientifique.

Karavanić, I. (2016). Early Upper Palaeolithic osseous points from Croatia. In M. C. Langley (Ed.), *Osseous projectile weaponry: Towards an understanding of Pleistocene cultural variability* (pp. 43–54). Dordrecht: Springer.

Kellar, J. H. (1955). *The atlatl in North America* (Prehistory Research Series 3(3)). Indianapolis, IN: Indiana Historical Society.

Klein, R. G. (1999). *The human career: Human biological and cultural origins*. Chicago: The University of Chicago Press.

Klein, R. G. (2000). Archeology and the evolution of human behavior. *Evolutionary Anthropology: Issues News and Reviews, 9*(1), 17–36.

Klein, R. G. (2008). Out of Africa and the evolution of human behavior. *Evolutionary Anthropology: Issues, News, and Reviews, 17*(6), 267–281.

Kriegler, K. F. (1962). Die bearbeitete mammuthspeiche au Stillfried an der March, Niederösterreich. *Acta Praehistorica, 3*(4), 6–13.

Kroeber, A. L. (1925). *Handbook of the Indians of California*. Washington, DC: Bureau of American Ethnology Bulletin 78.

Langlais, M., Costamango, S., Laroulandie, V., Pétillon, J.-M., Discamps, E., Mallye, J.-B., et al. (2012). The evolution of Magdalenian societies in south-west France between 18,000 and 14,000 cal BP: Changing environments, changing tool kits. *Quaternary International, 272–273*, 138–149.

Langley, M. C. (2015). Investigating maintenance and discard behaviours for osseous projectile points: A middle to late Magdalenian (c. 19,000–14,000 cal. BP) example. *Journal of Anthropological Archaeology, 40*, 340–360.

Langley, M. C., Clarkson, C., & Ulm, S. (2011). From small holes to grand narratives: The impact of taphonomy and sample size on the modernity debate in Australia and New Guinea. *Journal of Human Evolution, 61*, 197–208.

Langley, M. C., & Street, M. (2013). Long range inland-coastal networks during the late Magdalenian: Evidence for individual acquisition of marine resources at Andernach-Martinsberg, German Central Rhineland. *Journal of Human Evolution, 64*, 457–465.

Langley, M. C., Pétillon, J.-M., & Christensen, M. (2016). Diversity and evolution of osseous hunting equipment during the Magdalenian (21,000–14,000 cal. BP). In M. C. Langley (Ed.), *Osseous projectile weaponry: Towards an understanding of Pleistocene cultural variability* (pp. 143–159). Dordrecht: Springer.

Lartet, E. (1861). Nouvelles recherches sur la coexistence de l'homme et des grands mammiferes fossiles. *Annales des Sciences Naturelles, 15*, 177–253.

Lartet, E., & Christy, H. (1875). *Reliquae aquitanicae: Being contribution to the archaeology and paleontology of Perigord and adjoining provinces of Southern France*. London: Ruper Jones.

Lave, J., & Wenger, E. (1991). *Situated learning: Legitimate peripheral participation*. Cambridge, England: Cambridge University Press.

Lee, R. B., & Devore, I. (1968). *Man the hunter*. Chicago: Aldine.

Lévi-Strauss, C. (1969). *Les strutures élémentaires de la parenté*. Paris: Du Gruyter.

Lombard, M. (2011). Quartz-tipped arrows older than 60 ka: Further use-trace evidence from Sibudu, KwaZulu-Natal, South Africa. *Journal of Archaeological Science, 38*, 1918–1930.

Lombard, M., & Phillipson, L. (2010). Indications of bone and stone-tipped arrow use 64 000 years ago in KwaZulu-Natal, South Africa. *Antiquity, 84*, 635–648.

Lowie, R. H. (1935). *The Crow Indians*. New York: Farrar & Rinehart.

Mitchell, P. (2002). *The archaeology of Southern Africa*. Cambridge: Cambridge University Press.

Monod, T., & Mauny, R. (1957). Decouverte de nouveaux instruments en os dans l'Ouest Africain. In J. D. Clark & S. Cole (Eds.), *Third Pan-African congress on prehistory* (pp. 242–247). London: Chatto and Windus.

Nadel, D., & Zaidner, Y. (2002). Upper Pleistocene and mid-Holocene net sinkers from the Sea of Galilee, Israel. *Journal of the Israel Prehistoric Society, 32*, 49–71.

Nassaney, M. S., & Pyle, K. (1999). The adoption of the bow and arrow in Eastern North America: A view from Central Arkansas. *American Antiquity, 64*, 243–263.

O'Brien, M. J., Lyman, R. L., Buchanan, B., & Collard, M. (2016). A review of late Pleistocene North American Bone and Ivory Tools. In M. C. Langley (Ed.), *Osseous projectile weaponry: Towards an *understanding of Pleistocene cultural variability* (pp. 221–235). Dordrecht: Springer.

O'Connell, J. F. (1995). Ethnoarchaeology needs a general theory of behavior. *Journal of Archaeological Research, 3*(3), 205–255.

O'Connor, S., Ono, R., & Clarkson, C. (2011). Pelagic fishing at 42,000 years before the present and the maritime skills of modern humans. *Science, 334*, 1117–1121.

Opperman, H. (1987). *The Later Stone Age of the Drakensberg Range and its foothills*. Oxford: Archaeopress.

Otte, M. (1997). Contacts trans-méditerranéens au Paléolithique. In J. M. Fullola & N. Soler (Eds.), *El mon Mediterrani després del Pleniglacial* (pp. 29–39). Girona, Spain: Museu d'Arqueolgia de Catalunya.

Pasda, C. (2001). Das Knochengerät vom spätpaläolithischen Fundplatz Kleinkiesow in der Niederlausitz. In B. Gehlen, M. Heinen, & A. Tillmann (Eds.), *Zeit-Räume. gedenkschrift für Wolfgang Taute* (pp. 397–408). Bonn, Germany: Habelt-Verlag.

Perera, N., Roberts, P., & Petraglia, M. (2016). Bone technology in South Asia from late Pleistocene rockshelter deposits in Sri Lanka. In M. C. Langley (Ed.), *Osseous projectile weaponry: Towards an understanding of Pleistocene cultural variability* (pp. 173–188). Dordrecht: Springer.

Peter, H. (1986). *Wesen und bedeutung des bumerangs*. Vienna: Braumüller.

Pétillon, J.-M. (2008). First evidence of a whale-bone industry in the western European Upper Paleolithic: Magdalenian artifacts from Isturitz (Pyrénées-Atlantiques, France). *Journal of Human Evolution, 54*, 720–726.

Pétillon, J.-M. (2013). Circulation of whale-bone artifacts in the northern Pyrenees during the late Upper Paleolithic. *Journal of Human Evolution, 65*, 525–543.

Pétillon, J.-M., Langlais, M., Kuntz, D., Normand, C., Barshay-Szmidt, C., Costamagno, S., et al. (2015). The human occupation of the northwestern Pyrenees in the Late Glacial: New data from the Arudy basin, lower Ossau valley. *Quaternary International, 364*, 126–143.

Petit-Marie, N., Celles, N. C., Commelin, D., Delibrias, G., & Raimbault, M. (1983). The Sahara in Northern Mali: Man and his environment between 10,000 and 3,500 years BP. *African Archaeological Review, 1*, 105–126.

Peyrony, D. (1909). Station préhistorique du Ruth (Dordogne). *Revue de l'Ecole d'Anthropologie de Paris, 29*, 156–176.

Phillipson, D. W. (1977). Lowasera. *Azania, 12*, 1–32.

Piana, E. L., & Orquera, L. A. (2009). The southern top of the world: The first peopling of Patagonia and Tierra del Fuego and the cultural endurance of the Fuegian sea-nomads. *Arctic Anthropology, 46*, 103–117.

Piette, E. (1895). Etudes d'Ethnographie Prehistorique. *L'Anthropologie, 6*, 276–292.

Piper, P. J., & Rabett, R. J. (2009). Hunting in a tropical rainforest: Evidence from the terminal Pleistocene at Lobang, Hangus, Niah Caves, Sarawak. *International Journal of Osteoarcheology, 19*, 551–565.

Pittioni, R. (1954). *Urgeschichte des Österreichischen raumes*. Vienna: Franz Deuticke Publisher.

Potts, R. (1998). Variability selection and hominid evolution. *Evolutionary Anthropology, 7*, 81–96.

Rabett, R. J. (2005). The early exploitation of Southeast Asian mangroves: Bone technology from caves and open sites. *Asian Perspectives, 44*, 154–179.

Rabett, R. J., Barker, G., Barton, H., Hunt, C., Lloyd-Smith, L., Paz, V., et al. (2013). Landscape transformations and human responses c.11,500-c.4500 years ago. In G. Barker (Ed.), *Rainforest foraging and farming in Island Southeast Asia* (pp. 217–253). Cambridge, England: McDonald Institute Monographs.

Radcliffe-Brown, A. R. (1931). The social organisation of Australian tribes. *Oceania, 1*, 34–63, 322–341, 426–456.

Robbins, L. H. (1974). *The Lothagam Site: A late stone age fishing settlement in the Lake Rudolf Basin, Kenya* (Michigan State University Museum, Anthropology Series 1(2)). Ann Arbor, MI: Publications of the Museum-Michigan State University.

Robbins, L. H., Murphy, M. L., Stewart, K. M., Campbell, A. C., & Brook, G. A. (1994). Barbed bone points, paleoenvironment, and the antiquity of fish exploitation in the Kalahari Desert, Botswana. *Journal of Field Archaeology, 21*, 257–264.

Sackett, J. R. (1985). Style and ethnicity in the Kalahari: A reply to Wiessner. *American Antiquity, 50*, 154–159.

Shea, J. J. (2011). *Homo sapiens* is as *Homo sapiens* was. *Current Anthropology, 52*, 1–35.

Stodiek, U. (1993). *Zur technologie der Jungpalaolitischen speerschleuder. Eine studie auf der basis archäologischer, ethnologischer und experimenteller erkenntnisse*. Tübingen, Germany: Tübinger Monographien zur Urgeschichte 9.

Stordeur, D. (1991). Le Natoufien et son évolution à travers les artefacts en os. In O. Bar-Yosef & F. R. Valla (Eds.), *The Natufian culture in the Levant* (pp. 467–482). Ann Arbor, MI: International Monographs in Prehistory.

Straus, L. G. (1993). Upper Paleolithic hunting tactics and weapons in Western Europe. In G. L. Peterkin, H. M. Bricker, & P. Mellars (Eds.), *Hunting and animal exploitation in the Later Paleolithic and Mesolithic of Eurasia* (pp. 83–93). Washington, DC: American Anthropological Association.

Straus, L. G. (2001). Africa and Iberia in the Pleistocene. *Quaternary International, 75*, 91–102.

Street, M., Baales, M., Cziesla, E., Hartz, S., Heinen, M., Jöris, O., et al. (2002). Final Paleolithic and Mesolithic research in reunified Germany. *Journal of World Prehistory, 15*, 365–453.

Szmidt, C., Pétillon, J.-M., Cattelain, P., Normand, C., & Schwab, C. (2009). Premières dates radiocarbone pour le Magdalénien d'Isturitz (Pyrénées Atlantiques). *Bulletin de la Société Préhistorique Française, 106*, 588–592.

Taçon, P. S. C., & Chippendale, C. (1994). Australia's ancient warriors: Changing depictions of fighting in the rock art of Arnhem Land, N.T. *Cambridge Archaeological Journal, 4*, 211–248.

Tejero, J.-M. (2016). Spanish Aurignacian projectile points: An example of the first European palaeolithic hunting weapons in osseous materials. In M. C. Langley (Ed.), *Osseous projectile weaponry: Towards an understanding of Pleistocene cultural variability* (pp. 55–69). Dordrecht: Springer.

Tisnerat-Laborde, N., Valladas, H., & Ladier, E. (1997). Nouvelles datations carbone 14 en SMA pour le Magdalénien supérieur de la vallée de l'Aveyron. *Préhistoire Ariégeoise, 52*, 129–136.

Tostevin, G. B. (2007). Social intimacy, artefact visibility and acculturation. Models of Neanderthal - Modern human interaction. In P. Mellars, K. V. Boyle, O. Bar-Yosef, & C. Stringer (Eds.), *Rethinking the human revolution* (pp. 341–357). Cambridge, England: McDonald Institute for Archaeological Research.

Valde-Nowak, P., Nadachowski, A., & Wolsan, M. (1987). Upper Paleolithic boomerang made of mammoth tusk in south Poland. *Nature, 329*, 436–438.

Vanhaeren, M., & d'Errico, F. (2005). Grave goods from the Saint-Germain-la-Riviere burial: Evidence for social inequality in the Upper Palaeolithic. *Journal of Anthropological Archaeology, 24*, 117–134.

Villa, P., & d'Errico, F. (2001). Bone and ivory points in the lower and middle Paleolithic of Europe. *Journal of Human Evolution, 41*, 69–112.

Villa, P., Soriano, S., Teyssandier, N., & Wurz, S. (2010). The Howiesons Poort and MSA III at Klasies River main site, Cave 1A. *Journal of Archaeological Science, 37*, 630–655.

Villaverde, V., & Roman, D. (2005–2006). Los arpones del Magdaleniense superior de la Cova de les Cendres y su valoración en el contexto del Magdaleniense mediterráneo. *Munibe, 57*(2), 207–225.

Villaverde, V., Aura, J. E., Borao, M., & Roman, D. (2016). Upper palaeolithic bone and Antler projectiles in the Spanish Mediterranean Region: The Magdalenian period. In M. C. Langley (Ed.), *Osseous projectile weaponry: Towards an understanding of Pleistocene cultural variability* (pp. 109–130). Dordrecht: Springer.

Waguespak, N. M., Surovell, T. A., Denoyer, A., Dallow, A., Savage, A., Hyneman, J., et al. (2009). Making a point: Wood-versus stone tipped projectiles. *Antiquity, 83*, 786–800.

Wendt, W. E. (1966). Two prehistoric archaeological sites in Egyptian Nubia. *Postilla, 102*, 1–49.

Weniger, G. C. (1987). Magdalenian settlement pattern and subsistence in Central Europe: The southwestern and central German cases. In O. Soffer (Ed.), *The Pleistocene Old World: Regional perspectives* (pp. 201–215). New York: Plenum Press.

Weniger, G. C. (1992). Function and form: An ethnoarchaeological analysis of barbed points from northern hunter-gatherers. In F. Audouze (Ed.), *Ethnoarchéologie: Justification, problèmes, limites* (pp. 257–268). Juan-les-Pins, France: XIIe Rencontres Internationales d'Archéologie et d'Historie d'Antibes, Centre National de Recherches Archéologiques en Belgique.

Weniger, G. C. (1995). *Widerhaken Spitzen des Magdalenien Westeuropas. Ein Vergleich mit Ethnohistorischen Jägergruppen Nordamerikas*. Mainz.

Weniger, G.-C. (2000). Magdalenian barbed points: Harpoons, spears and arrowheads. In C. Bellier, P. Cattelain, & M. Otte (Eds.), *La chasse dans la préhistoire/Hunting in prehistory* (pp. 79–87). Bruxelles: Société Royale Belge d'Anthropologie et de Préhistoire/Service de Préhistoire de l'Universitéde Liège/Centre d'Études et de Documentation Archéologiques.

Wiessner, P. (1983). Style and social information in Kalahari San projectile points. *American Antiquity, 48*, 253–276.

Wobst, M. H. (1977). Stylistic behavior and information exchange. In C. E. Cleland (Ed.), *For the director: Research essays in honor of James B. Griffin* (pp. 317–342). Ann Arbor, MI: Museum of Anthropology, University of Michigan.

Wobst, H. M. (1978). The archaeo-ethnology of hunter-gatherers or the tyranny of ethnographic record in archaeology. *American Antiquity, 43*, 303–309.

Yellen, J. E. (1998). Barbed bone points: Tradition and continuity in Saharan and Sub-Saharan Africa. *African Archaeological Review, 15*, 173–198.

Yellen, J. E., Brooks, A. S., Cornelbsen, E., Mehlman, M., & Stewart, K. (1995). A Middle Stone Age worked bone industry from Katanda, Upper Semliki Valley, Zaire. *Science, 268*, 553–556.

Yohe, R. M. (1998). The introduction of the bow and arrow and lithic resource use at Rose Spring (CA-INY-372). *Journal of California and Great Basin Anthropology, 20*, 26–52.

Zilhão, J., & d'Errico, F. (1999). The chronology and taphonomy of the earliest Aurignacian and its implications for the understanding of Neandertal extinction. *Journal of World Prehistory, 13*, 1–68.

Part I
Africa

Chapter 2
Osseous Projectile Weaponry from Early to Late Middle Stone Age Africa

Lucinda Backwell and Francesco d'Errico

Abstract Discussion about early projectile technology typically includes criteria used to distinguish artefacts used as hafted points from those employed for other purposes, associated faunal and lithic assemblages, palaeoenvironment, age of the material, associated hominins and their cognitive capacities, criteria used to identify complex technology and cognition, and how innovative technologies might have developed and spread. Here we summarize what is known about osseous weaponry in the African Middle Stone Age, and discuss the implications of these items for the origin(s) of modern cognitive complexity. Results indicate the use of bone spear points in the Aterian and Still Bay, and bone-tipped arrowheads in the Howiesons Poort and the Early Later Stone Age. The appearance and disappearance of projectile technology suggests that it likely emerged more than once, as an adaptation to local environments, rather than being the outcome of a process in which technology advanced in step with developing cognition.

Keywords Arrow • Bone • Howiesons Poort • Projectile point • Still Bay • Tusk

L. Backwell (✉)
Evolutionary Studies Institute and DST/NRF Centre of Excellence in Palaeosciences and School of Geosciences, University of the Witwatersrand, Private Bag 3, Wits 2050, Johannesburg, South Africa
e-mail: lucinda.Backwell@wits.ac.za

F. d'Errico
CNRS UMR 5199, PACEA, Université de Bordeaux, Bâtiment B18, Allée Geoffroy Saint Hilaire, CS 50023, 33615 Pessac Cedex, France

Evolutionary Studies Institute and DST/NRF Centre of Excellence in Palaeosciences and School of Geosciences, University of the Witwatersrand, Private Bag 3, Wits 2050, Johannesburg, South Africa
e-mail: francesco.derrico@u-bordeaux.fr

Introduction

The view that the development of technology was a gradual process that proceeded in parallel with biological evolution and complex cognition has been challenged in recent years by evidence from Middle Stone Age (MSA) sites in Africa. As Wadley (2006) points out, a lot has changed since Isaac (1974) referred to African Middle Pleistocene archaeology as the "muddle in the middle". Since then a dozen MSA sites have been newly dated, and the cultural sequence is now firmly understood in southern Africa to include: pre-Still Bay, Still Bay, Howiesons Poort, post-Howiesons Poort, late and final MSA phases. The situation has also significantly improved in northern Africa, an area in which the earliest cultural horizons preserving unambiguous evidence of bone tool utilization belong to the Aterian, an industry that stratigraphically follows a local Mousterian and whose time and geographic range is now much better known than it was just a decade ago (Barton et al. 2009; Richter et al. 2010; Dibble et al. 2013; Scerri 2013). Other areas of Africa, however, have not benefitted from a comparable research effort, and some MSA cultural adaptations remain insufficiently characterized and poorly dated (Barham and Mitchell 2008).

Bone tools have been particularly valuable in tracking cultural changes in time, identifying regional patterns, characterizing technical systems, and inferring the degree of complexity of cultural adaptations outside Africa (Breuil 1912). Although we are only at an early stage of that process in Africa, the identification of formal bone tools in MSA deposits has made this possible, and attempts at following this path already exist (d'Errico et al. 2012b). Formal bone tools are those that are cut, carved or polished, to form points, awls, borers, and so forth (Klein 1999). The use of bone, and its shaping into task-specific tools, is among the list of traits identified as characteristic of modern human behavior (McBrearty and Brooks 2000). As such, the early appearance

Michelle C. Langley (ed.), *Osseous Projectile Weaponry*, Vertebrate Paleobiology and Paleoanthropology,
DOI 10.1007/978-94-024-0899-7_2

of formal bone tools in the African MSA—together with pigments (Watts 2009), engravings (Henshilwood et al. 2009; Texier et al. 2010), personal ornaments (d'Errico et al. 2009a) and other forms of "modern" behavior, for example heat treatment of stone for knapping (Brown et al. 2009; Mourre et al. 2010) and hafting tools with compound adhesives (Wadley et al. 2009)—has been used to support the "Out of Africa" hypothesis, which postulates a causal connection between the origin of modern humans in Africa around 200 ka, synchronous with the gradual emergence of modern culture. This model predicts a steady accrual of cultural innovations in Africa that culminated in the increase and spread of modern humans, and the rapid replacement of archaic hominins in Africa and Eurasia by about 30 ka.

The evolution of tool-making from simple to composite, and the diversification of raw materials utilized, suggest to many that complex cognition, and perhaps some sort of language with a complex structure, had to have been in place in order for this to have happened (Ambrose 2010; Wadley 2010a, b; Sterelny 2012). It is for these reasons that evidence of hafting—in particular the manufacture of bone and ivory points, using techniques such as scraping, grinding, grooving and polishing—is considered good evidence of complex cognition and modern human behavior. However, if the Out of Africa scenario is correct, one would not expect to find behaviors considered specific to *Homo sapiens sapiens* associated with archaic populations outside of Africa. The fact that Neanderthals exhibited many of these complex behaviors (funerary practices, complex hafting techniques, bone tool manufacture, personal ornamentation, pigment use) before or at the very moment of contact with modern humans, contradicts this theory (Koller et al. 2001; Pettitt 2002; d'Errico 2003; d'Errico et al. 2003; Mazza et al. 2006; Soressi and d'Errico 2007; Zilhão et al. 2009; Caron et al. 2011; but see Bar-Yosef and Bordes 2010). In addition, contrary to showing a gradual increase in innovations, the archaeological record shows a discontinuous pattern in cultural evolution, with innovations appearing, disappearing and reappearing again in different forms, indicating regional cultural traditions and discontinuity in cultural transmission (Villa et al. 2005; Jacobs et al. 2008a; d'Errico and Vanhaeren 2009; d'Errico et al. 2009b; d'Errico and Stringer 2011; Lombard and Parsons 2011). Changes in mechanisms of cultural transmission (Hovers and Belfer-Cohen 2006), climate (Ambrose 1998a; Lahr and Foley 1998; Henshilwood 2008; d'Errico et al. 2009a; d'Errico and Banks 2013) and demography (Shennan 2001; Henrich 2004; Powell et al. 2009) are proposed to account for the discontinuous pattern, but how much each of these factors potentially contributed to the process, and if so, to what degree they were inter-related, remains unclear.

Considering that early hominin bone tools, dated to between 2 and 1 million years ago, and associated with *Homo ergaster* and robust australopithecines in the same deposits, record evidence of intentional shaping through grinding and knapping (d'Errico and Backwell 2003; Backwell and d'Errico 2004, 2005), one may wonder how 'formal' a bone tool must be to tell us something about the identity and cognitive abilities of its maker and user. Most of the techniques used to manufacture bone tools do not require a particularly high level of dexterity or cognition, nor do they appear difficult to transmit from one generation to another, even without language. Researchers propose a cognitive attribution based on the amount of finishing to an end product, or type and placement of wear observed, and consider many steps in the process to be synonymous with cognitive sophistication, but the truth is that bone tools themselves do not always provide criteria by which to judge a degree of cognition. It is the thought processes involved in achieving the intended outcome that gauge behavioral complexity, and these are more difficult to document and evaluate. Among the archaeological tools listed as reflecting modernity (points, awls and borers) are minimally modified forms that do not fall within the standardized range of types and dimensions one might expect from anatomically and behaviorally modern people.

The MSA archaeological record, as with ethnographic accounts, shows the use of bone tools to be highly variable, ranging from absent in some societies, to minimally modified and highly sophisticated and specialized in others. In this milieu it is difficult to assess the nature and cultural significance of archaeological bone tools, let alone identify the maker and user. An extremely small sample of modern and archaic African middle and late Pleistocene hominin remains renders our knowledge of the extent of hominin diversity, distribution, first and last appearances, and associations with lithic industries and each other, virtually nil (Barham and Mitchell 2008; Backwell et al. 2014). It is widely accepted that by 60 ka all people in southern Africa were anatomically modern (e.g., McBrearty and Brooks 2000; Grün and Beaumont 2001; Deacon and Wurz 2005; Marean and Assefa 2005), but with the fossil evidence available, and the fact that late MSA lithic assemblages are characterized by being highly variable (Wurz 2002; Wadley 2005), we really do not know who was on the landscape, and what they were making and using. Tracing the origin and transmission of innovative behavior is all the more difficult when it is unclear whether there was an accretional emergence of the modern human morphotype from an archaic ancestor with a pan-African distribution (Bräuer 2008; Pearson 2008), or whether there was a more punctuated appearance, possibly a speciation event from a geographically-restricted subpopulation of archaic humans (Stringer 2002). It is against this backdrop that we discuss what is known about osseous projectile weaponry in Africa during the MSA, and discuss the implications of this evidence for the origin of modern cognitive complexity.

Fig. 2.1 Map showing African Middle Stone Age sites mentioned in the text. *Black dots* indicate MSA sites with bone points

Osseous Projectile Technologies

Given the small sample of osseous tools known from the MSA, and the fact that attributing a clear function is problematic in a number of cases, the word "projectile" is used here in its broadest sense, to incorporate all forms of bone points, from bone-tipped thrusting spears to hand-delivered and mechanically projected weaponry. The oldest evidence for the production of 'formal' bone tools comes from eight African MSA sites (Fig. 2.1). Most of the finds are unique, or consist of small collections of objects, in a number of cases of uncertain stratigraphic provenance and chronological attribution. Three bone objects from Broken Hill (Kabwe) in Zambia (Clark et al. 1947), tentatively attributed to the early MSA (Barham et al. 2002), and thought to be associated with *Homo heidelbergensis* (elsewhere named *Homo rhodesiensis*), comprise a pointed implement and two bone flakes with traces of scraping and polish due to use. The uncertain provenance of these objects within the Broken Hill cave system makes it difficult to draw definitive conclusions about their age and significance (McBrearty and Brooks 2000). A point reported from Mumbwa Cave in Zambia is considered doubtful. The barbed and unbarbed bone points from the Katanda sites in the Semliki Valley, Democratic Republic of the Congo (formerly Zaïre), are at present the oldest known points from Africa that were clearly made to be hafted (Fig. 2.2). The layer from which they originate has been attributed an age of ca. 90 ka (Brooks et al. 1995; Yellen et al. 1995; Yellen 1998; Brooks et al. 2006). Although considered by some as possibly younger (Ambrose 1998b; Klein 1999, 2000, 2008), more recent dating of the site confirms an old age, at least in excess of 60–70 ka, and certainly no younger than 50 ka (Feathers and Migliorini 2001). Given the uniqueness of these kinds of artefacts, which predate

Fig. 2.2 Selection of barbed bone points from Katanda, Democratic Republic of the Congo, dated to c. 90 ka, and interpreted as harpoons used for fishing. (After Backwell and d'Errico 2014)

well-documented Later Stone Age (LSA) harpoons by 50,000 years or more, it is understandable that the age estimates of these objects have been challenged.

The presence of bone tools in the North African Aterian was reported more than 20 years ago by El Hajraoui (1994), from the site of El Mnasra in Morocco. Until now, the significance of the material has not been fully realized, partly due to a scarcity of contemporaneous material elsewhere in Africa at the time, and largely because the pieces were fragmentary, and presented in the form of line drawings, which renders independent assessment of the nature of the material difficult. Recent publications by El Hajraoui and Debénath (2012), and Campmas and colleagues (2015) convincingly demonstrate their identification as worked bone artefacts. Nine objects come from layer 5 of El Mnasra, dated by optically stimulated luminescence (OSL) to 107 ± 6.6 and 106 ± 6.6 ka (Jacobs et al. 2012). Most of them consist of complete or longitudinally split ribs of large herbivores, thinned by scraping and grinding. The most complete specimen, a split rib, has been shaped by grinding its medullary aspect and edges. One end was additionally thinned by scraping to produce an elongated slender tip, and the outer surface of the rib marked with incisions (Fig. 2.3, left). A mesial fragment of a similar object, shaped with the same techniques, features on one side a curved protuberance (Fig. 2.3, right). This second object is interpreted by El Hajraoui and Debénath (2012) as a broken spear point. Although these objects may have been used for such a function, their fragmentary state, which prevents analysis of the tool tip, makes it difficult to establish that with confidence.

Fig. 2.3 Bone implements from layer 5 at El Mnasra, Morocco, interpreted as spear points. (Modified after El Hajraoui and Debénath 2012)

It is, however, noteworthy that some spear points from Katanda, probably made from ribs of very large mammals and clearly used as spear points, as indicated by the presence of barbs (Fig. 2.2), are reminiscent of the objects found at El Mnasra, which supports the identification of the El Mnasra artefacts as spear points. If the function and age of these artefacts are confirmed by new discoveries, the Aterian might represent an African cultural adaptation in which osseous projectile points were elaborated for the first time, emerging during the interglacial conditions of MIS5 (Campmas et al. 2015). El Mnasra and Katanda may represent the only known examples of a phase in which large spear points were used (including flat specimens made out of ribs), perhaps for

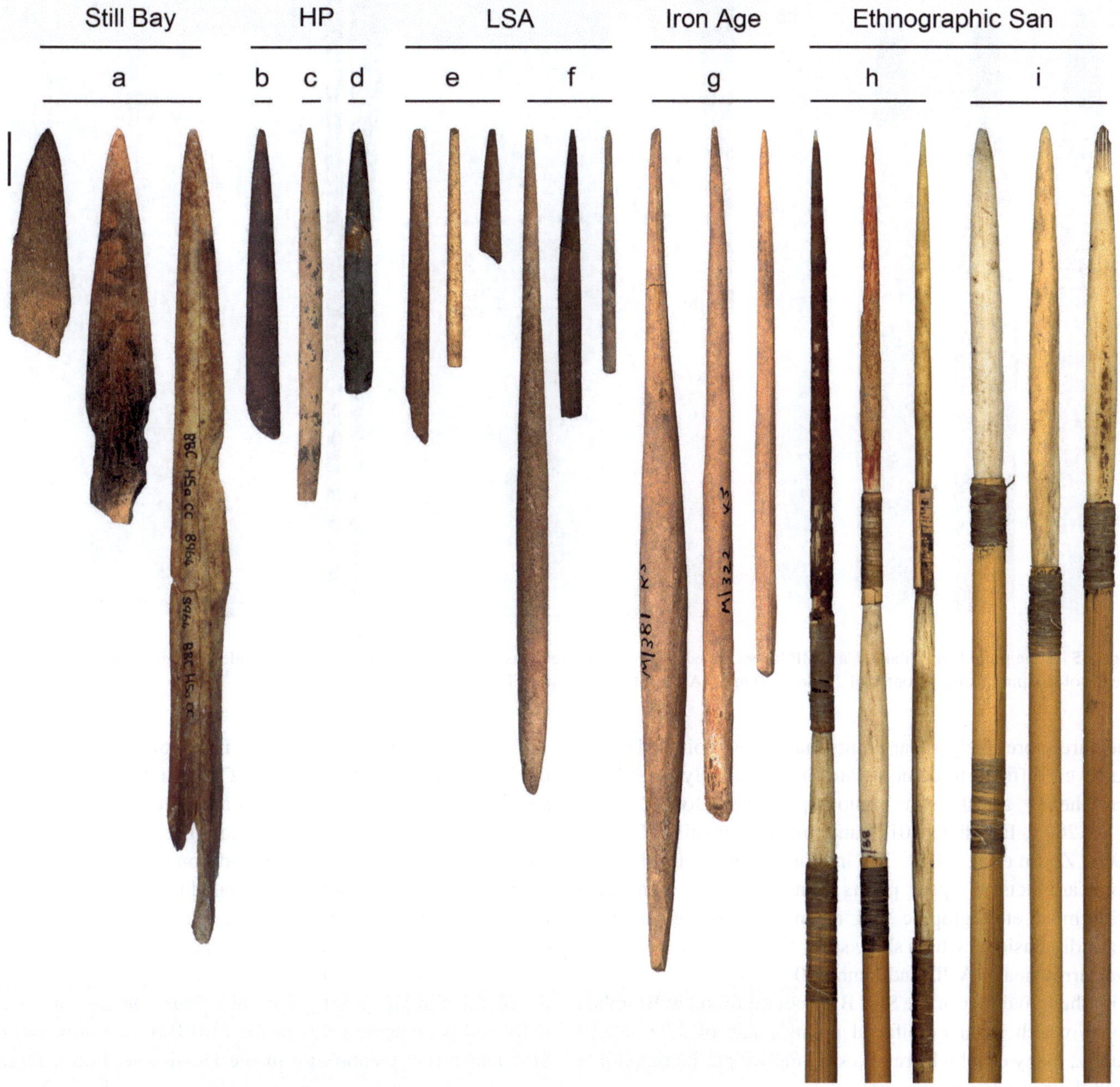

Fig. 2.4 Bone points from MSA deposits at Blombos Cave (**a**), Peers Cave (**b**), Klasies River (**c**) and Sibudu Cave (**d**); Later Stone Age layers at Rose Cottage Cave (**e**) and Jubilee Shelter (**f**), and an Iron Age occupation at Mapungubwe, Zimbabwe (**g**). Two types of traditional San bone arrowheads are shown: reversible poison-coated types with linkshafts (**h**) and robust fixed types without linkshafts and poison (**i**). Note the robusticity of the Blombos points, interpreted as spear heads (**a**) compared with those from later contexts, and the reduction in point size through time. Also note the similarity between the Sibudu specimen (**d**) and arrowheads of the un-poisoned fixed type used by San hunters (**i**). Scale = 10 mm. (After Backwell et al. 2008)

harpooning large fish, as suggested by Yellen et al. (1995) for the Katanda specimens.

The oldest formal bone tools from southern Africa are presently found in the Still Bay (SB) layers at Blombos Cave (Fig. 2.4a) dated to 75 ka (d'Errico and Henshilwood 2007; Henshilwood et al. 2009), and in pre-Still Bay layers at Sibudu Cave (d'Errico et al. 2012b) dated to 72.5 ± 2.0 ka (Jacobs et al. 2008a, b). A point tip, a mesial fragment, an almost complete spear point and a tanged bone point are reported from M1 and M2 layers at Blombos Cave. A single massive point, different from those found in the MSA and LSA layers at Blombos Cave, was recovered in the dune sand layer, with an age of ~70 ka (Jacobs et al. 2003).

The morphological variability in the bone points from Blombos Cave, and the size and weight of the three almost complete specimens and a proximally broken tip, suggests

Fig. 2.5 Bone point from Sibudu Cave HP layers, dated >61 ka, interpreted as an arrowhead, showing clear evidence of intentional shaping by means of scraping using a stone tool. Scales = 10 mm. (After Backwell et al. 2008)

they are more likely spear points than arrow points. This is, however, difficult to demonstrate since relatively large bone arrowheads are known ethnographically (Bosc-Zanardo et al. 2008; Bradfield 2012) and archaeologically (Guthrie 1983; Zhilin et al. 2014). The interpretation of the Blombos bone artefacts as spear points is none-the-less consistent with most ethnographic and recent archaeological stone point dimensions, which show spear tips to be 5 times larger than arrowheads (Villa and Lenoir 2006). It is also consistent with the remainder of the Still Bay tool kit found at Blombos Cave, which includes bifacial points made of silcrete and quartz, many of which are of a size and weight incompatible with their use as arrow points.

A bone point from Peers (Skildergatkop) Cave (Fig. 2.4b) was retrieved from either the Howiesons Poort (HP) or Still Bay layers at the site (Peers 1929). A study of carbon-nitrogen ratios in the Peers point, and a sample of MSA and LSA faunal remains from this site, confirms that the point originates from MSA layers (d'Errico and Henshilwood 2007). A bone point from the lowest HP levels at Klasies River (Fig. 2.4c) in layer 19 of Shelter 1a, dated to c. 70–65 ka (Miller et al. 1999; Vogel 2001) is described by Singer and Wymer (1982) as similar in color to MSA bone from the same level. Pointed MSA bone tools from Sibudu Cave include a large bone point from layer GS (Figs. 2.4d and 2.5), in HP layers with age estimates >61 ka (Backwell et al. 2008; Jacobs et al. 2008b, c).

Comparative microscopic and morphometric analysis of the large bone point from Sibudu Cave with bone tools from southern African Middle and Later Stone Age deposits (Fig. 2.4e, f), an Iron Age occupation (Fig. 2.4g), San hunter-gatherer toolkits (Fig. 2.4h, i), and bone tools used experimentally in a variety of tasks, revealed that it is most similar to arrow points from LSA, Iron Age, and historical San sites (Fig. 2.6). This is interpreted by Backwell and colleagues (2008), together with the extreme symmetry recorded in the tip of the Sibudu point, as a shift from the use of hand-delivered bone spearheads in the Still Bay (at Blombos) to bow and arrow technology in the Howiesons Poort, represented at Sibudu Cave, and probably Klasies River Mouth and Peers Cave. Table 2.1 provides contextual information on the bone points analyzed. The Sibudu bone point also falls within the morphological variability of a type of unpoisoned fixed bone arrow point used by Bushmen for hunting small game and birds (Fig. 2.4i), which is in accordance with the associated fossil fauna, represented mostly by small forest antelope (Plug 2004; Clark and Plug 2008; Wadley 2010a, b).

Use-wear and residue analysis of lithics from Sibudu Cave show that many segments from HP layers also have ochre and plant adhesive traces on their curved backs where they would have been hafted to shafts or handles (Lombard 2006, 2008). Some lack ochre, and instead have fat mixed with plant material (Wadley et al. 2009). Design, impact

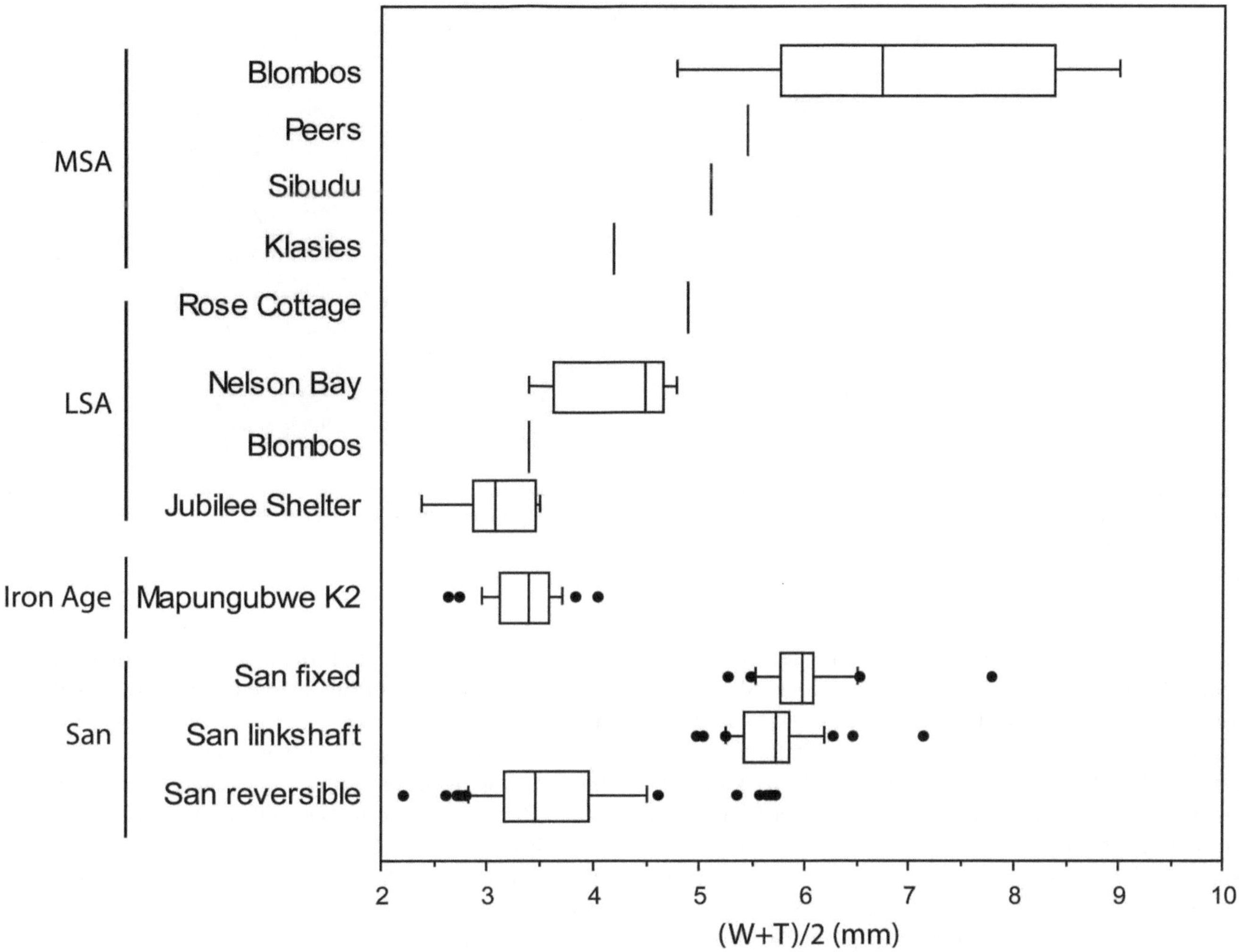

Fig. 2.6 Variation in size at 30 mm from the tips of bone points from MSA, LSA, and Iron Age sites, and ethnographic San hunting kits. Note the difference in standardization between the specimens from Blombos Cave, interpreted as spear heads, and those from later contexts. (After Backwell et al. 2008)

Table 2.1 Contextual information on bone points analyzed from southern Africa

Site	Cultural attribution	Age	*n*	Museum/institution	Reference
Blombos Cave	MSA (SB)	~84–72 ka	3	Iziko	Henshilwood et al. (2001); Jacobs et al. (2006); d'Errico and Henshilwood (2007)
Klasies River Mouth	MSA (HP)	66–45 ka	1	Iziko	Singer and Wymer (1982); Deacon and Wurz (1996); Wurz (1999)
Sibudu Cave	MSA (HP)	>61 ka	1	Wits	Wadley (1987, 2006); Wadley and Jacobs (2004, 2006)
Peers Cave	MSA (SB/HP)	75–50 ka	1	Iziko	d'Errico and Henshilwood (2007)
Nelson Bay Cave	Later Stone Age (CW)	9000–5300 BP	5	Iziko	Inskeep (1987); Deacon (1982, 1984b)
Jubilee Shelter	Later Stone Age (IW)	8500–5200 BP	7	Wits	Wadley (1989, 1993)
Rose Cottage Cave	Later Stone Age (IW)	4000–2000 BP	1	Wits	Sampson (1974); Wadley (1993, 2000)
Mapungubwe	Iron Age	AD 200–present	25	Wits	Voigt (1983)
Namibia (Kalahari)	Modern San hunter-gatherers	Historical times	117	Museum Africa	Wanless (2007)

SB Still Bay, *HP* Howiesons Poort, *CW* Coastal Wilton, *IW* Interior Wilton, *Wits* University of the Witwatersrand

Fig. 2.7 Split warthog (*P. aethiopicus*) or bushpig (*P. larvatus*) lower canine (tusk) from layer 1BS Lower B–C at Border Cave, dated to between 43 and 42 ka. It shows scraping to produce a lanceolate shape, and on one end has deep transverse incisions, probably to facilitate hafting. Scale = 1 cm. (After d'Errico et al. 2012a)

fractures and residues suggest that some segments from Sibudu Cave and nearby Umhlatuzana Rock Shelter are likely to have functioned as arrowheads (Lombard and Pargeter 2008; Wadley 2008; Lombard and Phillipson 2010; Lombard 2011), supporting an early date of >61 ka for the origin of bow and arrow technology in sub-Saharan Africa. Impact fractures and animal residues on points from post-HP layers at Sibudu Cave suggest their use as spear tips (Lombard 2004, 2005; Villa et al. 2005). A warthog (*Phacochoerus aethiopicus*) or bushpig (*Potamochoerus larvatus*) tusk from Border Cave, dated 43–42 ka, and scraped into a lanceolate shape, supports the use of spears in the late MSA (Fig. 2.7). This may either imply the loss of bow and arrow technology for about 20,000 years, until it appeared again at Border Cave in the Early LSA (Fig. 2.8), between 44 and 43 ka (d'Errico et al. 2012a), or a continuity that has remained, for the moment, archaeologically invisible. The former hypothesis is consistent with the fact that, contrary to the lithic technology at Border Cave, which shows a gradual evolution from about 56 ka towards the LSA (Villa et al. 2012), a suite of organic artefacts similar to LSA and modern Kalahari San material culture, including probable arrowheads, appear quite abruptly, highlighting an apparent mismatch in rates of cultural change.

Detailed analyses show that MSA bone tool production methods follow a sequence of deliberate technical choices, starting with blank production, the use of various shaping methods, and the final finishing of the artefacts to produce projectile points and other tool types (d'Errico et al. 2012b). Tool production processes in the MSA conform to generally accepted descriptions of 'formal' techniques of bone tool manufacture, and apart from a large flat point from Still Bay layers at Blombos Cave, careful scraping is typical on MSA bone points, while LSA bone points evidence shaping through scraping and grinding. Bone points dated to between 39 and 10 ka are known at four southern African sites: Border Cave, Boomplaas, Nelson Bay Cave and Bushman Rock Shelter. Thereafter they become widespread in southern Africa and common in LSA sites (Deacon 1984a; Deacon and Deacon 1999; Wadley 1993).

Discussion

When observed in a broader context, African MSA bone tools in general, and spear points in particular, begin to shed light on previously undetected regional variations in bone tool technology and utilization. The possible bone spear points from Morocco are remarkably different in technology and shape from the Katanda harpoons, the relatively robust Blombos spears, the slender Sibudu specimen, the probable point made of a bushpig tusk from Border Cave and the thin San-like arrow points from Border Cave and contemporary sites.

An apparent mismatch appears when comparing the cultural affiliation of sites based on lithics and the presence/absence of bone tools, or the associated bone artefact types. Bone tools in the form of thinned ribs, modified long bone shafts and retouchers only occur at a few early Aterian sites

Fig. 2.8 Bone point from Border Cave, shaped by scraping, dated to the Early LSA, between 44 and 43 ka, and decorated with a spiral incision filled with red pigment (*top*), interpreted as a mark of ownership, as practiced by modern San hunters. Burnt bone point in three pieces, dated 43–42 ka, and shaped by intense scraping with a stone flake (*bottom*). Horizontal scales = 1 cm, vertical scales = 1 mm. (After d'Errico et al. 2012a)

located on the Atlantic coast of Morocco. We are left to wonder why no bone tools are found at the numerous more recent Aterian sites in the remainder of North Africa (Tixier 1967; Richter et al. 2010; Schwenninger et al. 2010), a number of which are in caves and shelters with reasonable preservation of bone. Bone tools occur in the pre-SB at Sibudu, but are absent in pre-SB layers elsewhere. They are numerous in the SB of Blombos Cave, but are absent in the SB of Diepkloof and Sibudu. Numerous bone tools are found in the HP and post-HP layers at Sibudu, but are absent, apart from four possible objects from Klasies, and from the many HP sites excavated so far in southern Africa, including the recently and meticulously excavated site of Diepkloof (Parkington et al. 2013). Sibudu bone tools are not only more varied in their conception, morphology, and the variety of tasks and material for which they were used, some categories such as splitting tools (scaled pieces, wedges) and smoothers are peculiar to this site, and straddle the HP and post-HP technocomplexes.

Such a pattern cannot be attributed to preservation factors, because well-preserved faunal assemblages were recovered from MSA sites with no bone artefacts. Raw material availability is also not a viable proposition, as a large variety of animal taxa and size are recorded at MSA sites. Site function cannot account for these differences, as many, if not all of the MSA sites discussed were places of habitation, and therefore where most of the subsistence and social activities were performed. Differences in available resources may of course have stimulated the creation of different bone tool traditions in different regions in response to local need, but we find the environmental explanation alone unsatisfactory. It does not explain why similar regional differences do not emerge in the lithic technology. Moreover, regional differences are now emerging in categories of material culture that are less linked to environment. Fragments of ostrich eggshells are ubiquitous at MSA sites, but engraved ostrich eggshells are only found in the HP layers of three sites, namely Diepkloof (Texier et al. 2010, 2013), Klipdrift (Henshilwood et al. 2014), and Apollo 11 (Vogelsang et al. 2010), and at Diepkloof they also occur in pre-HP layers. Although most designs used are found at both Diepkloof and Klipdrift, some only occur at one site, suggesting regional variation in style. Engraved hematite fragments, which are abundant in Blombos SB and pre-SB layers, and present in the pre-SB layers of Pinnacle Point (Watts 2010) and in the HP at Klein Kliphuis (Mackay and Welz 2008), are absent from SB and pre-SB layers at Diepkloof and Sibudu. Two *Conus* shells, intentionally perforated and coated with red pigment, one of which is associated with an infant burial, come from HP layers at Border Cave (Cooke et al. 1945; Beaumont and Bednarik 2013; d'Errico and Backwell 2016), dated to 74 ka (Grün et al. 2003). Shell beads are found at Blombos in the SB layers and at a number of sites in north Africa (Bouzouggar et al. 2007; d'Errico et al. 2009a, b) and the Near East (Mayer et al. 2009), but none is recorded at Diepkloof in the same cultural horizons, and the few possible shell beads from Sibudu SB layers (d'Errico et al. 2008) belong to a different taxon in spite of the availability at Sibudu of the species used at Blombos. The Katanda bone harpoons, which exhibit an unparalleled precocious technological sophistication in the central African region, probably reflect the same trend, which is the localized emergence and loss of a significant innovation.

The question of the emergence of cultural modernity has generally been approached by analyzing the archaeological record in search of behaviors considered as comparable to our own. Personal ornaments, engravings, projectile weaponry etc. represent items of complexity, either in the level of cognition required to conceive of and manufacture them, or in the symbolic meaning they carry. When extracted from their original context, some items tell little of their once useful function, and say nothing about their symbolic significance. The truth is that we have little idea about what happened in Africa between 60 and 30 ka, and it is at present difficult to assess the relationship, if any, between the suite of cultural adaptations recorded before 60 ka and those that emerged with the Later Stone Age.

Although they may well have been used in a comparable framework, the possibility exists that in the deep past such cultural items played a completely different role from the one attributed to them by archaeologists. In light of this, making inferences about MSA societies based on what we know about modern hunter-gatherers from southern Africa is problematic. The ethnographic record can stretch only so far, with the longest connection to modern culture as we know it, represented at Border Cave in South Africa at 44 ka (d'Errico et al. 2012a). Here the suite of complex and varied technical and symbolic items that characterize more recent LSA and historical San material culture enter the archaeological record abruptly, including bone points identical to San poisoned arrow points, one of which is incised with a mark of ownership filled with red ochre (Fig. 2.8), ostrich eggshell beads, a digging stick, and items associated with hunting, such as a wooden poison applicator and lump of beeswax.

Exploring whether snares and traps were used in the MSA (Wadley 2010a, b) is a good example of how the archaeological record may be mute in its hard evidence of innovative behavior. Take for example the ingenious ostrich trap made by traditional San hunters, which comprises about six minimally modified short sticks and a rope made from plant fibers. If preserved in a MSA deposit, the chances of identifying the items as a trap, and as part of one implement, are highly unlikely, and assigning an advanced level of cognition to the maker extremely low. The Paleolithic bone tool record shows that hominin technological evolution advanced in a nonlinear manner, and that from the outset bone tools exhibit signs of innovation, manifest as implements intentionally modified through knapping and grinding. The identification of a discontinuous pattern, with innovations appearing and disappearing, or being associated in a way that does not match the expected trend, supports the view that bone and complex lithic technology do not necessarily represent reliable hallmarks of "modern behavior" and cannot be attributed an unequivocal evolutionary significance. The discontinuous pattern shows that what we perceive today as modern behavior is the result of nonlinear trajectories, which may be better understood when documented at a regional scale (Hovers and Belfer-Cohen 2006; Hovers 2009; d'Errico and Stringer 2011; Shea 2011; d'Errico et al. 2012b; Villa et al. 2012).

The small number of post-200 ka sites that have been properly excavated in Africa, or dated using recent standards, together with a lack of coherence in hominin taxonomic attribution and regional palaeoenvironmental records, make it difficult to establish whether sampling bias, erosion and destructive taphonomic processes, discontinuity in cultural transmission, or the behavior of different hominin taxa account for the variability that characterizes MSA assemblages, the irregular manner in which they occur, and what appears to be the emergence of modern human behavior on the southern tip of the continent.

At this juncture we would argue that attempts at identifying the origin of behavioral modernity are premature and inherently biased, purely because current data derived from inland open air and coastal or near-coastal cave contexts are not really comparable. Cave sites have been preferentially studied over open air sites, which suffer the effects of erosion and a range of destructive taphonomic processes, presenting problems with dating, association and representation. Unlike open air sites that generally served a specific short-lived subsistence-related function, cave sites may preserve a range of cultural manifestations of different aspects of daily life over a significant period of time. Apart from symbolic items that may have been lost, one would not expect to find evidence of symbolic behavior in the context of a carcass processing site, just as one would not expect the distribution of MSA sites to fall within the borders of modern geopolitical divisions. Very few MSA sites in caves or rock shelters are recorded for Namibia and Botswana (Lane et al. 1998), but we know from field work that the landscape north of South Africa is littered with *in situ* MSA artefact knapping sites.

Of the few rock shelter sites studied, Apollo 11 in Namibia has yielded the oldest *art mobilier*, and White Paintings Rock Shelter in Botswana has recently yielded ancient bone arrow points, OSL dated from their association with sediments to between 37 and 35 ka (Robbins et al. 2012). Good examples of innovative material culture are recorded at five inland sites in southern Africa, namely fire at Wonderwerk (Berna et al. 2012), *art mobilier* at Apollo 11 (Wendt 1976), compound adhesives at Sibudu (Wadley et al. 2009), and ostrich eggshell beads at Cave of Hearths (Mason 1993) and Bushman Rock Shelter (Plug 1982), so we know that complex human cognition was widespread.

In order to explain the discontinuous pattern in hominin technological evolution, and the trans-species phenomenon of bone tool utilization in prehistory, we need to evoke social, demographic, and climatic factors, and their potential impact on similar innovations among geographically dispersed populations in Africa and Eurasia. Until such time that more MSA sites in the interior are excavated and properly dated, the deficiency in data available on inland MSA populations will continue to hinder meaningful comparison between the cultural technology and cognitive abilities of contemporaneous coastal or near-coastal dwellers, and bias attempts at identifying the geographic or taxonomic origin(s) of innovative behavior and complex modern human cognition. Based on the archaeological evidence at hand, and in particular the variability in technology, size and function that characterizes the handful of bone tools found at MSA sites, in our view it is likely that osseous projectile technology emerged and evolved more than once in human evolution, as an adaptation to local environments, rather than as the outcome of a protracted process in which bone tool technology advanced in step with the development of complex cognition. We see the need to consider each instance of bone use as an independent cultural adaptation to environmental conditions. This view seeks to provide best-fit explanations for the role of a bone technology within a specific subsistence strategy, and does not assume gradual patterns of evolution in technology.

Acknowledgments We thank Michelle Langley for inviting us to contribute a chapter to this monograph. We also thank A. Aomar, M. El Hajraoui, E. Campmas, R. Nespulet and B. Leprêtre (photographer) for granting permission to use photographs presented in Fig. 2.3. This research was supported by grants from: European Research Council (FP7/2007/2013, TRACSYMBOLS 249587); Origin of Man, Language and Languages Program of the European Science Foundation; Project Origines II, Aquitaine Region; South Africa/France Scientific Cooperation Agreement; Institut Français d'Afrique du Sud; Evolutionary Studies Institute and DST/NRF Centre of Excellence in Palaeosciences, University of the Witwatersrand; Ernest Oppenheimer Memorial Trust.

References

Ambrose, S. H. (1998a). Late Pleistocene human population bottlenecks, volcanic winter, and differentiation of modern humans. *Journal of Human Evolution, 34*, 623–651.

Ambrose, S. H. (1998b). Chronology of the Later Stone Age and food production in East Africa. *Journal of Archaeological Science, 25*, 337–392.

Ambrose, S. H. (2010). Coevolution of composite-tool technology, constructive memory, and language: Implications for the evolution of modern human behavior. *Current Anthropology, 51*(S1), S135–S147.

Backwell, L. R., & d'Errico, F. (2004). The first use of bone tools: A reappraisal of the evidence from Olduvai Gorge, Tanzania. *Palaeontologia Africana, 40*, 95–158.

Backwell, L. R., & d'Errico, F. (2005). The origin of bone tool technology and the identification of early hominid cultural traditions. In F. d'Errico & L. R. Backwell (Eds.), *From tools to symbols. From early hominids to modern humans* (pp. 238–275). Johannesburg: Wits University Press.

Backwell, L. R., & d'Errico, F. (2014). Palaeolithic bone tools. In C. Smith (Ed.), *Encyclopedia of global archaeology* (Vol. 2, pp. 950–962). New York: Springer.

Backwell, L. R., d'Errico, F., & Wadley, L. (2008). Middle Stone Age bone tools from the Howiesons Poort layers, Sibudu Cave, South Africa. *Journal of Archaeological Science, 35*(6), 1566–1580.

Backwell, L., McCarthy, T., Wadley, L., Henderson, Z., Steininger, C., deKlerk, B., et al. (2014). Multiproxy record of late Quaternary climate change and Middle Stone Age human occupation at Wonderkrater, South Africa. *Quaternary Science Reviews, 99*, 42–59.

Barham, L., & Mitchell, P. (2008). *The first Africans: African archaeology from the earliest tool makers to most recent foragers*. Cambridge: Cambridge University Press.

Barham, L. S., Llona, A. C. P., & Stringer, C. B. (2002). Bone tools from Broken Hill (Kabwe) cave, Zambia, and their evolutionary significance. *Before Farming, 2002*(2), 1–16.

Barton, R. N. E., Bouzouggar, A., Collcutt, S., Schwenninger, J.-L., & Clark-Balzan, L. (2009). OSL dating of the Aterian levels at Dar es-Soltan I (Rabat, Morocco) and implications for the dispersal of modern *Homo sapiens*. *Quaternary Science Reviews, 28*, 1914–1931.

Bar-Yosef, O., & Bordes, J.-G. (2010). Who were the makers of the Châtelperronian culture? *Journal of Human Evolution, 59*, 586–593.

Berna, F., Goldberg, P., Horwitz, L. K., Brink, J., Holt, S., Bamford, M., et al. (2012). Microstratigraphic evidence of *in situ* fire in the Acheulean strata of Wonderwerk Cave, Northern Cape Province, South Africa. *Proceedings of the National Academy of Sciences, 109*(20), E1215–E1220.

Beaumont, P. B., & Bednarik, R. G. (2013). Tracing the emergence of palaeoart in sub-Saharan Africa. *Rock Art Research: The Journal of the Australian Rock Art Research Association (AURA), 30*(1), 33–54.

Bosc-Zanardo, B., Bon, F., & Fauvelle-Aymar, F. X. (2008). Bushmen arrows and their recent history: Crossed outlooks of historical, ethnological and archaeological sources. In Projectile weapon elements from the Upper Palaeolithic to the Neolithic. Proceedings of session C83, 15th World Congress UISPP, Lisbon, September 4–9, 2006. *P@lethnologie, 1*, 341–357.

Bouzouggar, A., Barton, N., Vanhaeren, M., d'Errico, F., Collcutt, S., & Higham, T. (2007). 82,000-year-old shell beads from North Africa and implications for the origins of modern human behavior. *Proceedings of the National Academy of Sciences, 104*(24), 9964–9969.

Bradfield, J. (2012). Macrofractures on bone-tipped arrows: Analysis of hunter-gatherer arrows in the Fourie Collection from Namibia. *Antiquity, 86*(334), 1179–1191.

Bräuer, G. (2008). The origin of modern anatomy: By speciation or intraspecific evolution? *Evolutionary Anthropology: Issues, News, and Reviews, 17*(1), 22–37.

Breuil, H. (1912). Les subdivisions du Paléolithique supérieur et leur signification. *XIV Congrés International d'Anthropologie et d'Archéologie Préhistoriques, 1*, 165–238.

Brooks, A. S., Helgren, D. M., Cramer, J. S., Franklin, A., Hornyak, W., Keating, J. M., et al. (1995). Dating and context of three Middle Stone Age sites with bone points in the Upper Semliki Valley, Zaïre. *Science, 268*, 548–553.

Brooks, A. S., Nevell, L., Yellen, J. E., & Hartman, G. (2006). Projectile technologies of the MSA: Implications for modern human origins. In E. Hovers & S. L. Kuhn (Eds.), *Transitions before the transition: Evolution and stability in the Middle Palaeolithic and Middle Stone Age* (pp. 233–255). New York: Springer.

Brown, K. S., Marean, C. W., Herries, A., Jacobs, Z., Tribolo, C., Braun, D., et al. (2009). Fire as an engineering tool of early modern humans. *Science, 325*(5942), 859–862.

Campmas, E., Michel, P., Costamagno, S., Amani, F., Stoetzel, E., Nespoulet, R., et al. (2015). Were Upper Pleistocene human/non-human predator occupations at the Témara caves (El Harhoura 2 and El Mnasra, Morocco) influenced by climate change? *Journal of Human Evolution, 78*, 122–143.

Caron, F., d'Errico, F., Del Moral, P., Santos, F., & Zilhão, J. (2011). The reality of Neandertal symbolic behavior at the Grotte du Renne, Arcy-sur-Cure, France. *PLoS ONE, 6*(6), e21545.

Clark, J. D., Oakley, K. P., Wells, L. H., & McClelland, J. A. C. (1947). New studies on Rhodesian Man. *Journal of the Royal Anthropological Society, 77*, 4–33.

Clark, J. L., & Plug, I. (2008). Animal exploitation strategies during the South African Middle Stone Age: Howiesons Poort and post-Howiesons Poort fauna from Sibudu Cave. *Journal of Human Evolution, 54*(6), 886–898.

Cooke, H. B. S., Malan, B. D., & Wells, L. H. (1945). Fossil man in the Lebombo Mountains, South Africa: The 'Border Cave', Ingwavuma district, Zululand. *Man, 45*(3), 6–13.

Deacon, H. J., & Deacon, J. (1999). *Human beginnings in South Africa: Uncovering the secrets of the Stone Age*. Walnut Creek, CA: Rowman Altamira.

Deacon, H. J., & Wurz, S. (1996). Klasies River main site, cave 2: A Howiesons Poort occurrence. In G. Pwiti & R. Soper (Eds.), *Aspects of African archaeology* (pp. 213–218). Harare: University of Zimbabwe.

Deacon, H. J., & Wurz, S. (2005). *A Late Pleistocene archive of life at the coast, Klasies River*.

Deacon, J. (1982). *The Later Stone Age in the Southern Cape, South Africa*. Ph.D. thesis, University of Cape Town.

Deacon, J. (1984a). *The Later Stone Age of southernmost Africa*. Oxford: Archaeopress.

Deacon, J. (1984b). Later Stone Age people and their descendants in southern Africa. In R. G. Klein (Ed.), *Southern African prehistory and paleoenvironments*. Rotterdam: A.A. Balkema.

Dibble, H. L., Aldeias, V., Jacobs, Z., Olszewski, D. I., Rezek, Z., Lin, S. C., et al. (2013). On the industrial attributions of the Aterian and Mousterian of the Maghreb. *Journal of Human Evolution, 64*(3), 194–210.

d'Errico, F. (2003). The invisible frontier. A multiple species model for the origin of behavioral modernity. *Evolutionary Anthropology, 12*, 188–202.

d'Errico, F., & Backwell, L. R. (2003). Possible evidence of bone tool shaping from the early hominid site of Swartkrans, South Africa. *Journal of Archaeological Science, 30*, 1559–1576.

d'Errico, F., & Backwell, L. (2016). Earliest evidence of personal ornaments associated with burial: the Conus shells from Border Cave. *Journal of Human Evolution, 93*, 91–108.

d'Errico, F., Backwell, L. R., Villa, P., Degano, I., Lucejko, J., Bamford, M., et al. (2012a). Early evidence of San material culture represented by organic artifacts from Border Cave, South Africa. *Proceedings of the National Academy of Sciences, 109*(33), 13214–13219.

d'Errico, F., Backwell, L. R., & Wadley, L. (2012b). Indentifying regional variability in Middle Stone Age bone technology. The case of Sibudu Cave. *Journal of Archaeological Science, 39*, 2479–2495.

d'Errico, F., & Henshilwood, C. (2007). Additional evidence for bone technology in the southern African Middle Stone Age. *Journal of Human Evolution, 52*(2), 142–63.

d'Errico, F., Julien, M., Liolios, D., Vanhaeren, M., & Baffier, D. (2003). Many awls in our argument. Bone tool manufacture and use in the Châtelperronian and Aurignacian levels of the Grotte du Renne at Arcy-sur-Cure. In J. Zilhão & F. d'Errico (Eds.), *The chronology of the Aurignacian and of the transitional technocomplexes. Dating, stratigraphies, cultural implications* (pp. 247–271). Lisbon: Instituto Português de Arqueologia.

d'Errico, F., & Stringer, C. B. (2011). Evolution, revolution or saltation scenario for the emergence of modern cultures? *Philosophical Transactions of the Royal Society B: Biological Sciences, 366*(1567), 1060–1069.

d'Errico, F., & Vanhaeren, M. (2009). Earliest personal ornaments and their significance for the origin of language debate. In R. Botha & C. Knight (Eds.), *The cradle of human language* (Vol. 2, pp. 16–40). Oxford: Oxford University Press.

d'Errico, F., Vanhaeren, M., Barton, A., Bouzouggar, A., Mienis, H., Richter, D., et al. (2009a). Additional evidence on the use of personal ornaments in the Middle Paleolithic of North Africa. *Proceedings of the National Academy of Sciences, 106*, 16051–16056.

d'Errico, F., Vanhaeren, M., Henshilwood, C., Lawson, G., Maureille, B., Gambier, D., et al. (2009b). From the origin of language to the diversification of language. What can archaeology and palaeoanthropology say? In F. d'Errico & J.-M. Hombert (Eds.), *Becoming eloquent. Advances on the emergence of language, human cognition, and modern cultures* (pp. 13–68). Amsterdam: Benjamins.

d'Errico, F., Vanhaeren, M., & Wadley, L. (2008). Possible shell beads from the Middle Stone Age layers of Sibudu Cave, South Africa. *Journal of Archaeological Science, 35*(10), 2675–2685.

d'Errico, F., & Banks, W.W. (2013). Identifying mechanisms behind Middle Paleolithic and Middle Stone Age cultural trajectories. *Current Anthropology 54*, S371–S387

El Hajraoui, A. M., & Debénath, A. (2012). El Mnasra - Chapter XXIV. L'industrie osseuse. Préhistoire de la région de Rabat-Témara. In A. M. El Hajraoui, R. Nespoulet, A. Debénath, & H. Dibble (Eds.), *Villes et sites d'archéologie Marocaine, Royaume du Maroc* (Vol. III, pp. 179–188). Rabat: Ministère de la Culture et Institut National des Sciences de l'Archéologie et du Patrimoine.

El Hajraoui, M. A. (1994). L'Industrie osseuse Atérienne de la Grotte d'el Mnasra. *Préhistoire et Anthropologie Médetiterranéenne, 3*, 91–94.

Feathers, J. K., & Migliorini, E. (2001). Luminescence dating at Katanda—a reassessment. *Quaternary Science Reviews, 20*, 961–966.

Guthrie, R. D. (1983). Osseous projectile points: Biological considerations affecting raw material selection and design among Palaeolithic and Paleoindian peoples. In J. Clutton-Brook & C. Grigson (Eds.), *Animals and archaeology, British Archaeological Reports International Series 163* (pp. 273–294). Oxford: Archaeopress.

Grün, R., & Beaumont, P. (2001). Border Cave revisited: A revised ESR chronology. *Journal of Human Evolution, 40*(6), 467–482.

Grün, R., Beaumont, P., Tobias, P. V., & Eggins, S. (2003). On the age of Border Cave 5 human mandible. *Journal of Human Evolution, 45*, 155–167.

Henshilwood, C. S. (2008). Winds of change: Palaeoenvironments, material culture, and human behaviour in the late Pleistocene (77 ka–48 ka ago) in the Western Cape Province, South Africa. *South African Archaeological Society Goodwin Series, 10*, 1–17.

Henshilwood, C., d'Errico, F., Marean, C., Milo, R., & Yates, R. (2001). An early bone tool industry from the Middle Stone Age at Blombos Cave, South Africa: Implications for the origins of modern human behaviour, symbolism and language. *Journal of Human Evolution, 41*, 632–678.

Henshilwood, C., d'Errico, F., & Watts, I. (2009). Engraved ochres from the Middle Stone Age levels at Blombos Cave, South Africa. *Journal of Human Evolution, 57*, 27–47.

Henshilwood, C. S., van Niekerk, K. L., Wurz, S., Delagnes, A., Armitage, S. J., Rifkin, R. F., et al. (2014). Klipdrift shelter, southern Cape, South Africa: Preliminary report on the Howiesons Poort layers. *Journal of Archaeological Science, 45*, 284–303.

Henrich, J. (2004). Demography and cultural evolution: Why adaptive cultural processes produced maladaptive losses in Tasmania. *American Antiquity, 69*(2), 197–221.

Hovers, E. (2009). The Middle-to-Upper Paleolithic transition: What news? In M. Camps & P. Chauhan (Eds.), *Sourcebook of Paleolithic transitions: Methods, theories, and interpretations* (pp. 455–462). New York: Springer.

Hovers, E., & Belfer-Cohen, A. (2006). "Now you see it, now you don't"—Modern human behavior in the Middle Paleolithic. In E. Hovers & S. L. Kuhn (Eds.), *Transitions before the transition: Evolution and stability in the Middle Paleolithic and Middle Stone Age* (pp. 295–304). New York: Springer.

Inskeep, R. R. (1987). *Nelson Bay cave, Cape Province, South Africa. The Holocene levels. British Archaeological Reports International Series 357*. Oxford: Archaeopress.

Isaac, G. L. (1974). Sorting out the muddle in the middle: An anthropologist's post-conference appraisal. In K. W. Butzer & G. L. Isaac (Eds.), *After the australopithecines: Strategy, ecology and culture change in the Middle Pleistocene* (pp. 875–877). The Hague: Mouton.

Jacobs, Z., Duller, G. A. T., & Wintle, A. G. (2003). Dating of dune sand from Blombos Cave, South Africa: II—single grain data. *Journal of Human Evolution, 44*, 613–625.

Jacobs, Z., Duller, G. A. T., Wintle, A. G., & Henshilwood, C. S. (2006). Extending the chronology of deposits at Blombos Cave, South Africa, back to 140 ka using optical dating of single and multiple grains of quartz. *Journal of Human Evolution, 51*, 255–273.

Jacobs, Z., Roberts, R. G., Galbraith, R. F., Deacon, H. J., Grun, R., Mackay, A., et al. (2008a). Ages for the Middle Stone Age of southern Africa: Implications for human behavior and dispersal. *Science, 322*, 733–735.

Jacobs, Z., Roberts, R. G., Nespoulet, R., El Hajraoui, M. A., & Debénath, A. (2012). Single-grain OSL chronologies for Middle Palaeolithic deposits at El Mnasra and El Harhoura 2, Morocco: Implications for Late Pleistocene human-environment interactions along the Atlantic coast of northwest Africa. *Journal of Human Evolution, 62*(3), 377–394.

Jacobs, Z., Wintle, A. G., Duller, G. A. T., Roberts, R. G., & Wadley, L. (2008b). New ages for the post-Howiesons Poort, late and final Middle Stone Age at Sibudu Cave, South Africa. *Journal of Archaeological Science, 35*, 1790–1807.

Jacobs, Z., Wintle, A. G., Roberts, R. G., & Duller, G. A. T. (2008c). Equivalent dose distributions from single grains of quartz at Sibudu, South Africa: Context, causes and consequences for optical dating of archaeological deposits. *Journal of Archaeological Science, 35*, 1808–1820.

Klein, R. G. (1999). *The human career: Human biological and cultural origins*. Chicago: University of Chicago Press.

Klein, R. G. (2000). Archeology and the evolution of human behavior. *Evolutionary Anthropology: Issues News and Reviews, 9*(1), 17–36.

Klein, R. G. (2008). Out of Africa and the evolution of human behavior. *Evolutionary Anthropology: Issues, News, and Reviews, 17*(6), 267–281.

Koller, J., Baumer, U., & Mania, D. (2001). High-tech in the Middle Palaeolithic: Neandertal manufactured pitch identified. *European Journal of Archaeology, 4*, 385–397.

Lahr, M. M., & Foley, R. A. (1998). Towards a theory of modern human origins: Geography, demography, and diversity in recent human evolution. *Year Book of Physical Anthropology, 41*, 137–176.

Lane, P. J., Reid, A., & Segobye, A. (Eds.). (1998). *Ditswa Mmung. The archaeology of Botswana*. Pula Press and Botswana Society.

Lombard, M. (2004). Distribution patterns of organic residues on Middle Stone Age points from Sibudu Cave, KwaZulu-Natal, South Africa. *South African Archaeological Bulletin, 59*, 37–44.

Lombard, M. (2005). Evidence for hunting and hafting during the Middle Stone Age at Sibudu Cave, KwaZulu-Natal, South Africa: A multi-analytical approach. *Journal of Human Evolution, 48*, 279–300.

Lombard, M. (2006). Direct evidence for the use of ochre in the hafting technology of Middle Stone Age tools from Sibudu Cave. *Southern African Humanities, 18*(1), 57–67.

Lombard, M. (2008). Finding resolution for the Howiesons Poort through the microscope: Micro-residue analysis of segments from Sibudu Cave, South Africa. *Journal of Archaeological Science, 35*(1), 26–41.

Lombard, M. (2011). Quartz-tipped arrows older than 60 ka: Further use-trace evidence from Sibudu, KwaZulu-Natal, South Africa. *Journal of Archaeological Science, 38*, 1918–1930.

Lombard, M., & Pargeter, J. (2008). Hunting with Howiesons Poort segments: Pilot experimental study and the functional interpretation of archaeological tools. *Journal of Archaeological Science, 35*, 26–41.

Lombard, M., & Parsons, I. (2011). What happened to the human mind after the Howiesons Poort? *Antiquity, 85*, 1–11.

Lombard, M., & Phillipson, L. (2010). Indications of bow and stone-tipped arrow use 64000 years ago in KwaZulu-Natal, South Africa. *Antiquity, 84*, 1–14.

Mackay, A., & Welz, A. (2008). Engraved ochre from a Middle Stone Age context at Klein Kliphuis in the Western Cape of South Africa. *Journal of Archaeological Science, 35*(6), 1521–1532.

Marean, C. W., & Assefa, Z. (2005). The Middle and Upper Pleistocene African record for the biological and behavioral origins of modern humans. In A. B. Stahl (Ed.), *African archaeology* (pp. 93–129). New York: Blackwell.

Mason, R. J. (1993). The Cave of Hearths, Makapansgat, Transvaal, South Africa 1937-1988. *Anthropologie, 97*(1), 85–95.

Mayer, D. E. B. Y., Vandermeersch, B., & Bar-Yosef, O. (2009). Shells and ochre in Middle Paleolithic Qafzeh Cave, Israel: Indications for modern behavior. *Journal of Human Evolution, 56*(3), 307–314.

Mazza, P., Martini, F., Sala, B., Magi, M., Colombini, M. P., Giachi, G., et al. (2006). A new Palaeolithic discovery: Tar-hafted stone tools in a European Mid-Pleistocene bone-bearing bed. *Journal of Archaeological Science, 33*, 1310–1318.

McBrearty, S., & Brooks, A. S. (2000). The revolution that wasn't: A new interpretation of the origin of modern human behaviour. *Journal of Human Evolution, 39*, 453–563.

Miller, G. H., Beaumont, P. B., Deacon, H. J., Brooks, A. S., Hare, P. E., & Jull, A. J. T. (1999). Earliest modern humans in southern Africa dated by isoleucine epimerization in ostrich eggshell. *Quaternary Science Reviews, 18*, 1537–1548.

Mourre, V., Villa, P., & Henshilwood, C. S. (2010). Early use of pressure flaking on lithic artifacts at Blombos Cave, South Africa. *Science, 330*(6004), 659–662.

Parkington, J. E., Rigaud, J. P., Poggenpoel, C., Porraz, G., & Texier, P. J. (2013). Introduction to the project and excavation of Diepkloof Rock Shelter (Western Cape, South Africa): A view on the Middle Stone Age. *Journal of Archaeological Science, 40*(9), 3369–3375.

Pearson, O. M. (2008). Statistical and biological definitions of "anatomically modern" humans: Suggestions for a unified approach to modern morphology. *Evolutionary Anthropology, 17*, 38–48.

Peers, B. (1929). *Preliminary report on the archaeology of the Fish Hoek-Noord Hoek valley*. Unpublished manuscript, South African Museum, Cape Town.

Pettitt, P. B. (2002). The Neanderthal dead: Exploring mortuary variability in Middle Palaeolithic Eurasia. *Before Farming, 1*, 1–19.

Plug, I. (1982). Bone tools and shell, bone and ostrich eggshell beads from Bushman Rock Shelter (BRS), eastern Transvaal. *The South African Archaeological Bulletin, 37*(136), 57–62.

Plug, I. (2004). Resource exploitation: Animal use during the Middle Stone Age at Sibudu Cave, KwaZulu-Natal. *South African Journal of Science, 100*, 151–158.

Powell, A., Shennan, S., & Thomas, M. (2009). Late Pleistocene demography and appearance of modern human behavior. *Science, 324*, 1298–1301.

Richter, D., Moser, J., Nami, M., Eiwanger, J., & Mikdad, A. (2010). New chronometric data from Ifri n'Ammar (Morocco) and the chronostratigraphy of the Middle Palaeolithic in the Western Maghreb. *Journal of Human Evolution, 59*(6), 672–679.

Robbins, L. H., Campbell, A. C., Brook, G. A., Murphy, M. L., & Hitchcock, R. K. (2012). The antiquity of the bow and arrow in the Kalahari Desert: Bone points from White Paintings Rock Shelter, Botswana. *Journal of African Archaeology, 10*(1), 7–20.

Sampson, C. G. (1974). *The Stone Age archaeology of southern Africa*. New York: Academic Press.

Scerri, E. M. L. (2013). The Aterian and its place in the North African Middle Stone Age. *Quaternary International, 300*, 111–130.

Schwenninger, J.-L., Collcutt, S. N., Barton, N., Bouzouggar, A., Clark-Balzan, L., El Hajraoui, M. A., et al. (2010). A new luminescence chronology for Aterian cave sites on the Atlantic coast of Morocco. In E. A. A. Garcea (Ed.), *South-Eastern Mediterranean peoples between 130,000 and 10,000 years ago* (pp. 18–36). Oxford: Oxbow Books.

Shea, J. J. (2011). *Homo sapiens* is as *Homo sapiens* was. *Current Anthropology, 52*, 1–35.

Shennan, S. (2001). Demography and cultural innovation: A model and its implications for the emergence of modern human culture. *Cambridge Archaeological Journal, 11*, 5–16.

Singer, R., & Wymer, J. (1982). *The Middle Stone Age at Klasies River Mouth in South Africa*. Chicago: Chicago University Press.

Soressi, M., & d'Errico, F. (2007). Pigment, gravures, parures: les comportements symboliques contro-versés des Néandertaliens. In B. Vandermeersch & B. Maureille (Eds.), *Les Néandertaliens. Biologie et cultures* (pp. 297–309). Paris: CTHS.

Sterelny, K. (2012). Language, gesture, skill: The co-evolutionary foundations of language. *Philosophical Transactions of the Royal Society of London, 367*(1599), 2141–2151.

Stringer, C. (2002). Modern human origins: Progress and prospects. *Philosophical Transactions of the Royal Society of London, 357*, 563–579.

Texier, J.-P., Porraz, G., Parkington, J., Rigaud, J.-P., Poggenpoel, C., Miller, C., et al. (2010). A Howiesons Poort tradition of engraving ostrich eggshell containers dated to 60,000 years ago at Diepkloof Rock Shelter, South Africa. *Proceedings of the National Academy of Sciences, 107*, 7621–7622.

Texier, P.-J., Porraz, G., Parkington, J., Rigaud, J.-P., Poggenpoel, C., & Tribolo, C. (2013). The context, form and significance of the MSA engraved ostrich eggshell collection from Diepkloof Rock Shelter, Western Cape, South Africa. *Journal of Archaeological Science, 40*, 3412–3431.

Tixier, J. (1967). Procédés d'analyse et questions de terminologie concernant l'étude des ensembles industriels du Paléolithique récent et de l'Epipaléolithique dans l'Afrique du nord-ouest. In W. W. Bishop & J. D. Clark (Eds.), *Background to evolution in Africa* (pp. 771–820). Chicago: University of Chicago Press.

Villa, P., Delagnes, A., & Wadley, L. (2005). A late Middle Stone Age artefact assemblage from Sibudu (KwaZulu-Natal): Comparisons with the European Middle Palaeolithic. *Journal of Archaeological Science, 32*, 399–422.

Villa, P., & Lenoir, M. (2006). Hunting weapons of the Middle Stone Age and the Middle Palaeolithic: Spear points from Sibudu, Rose Cottage and Bouheben. *Southern African Humanities, 18*(1), 89–122.

Villa, P., Soriano, S., Tsanova, T., Degano, I., Higham, T. F., d'Errico, F., et al. (2012). Border Cave and the beginning of the Later Stone Age in South Africa. *Proceedings of the National Academy of Sciences, 109*(33), 13208–13213.

Vogel, J. C. (2001). Radiometric dates for the Middle Stone Age in South Africa. In P. V. Tobias, M. A. Raath, J. Moggi-Cecchi, & G. A. Doyle (Eds.), *Humanity from African naissance to coming millennia* (pp. 261–268). Florence: Firenze University Press.

Vogelsang, R., Richter, J., Jacobs, Z., Eichhorn, B., Linseele, V., & Roberts, R. (2010). New excavations of Middle Stone Age deposits at Apollo 11 rockshelter, Namibia: Stratigraphy, archaeology, chronology and past environments. *Journal of African Archaeology, 8*, 185–210.

Voigt, E. A. (1983). Mapungubwe. An archaeological interpretation of an Iron Age community. *Transvaal Museum Monograph* no. 1.

Wadley, L. (1987). *Later Stone Age hunters and gatherers of the southern Transvaal: Social and ecological interpretations. British Archaeological Reports International Series 380*. Oxford: Archaeopress.

Wadley, L. (1989). Legacies of the Later Stone Age. *South African Archaeological Society, Goodwin Series, 6*, 42–53.

Wadley, L. (1993). The Pleistocene Later Stone Age south of the Limpopo River. *Journal of World Prehistory, 7*, 243–296.

Wadley, L. (2000). The early Holocene layers of Rose Cottage Cave, eastern Free State: Technology, spatial patterns and environment. *South African Archaeological Bulletin, 55*, 18–31.

Wadley, L. (2005). A typological study of the final Middle Stone Age stone tools from Sibudu Cave, KwaZulu-Natal. *South African Archaeological Bulletin, 60*(182), 51–63.

Wadley, L. (2006). Partners in grime: Results of multi-disciplinary archaeology at Sibudu Cave. *Southern African Humanities, 18*(1), 315–341.

Wadley, L. (2008). The Howieson's Poort industry of Sibudu Cave. *Goodwin Series, 10*, 122–132.

Wadley, L. (2010a). Compound-adhesive manufacture as a behavioral proxy for complex cognition in the Middle Stone Age. *Current Anthropology, 51*(S1), S111–S119.

Wadley, L. (2010b). Were snares and traps used in the Middle Stone Age and does it matter? A review and a case study from Sibudu, South Africa. *Journal of Human Evolution, 58*(2), 179–192.

Wadley, L., Hodgskiss, T., & Grant, M. (2009). Implications for complex cognition from the hafting of tools with compound adhesives in the Middle Stone Age, South Africa. *Proceedings of the National Academy of Sciences, 106*(24), 9590–9594.

Wadley, L., & Jacobs, Z. (2004). Sibudu Cave KwaZulu-Natal: Background to the excavations of Middle Stone Age and Iron Age occupations. *South African Journal of Science, 100*, 145–150.

Wadley, L., & Jacobs, Z. (2006). Sibudu Cave: Background to the excavations and dating. *Southern African Humanities, 18*(1), 1–26.

Wanless, A. (2007). *The silence of colonial melancholy: The Fourie collection of Khoisan ethnologica*. Ph.D. thesis, University of the Witwatersrand.

Watts, I. (2009). Red ochre, body-painting, and language: Interpreting the Blombos ochre. In R. Botha & C. Knight (Eds.), *The cradle of language* (pp. 62–92). Oxford: Oxford University.

Watts, I. (2010). The pigments from Pinnacle Point Cave 13B, Western Cape, South Africa. *Journal of Human Evolution, 59*, 392–411.

Wendt, W. E. (1976). '*Art mobilier*' from the Apollo 11 Cave, South West Africa: Africa's oldest dated works of art. *South African Archaeological Bulletin, 31*, 5–11.

Wurz, S. (1999). The Howiesons Poort backed artefacts from Klasies River: An argument for symbolic behaviour. *South African Archaeological Bulletin, 54*, 38–50.

Wurz, S. (2002). Variability in the Middle Stone Age lithic sequence, 115,000–60,000 years ago at Klasies River, South Africa. *Journal of Archaeological Science, 29*(9), 1001–1015.

Yellen, J. E. (1998). Barbed bone points: Tradition and continuity in Saharan and sub-Saharan Africa. *African Archaeological Review, 15*, 173–198.

Yellen, J. E., Brooks, A. S., Cornelissen, E., Mehlman, M. J., & Stewart, K. (1995). A Middle Stone Age worked bone Industry from Katanda, Upper Semliki Valley, Zaïre. *Science, 268*, 553–556.

Zhilin, M. G., Savchenko, S., Nikulina, E., Schmölcke, U., Hartz, S., & Terberger, T. (2014). Eleven bone arrowheads and a dog coprolite – the Mesolithic site of Beregovaya 2, Urals region (Russia). *Qüartar, 61*, 165–187.

Zilhão, J., Angelucci, D. E., Badal-García, E., d'Errico, F., Daniel, F., Dayet, L., et al. (2009). Symbolic use of marine shells and mineral pigments by Iberian Neandertals. *Proceedings of the National Academy of Sciences, 107*, 1023–1028.

Chapter 3
Bone Point Functional Diversity: A Cautionary Tale from Southern Africa

Justin Bradfield

Abstract In this chapter I present the results of a use-trace study conducted on 357 pointed bone tools from terminal Pleistocene and Holocene assemblages in southern Africa. All the bone points considered here conform to the morphological criteria of projectile arrow heads, as defined by analogy to historic Bushman arrows. Use-wear and residue traces consistent with wood-working and hide processing reveal that not all bone points functioned as projectile armatures in the past. Functional diversity is evident only during the last 6000 years. Bone points from the Pleistocene are routinely subject to rigorous use-wear analyses to establish their function, yet it is often taken for granted that similar tools found in the more recent Holocene were used as projectile tips. This paper cautions against the specious imputation to projectile technology of all bone points based solely on morphometric criteria.

Keywords Artefact function • Arrow points • Microscopy • Use-traces

Introduction

In southern Africa, as in the rest of the world, Stone Age societies are understood largely in terms of their technology. The ways in which we frame our research and understanding of these past societies are based almost exclusively on stone tools and ceramics, yet these materials represent only a small portion of traditional hunter-gatherer paraphernalia and do not necessarily reflect the complexity of cultural adaptation and technological achievements of past societies (Hayden 1979; Binford 1981; O'Connor et al. 2014). Studying components of past technological systems in isolation risks creating a distorted image of the pasts of past societies (Hayden 1979; van Gijn 2007).

Although osseous technology may have been a significant aspect of past societies, it constitutes a comparatively minor component of the archaeological record. Bone is dependent for its survivability on a host of factors (Berner 1971; Olsen 2007; Choyke and Daróczi-Szabó 2010), which means that the archaeological record of bone-tool use will be inevitably incomplete and skewed in favor of those few sites and regions that are best suited for organic preservation. Bone can survive for great lengths of time in suitable environments. For example, they have been found in ~1.5 Ma deposits associated with *Australopithecus* and *Paranthropus* remains (Backwell and d'Errico 2001, 2008) and in a few Middle Stone Age (c. 300–20 ka) sites in southern Africa, where some examples are thought to represent, together with small stone segments, early evidence for the use of mechanically projected weapon systems like the bow and arrow (d'Errico and Henshilwood 2007; Backwell et al. 2008; Lombard and Phillipson 2010; Bradfield and Lombard 2011; Lombard 2011). But it is not until the Holocene that pointed bone tools occur with any degree of regularity in this region. The extent to which this record is a reflection of taphonomy or technological choice is still poorly understood.

In this chapter, I examine the class of bone artifacts known simply as 'bone points'. Bone points are often assumed to be projectile weapon tips based on morphological analogy with recent San/Bushman arrows (e.g., Schweitzer 1979; Bradfield 2014). However, as studies in other parts of the world have shown, a high degree of functional variability may be seen within morphological tool classes (e.g., Chomko 1975; Becker 2001; St-Pierre 2007). Attributing projectile function based solely on the shape of an object is a misleading practice and can have adverse implications for our understanding of past techno-cultural variability. Here I present the results of a use-trace study on a collection of 357 pointed bone artifacts from eight southern African archaeological sites spanning the last 18,000 years. Morphological residue

J. Bradfield (✉)
Evolutionary Studies Institute, University of the Witwatersrand, Johannesburg, South Africa
e-mail: jbradfield8@gmail.com

Michelle C. Langley (ed.), *Osseous Projectile Weaponry*, Vertebrate Paleobiology and Paleoanthropology,
DOI 10.1007/978-94-024-0899-7_3

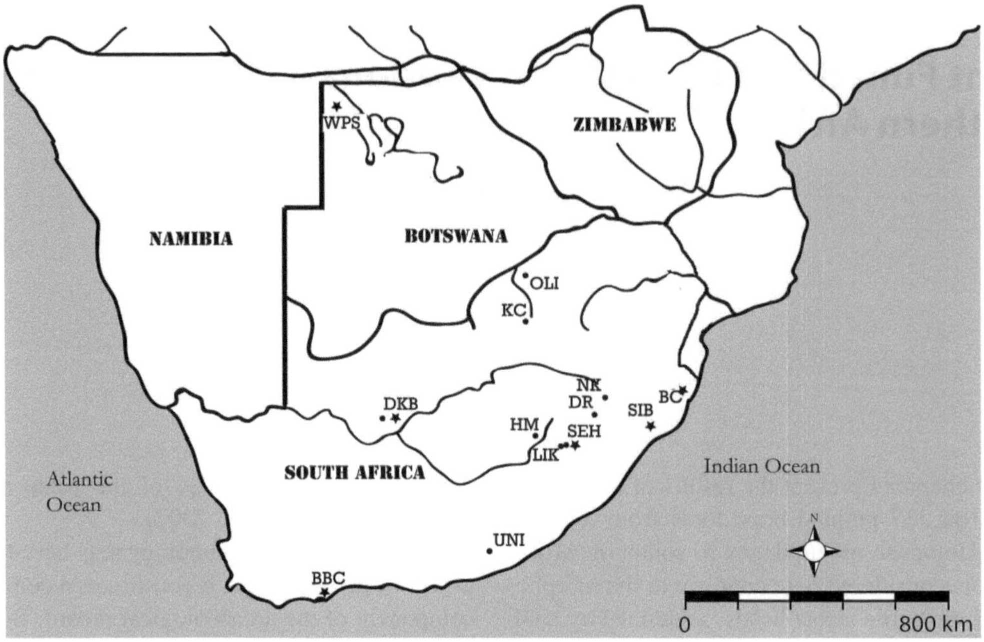

Fig. 3.1 Map of southern Africa showing archaeological sites referred to in the text. Starred sites represent Pleistocene assemblages while dots represent Holocene assemblages. BBC—Blombos Cave; BC—Border Cave; DKB—Dikbosch; DR—Driel; HM—Ha Makotoko; KC—Kruger Cave; LIK—Likoaeng; NK—Nkupe; OLI—Oliboomspoort; SEH—Sehonghong; SIB—Sibudu; UNI—Uniondale; WPS—White Paintings Shelter

analysis, use-wear, and macrofracture analysis were conducted on bone tools that meet the morphological criteria of arrow points in the southern African typological system (Bradfield 2014). In particular, the variable functions within the bone point tool class from two terminal Pleistocene sites and nine Holocene sites are outlined and discussed (Fig. 3.1).

Background

The Pleistocene of southern Africa is a period from which few bone tools survive (Fig. 3.2). The earliest manifestation of purported bone projectile components in southern Africa comes from Sibudu Cave, where a number of specialized bone tools have been recovered and a local tradition of bone tool manufacture identified from ~61 to 72 ka levels (Backwell et al. 2008; d'Errico et al. 2012a; Backwell and d'Errico 2016). After the Howieson's Poort industry (~66–58 ka), there appears to have been a hiatus of nearly 40 kyr in bone tool manufacture, during which time bone may have been substituted for more perishable materials like wood (sensu O'Connor et al. 2014). Thus far bone points have been found at only two sites in southern Africa dating to this period, namely, White Paintings Shelter and Border Cave (Fig. 3.1). At both sites decoration on the points has been interpreted as indicating the antiquity of recent hunter-gatherer material culture and hunting-ritual practices akin to those documented in the 1960s (d'Errico et al. 2012b; Robbins et al. 2012). The dimensions of these bone points, which are in keeping with the range of historic and ethnographic examples, are cited as additional evidence that the points must have been intended to be reversible and poisoned, and therefore represent the same social organization, world view, and symbolic systems as ethnographically documented hunter-gatherers in the Kalahari (Robbins et al. 2012; also see Plug 2012).

At a number of terminal Pleistocene sites in southern Africa, stone bladelet-based assemblages have been interpreted as signaling the use of composite tools with multiple lithic insets hafted around a bone or wood shaft (Clark 1977; Lombard and Parsons 2008). Several hafting configurations have been suggested based on ethnographic arrow collections housed in various museums (Lombard and Parsons 2008), although only a transversally hafted example has been recovered from an archaeological context (Fig. 3.3; Binneman 1994). Bone points occur infrequently during the terminal Pleistocene, although they are present at a number of sites (Bradfield 2014) where they invariably have been interpreted as arrow points (e.g., Humphreys and Thackeray 1983; Mitchell 1995). It is not until 12 ka, however, that bone tools occur more regularly in archaeological deposits (Deacon and Deacon 1999; Mitchell 2002).

The terminal Pleistocene/Holocene boundary in southern Africa is marked by a number of environmental and techno-

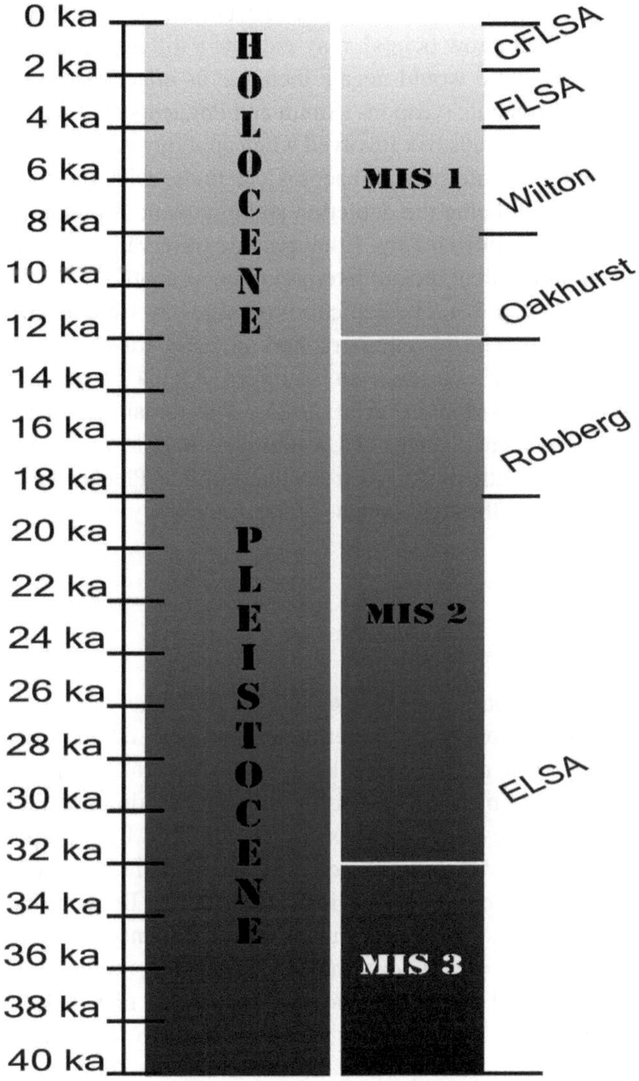

Fig. 3.2 Timeline showing chronology of southern African Later Stone Age technocomplexes according to Lombard et al. (2012). *Abbreviations*: ELSA = early Later Stone Age; FLSA = final Later Stone Age; CFLSA = ceramic final Later Stone Age

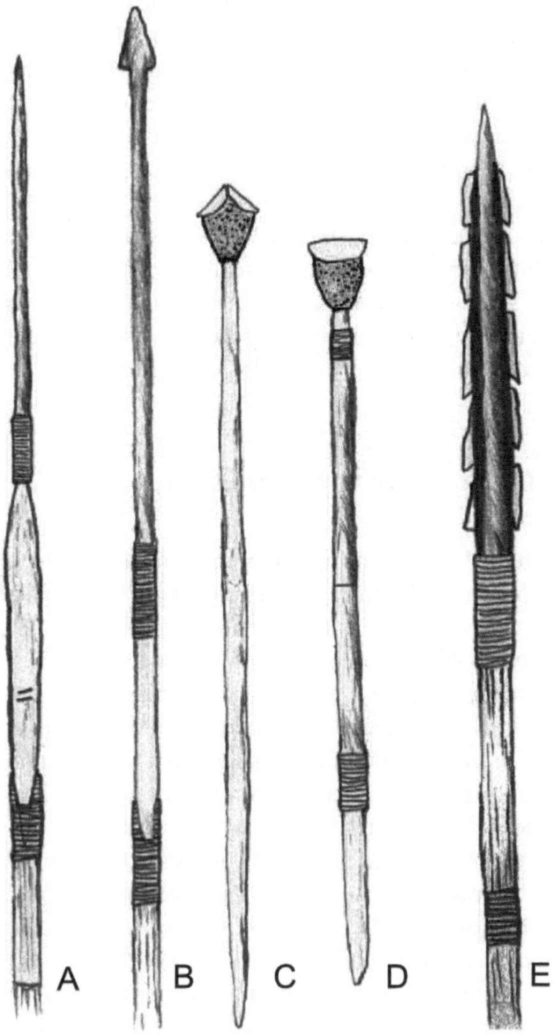

Fig. 3.3 Reproduction of four types of southern African Bushman arrows employing a bone component (after Bradfield 2015a). (**a**) Reversible bone point with link-shaft; (**b**) tanged bone arrowhead; (**c**, **d**) stone segment mounted on the end of a bone point (sensu Goodwin 1945); (**e**) hypothetical reconstruction of bone arrowhead with stone insets mounted down the side (after Lombard and Parsons 2008)

logical changes. The period after 12 ka saw a steady shift from open grassland habitats to more mesic environments in much of the sub-continent (Deacon 1978). This shift was paralleled by a concomitant focus on more predictable subsistence resources such as geophytes, shellfish and tortoises, and by the replacement of large gregarious grazers with smaller non-gregarious species (Deacon 1976; Mazel 1989; Mitchell 2002). At most archaeological sites dating to this period there is a decrease in frequency of backed microliths and an increase in frequency of stone scrapers (Deacon 1976; Mazel 1989; Mitchell 2002). This change is thought to reflect a technological shift from lithic-based multi-barbed spears to arrows tipped with bone points (Wadley 1987). This evidence is supported by an accompanying change in stone flaking technique, raw material preference and an increase in the size of untrimmed flakes and scrapers (Klein 1972; Deacon 1984; Inskeep 1987; Mitchell 1988). This technological change probably was precipitated by a shift in environmental conditions 2–4 ka earlier, which resulted in the disappearance of grassland, a rise in sea level and consequently the extinction of certain mega-faunal species (Klein 1972; Deacon 1984). Whether for taphonomic or subsistence reasons, this period is also marked by increased visibility of bone points in the archaeological record (Mitchell 2000, 2002). The decrease in microliths associated with increased frequencies of bone points suggests a reliance on simple bone point technology, without the addition of stone inserts.

During the historic period a number of arrow types employing a bone component were observed among the

Bushmen, some of which are reproduced here (Fig. 3.3; Bradfield 2015a). Lichtenstein (1930 [1812]) observed that different arrows were used by Bushmen for different purposes, while a Bushman informant once told ethno-linguist Lucy Lloyd that certain arrows were used to hunt specific game (Goodwin 1945:439). Different arrows could also indicate different hunting techniques (Deacon 1976) or cross-cultural relationships (Clark 1977). The most common Bushman arrow type is that which is tipped by a simple bone point (Fig. 3.3; Goodwin 1945; Bradfield 2015a), and it is this arrow type that is generally thought to be represented by the bone points in the archaeological record.

Equating Form with Function

The term 'bone point' is a morpho-functional category that has come to assume a degree of notoriety in functional studies, as it can veil all manner of pointed bone tools (Bradfield 2015b). While the definition of the bone point was based on morphology, the ascribed function was, at best, based on historical eye-witness descriptions (e.g., Sparrman 1977 [1786]), and at worst, simply on intuition, because "direct evidence for function is lacking" (see Schweitzer 1979:139). In southern African archaeology it is conventional to refer to bone tools by assumed functional names, which include terms such as awls, link-shafts, arrows, needles and spears (e.g., Goodwin and Van Riet Lowe 1929; Deacon and Deacon 1999; Mitchell 2002). The probable functions of the various pointed bone artifacts are based primarily on analogy with ethnographic tools of similar morphology. Equating form with function can be misleading, however, as it denies the possibility of multiple or alternative uses and ignores the morphological variability within particular tool classes (St-Pierre 2007).

In particular, how we lump bone points into morpho-functional categories is an inherent problem. Bone points have an air of uniformity all over the world (LeMoine 2001) and, while only a limited amount of variability is possible while still allowing for optimal functionality (Knecht 1997), such variation is equally likely to reflect social choices (Guthrie 1983; LeMoine 2001). Through use-trace analyses we see that bone points, usually assumed to be components of arrows, can encompass a large variety of quite different implements and comprise a broad variety of shapes (Becker 2001). A similar conclusion has been reached for 'awls', one of the most ambiguous categories among archaeological bone tools (Legrand and Sidéra 2007; Olsen 2007). Chomko's (1975) study revealed that awls within a single morphological type occasionally displayed evidence of divergent uses, while awls of different morphological types displayed evidence of similar uses. In South Africa, the curvature of some worked bone pieces that would otherwise fall into the same typological category defined by Goodwin and Van Riet Lowe (1929) as 'bone points' [read arrow points], may indicate a different function, as the curvature would negate their use as effective components of projectile weapons (Smith and Poggenpoel 1988).

There is always risk involved when applying ethnographic analogues to interpret archaeological finds, with the unintended result being the depiction and treatment of prehistory as linear and unchanging. Ethnographic observations should inform rather than dictate interpretation, as southern African rock art has shown. The depictions of large bows and fletched arrows, not seen by ethnographers or early travelers, may indicate their existence in southern Africa in the pre-colonial past (Manhire et al. 1985). The remainder of this chapter looks at changes and continuities in bone point technology in southern Africa from the terminal Pleistocene to the Holocene through the lens of use-trace analyses.

Methods

Modern interpretations of ancient bone tool functions rely on microscopic analysis, experimentation and ethnographic analogy (Olsen 2007; Bradfield 2015c; Évora 2015). The most reliable means of assessing the past function/s of tools is through use-trace analyses, which can shed light on past activities for which no direct evidence remains (e.g., hide working, basketry and weaving; Stone 2013). Bone surface modification constitutes a crucial line of evidence for investigating issues as diverse as site formation, taphonomic processes and ritual behaviors (Cook 1986; Fisher 1995; Russell 2001; Choyke and Daróczi-Szabó 2010). Here I briefly present the methods I used to analyze the bone tools.

All manufacturing and use-related traces were recorded using a Celestron® handheld digital microscope (model #44302-A) at 10×–50× and 100×–150× magnification. In some cases the specimens were further analyzed using an Olympus binocular light microscope (model #SZX16) with magnifications of between 10× and 110×. Polish was recorded and described in terms of its luster, extent, orientation and placement in relation to other traces. In most instances (but not all), polish develops along a linear spectrum—the more intense the polish the longer the duration of use and vice versa. Next, the use-related striations were recorded and described. Following the rule of superimposition, striations overlying polish or other striations were interpreted as use-related. The direction, orientation and shape of striations may all yield information about the probable function of a particular pointed bone tool.

Ancient micro-residues as well as modern contaminants were identified in the first instance using the Celestron® microscope. The residues were photographed and their placement and orientation on the tools were recorded. Where it was felt that the residue/substance required further investigation, the residue was lifted using a cellulose-based adhesive peel and analyzed under laboratory conditions using an Olympus BX51M light microscope with polarizing and cross-polarizing capabilities and using Bright- and Dark-Field illumination. Magnifications ranged from 50× to 500×, although, typically, 200× proved sufficient for most residue interpretations. Residues were interpreted based on their morphological traits (see Lombard and Wadley 2007; Högberg et al. 2009; Bradfield 2015b).

Often the clearest and most obvious traces found on bone tools result from manufacture rather than use. Two types of manufacturing techniques were identified on the southern African sample examined: (1) scraping parallel to the long axis, and (2) abrasive grinding diagonal to the long axis of the bone. At most sites both techniques appear to have been used, although there is a tendency for older bone points to favor longitudinal scraping rather than coarse abrasive diagonal grinding—but this is not true for all sites. The orientation of the striations, I suggest, is an indicator of the dextrality of the maker (Bradfield 2015b, submitted). Use-wear and residue indicators support diagnostic impact fracture results in 80 % of cases.

Use-Trace Indicators of Bone Point Functional Diversity

Use-wear consisted primarily of polish and striations, although other features such as pitting and general surface topography were also considered. On the vast majority of specimens examined, use-wear features were fairly uninformative, meaning that they were insufficiently developed to isolate a specific activity. Poorly formed, indistinct use-wear is nevertheless consistent with hunting. Use-wear becomes diagnostic only after prolonged, fricative contact—unlikely to be occasioned through hunting (LeMoine 1994; Buc and Loponte 2007). However, a few bone points from the Holocene deposits at six archaeological sites displayed signs of use-wear that were diagnostic of hide working (n=4) and wood working (n=11; Bradfield 2014; 2015b). These activities were inferred based on the presence of characteristic use-wear and residues (see LeMoine 1994; Francis 2002; Lombard 2008). Figure 3.4 presents examples of these features. Because of the length of time needed for distinctive use-wear to accrue, we can be sure that these traces did not occur through incidental contact during hunting. Use-wear and residue indicators support diagnostic impact fracture results in 80 % of cases.

Possible hafting residues, in the form of gum, resin and woody parenchyma cells, were found only on bone tools younger than 4 ka. Use-wear evidence suggestive of wood-working is present during the same period. Animal and plant residues are rare in the Robberg and Oakhurst technocomplexes, but are prevalent on bone tools during the last 6 ka. Rodent hairs were recovered embedded in possible poison residues from two bone points, perhaps supporting faunal data that indicate a more pronounced subsistence focus on smaller animals during the Holocene compared with that of the terminal Pleistocene. Putative poison residues were identified based on their consistency and placement, and were present on tools from the mid-to late-Wilton technocomplex onwards. It is still uncertain to what extent this pattern is a reflection of taphonomic conditions. The concomitant increase in tool width during this period, however, unaffected by taphonomy, would seem to argue in favor of their integrity.

Apart from residues and use-wear, a number of other features were noted on the bone tools from the later Holocene assemblages that are conspicuous by their absence on the Pleistocene bone tools. Evidence of deliberate snapping of the distal ends of bone points and the presence of circumferential chipping around the break facet may indicate the attachment of an additional element, such as a metal collar, similar to those seen among historic arrow collections (Bradfield 2015a). In addition, there is evidence that some of the pieces from the Holocene levels were reused after they broke—also absent in Pleistocene levels.

Conclusion

Although I have focused here on formally fashioned bone tools, prehistoric osseous technology was not necessarily limited to these 'formal' tools. Informal stone and bone flakes are just as likely to have been used as formally retouched tools (see Stow 1905; Plug 2012). Having said this, at Sibudu various specialized tool types have been identified in Pleistocene contexts, some of which appear to have Holocene counterparts and which appear to have been used in a variety of tasks and on different materials (d'Errico et al. 2012a). The intermittent nature of these artifacts does not necessarily mean that the technology was lost or abandoned; bone could be substituted for more perishable materials such as wood (see O'Connor et al. 2014). As an arrow armature, wood is just as effective at penetrating animal skin as bone or stone (Waguespack et al. 2009). Indeed, wooden arrows have been recovered from mid-Holocene deposits (Manhire 1993), and are present in the twentieth-century Fourie collection, housed in Museum Africa. It has been said, with reference to bone technology in southern Africa, that the similarity of bone points through time could signal the antiquity of certain behaviors or practices (Plug 2012). Indeed, bone technology present at Border Cave at roughly 40 ka, has been interpreted as indicating the early emergence of quintessential Bushman culture (d'Errico et al. 2012b).

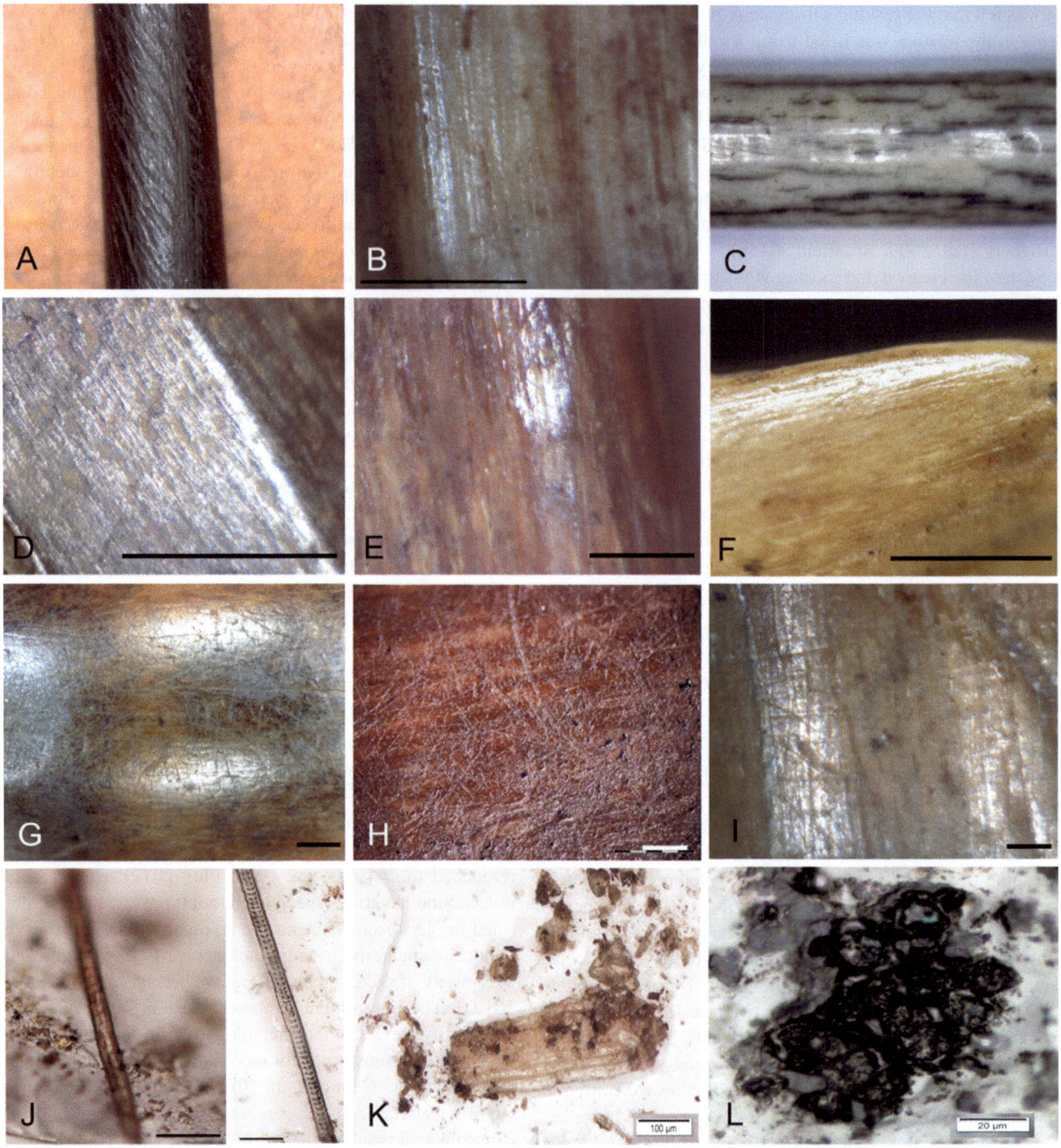

Fig. 3.4 Examples of use-trace indicators discussed in the text. (**a**) Diagonal grinding striations on 6480 IC2 from Dikbosch; (**b**) longitudinal scraping striations on SF105 from Likoaeng; (**c**) chattermarks on 1886.1.498.2 from the Dunn collection housed at the Pitt Rivers Museum; (**d**) hide-working use-wear on KC313 from Kruger Cave; (**e**) hide-working use-wear on SF598SA from Sehonghong (**f**) hide-working use-wear on KC865 from Kruger Cave; (**g**, **h**) wood-working use-wear on KC2688 from Kruger Cave (**i**) wood-working use-wear on R11 WA3A from Nkupe; (**j**) rodent hair strand on 2829DB34 from Likoaeng; (**k**) woody parenchyma tissue on SF095DC from Sehonghong; (**l**) blood residue on SF055GWA from Sehonghong. Scale bars represent 1 mm unless otherwise specified

A similar picture is emerging at White Paintings Shelter, where decoration on bone points has been imputed to the antiquity of Bushman hunting practices (Robbins et al. 2012), and by extension, Bushman culture. I am skeptical about this point, as decorations can mean different things to different people at different times. The presented argument assumes that decorations were applied to arrows for the same reasons over extensive periods, However, just as cultures are constantly changing and evolving into new forms, so too might different cultures have existed in the past (Barnard 1992). Decorations on bone arrow components do not necessarily indicate marks of ownership, nor do they imply !Kung-like meat distribution practices (sensu Marshall 1976; Lee 1979). Nor does the similarity in tool form necessarily suggest cultural continuity (see Hodder and Hutson 2003). The uniformity of bone tools, both in finished form and manufacturing technique, is not unique to southern Africa. Bone is optimally shaped using the groove-and-splinter technique, followed by grinding against an abrasive surface—this technique being utilized almost universally (e.g., Clark and Thompson 1953; Semenov 1964; Newcomer 1974; Morrison 1986; Smith and Poggenpoel 1988; Knecht 1997; Choyke and Bartosiewicz 2001; St-Pierre 2007; Legrand-Pineau et al. 2010; Rabett and Piper 2012a, b). While cortical bone can be flaked similarly to stone, it is ill-suited to produce a sharp edge (Johnson 1985; Fisher 1995), and so a pointed shape is the most logical outcome for a bone-tipped hunting weapon. In other words, there is a limited amount of variability in bone tool shape if penetrative function is to be maintained (Knecht 1997).

A more profitable line of investigation for bone tool technology is determining the uses to which individual bone tools were put. The results of this use-trace study of 357 bone points revealed a wider range of functions for pointed bone artifacts morphologically akin to arrow points. Evidence of wood or plant working, as well as prolonged hide working, was evident on some specimens, albeit the vast minority, that would otherwise have been interpreted as hunting weapons if only morphology had been considered (Bradfield 2015b). These specimens all came from the Holocene assemblages. Unsurprisingly, the majority of specimens conform to hunting related use-wear/residues, although most of the micro-wear features are from manufacture. No use-trace features such as invasive polish or developed striations that would contra-indicating hunting were found on the Oakhurst and Robberg samples. Thus it appears that bone points were more functionally versatile during the last 2 ka than previously thought.

There does not appear to be a marked 'innovative technological production' or clear linear evolution of bone point design and manufacture in response to known environmental or demographic fluctuations or concomitant with changing lithic technocomplexes (sensu Bousman 2005). Bone points in southern Africa start displaying variability only during the last 6 ka while manufacturing techniques remain much the same for the 18 ka covered in my study (Bradfield 2014). The most noticeable changes occur during the Wilton and after the widespread adoption of metal as a component of the arrows during the ceramic final Later Stone Age (also see Deacon 1992). In other words, changes in bone points seem to occur during rather than at the boundaries of lithic technocomplexes.

It is generally recognized that organic technology played a major role in hunter-gatherer societies from early on (see Hayden 1979; Binford 1981; St-Pierre and Walker 2007; d'Errico et al. 2012a). Bone tools, especially completely ground and polished bone points, are highly curated artifacts, taking much longer to make than stone tools (Knecht 1997), suggesting that they may have had a value, either intrinsic or symbolic, to the people who made and owned them (sensu Hurcombe 2007). My own work, conducted primarily on Holocene assemblages, has shown that not all bone points functioned as projectile armatures (Bradfield 2014; 2015b). While more and more evidence is emerging that bow hunting was established very early on in southern Africa (e.g., Lombard and Phillipson 2010; Lombard 2011; d'Errico et al. 2012b; Robbins et al. 2012), we must be careful not to impute too much based solely on morphological similarity or the simple presence of decorations. In so doing, we risk missing the interesting range of variation in bone point function.

Bone and other organic artifacts, however, will always remain under-represented in the archaeological record and our understanding of the diversity of bone tools will remain meager until such time as they receive the same attention as stone tools. Use-wear and residue studies have enormous potential for understanding the diversity of past functions of bone artifacts, but are seldom included in analyses. Similar to the lithic component of archaeological sites, bone tools deserve closer scrutiny. It is to be hoped that the papers in this volume go some way towards achieving this aim.

Acknowledgements My thanks go to Michelle Langley for inviting me to contribute to this volume. I also wish to thank the South African National Department of Arts and Culture, the Palaeontological Scientific Trust (PAST) and its Scatterlings of Africa programs, the South African Archaeological Society's Kent Bequest, the Oppenheimer Fund, and the University of Johannesburg for their financial support. My thanks also to the numerous individuals and institutions for granting me access to material. Finally, I thank my reviewers for their constructive criticisms of an earlier draft of this manuscript.

References

Backwell, L., & d'Errico, F. (2001). Evidence of termite foraging by Swartkrans early hominids. *Proceedings of the National Academy of Sciences of the United States of America, 98*, 1358–1363.

Backwell, L., & d'Errico, F. (2008). Early bone tools from Drimolen, South Africa. *Journal of Archaeological Science, 35*, 2880–2894.

Backwell, L., d'Errico, F., & Wadley, L. (2008). Middle Stone Age bone tools from the Howiesons Poort layers, Sibudu Cave, South Africa. *Journal of Archaeological Science, 35*, 1566–1580.

Backwell, L. R., & d'Errico, F. (2016). Osseous projectile weaponry from early to late Middle Stone Age Africa. In M. C. Langley (Ed.), *Osseous projectile weaponry: Towards an understanding of Pleistocene cultural variability* (pp. 15–29). Dordrecht: Springer.

Barnard, A. (1992). *Hunters and herders of Southern Africa: A comparative ethnography of the Khoisan peoples*. Cambridge: Cambridge University Press.

Becker, C. (2001). Bone points – No longer a mystery? Evidence from the Slavic urban fortification of Berlin-Spandau. In A. Choyke & L. Bartosiewicz (Eds.), *Crafting bone: Skeletal technologies through time and space. British Archaeological Reports International Series 937* (pp. 129–146). Oxford: Archaeopress.

Berner, R. (1971). *Principles of chemical sedimentology*. New York: McGraw Hill.

Binford, L. R. (1981). *Bones: Ancient men and modern myths*. New York: Academic Press.

Binneman, J. (1994). A unique stone-tipped arrowhead from Adam's Kranz Cave, Eastern Cape. *Southern African Field Archaeology, 3*, 58–60.

Bousman, B. (2005). Coping with risk: Later Stone Age technological strategies at Blydelontein Rock Shelter, South Africa. *Journal of Anthropological Archaeology, 24*, 193–226.

Bradfield, J. (2014). *Bone point technology in southern Africa*. Unpublished Ph.D. thesis, University of Johannesburg, Johannesburg.

Bradfield, J. (2015a). Identifying bone-tipped arrow types in the archaeological record: The contribution of use-trace studies. *Journal of African Archaeology, 13*, 135–147.

Bradfield, J. (2015b). Pointed bone tool technology in southern Africa: results of use-trace analyses. *Southern African Humanities, 27*, 1–27.

Bradfield, J. (2015c). Use-trace analysis on bone tools: a brief overview of four methodological approaches. *South African Archaeological Bulletin, 70*, 3–14.

Bradfield, J., & Lombard, M. (2011). A macrofracture study of bone points used in experimental hunting with reference to the South African Middle Stone Age. *South African Archaeological Bulletin, 66*, 67–76.

Buc, N., & Loponte, D. (2007). Bone tool types and microwear patterns: Some examples from the Pampa region, South America. In C. G. St-Pierre & R. B. Walker (Eds.), *Bones as tools: Current methods and interpretations in worked bone studies. British Archaeological Reports International Series 1622* (pp. 143–157). Oxford: Archaeopress.

Chomko, S. (1975). Bone awls and utilized antler tines from Arnold Research Cave, 23CY64, Missouri. *The Plains Anthropologist, 20*, 27–40.

Choyke, A., & Bartosiewicz, L. (2001). *Crafting bone: Skeletal technologies through time and space. Proceedings of the 2nd meeting of the (ICAZ) Worked Bone Research Group, Budapest. British Archaeological Reports International Series 937*. Oxford: Archaeopress.

Choyke, A., & Daróczi-Szabó, M. (2010). The complete and usable tool: Some life histories of prehistoric bone tools in Hungary. In A. Legrand-Pineau & I. Sidéra (Eds.), *Ancient and modern bone artefacts from America to Russia: Cultural, technological and functional signature. British Archaeological Reports International Series 2136* (pp. 235–248). Oxford: Archaeopress.

Clark, G., & Thompson, M. (1953). The groove and splinter technique of working antler in Upper Palaeolithic and Mesolithic Europe with special reference to the material from Star Carr. *Proceedings of the Prehistoric Society, 19*, 148–160.

Clark, J. D. (1977). Interpretations of prehistoric technology from ancient Egyptian and other sources. Part II: Prehistoric arrow forms in Africa as shown by surviving examples of the traditional arrows of the San Bushmen. *Paleorient, 3*, 127–150.

Cook, J. (1986). The application of scanning electron microscopy to taphonomic and archaeological problems. In D. Roe (Ed.), *Studies in the Upper Palaeolithic of Britain and Northwest Europe. British Archaeological Reports International Series 296* (pp. 143–163). Oxford: Archaeopress.

Deacon, H. (1976). *Where hunters gathered: A study of Stone Age people in the Eastern Cape*. South African Archaeological Society Monographs 1. Claremont: South African Archaeological Society.

Deacon, J. (1978). Changing patterns in the Pleistocene/early Holocene prehistory of southern Africa as seen from the Nelson Bay Cave stone artefact sequence. *Quaternary Research, 10*, 84–111.

Deacon, J. (1984). *The Later Stone Age of southernmost Africa. Cambridge Monographs in African Archaeology 12. British Archaeological Reports International Series 213*. Oxford: Archaeopress.

Deacon, J. (1992). Arrows as agents of belief among the /Xam bushmen. In *Margaret Shaw Lecture, 3*. South African Museum: Cape Town.

Deacon, H., & Deacon, J. (1999). *Human beginnings in South Africa*. Cape Town: David Philip.

d'Errico, F., Backwell, L., & Wadley, L. (2012). Identifying regional variability in Middle Stone Age bone technology. The case of Sibudu Cave. *Journal of Archaeological Science, 39*, 2479–2495.

d'Errico, F., Backwell, L., Villa, P., Deganog, I., Lucejkog, J., Bamford, et al. (2012b). Early evidence of San material culture represented by organic artefacts at Border Cave, South Africa. *Proceedings of the National Academy of Sciences, 109*, 13214–13219.

d'Errico, F., & Henshilwood, C. (2007). Additional evidence for bone technology in the southern African Middle Stone Age. *Journal of Human Evolution, 52*, 142–163.

Évora, M. (2015). Use-wear methodology on the analysis of osseous industries. In J. Marreiros, N. Bicho, & J. Gibaja (Eds.), *Use-wear and residue analysis in archaeology, Manuals in archaeological method, theory and technique* (pp. 159–170). Switzerland: Springer International Publishing.

Fisher, J. (1995). Bone surface modifications in zooarchaeology. *Journal of Archaeological Method and Theory, 2*, 7–68.

Francis, V. (2002). Twenty interesting points: An analysis of bone artefacts from Platypus Rockshelter. *Queensland Archaeological Review, 13*, 63–70.

Goodwin, A. J. H. (1945). Some historical Bushmen arrows. *South African Journal of Science, 61*, 429–443.

Goodwin, A. J. H., & Van Riet Lowe, C. (1929). The Stone Age cultures of South Africa. *Annals of the South African Museum, 27*, 1–289.

Guthrie, R. D. (1983). Osseous projectile points: Biological considerations affecting raw material selection and design among Palaeolithic and Paleoindian peoples. In J. Clutton-Brook & C. Grigson (Eds.), *Animals and archaeology, British Archaeological Reports International Series 163* (pp. 273–294). Oxford: Archaeopress.

Hayden, B. (1979). *Palaeolithic reflections: Lithic technology and ethnographic excavation among Australian Aborigines*. Canberra: Australian Institute for Aboriginal Studies.

Hodder, I., & Hutson, S. (2003). *Reading the past: Current approaches to interpretation in archaeology* (3rd ed.). Cambridge: Cambridge University Press.

Högberg, A., Puseman, K., & Yost, C. (2009). Integration of use-wear with protein residue analysis – A study of tool use and function in the south Scandinavian Early Neolithic. *Journal of Archaeological Science, 36*, 1725–1737.

Humphreys, A. J. B., & Thankeray, A. (1983). *Ghaap and Gariep: Later Stone Age studies in the Northern Cape*. The South African Archaeological Society Monograph series 2.

Hurcombe, L. (2007). *Archaeological artefacts as material culture*. London: Routledge.

Inskeep, R. R. (1987). *Nelson Bay Cave, Cape Province, South Africa: The Holocene levels. British Archaeological Reports International Series, 351*. Oxford: Archaeopress.

Johnson, E. (1985). Current developments in bone technology. *Advances in Archaeological Method and Theory, 8*, 157–235.

Klein, R. G. (1972). The late quaternary mammalian fauna of Nelson Bay Cave (Cape Province, South Africa): Its implications for megafaunal extinctions and environmental and cultural change. *Quaternary Research, 2*, 135–142.

Knecht, H. (1997). Projectile points of bone, antler and stone: Experimental explorations of manufacture and use. In H. Knecht (Ed.), *Projectile technology* (pp. 193–212). London: Plenum Press.

Lee, R. (1979). *The !Kung San: Men, women and work in a foraging society*. Cambridge: Cambridge University Press.

Legrand, A., & Sidéra, I. (2007). Methods, means and results when studying European bone industries. In C. G. St-Pierre & R. B. Walker (Eds.), *Bones as tools: Current methods and interpretations in worked bone studies. British Archaeological Reports International Series 1622* (pp. 67–78). Oxford: Archaeopress.

Legrand-Pineau, A., Sidéra, I., Buc, N., David, E., & Scheinsohn, V. (Eds.). (2010). *Ancient and modern bone artefacts from America to Russia: Cultural, technological and functional signature. British Archaeological Reports International Series 2136*. Oxford: Archaeopress.

LeMoine, G. (1994). Use wear on bone and antler tools from the Mackenzie Delta, Northwest Territories. *American Antiquity, 59*, 316–334.

LeMoine, G. (2001). Skeletal technology in context: An optimistic overview. In A. Choyke & L. Bartosiewicz (Eds.), *Crafting bone: Skeletal technologies through time and space. British Archaeological Reports International Series 937* (pp. 1–6). Oxford: Archaeopress.

Lichtenstein, H. (1930) [1812]. *Travels in Southern Africa in the years 1803, 1804, 1805 and 1806*. Cape Town: Van Riebeeck Society No. 11.

Lombard, M. (2008). Finding resolution for the Howiesons Poort through the microscope: Micro-residue analysis of segments from Sibudu Cave, South Africa. *Journal of Archaeological Science, 35*, 26–41.

Lombard, M. (2011). Quartz-tipped arrows older than 60 ka: Further use-trace evidence from Sibudu, KwaZulu-Natal, South Africa. *Journal of Archaeological Science, 38*, 1918–1930.

Lombard, M., & Parsons, I. (2008). Blade and bladelet function and variability in risk management during the last 2000 years in the Northern Cape. *South African Archaeological Bulletin, 63*, 18–27.

Lombard, M., & Phillipson, L. (2010). Indications of bow and stone-tipped arrow use 64,000 years ago in KwaZulu-Natal, South Africa. *Antiquity, 84*, 1–14.

Lombard, M., & Wadley, L. (2007). The morphological identification of micro-residues on stone tools using light microscopy: Progress and difficulties based on blind tests. *Journal of Archaeological Science, 34*, 155–165.

Lombard, M., Wadley, L., Deacon, J., Wurz, S., Parsons, I., Mohapi, L., et al. (2012). South African and Lesotho Stone Age sequence updated. *South African Archaeological Bulletin, 67*, 123–144.

Manhire, T. (1993). A report on the excavations at Faraoskop Rock Shelter, in the Graafwater District of the south-western Cape. *South African Field Archaeology, 2*, 3–23.

Manhire, T., Parkington, J., & Yates, R. (1985). Nets and fully recurved bows: Rock paintings and hunting methods in the Western Cape, South Africa. *World Archaeology, 17*, 161–174.

Marshall, L. (1976). *The !Kung of the Nyae-Nyae*. Cambridge: Harvard University Press.

Mazel, A. (1989). People making history. *Natal Museum Journal of Humanities, 1*, 96–159.

Mitchell, P. J. (1988). *The early microlithic assemblages of Southern Africa. British Archaeological Reports 388*. Oxford: Archaeopress.

Mitchell, P. (1995). Revisiting the Robberg: New results and a revision of old ideas at Sehonghong Rock Shelter, Lesotho. *South African Archaeological Bulletin, 50*, 28–38.

Mitchell, P. (2000). The organization of Later Stone Age lithic technology in the Caledon Valley, southern Africa. *African Archaeological Review, 17*, 141–176.

Mitchell, P. (2002). *The archaeology of Southern Africa*. Cambridge: Cambridge University Press.

Morrison, D. (1986). Inuit and Kutchin Bone and Antler Industries in Northwestern Canada. *Canadian Journal of Archaeology, 10*, 107–125.

Newcomer, M. (1974). Study and replication of bone tools from Ksar Akil (Lebanon). *World Archaeology, 6*, 138–153.

O'Connor, S., Robertson, G., & Aplin, K. (2014). Are osseous artefacts a window to perishable material culture? Implications of an unusually complex bone tool from the Late Pleistocene of East Timor. *Journal of Human Evolution, 67*, 108–119.

Olsen, S. (2007). Conclusions: Bone artifacts and their importance to archaeology. In C. G. St-Pierre & R. B. Walker (Eds.), *Bones as tools: Current methods and interpretations in worked bone studies. British Archaeological Reports International Series 1622* (pp. 175–182). Oxford: Archaeopress.

Plug, I. (2012). Diversity and applications: Some bone tools from the past to the present in southern Africa. In K. Seetah & B. Gravina (Eds.), *Bones for tools – tools for bones: The interplay between objects and objectives* (pp. 87–96). Cambridge: McDonald Institute Monographs XII.

Rabett, R., & Piper, P. (2012a). Eating your tools: Early butchery and craft modification of primate bones in tropical southeast Asia. In K. Seetah & B. Gravina (Eds.), *Bones for tools – tools for bones: The interplay between objects and objectives* (pp. 131–142). Cambridge: McDonald Institute Monographs XII.

Rabett, R., & Piper, P. (2012b). The emergence of bone technologies at the end of the Pleistocene in southeast Asia: Regional and evolutionary implications. *Cambridge Archaeological Journal, 22*, 37–56.

Robbins, L., Campbell, A., Brook, G., Murphy, M., & Hitchcock, R. (2012). The antiquity of the bow and arrow in the Kalahari Desert: Bone points from White Paintings Rock Shelter, Botswana. *Journal of African Archaeology, 10*, 7–20.

Russell, N. (2001). Neolithic relations of production: insights from the bone tool industry. In A. Choyke & L. Bartosiewicz (Eds.), *Crafting bone: Skeletal technologies through time and space. Proceedings of the 2nd meeting of the (ICAZ) Worked Bone Research Group, Budapest. British Archaeological Reports International Series 937* (pp. 271–280). Oxford: Archaeopress.

Schweitzer, F. (1979). Excavations at Die Kelders, Cape Province, South Africa: The Holocene deposits. *Annals of the South African Museum, 78*, 101–233.

Semenov, S. (1964). *Prehistoric technology*. Bath: Adams and Dart.

Smith, A., & Poggenpoel, C. (1988). The technology of bone tool fabrication in the South-Western Cape, South Africa. *World Archaeology, 20*, 103–114.

Sparrman, A. (1977) [1786]. *A voyage to the Cape of Good Hope 1772 to 1776*. Cape Town: Van Riebeek Society.

Stone, E. (2013). The identification of perishable technologies through use-wear on osseous tools: Wear patterns on historic and contemporary tools as a standard for identifying raw materials worked in the Late Upper Palaeolithic. In A. Choyke & S. O'Connor (Eds.), *From these bare bones: Raw materials and the study of worked osseous objects* (pp. 28–35). Oxford: Oxbow Books.

Stow, G. W. (1905). *The native races of South Africa: A history of the intrusion of the Hottentots and Bantu into the hunting grounds of the Bushmen, the Aborigines of the country*. London: Swan Sonnenschein and Co., Ltd.

St-Pierre, C. (2007). Bone awls of the St. Lawrence Iroquoians: a microwear analysis. In C. G. Gates St-Pierre & R. B. Walker (Eds.), *Bones as tools: Current methods and interpretations in worked bone studies. British Archaeological Reports International Series 1622* (pp. 107–118). Oxford: Archaeopress.

St-Pierre, C. G., & Walker, R. B. (Eds.). (2007). Bones as tools: Current methods and interpretations in worked bone studies. British Archaeological Reports International Series 1622. Oxford: Archaeopress.

van Gijn, A. (2007). The use of bone and antler tools: two examples from the Late Mesolithic in the Dutch Coastal Zone. In C. G. St-Pierre & R. B. Walker (Eds.), *Bones as tools: Current methods and interpretations in worked bone studies. British Archaeological Reports International Series 1622* (pp. 81–92). Oxford: Archaeopress.

Wadley, L. (1987). *Later Stone Age hunter-gatherers of the Southern Transvaal: Social and ecological interpretation. Cambridge Monographs in African Archaeology 25. British Archaeological Reports International Series 380*. Oxford: Archaeopress.

Waguespack, N., Surovell, T., Denoyer, A., Dallow, A., Savage, A., Hyneman, J., et al. (2009). Making a point: Wood- versus stone-tipped projectiles. *Antiquity, 83*, 786–800.

Part II
Europe

Chapter 4
Early Upper Paleolithic Osseous Points from Croatia

Ivor Karavanić

Abstract This chapter discusses Croatian sites that contain early Upper Paleolithic osseous points and alternative interpretations of this evidence. At Vindija and Velika pećina, sites in the region of Hrvatsko zagorje (northwestern Croatia), split base and massive base (Mladeč) osseous points were found in early Upper Paleolithic contexts associated with a very limited number of lithic finds. The unusual association of Neanderthal remains with Upper Palaeolithic osseous points in Vindija level G1 has been explained either as a result of stratigraphic mixing, or as a true cultural assemblage. Further south at Bukovac pećina, in the region of Gorski kotar, another point was found. The base of this point is missing, but it was probably massive in section. A small split-base point, similar to the points found in Franco-Cantabrian Magdalenian contexts, was found at Šandalja II on the Istrian peninsula. Osseous points from all of these sites mark the first appearance of osseous technology in the different regions of Croatia.

Keywords Neanderthals • Early Modern Humans • Mladeč point • Split-base point • Middle Paleolithic

Introduction

The sites of Croatia are known worldwide in prehistoric archaeology owing to their important finds of Paleolithic industries and/or fossil human remains. Vindija and Velika pećina, situated in northwestern (continental) Croatia (Fig. 4.1), contain both Middle and Upper Paleolithic stratigraphic units which play an important role in the debate on the Middle/Upper Paleolithic transition. At both of these sites, early Upper Paleolithic osseous points were found and these artefacts have become central to the cultural development discourse. An Upper Paleolithic osseous point was also found at Bukovac pećina, situated further south in the region of Gorski kotar (Fig. 4.1). In contrast to the situation in northwestern Croatia, not a single site from the Eastern Adriatic region contains a stratigraphic sequence that includes both Middle and Upper Paleolithic levels, although there are sites with either late Mousterian or Aurignacian finds. The only site from the eastern Adriatic region containing a bone point found in an early Upper Paleolithic context is Šandalja II, situated on the Istrian peninsula (southwestern Croatia, Fig. 4.1). The problems concerning chronology and function of these points, industrial affiliation and association with fossil human remains (Vindija) are presented and discussed in this chapter. Comparison with other sites from Slovenia and abroad is also presented.

Sites

Vindija

Site Location and History of Excavations

Vindija cave is situated in Hrvatsko zagorje (northwestern Croatia), 2 km west of the village of Donja Voća, and 20 km west of the center of Varaždin (Fig. 4.1). Its entrance lies in a narrow gorge on the southwestern slope of Križnjakov vrh, 275 m above sea level. The cave is more than 50 m deep, up to 28 m wide, and more than 10 m high (Fig. 4.2). Vuković (1950), who visited the site in 1928, excavated the cave for more than 30 years. Malez (1975) started systematic excavations at Vindija in 1974, and he directed excavations until 1986. During this later period, most of the Paleolithic archaeological finds and all of the fossil human remains were collected.

I. Karavanić (✉)
Faculty of Humanities and Social Sciences, Department of Archaeology, University of Zagreb, Ivana Lučića 3, HR-10000 Zagreb, Croatia
e-mail: ikaravan@ffzg.hr

Michelle C. Langley (ed.), *Osseous Projectile Weaponry*, Vertebrate Paleobiology and Paleoanthropology,
DOI 10.1007/978-94-024-0899-7_4

Fig. 4.1 Location of Croatian sites mentioned in the text

Stratigraphy, Chronology, and Hominins

The stratigraphic profile is approximately 9 m high and comprises about 20 strata, which, according to Malez and Rukavina (1979), cover the period from the onset of the Riss glaciation (oxygen isotope stage 6 or earlier) through to the Holocene.

The G complex, comprising five stratigraphic levels, numbered G1 (top) through G5, produced the Neanderthal skeletal remains from the site. Level G3 contained approximately 100 fragmentary Neanderthal skeletal remains associated with a late Mousterian industry. Neanderthal remains from level G3 show distinct changes in facial morphology when compared to earlier Neanderthals (see Wolpoff et al. 1981; Smith 1984; Wolpoff 1999). These remains were dated to over 42 ka by radiocarbon AMS (Krings et al. 2000) and 4 years later to about 38 ka by the same method (Serre et al. 2004). There is also another AMS radiocarbon determination on a Neanderthal bone from unit G (level unknown) that yielded results of about 44 ka (Green et al. 2010; for additional dates see Wild et al. 2001; Ahern et al. 2004: Table 1). A series of human skeletal remains was also recovered from level G1, and diagnostic morphology of these specimens identifies the remains as Neanderthal (Smith and Ahern 1994; Smith et al. 1999). Several different radiocarbon dates on bone samples from this level have been obtained (see Ahern et al. 2004: Table 1). The most important are the direct dates from the Neanderthal skeletal remains. These bones were first dated to 28 and 29 ka, respectively (Smith et al. 1999). The same samples were re-dated, using a more accurate technique, to about 33 ka (Higham et al. 2006), which corresponds well with one of the previous dates obtained on animal bone (Karavanić 1995). Results from stable isotope analysis show that Vindija G1 Neanderthals were top-level carnivores, obtaining almost all of their dietary protein from animal sources (Richards et al. 2000; Karavanić and Patou-Mathis 2009). In this aspect, the Vindija people are similar to Neanderthal populations in Western Europe (Bocherens and Drucker 2006).

Fig. 4.2 View from inside Vindija Cave. Photo I. Karavanić

Radiocarbon dating of cave bear bones found in level Fd/d yielded an age of about 26 ka BP (Obelić et al. 1994), while radiocarbon dating of a sample of charcoal found between levels Fd and Fd/d (Malez and Rukavina 1979) or in the level F/d (Malez 1988) yielded an age of about 27 ka. Three isolated undiagnostic hominin teeth were found in level Fd, while a posterior fragment of left parietal originated from the contact of Aurignacian levels Fd and Fd/d (Smith et al. 1985). Radiocarbon dating of cave bear bones from level E produce an age of about 18 ka (Karavanić 1995), while direct dating of human remains from level D produced an Holocene age (M. Richards, personal communication).

Lithic Industries

In the lower Mousterian levels, tools were found which were produced from local raw materials (Kurtanjek and Marci 1990; Blaser et al. 2002) using the Levallois technique. In contrast, the Levallois technique was not employed in Level G3, where local raw materials (chert, quartz, tuff, etc.) were also utilized. The Late Mousterian industry from the level G3 is dominated by side scrapers, notched pieces and denticulates, but also includes some Upper Paleolithic types (e.g., end scrapers). In addition to flake technology, Level G3 also includes evidence of bifacial and blade technology (Karavanić and Smith 1998). As in Level G3, the mixture of Middle and Upper Paleolithic typological characteristics is also present in the stone tool assemblage from Level G1, where osseous points and Neanderthal remains were found. Although this level includes a lithic industry of poor quality, the finds suggest a continuation of the Mousterian technological and typological tradition (excluding the Levallois technique); osseous tools from the same level are typical of the Upper Paleolithic. Although an unusual association of Neanderthal remains and Upper Paleolithic osseous points in Level G1 can be explained as a result of stratigraphic mixing (Zilhão and d'Errico 1999; Bruner 2009; Zilhão 2009), it may also represent an original cultural assemblage (Karavanić and Smith 1998, 2000, 2011). This problem will be discussed later in more detail. A small number of typical Aurignacian stone tool types was found in lower complex F and in the G/F interface. The small archaeological assemblages from these stratigraphic units and level G1 may suggest very short occupations by mobile groups of Paleolithic hunters. Industry from later levels of complex F and level E may represent the Epigravettian.

Osseous Industry

The typology of the osseous tools from Vindija were published more than 20 years ago (Malez 1988; Karavanić 1994). More recent analyses of these artefacts were primarily based on taphonomy and the characteristics of the raw material, and secondarily on technology (Karavanić and Patou-Mathis 2009). These tools, recovered from level G1, are typical of the early Upper Paleolithic, particularly the split-base point, as well as two massive-base points ("Mladeč points") and massive base point fragments (Fig. 4.3). The distal parts are missing on both split-base (Fig. 4.3, no. 1) and massive base points (Fig. 4.3, no. 3), while the tip is missing on another massive base point with this specimen also having a damaged base (Fig. 4.3, no. 2). There are also several osseous tool fragments including a basal section (Fig. 4.3, no. 4). These fragments and basal section are made on antler as well as one point (Fig. 4.3, no. 3) while other points are made on bone (S. Radović, personal communication).

A bear baculum with engraved circumferential markings, and a so-called "bone button" are designated as deriving from this level. The latter is produced by cave bear activity while markings on the former object (Karavanić and Smith 1998) could also be a result of natural processes and not human activity (Karavanić and Patou-Mathis 2009). Furthermore, new analyses (Karavanić and Patou-Mathis 2009) show that some "retouchers" from complex G (Karavanić and Šokec 2003; Ahern et al. 2004) are in fact pseudo-tools. Marks on some of those artifacts are supposed to be of recent age, as the incisions contain no traces of patina. Although the bear baculum has been attributed to level G1 (Malez 1988), a note associated with this specimen suggests that it may in reality have come from the upper part of G3, which would make it even older (Karavanić and Smith 1998).

A bone awl and several "bone buttons" (products of animal activity) are marked with "F" only. The F complex was subsequently divided into several levels marked Fd/d, Fd, Fd/s, Fs and Fg (top). As in level G1, points with a massive base are also present in Fd/d+G1 interface (Fig. 4.3, nos. 5 and 6), level Fd/d (Fig. 4.3, nos. 7 and 8) and E/F interface. Massive base fragments were also found in level F/s together with distal (Fig. 4.3, no. 9) and mesial fragments. Some of these points (Fig. 4.3, nos. 7 and 9) are probable made on antler as well as some fragments. Sagaie fragments are present in upper levels (see Malez 1988; Karavanić 1994, 1995).

Velika Pećina

Site Location and History of Excavations

Velika pećina, another important Paleolithic site in northwestern Croatia, is situated between the sites of Krapina and Vindija, near the village of Goranec on Ravna Gora (Fig. 4.1). The cave is 25 m deep. Excavations were conducted initially by M. Malez in 1948, with subsequent excavations begun in 1957 and, with some interruptions, lasting until 1979.

Stratigraphy, Chronology and Hominins

Stratigraphy at this site consists of 16 defined levels, which are in some parts of the cave over 10 m deep, ranging from the end of the Riss glaciation (oxygen isotope stage 6 or earlier) through to the Holocene (Malez 1979). Radiocarbon dating of the sample from level i yielded an age of about 34 ka (Malez and Vogel 1970).

The human frontal bone from Velika pećina Level j, generally considered one of the earliest finds of early modern Europeans, has been directly dated by AMS radiocarbon to ca. 5 ka (Smith et al. 1999). This result removes the frontal bone from the list of finds of the early Modern Human record in Europe.

Lithic Industries

The lower levels (levels p to k) yielded a Mousterian industry (Malez 1979), though Malez (1967:28) attributed the artifacts from the lower part of Level k to the Mousterian and those from the upper part tentatively to the proto-Aurignacian or to the Mousterian. Reanalysis of the artifacts from Level k did not provide any convincing reason to recognize two different industries. Revision of lithics from this site show that the lower part of Level k probably does belong to the Mousterian (as well as the finds from lower levels), while the upper part may contain only pseudo-tools (Karavanić and Smith 1998). All tools are small and similar to the so-called Micromousterian. Only one stone tool originates from level j, a blade with retouched edges and a notch. Seven stone tools and one bladelet core were found in level i. While these tools include some Upper Paleolithic types, Middle Paleolithic types (three side scrapers) are also represented (Karavanić and Smith 1998). The osseous points from this same level strongly suggest an early Upper Paleolithic affiliation.

Stone tools are also found in later levels. A small number of artifacts in all Paleolithic levels of Velika pećina suggests very short occupations of the site during several episodes from the Middle till late Upper Paleolithic (Malez 1967, 1979).

Osseous Industry

Early Upper Paleolithic osseous tools at this site consist of four points from level i (Fig. 4.4, nos. 1, 2, 3 and 4) two of which (Fig. 4.4, nos. 1 and 2) probably had split bases (the

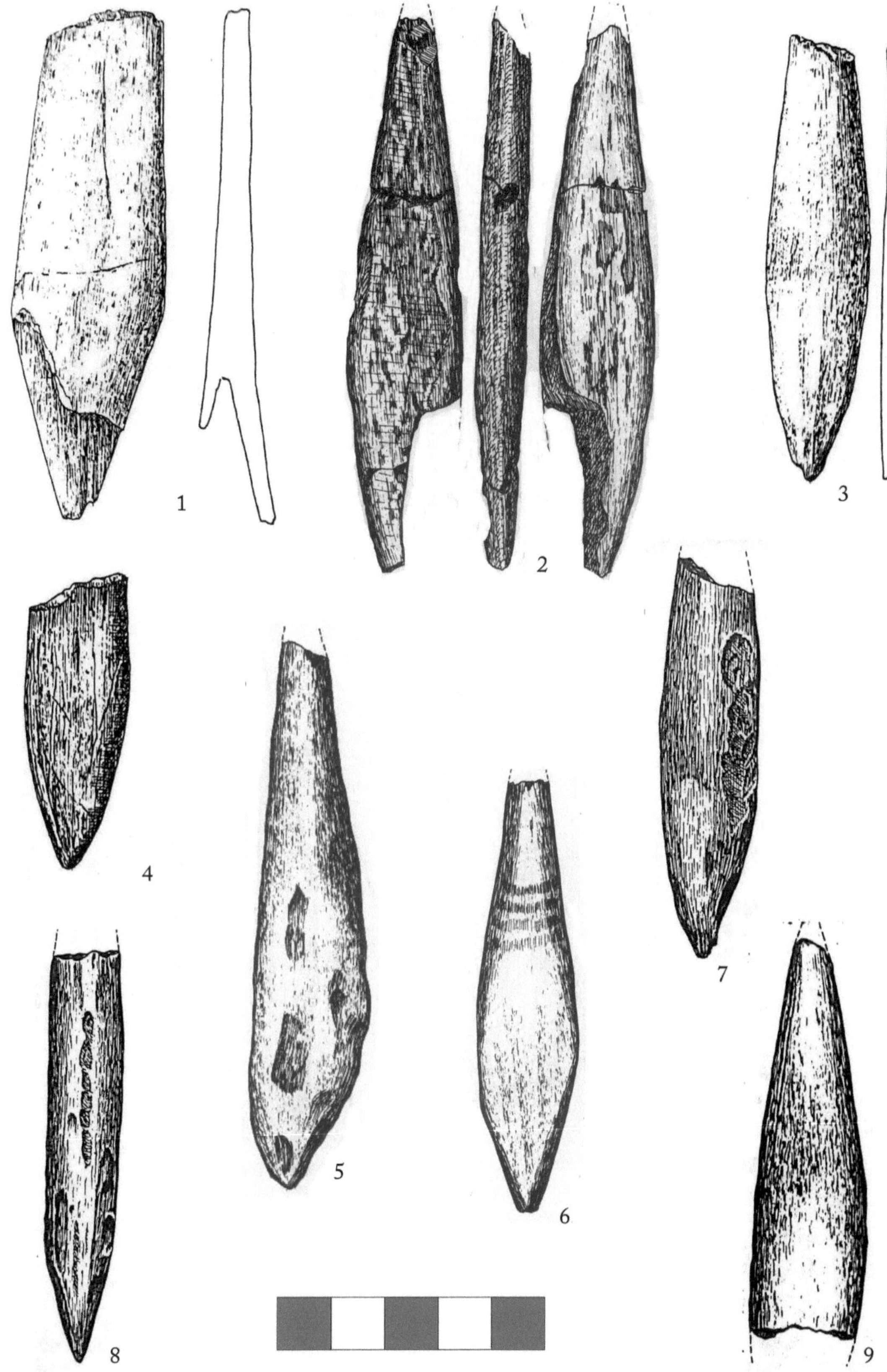

Fig. 4.3 Osseous points from Vindija. Level G1: 1. Split-base point, 2 and 3. Massive-base points, 4. Massive-base fragment (modified after Karavanić and Smith 1998: Fig. 8, nos. 1, 2 and 9). Fd/d interface: 5 and 6. Massive-base points (after Malez 1988: Fig. 6, no. 1a and Fig. 4, no. 1a). Level Fd/d: 7 and 8. Massive-base points (after Malez 1988: Fig. 4, nos. 2a and 4a). Level Fd/s:9. point distal fragment (after Malez 1988: Fig. 5, no. 1c)

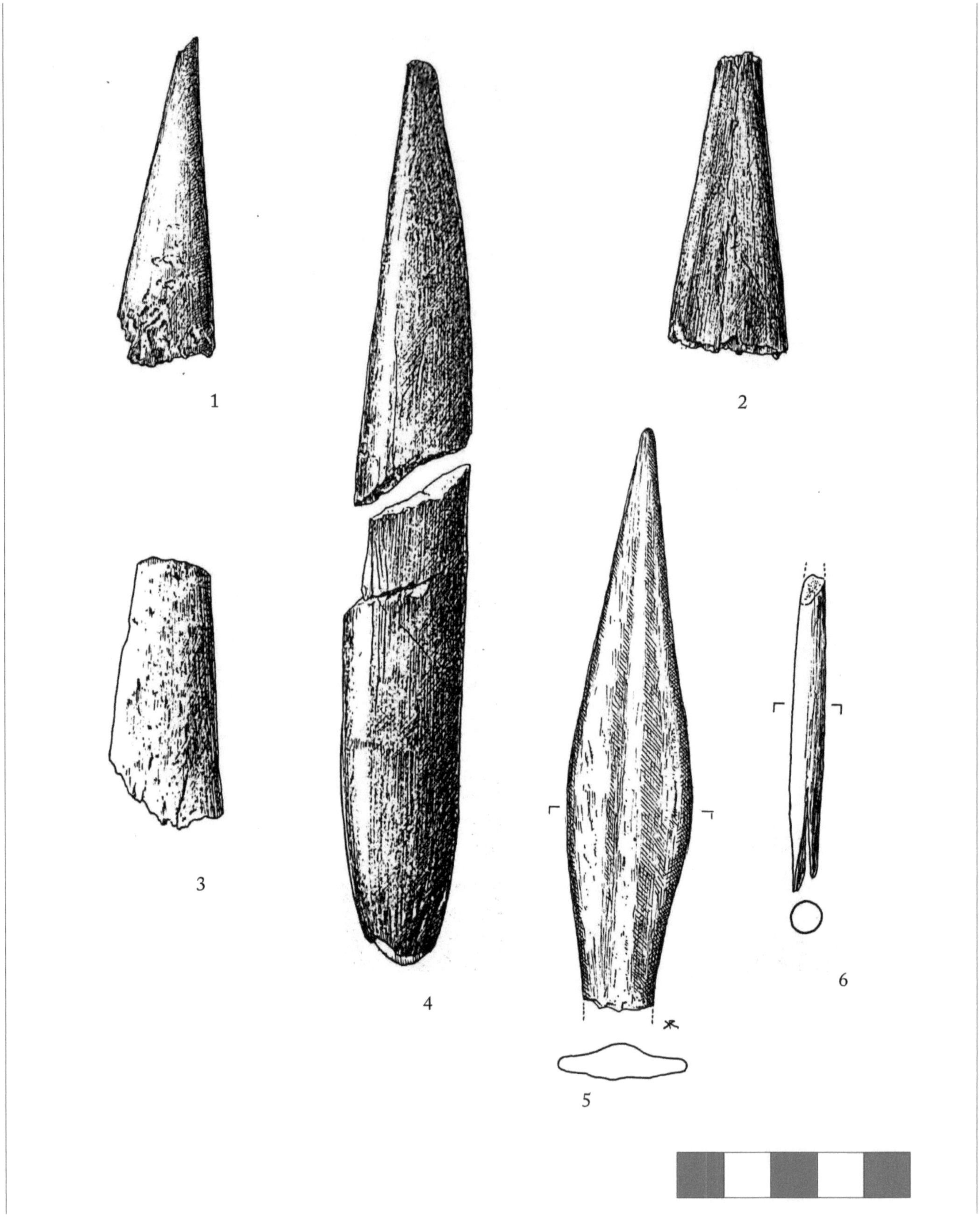

Fig. 4.4 Osseous points from Velika pećina Level i: 1–3. probable split-base points, 4. Massive-base point (after Karavanić and Smith 1998: Fig. 10, nos. 4, 6, 7 and 8); 5. probable massive-base point from Bukovac pećina (Drawing by Krešimir Rončević); 6. Split-base point from Šandalja II (Drawing by Krešimir Rončević)

bases are damaged), one mesial fragment and one massive base point (Fig. 4.4, no. 4). Originally Malez (1967) attributed the base of this massive base point to level h, but it fits with a distal fragment from level i (Karavanić and Smith 1998). However, there is no evidence of mixed layers at Velika pećina. These two fragments of the same point are made on antler which was probable also used for manufacture of another piece (Fig. 4.4, no. 2), while other points are made on bone (S. Radović, personal communication). Bone tools are also found in later levels of Velika pećina but they are (sagaie fragments) typologically different from those belonging to the early Upper Paleolithic (see Malez 1967).

Bukovac Pećina

Site Location and History of Excavations

Bukovac pećina is located in Croatia's Gorski kotar region (Fig. 4.1), southeast of the town of Lokve on the northwestern slopes of Sleme Hill (Malez 1979). It is situated in a mountain region within the border zone between the Mediterranean and continental zones of Croatia, closer to the Adriatic than to the Hrvatsko zagorje sites. The cave was first test excavated by T. Kormos (1912) and L. Szilágy in 1911 (Malez 1979). A trench excavated in the front of the cave yielded no significant discoveries, but a test pit deeper inside the cave resulted in the recovery of faunal remains and an osseous point (Fig. 4.4, no. 5). Malez (1979) excavated the cave during 1970s. I. Janković has directed excavations from 2010 till 2014 (Fig. 4.5).

The Osseous Point

The point has been assigned to different cultures (Malez 1979), but today the overriding view is that it belongs to the Aurignacian or Olschewian (Malez 1979; Montet-White 1996; Horusitzky 2004). It is probable made on antler. The base of the point is missing, but based on the sudden thinning of the widest part it can be argued that it was a so-called Mladeč point (Fig. 4.4, no. 5). During the 1970s, excavations by Malez (1979), yielded no further artefacts. Therefore, based solely on the single osseous point, assignment of the industry to the early Upper Paleolithic is tenuous although likely. One of the major aims of the excavation under the direction of I. Janković in 2010 (Janković et al. 2011), was to determine the layer from which this find comes, based on the stratigraphy by Kormos (1912), and obtain material for dating. Thus far, a single date confirms the Aurignacian timeframe (I. Janković, personal communication).

Fig. 4.5 Excavation at Bukovac pećina. Photo I. Karavanić

Šandalja II

Site Location and History of Excavations

During mining in the quarry near the city of Pula on Istrian peninsula in 1961, a cavern containing Quaternary sediments was exposed, and named Šandalja I, while in the next year (1962) a second cavern was found and named Šandalja II (Fig. 4.1). Both are, most likely, part of a single, larger underground complex (Malez 1979). Therefore Šandalja II refers to the part of the cave that yielded the Upper Paleolithic finds, while the bone breccia of the Villafranchian age is referred to as Šandalja I (Karavanić 1999; D. Rukavina, personal communication). The Šandalja II site was excavated by M. Malez on 22 occasions, between 1962 and 1989 (Miracle 1995).

Stratigraphy and Chronology

Basic stratigraphy of the site is over 8 m thick and has been divided into 8 layers (A-H) (Malez 1963, 1964, 1979). They are composed of compact or sandy clay with large or small rock fragments, containing Upper Paleolithic (Aurignacian and Epigravettian) lithics, fauna, and human remains (Malez 1979; Brajković 2000). Layers G, F, and E have provided radiocarbon dates between 28 and 23 ka (Malez and Vogel 1969; Srdoč et al. 1979). Material in layers H and G/H is scarce and attributable to the early Upper Paleolithic. Layers G, F and E yielded Aurignacian lithic material, layer D contained both Aurignacian and Epigravettian material while levels of C and B complexes belong to the Epigravettian (Karavanić 1999). Human remains were only found in late Epigravettian level B/s (Malez 1972; Janković et al. 2012).

Lithic Industries

Debitage in Aurignacian levels, mainly produced on local grey chert, are often patinated. Flakes are most common, while bladelets are more numerous than blades in all Aurignacian layers except level F (Karavanić 2003, 2009). Both blades and bladelets have been produced by direct percussion using a soft hammer. Flakes from retouching are very rare. This could be the result of the excavation method given that sieving was not practiced, and thus, may not reflect a true lack of retouching at the site itself. The large number of chunks is easily explained by the use of the local grey chert, which breaks irregularly. The very small percentage of tools could be explained either by their production elsewhere or, alternatively, by hominins having taken them from the site.

Nosed and carinated end scrapers are quite common, while Aurignacian blades are missing (Karavanić 2003, 2009). Side scrapers and notches are present in significant quantities. Dufour bladelets are missing from the sample but it is not clear whether this reflects a real situation at the site or the fact that the sediment was not sieved. On the other hand, Epigravettian layers that were excavated using the same archaeological methods have yielded numerous smaller finds (for example, the backed bladelets), therefore making it likely that the Dufour bladelets would have been collected, if present in the layers. Although Dufour bladelets are missing, the lithic industry of the stratigraphic units F, E/F and F represents the Aurignacian (Karavanić 2003, 2009).

Osseous Industry

The most common bone tools in these units are awls, while four pierced animal teeth from the Aurignacian layers represent decorative items and represent symbolic behavior (see Karavanić 2003:Fig. 9). On two of these artefacts, the root was thinned by scraping before the actual piercing was done. The root of the third tooth was damaged, while the fourth tooth (although damaged), also shows deliberate thinning of the root. A small bone point with a split base was found at Šandalja II (layer H) and is more similar to examples from the Franco-Cantabrian Magdalenian (L. G. Straus, personal communication) than to the Aurignacian types (Fig. 4.4, no. 6). In addition to the above mentioned point, a few osseous tools and fragments derived from Aurignacian levels (distal part of point or awl, awl, spatula or chisel, a tool with circular cross section, a proximal part of a bone tool, a probable fragment of a bone awl, distal fragment of decorated bone point with a broken tip), there are also pierced animal teeth, bone points, awls, bone personal ornaments and bone engraved pieces found in Epigravettian levels (Malez 1987; Karavanić 1999, 2003).

Discussion

In summary, early Upper Paleolithic osseous points are present at four sites in Croatia, but it is the hominin finds and their association with these points that pose the most interesting and puzzling problem. In Vindija level G1, a split-base point and massive base points were found together with Neanderthal remains and some mixture of Middle and Upper Paleolithic stone tool typological characteristics is present in the lithic assemblage. A leaf-shaped bifacial stone piece, like those found in the Szeletien, has also been recovered from this level. This piece (a point) is thin and very finely worked on both faces. However, it is likely that some of the lithic material from G1 (e.g. Vi 1061, Vi 3383) are pseudo-tools, as argued recently by Zilhão (2009). These pieces are probably flakes with pseudo-retouch (edge modification caused by post-depositional processes which looks like retouch). The presence of pseudo-tools and the results of refitting (Bruner 2009; Zilhão 2009) confirms that there was some mixing of different layers, and that the presence of certain Upper Paleolithic lithic tool types made on high quality chert from G1 and G3 levels might be explained as a result of this mixing (Karavanić and Smith 2011). Different authors have long recognized that both bioturbation and cryoturbation occurred at Vindija and likely resulted in mixing of elements from different layers in some parts of the cave (Malez and Rukavina 1975; Smith 1984; Kozlowski 1996; d'Errico et al. 1998; Karavanić and Smith 1998). However, they are not seen uniformly throughout the site, and the area where many of the relevant finds are derived do not show evidence of disturbance. In light of the documented disturbance of layers, the Olschewian hypothesis as the transitional industry of the G1 layer (Karavanić 2000, 2007) is not likely. It is more probable that Middle and Upper Paleolithic typological characteristics of the G1 stone tool assemblage resulted from mixing of the material between levels than from a specific transitional industry (Karavanić and Smith 2011).

However, the problem of the association of Neanderthal remains and Upper Paleolithic osseous points is still open. Results of taphonomic analysis show that the preservation of osseous tools from G1 is similar to that of the bone remains of large mammals and humans from this same level, suggesting they all derive from the same context (Karavanić and Patou-Mathis 2009). These osseous points and Neanderthal remains do not show trampling traces (except for the base fragment of the Mladeč point – Vi 2510) and it should be noted that distinctive reddish sediment typical of the G1 layer was imbedded in the Mladeč type osseous point Vi 3439 and in Neanderthal skeletal remains from G1, thus proving that they were found in the same level.

Therefore, Straus (1999), Montet-White (1996), Karavanić and Smith (1998), Ahern and colleagues (2004), and Janković and colleagues (2006, 2011) see the unusual G1 associations in the context of a more complex pattern that characterizes the Middle/Upper Paleolithic transition in this region while Svoboda (2001, 2006) noted some similarities between the G1 layer of Vindija and Szeletian industry. While Pacher (2010) correctly pointed out the lack of attributable elements required to define Olschewian as an initial Upper Paleolithic industry, her suggestion that fossil human remains from Vindija level G1 are not Neanderthals has no foundation. Even though the human remains are very fragmented, as she properly noted, the anatomical characteristic clearly indicate an attribution to Neanderthals (with some Modern Human characteristics), which was published in numerous scientific papers and books (e.g., Karavanić and Smith 1998; Wolpoff 1999; Cartmill and Smith 2009).

In contrast to Zilhão (2009: Table 2), who sees the G1 layer material as mix of Szeletian, Aurignacian I and II, and material from Fd/d layer as Aurignacian II or III/IV, Karavanić and Smith (2011) offered two possible explanations. The first possibility is that the lithic industry of G1 represents Mousterian (see also Kozlowski 1996) without levallois technology. A bifacial stone point made on nonlocal Hungarian red radiolarite (Montet-White 1996; Biró and Markó 2007) is seen as an import, a result of the contact of various Neanderthal groups (if the Szeletian was produced by Neanderthals) or of contact between Neanderthals and early Modern Humans (if the Szeletian was produced by early Modern Humans) from northwestern Croatia and Hungary. The Upper Paleolithic elements, especially the osseous points, and possibly some lithic types, may be the result of contact (exchange or acculturation) between Neanderthals and early anatomically modern groups associated with Aurignacian. The second possibility is that although the lithic industry is Mousterian and the aforementioned stone bifacial point is imported, the presence of the osseous points and Upper Paleolithic lithic tools in the G1 level is a result of mixing with the upper layers of the site (Karavanić and Smith 2011). If this is the case, then the industry present in stratigraphic levels Fd/d and Fd is Aurignacian (as suggested by Karavanić in 1995 and Kozlowski in 1996). However, due to the variability of the Aurignacian industry (Kozlowski and Otte 2000; Teyssandier et al. 2009), the low percentage of typical Aurignacian stone tools in Vindija, and the fact that lithic industry is typologically different from the Aurignacian known in French sites, we are not comfortable using the terms Aurignacian I, II, III/IV for the Vindija assemblage (see also Miracle 1998).

Equivalents of Vindija and Velika pećina osseous points have been found at Mokriška jama, Potočka zijalka and Divje babe I, Alpine Palaeolithic sites in Slovenia (Brodar and Osole 1979; Brodar and Brodar 1983; Turk and Kavur 1997) and at many Central European sites (Albrecht et al. 1972). Furthermore, the combination of osseous tools similar to those from Vindija and small numbers of undiagnostic (i.e. non-Aurignacian) lithic artifacts has also been identified in the early Upper Paleolithic levels of Velika pećina, as well as Mokriška jama (Brodar and Osole 1979) and Divje babe I (Turk and Kavur 1997). Therefore, another possibility is that the lack of more typical Aurignacian stone tools at Vindija (and other sites) is the result of some type of functional specialization connected to regional specific hunting activity (cf. Hahn 1977). The small quantities of artifacts found at these sites also suggest that they may have been occupied only for short episodes. However, another site, Potočka zijalka (Brodar and Brodar 1983; Pacher et al. 2004), contained an abundant stone and osseous industry including typical Aurignacian tools absent in other assemblages (except end scraper on an Aurignacian blade in the Vindija G1 assemblage, as well as the same type of tool and Aurignacian blade, flat-nosed end scraper, etc. in Vindija "G/F interface" and unit F assemblages). Despite this minor variation, all these Croatian and Slovenian sites are similar in having yielded early Upper Paleolithic osseous points along with some stone tool.

Two osseous point types have been identified in some of those assemblages (i.e., so called massive-base and split-base bone points). The functional usefulness of the Vindija split-base point might be questionable. This point is very thin and wide with flat section. The distal part is missing (broken) and it seems that it was not recycled. The basal flanges are fragile (one part of a flange is missing) and the overall dimensions (31.1 × 5.6 mm) suggest a structurally weak point (Karavanić and Smith 1998). The points from Mokriška jama are distinguished by an oval-flat section similar to that of the split-base point found at Vindija and the points from Velika pećina. On the other hand, the points from Potočka zijalka are mainly thick and oval, like the massive-base points from Vindija.

Although direct dating of the osseous points from Vindija and Velika pećina failed (Smith et al. 1999), an age of 34 ka was determined for the "i" layer of Velika pećina (Malez and Vogel 1970). Thus, the same age can be assumed for the osseous points from this same layer of the same site (Malez and Vogel 1970). An osseous point (most likely with a split base) from Divje babe I (Slovenia) comes from a layer that has been dated to about 35 ka (Nelson 1997). This point was directly dated to about 30 ka (Moreau et al. 2015) while points from Potočka zijalka and Mokriška jama (Slovenia) are dated to between 35 and 29 ka (Hofreiter and Pacher 2004; Moreau et al. 2015). Likewise, the Mladeč type points from the Mamutova cave in Poland, near Krakow, date to between 33 and 32 ka (Wojtal 2007), while oldest osseous projectile points from Hungary were dated back to 37/38 ka (Davies and Hedges (2008–2009). Early Upper Paleolithic points from German sites have been dated to between 32 and ca. 27 ka (Conard and Bolus 2003; Bolus and Conard 2006; Conard and Bolus 2008), and the (proto) Aurignacian split-base points from Trou de la Mère Clochette in northeastern France have been dated to between 33 and 35 ka (Szmidt et al. 2010). Although some of these sites are geographically quite distant from Vindija, it should also be noted that some of their dates are older than the Vindija Neanderthals, while others are younger. Although we do not have direct dates on the points themselves, dates from comparable archaeological layers suggest that the bone points from Velika pećina are older than, or contemporaneous with, the Vindija Neanderthals. If we adhere to the generally accepted view that such points are associated with only Modern Humans, this association raises the question of possible interactions between these groups.

An Upper Paleolithic osseous point was found at Bukovac pećina. Thus far, a single date of the level from which the find comes, confirms the Aurignacian timeframe (I. Janković, personal communication). From the eastern Adriatic, only a single bone point has been found, and comes from layer H at the site of Šandalja II, in Istria. It is relatively small compared to the points from Central Europe and has a split-base and rounded cross section. It is similar to points from the Franco-Cantabrian Magdalenian (L.G. Straus, personal communication); and based on the recent date for the layer F at Šandalja II, it should be older than 32 ka (M. Richards, personal communication), if it did not originally come from one of the Epigravettian layers.

It is still questionable whether we should explain the early Upper Paleolithic osseous points from Vindija level G1 as a result of contact (exchange or acculturation) between Neanderthals and early anatomically modern groups or as a result of stratigraphic mixing. Direct radiocarbon dates on Vindija G1 Neanderthals indicate that they inhabited northwestern Croatia during the same period when the developed technology of split and massive base osseous points were present in Central Europe. Therefore, the possibility that the last Neanderthals in Europe who occupied Vindija cave some 33 ka adopted this technology cannot be excluded.

Although the faunal remains from the sites presented in this chapter are studied from different perspectives (e.g., Malez 1979; Miracle 1995; Brajković 2000; Brajković and Miracle 2008; Karavanić and Patou-Mathis 2009), unfortunately there was no detailed technological study of the faunal assemblages from where these points came from so far. Such analysis will be of crucial importance in the future, by which we will be able to establish if these points were produced in situ or obtained from some other location, and for the better understanding of technology of such points in general.

However, there are notable differences in the early Upper Paleolithic stone and bone industry between continental and the Adriatic sites of Croatia (see Karavanić 2007, 2009). These differences might reflect different forms of adaptation of local Paleolithic populations induced by different climatic and environmental conditions, and might well be interpretable as providing further evidence of the complexity and "mosaic" nature of the early Upper Paleolithic in Europe.

Acknowledgments I am grateful to Michelle C. Langley for inviting me to contribute to the edited volume and for comments and editing of the paper. I am also grateful the following individuals for various forms of assistance relative to the information presented in the paper: James C. M. Ahern, Ivor Janković, Fred H. Smith, Lawrence G. Straus, Siniša Radović, Mike Richards and Darko Rukavina. This paper also benefited from the comments of two anonymous reviewers. Rajna Šošić-Klindžić designed Figs. 4.3 and 4.4. Figure 4.1 was downloaded from GinkgoMaps-Project, http://www.gingomaps.com, licensed under CC-BY-3.0 and modified by M. Vuković.

References

Ahern, J. C. M., Karavanić, I., Paunović, M., Janković, I., & Smith, F. H. (2004). New discoveries and interpretations of hominid fossils and artifacts from Vindija Cave, Croatia. *Journal of Human Evolution, 46*, 25–65.

Albrecht, G., Hahn, J., & Torke, W. G. (1972). *Merkmalanalyse von Geschoßspitzen des mittleren Jungpleistozäns in Mittel- und Osteuropa. Archaeologica Venatoria*. Stuttgart, Germany: Verlag W. Kohlhammer.

Biró, K., & Markó, A. (2007). Stones "hit the road". In T. Kovács (Ed.), *Highways and byways in space and time* (pp. 14–19). Budapest, Hungary: Hungarian National Museum.

Blaser, F., Kurtanjek, D., & Paunović, M. (2002). L'industrie du site néandertalien de la grotte de Vindija (Croatie): Une révision des matières premières lithiques. *L'Anthropologie, 106*, 387–398.

Bocherens, H., & Drucker, D. (2006). Dietary competition between Neanderthals and Modern Humans: Insights from stable isotopes. In N. Conard (Ed.), *When Neanderthals and Modern Humans met* (pp. 129–143). Tübingen, Germany: Kerns Verlag.

Bolus, M., & Conard, N. J. (2006). Zur zeitstellung von geschosspitzen aus organischen materialien im späten Mittelpaläolithikum und Aurignacien. *Archäologisches Korrespondenzblatt, 36*, 1–15.

Brajković, D. (2000). Šandalja. *Histria Archaeologica, 29*(1998), 5–25.

Brajković, D., & Miracle, P. T. (2008). Middle Palaeolithic and early Upper Palaeolithic subsistence practices at Vindija cave, Croatia. In A. Darlas & D. Mihailović (Eds.), *The Palaeolithic of the Balkans* (pp. 107–116). Oxford, England: Archaeopress.

Brodar, S., & Brodar, M. (1983). *Potočka zijalka*. Ljubljana, Slovenia: Slovenian Academy of Sciences and Arts.

Brodar, M., & Osole, F. (1979). Paleolitske i mezolitske regije i kulture u Sloveniji. In A. Benac (Ed.), *Praistorija jugoslavenskih zemalja* (Vol. I, pp. 159–194). Sarajevo, Bosnia and Herzegovina: Svjetlost.

Bruner, K. M. (2009). *Testing stratigraphic integrity of Upper and Middle Paleolithic deposits in Vindija Cave (Croatia): A chipped stone refitting analysis*. Master theses, University of Kansas.

Cartmill, M., & Smith, F. (2009). *The human lineage*. Hoboken, NJ: Wiley-Blackwell.

Conard, N. J., & Bolus, M. (2003). Radiocarbon dating the appearance of Modern Humans and timing of cultural innovations in Europe: New results and new challenges. *Journal of Human Evolution, 44*, 331–371.

Conard, N. J., & Bolus, M. (2008). Radiocarbon dating the late Middle Paleolithic and the Aurignacian of the Swabian Jura. *Journal of Human Evolution, 55*, 886–897.

d'Errico, F., Zilhao, J., Julien, M., Baffier, D., & Pelegrin, J. (1998). Neanderthal acculturation in western Europe? A critical review of the evidence and its interpretation. *Current Anthropology, 39* (Suppl.), S1–S44.

Davies, W., & Hedges, R. (2008–2009). Dating a type site: Fitting Szeleta cave into its regional chronometric context. *Praehistoria, 9–10*, 35–45.

Green, R., Krause, J., Briggs, A., Maricic, T., Stenzel, U., Kircher, M., et al. (2010). A draft sequence of the Neandertal genome. *Science, 328*, 710–725.

Hahn, J. (1977). *Aurignacien. Das Ältere Jungpaläolithikum in Mittel- und Osteuropa*. (Fundamenta, Ser. A., Vol. 9). Köln, Germany: Böhlau.

Higham, T., Ramsey, C. B., Karavanić, I., Smith, F. H., & Trinkaus, E. (2006). Revised direct radiocarbon dating of the Vindija G1 Upper Paleolithic Neandertals. *Proceedings of the National Academy of Sciences of the United States of America, 103*(3), 553–557.

Hofreiter, M., & Pacher, M. (2004). Using ancient DNA to elucidate raw material origin of bone points from Potočka zijalka (Slovenia): Preliminary results. In M. Pacher, V. Pohar, & G. Rabeder (Eds.), *Potočka Zijalka: Palaeolontological and Archaeological Results of the Campaigns 1997–2000* (pp. 201–210). Vienna: Österreichischen Akademie der Wissenschaften.

Horusitzky, F. Z. (2004). Les artefacts en os et bois de cerf à Bukovac, Lokve (Croatie): Une seconde flûte possible? Relations entre les chasseurs de Lokve et les montagnards d'Olcheva au début du Paléolithique supérieur. *Arheološki Vestnik, 55*, 9–37.

Janković, I., Ahern, J. C. M., Karavanić, S., Stockton, T., & Smith, F. H. (2012). Epigravettian Human Remains and Artifacts from Šandalja II, Istria, Croatia. *Paleoanthropology, 2012*, 87–122.

Janković, I., Karavanić, I., Ahern, J. C. M., Brajković, D., Mauch Lenardić, J. I., & Smith, F. H. (2006). Vindija Cave and the Modern Human peopling of Europe. *Collegium Antropologicum, 30*, 457–466.

Janković, I., Karavanić, I., Ahern, J. C. M., Brajković, D., Mauch Lenardić, J., & Smith, F. H. (2011). Archaeological, paleontological and genomic perspectives on late European Neandertals at Vindija cave, Croatia. In S. Condemni & G.-C. Weniger (Eds.), *Continuity and discontinuity in the peopling of Europe: One hundred fifty years of Neanderthal study* (pp. 299–313). Vertebrate paleobiology and paleoanthropology. New York: Springer.

Janković, I., Karavanić, I., Ahern, J. C. M., & Mihelić, S. (2011). Pećina Bukovac. *Hrvatski Arheološki Godišnjak, 7*(2010), 477–479.

Karavanić, I. (1994). Gornjopaleolitičke kamene i koštane rukotvorine iz špilje Vindije. *Opuscula Archaeologica, 17*(1993), 53–163.

Karavanić, I. (1995). Upper Paleolithic occupation levels and late-occurring Neandertal at Vindija Cave (Croatia) in the context of Central Europe and the Balkans. *Journal of Anthropological Research, 51*, 9–35.

Karavanić, I. (1999). *Gornji paleolitik Šandalje II u okviru jadranske regije*. Ph.D. dissertation. University of Zagreb.

Karavanić, I. (2000). Olschewian and appearance of bone technology in Croatia and Slovenia. In J. Orschiedt & G.-Ch. Weniger (Eds.), *Neanderthals and Modern Humans – Discussing transition* (pp. 159–168). Wissenschaftliche Schriften des Neanderthal Museums, Bd. 2. Mettmann, Germany: Neanderthal Museum.

Karavanić, I. (2003). L'industrie aurignacienne de la grotte de Šandalja II (Istrie, Croatie) dans le contexte de la région de l'est de l'Adriatique. *L'Anthropologie, 107*, 577–602.

Karavanić, I. (2007). The Middle/Upper Paleolithic interface in Croatia. In J. Riel-Salvatore & G. A. Clark (Eds.), *New approaches to the study of Early Upper Paleolithic "transitional" industries in Western Eurasia: Transitions great and small* (pp. 75–89). Oxford, England: Archaeopress.

Karavanić, I. (2009). Adriatic coast of Croatia and its hinterland from 50 000 to 25 000 BP. In M. Camps & C. Szmidt (Eds.), *The Mediterranean from 50 000 to 25 000 BP: Turning points and new directions* (pp. 163–178). Oxford, England: Oxbow Books.

Karavanić, I., & Patou-Mathis, M. (2009). Middle/Upper Paleolithic interface in Vindija Cave (Croatia): New results and interpretations. In M. Camps & P. Chauhan (Eds.), *Sourcebook of Paleolithic transitions: Methods, theories and interpretations* (pp. 397–405). New York: Springer.

Karavanić, I., & Smith, F. H. (1998). The Middle/Upper Paleolithic interface and the relationship of Neanderthals and early Modern Humans in the Hrvatsko Zagorje, Croatia. *Journal of Human Evolution, 34*, 223–248.

Karavanić, I., & Smith, F. H. (2000). More on the Neanderthal problem: The Vindija case. *Current Anthropology, 41*, 838–840.

Karavanić, I., & Smith, F. H. (2011). Middle/Upper Paleolithic interface at Vindija Cave (Croatia) in the context of central Europe and the Adriatic. In A. P. Derevianko & M. V. Shunkov (Eds.), *Characteristic features of the Middle to Upper Paleolithic transition in Eurasia* (pp. 124–134). Novosibirsk, Russia: Institute of Archaeology and Ethnography SB RAS.

Karavanić, I., & Šokec, T. (2003). The Middle Paleolithic percussion or pressure flaking tools? The comparison of experimental and archaeological material from Croatia. *Prilozi Instituta za arheologiju u Zagrebu, 20*, 5–14.

Kormos, T. (1912). Die ersten Spuren des Urmenschen in Karst-Gebirge. *Földtani Közlöny, 52*, 97–104.

Kozlowski, J. K. (1996). Cultural context of the last Neanderthals and early Modern Humans in Central-Eastern Europe. In O. Bar-Yosef, L. L. Cavalli-Sforza, R. J. March, & M. Piperno (Eds.), *Proceedings of the XIII UISPP congress, the lower and the Middle Palaeolithic* (Vol. 5, pp. 205–218). Forlì, Italy: Abaco.

Kozlowski, J. K., & Otte, M. (2000). The formation of Aurignacian in Europe. *Journal of Anthropological Research, 56*, 513–534.

Krings, M., Capelli, C., Tschentscher, F., Geisert, H., Meyer, S., Von Haeseler, A., et al. (2000). A view of neandertal genetic diversity. *Nature Genetics, 26*, 144–146.

Kurtanjek, D., & Marci, V. (1990). Petrografska istraživanja paleolitskih artefakata spilje Vindije. *Rad Jugoslavenske Akademije Znanosti i Umjetnosti, 449*, 227–238.

Malez, M. (1963). Istraživanje pleistocenske stratigrafije i faune u 1962. godini. *Ljetopis Jugoslavenske Akademije Znanosti i Umjetnosti, 69*, 303–313.

Malez, M. (1964). Šandalja bei Pula - ein neuer und wichtiger paläolithischer fundort in Istrien. *Bulletin Scientifique du Conseil des Académies des Sciences et des Arts de la RSF Yougoslavie, 9*(6), 154–155.

Malez, M. (1967). Paleolit Velike pećine na Ravnoj gori u sjeverozapadnoj Hrvatskoj. *Arheološki Radovi i Rasprave, 4–5*, 7–68.

Malez, M. (1972). Ostaci fosilnog čovjeka iz gornjeg pleistocena Šandalje kod Pule (Istra). *Palaeontologia Jugoslavica, 12*, 1–39.

Malez, M. (1975). Die Höhle Vindija – eine neue Fundstelle fossiler Hominiden in Kroatien. *Bulletin Scientifique du Conseil des Académies des Sciences et des Arts de la RSF Yougoslavie, 20*(5–6), 139–141.

Malez, M. (1979). Nalazišta paleolitskog i mezolitskog doba u Hrvatskoj. In A. Benac (Ed.), *Praistorija jugoslavenskih zemalja* (Vol. I, pp. 227–276). Sarajevo, Bosnia and Herzegovina: Svjetlost.

Malez, M. (1987). Pregled paleolitičkih i mezolitičkih kultura na području Istre. In V. Jurkić (Ed.), *Arheološka istraživanja u Istri i Hrvatskom primorju* (Vol. I, pp. 3–47). Izdanja Hrvatskog arheološkog društva 11. Pula, Croatia: Hrvatsko Arheološko Društvo.

Malez, M. (1988). Prethistorijske koštane rukotvorine iz spilje Vindije (Hrvatska, Jugoslavija). *Radovi Zavoda za Znanstveni rad Jugoslavenske Akademije Znanosti i Umjetnosti, 2*, 217–252.

Malez, M., & Rukavina, D. (1975). Krioturbacijske pojave u gornjopleistocenskim naslagama pećine Vindije kod Donje Voće u sjeverozapadnoj Hrvatskoj. *Rad Jugoslavenske Akademije Znanosti i Umjetnosti, 371*(17), 245–265.

Malez, M., & Rukavina, D. (1979). Položaj naslaga spilje Vindije u sustavu članjenja kvartara šireg područja Alpa. *Rad Jugoslavenske Akademije Znanosti i Umjetnosti, 383*(18), 187–218.

Malez, M., & Vogel, J. C. (1969). Rezultati određivanja apsolutne starosti pleistocenskih naslaga Šandalje II kod Pule u Istri. *Geološki vjesnik, 22*, 121–133.

Malez, M., & Vogel, J. C. (1970). Die Ergebnisse der Radiokarbonanalisen der quartären Schichten der Velika Pećina in Nordwest-Kroatien. *Bulletin scientifique Conseil des Académies des Sciences et des Arts de la RSF de Yougoslavie, 15*(11–12), 390–391.

Miracle, P.T. (1995). *Broad-spectrum adaptations re-examined: Hunter-gatherer responses to late glacial environmental changes in the eastern Adriatic*. Ph.D. dissertation, The University of Michigan.

Miracle, P. T. (1998). The spread of modernity in Europe. In K. Omoto & P. V. Tobias (Eds.), *The origins and past of Modern Humans - towards reconciliation* (pp. 171–187). Singapore: World Scientific.

Montet-White, A. (1996). *Le Paléolithique en ancienne Yougoslavie*. Grenoble, France: Jérôme Millon.

Moreau, L., Odar, B., Higham, T., Horvat, A., Pirkmajer, D., & Turk, P. (2015). Reassessing Aurignacian of Slovenia: Techno-economic behaviour and direct dating of osseous projectile points. *Journal of Human Evolution, 78*, 158–180.

Nelson, D. E. (1997). Radiocarbon dating of bone and charcoal from Divje Babe I Cave. In I. Turk (Ed.), *Mousterian "Bone Flute" and other finds from Divje Babe I Cave site in Slovenia* (pp. 51–64). Ljubljana, Slovenia: Slovenian Academy of Sciences and Arts.

Obelić, B., Horvatinčić, N., Srdoč, D., Krajcar Bronić, I., Sliepčević, A., & Grgić, S. (1994). Rudjer Bošković Institute radiocarbon measurements XIII. *Radiocarbon, 36*(2), 303–324.

Pacher, M. (2010). Raw material analysis of Upper Paleolithic bone points and the invention of the Olschewian. In C. Neugebauer-Maresch & L. R. Owen (Eds.), *New aspects of the Central and Eastern European Upper Palaeolithic – Methods, chronology, technology and subsistence* (pp. 319–325). Vienna: Verlag der Österreichischen Akademie der Wissenschaften.

Pacher, M., Pohar, V., & Rabeder, G. (2004). *Potočka Zijalka: Palaeolontological and Archaeological Results of the Campaigns 1997–2000*. Vienna: Österreichischen Akademie der Wissenschaften.

Richards, M. P., Pettitt, P. B., Trinkaus, E., Smith, F. H., Paunović, M., & Karavanić, I. (2000). Neanderthal diet at Vindija and Neanderthal predation: The evidence from stable isotopes. *Proceedings of the National Academy of Science of the United States of America, 97*, 7663–7666.

Serre, D., Langaney, A., Chech, M., Teschler-Nicola, M., Paunović, M., Hofreiter, M., et al. (2004). No evidence of Neandertal mtDNA contribution to early Modern Humans. *PLoS Biology, 2*, 313–317.

Smith, F. H. (1984). Fossil hominids from the Upper Pleistocene of Central Europe and the origin of modern Europeans. In F. H. Smith & F. Spencer (Eds.), *The origin of Modern Humans: A world survey of the fossil evidence* (pp. 137–209). New York: Alan R. Liss.

Smith, F. H., & Ahern, J. C. (1994). Brief Communication: additional cranial remains from Vindija Cave, Croatia. *American Journal of Physical Anthropology, 93*, 275–285.

Smith, F. H., Boyd, C. D., & Malez, M. (1985). Additional Upper Pleistocene human remains from Vindija Cave, Croatia, Yugoslavia. *American Journal of Physical Anthropology, 68*, 375–388.

Smith, F. H., Trinkaus, E., Pettitt, P. B., Karavanić, I., & Paunović, M. (1999). Direct radiocarbon dates for Vindija G1 and Velika Pećina Late Pleistocene hominid remains. *Proceedings of the National Academy of Sciences of the United States of America, 96*, 12281–12286.

Srdoč, D., Sliepčević, A., Obelić, B., & Horvatinčić, N. (1979). Rudjer Bošković Institute Radiocarbon Measurements V. *Radiocarbon, 21*(1), 131–137.

Straus, L. G. (1999). The Neanderthal problem continued: Reply. *Current Anthropology, 40*, 352–355.

Svoboda, J. (2001). La question szélétienne. In D. Cliquet (Ed.), *Les industries a outils bifaciaux du paléolithique moyen d'Europe occidentale* (pp. 221–330). Liège, Belgium: ERAUL 98.

Svoboda, J. (2006). The Danubian gate to Europe: Patterns of chronology, settlement archaeology, demography of late Neandertals and early Modern Humans on the Middle Danube. In N. J. Conard (Ed.), *When Neanderthals and Modern Humans met* (pp. 233–267). Tübingen, Germany: Kerns Verlag.

Szmidt, C. C., Brou, L., & Jaccottey, L. (2010). Direct radiocarbon (AMS) dating of split-based points from the (Proto)Aurignacian of Trou de la Mère Clochette, Northeastern France. Implications for the characterization of the Aurignacian and the timing of technical innovations in Europe. *Journal of Archaeological Science, 37*, 3320–3337.

Teyssandier, N., Bon, F., & Bordes, J.-G. (2009). Within projectile range: Some thoughts on the appearance of the Aurignacian in Europe. *Journal of Anthropological Research, 66*, 209–229.

Turk, I., & Kavur, B. (1997). Survey and description of Paleolithic tools, fireplaces, and hearths. In I. Turk (Ed.), *Mousterian "bone flute" and other finds from Divje Babe I Cave Site in Slovenia* (pp. 119–156). Ljubljana, Slovenia: Slovenian Academy of Sciences and Arts.

Vuković, S. (1950). Paleolitska kamena industrija spilje Vindije. *Historijski Zbornik, 1–4*, 241–256.

Wild, E. M., Paunović, M., Rabeder, G., Steffan, I., & Steier, P. (2001). Age determination of fossil bones from the Vindija Neanderthal site in Croatia. *Radiocarbon, 43*, 1021–1028.

Wojtal, P. (2007). *Zooarchaeological studies of the Late Pleistocene sites in Poland*. Kraków, Poland: Institute of Systematics and Evolution of Animals, Polish Academy of Sciences.

Wolpoff, M. H. (1999). *Paleoanthropology* (2nd ed.). New York: McGraw-Hill.

Wolpoff, M. H., Smith, F. H., Malez, M., Radovčić, J., & Rukavina, D. (1981). Upper Pleistocene human remains from Vindija Cave, Croatia, Yugoslavia. *American Journal of Physical Anthropology, 54*, 499–545.

Zilhão, J. (2009). Szeletian, not Aurignacian: A review of the chronology and cultural associations of the Vindija G1 Neandertals. In M. Camps & P. Chauhan (Eds.), *Sourcebook of Paleolithic transitions* (pp. 407–426). New York: Springer.

Zilhão, J., & d'Errico, F. (1999). The Neanderthal problem continued: Reply. *Current Anthropology, 40*, 355–364.

Chapter 5
Spanish Aurignacian Projectile Points: An Example of the First European Paleolithic Hunting Weapons in Osseous Materials

José-Miguel Tejero

Abstract Projectile points constitute the main component of osseous equipment in the Spanish Aurignacian. Two different types follow one after the other chronologically: split-based points during the Early Aurignacian and then simple-based point during the evolved Aurignacian. With rare exceptions, antler is the chosen material to produce these projectile points. Contrary to bone work—which uses fragments recovered from food activities to make domestic tools—antler exploitation is unconnected to food activities and is instead driven by projectile production. This form of antler exploitation integrates, for the first time during the European Paleolithic, an organic material into the technical sphere. The limited availability of this material and the complex processes applied in its transformation, are reflected in the systematic shaping and resharpening of the projectile points. Issues surrounding these processes in Spanish Aurignacian split-based and simple-based points are outlined and discussed.

Keywords Upper Paleolithic • Split-based points • Simple/massive based points

Introduction

Hunting weapons manufactured from osseous raw materials first appeared in Europe during the Aurignacian, more specifically during its earliest stage (Early Aurignacian) about 40,000 years ago, though a few possible and controversial examples from the Proto-Aurignacian have been reported (i.e., Trou de la Mère Clochette, France: Szmidt et al. 2010; Fumane, Italy: Bertola et al. 2013). Prior to the advent of this techno-complex, the raw materials used for weapon manufacture were stone and/or wood—famous examples of wooden spears including those found at Schöningen, Germany (Thieme 1997) and Clacton-on-Sea, Great Britain (Oakley et al. 1977). On the African continent, evidence for the use of osseous materials to make presumed projectile points are older. Here we find several examples from Middle Stone Age (MSA) contexts, as well as barbed points from Katanga, Congo (Yellen et al. 1995; but see Klein 2009: 527–529 for criticism on their estimated age) and Broken Hill, Zambia (see Backwell and d'Errico 2016) to cite only a few of the most important finds.

On the European continent, the first osseous points (*pointes de sagaie*) are the split-based points which have long served to establish the periodization of the Aurignacian. The internal organization of this typo-technological tradition simultaneously relied on quantitative representation of lithic types and the nature of the osseous points, sometimes prioritizing the latter over the former. The predominance of typological classifications in early archaeology explains the great importance given to so-called '*fossil directeurs*' in bone and antler, which in Europe, are stratigraphically better defined than most stone tools. Initially, the rich sequences of southwestern France endowed by the excavations of D. Peyrony provided the references necessary to develop a more complex model than that originally proposed a few years earlier by H. Breuil. Peyrony's periodization of the Aurignacian has five phases, the first four found at the site of La Ferrassie and the last in Laugerie-Haute (Peyrony 1933, 1934). Each phase was set according to the presence of a specific bone/antler tool. The split-based point, being present in Early Aurignacian levels at numerous sites, was the first morpho-type recognized among the hunting weapons of the European Aurignacian (Fig. 5.1).

It is because of this strong chronological connection that these points have been brought into arguments put forward

J.-M. Tejero (✉)
CNRS ArScAn, Équipe Ethnologie Préhistorique, Paris, France

SERP Barcelona University, Barcelona, Spain
e-mail: jose-miguel.tejero@mae.cnrs.fr

Michelle C. Langley (ed.), *Osseous Projectile Weaponry*, Vertebrate Paleobiology and Paleoanthropology,
DOI 10.1007/978-94-024-0899-7_5

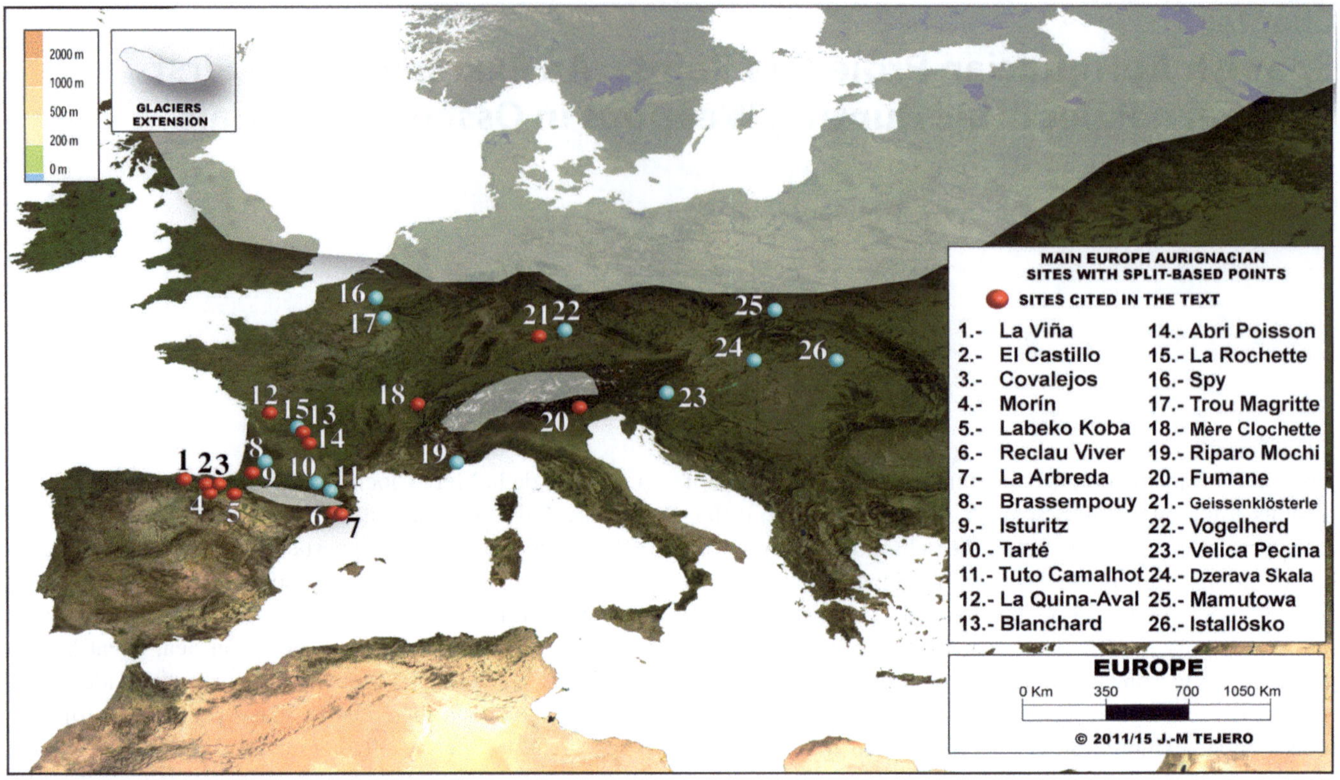

Fig. 5.1 Main Europe Aurignacian sites with split-based points

by defenders of a sharp break between the Middle and the Upper Paleolithic (i.e., Mellars 1989; Mellars and Stringer 1989; Klein 1995). These claims, however, are not entirely correct if we consider that there is earlier evidence for the use of hard animal raw materials in elements of symbolism. Pigments and seashells use, for instance, are both documented in Mousterian (e.g., Aircraft Cave in Spain: Zilhão et al. 2010), and Châtelperronian contexts (e.g., Grotte du Renne, France: Caron et al. 2011). Both of these cultures being associated with Neanderthals. Furthermore, the real break in osseous raw material working is not between the Mousterian and the Aurignacian, but between the Proto-Aurignacian and the Early Aurignacian and concerns—exclusively—antler working (Tejero 2014; Tejero and Grimaldi 2015). In all, this situation has caused split-based points to currently be one of the most studied osseous objects of the Paleolithic period.

Concerning the Spanish Aurignacian, projectile points constitute the main component of osseous equipment recovered. Here, two different types follow one another chronologically: split-based points (during the Early Aurignacian) and then simple-based or massive-based points in the following Evolved Aurignacian phase (Tejero 2010, 2013). From the analysis of these objects, and of the archaeological remains associated with their manufacture (blanks, waste, etc.), it was found that the raw material, debitage, and manufacturing schemes, were similar for both point types. The differences between split-based and simple-based points are instead limited to the proximal part, and necessarily by this fact, their hafting systems, though their mode of use may also differ.

In this chapter, I present the different aspects of osseous projectile points during the Spanish Aurignacian. But since, as we have seen, these objects are of great importance for the analysis of the chrono-cultural, economic, technical, and social aspects of the beginnings of the Upper Paleolithic, and because the Spanish corpus is numerically limited, I will attempt to integrate information from other main European sites currently being studied by myself in order to consider these technologies in a wider context.

The Raw Material for Hunting Equipment Manufacture

Although some early references to Spanish Aurignacian projectile points mention bone as the raw material for manufacture of these weapons (e.g., Bernaldo de Quirós 1982; Cabrera 1984), the fact is that almost all of these objects are made from deer antler, which is consistent with the available data for other parts of Europe (including southwest France and

Central Europe [Knecht 1991]). The only other raw material documented for making projectile points in Spain is mammoth (*Mammuthus primigenius*) ivory. One split-based point found in level Delta (Obermaier excavations) of El Castillo Cave (Fig. 5.2: 4) is the only such point in ivory found in Aurignacian contexts from all over Europe, though another point from Morín Cave (level 7) that does not conserve its base (and thus we cannot identify its morpho-type) has also been reported (Tejero 2010, 2013).

Although hunting weapons made from mammoth ivory are known in the Aurignacian archaeological record, most were documented in areas where the presence of this animal is relatively abundant during the initial Upper Paleolithic—such as Belgium (Otte 1995) or Central Europe (Hahn 1995).

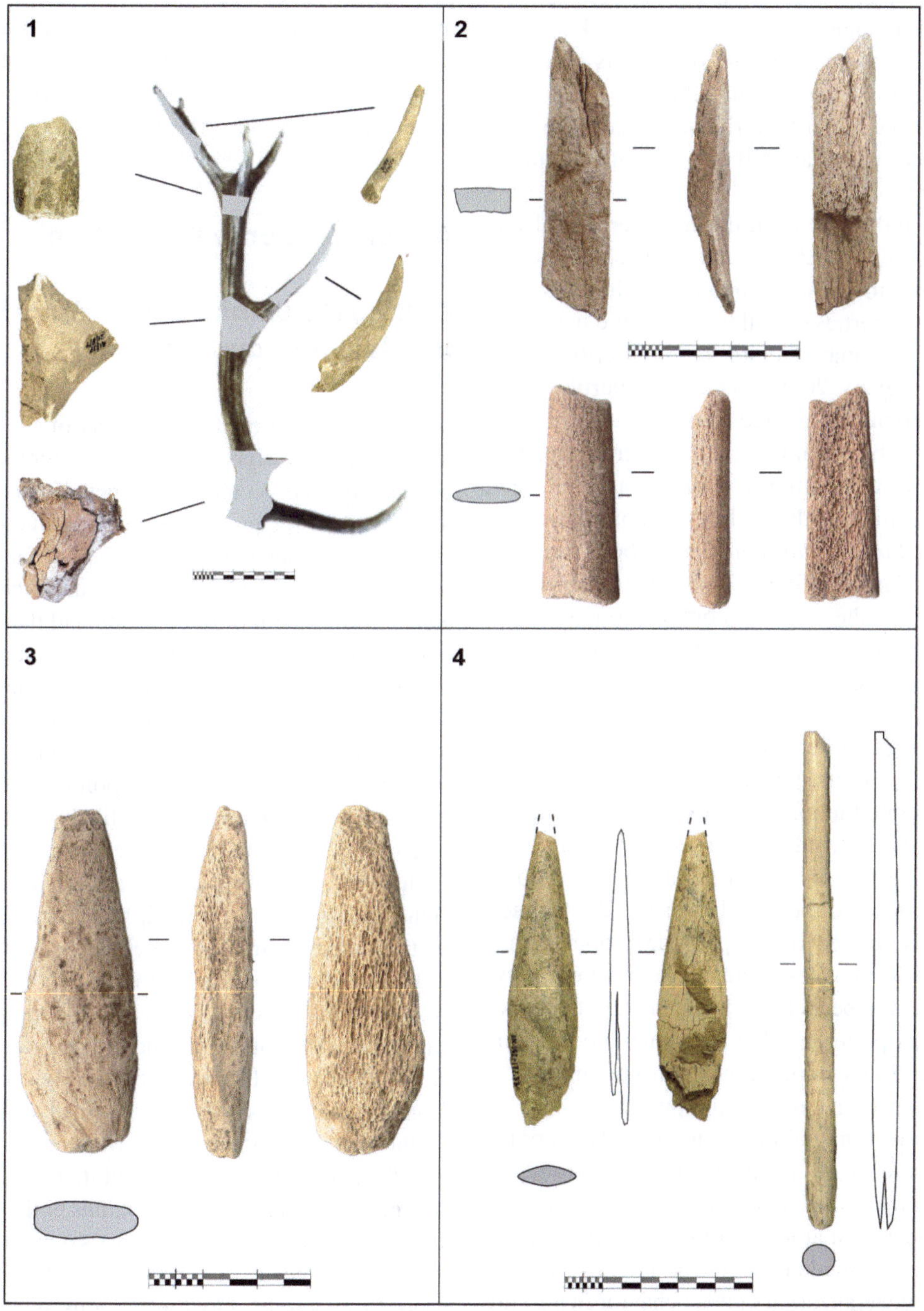

Fig. 5.2 Elements linked to the split-based points fabrication: (*1*) Wastes produced during antler preparation and secondary block production. El Castillo, level Delta. (*2*) Rough blank of the *baguette* type (*top*). Conde cave A/B. and a worked baguette blank (*bottom*). Cierro 5, 6. (*3*) Rough out of a projectile (*pointe de sagaie*) point. Conde Cave A/B. (*4*) Split-based point on antler (*left*) and ivory (*right*). Labeko Koba V. El Castillo Delta

In more southerly regions, and especially in Spain, the remains of this proboscidea are only marginally documented amongst the fauna of Aurignacian deposits (i.e., Labeko Koba: Altuna and Mariezkurrena 2000). Therefore, we can assume that the availability of this raw material was actually very limited. If we consider that the availability of deer antler was relatively high, that the properties of ivory are less suitable than antler or bone for the functional requirements of a projectile weapon (Christensen and Tejero 2015), and that other forms of ivory production in this area are not known (e.g., ivory portable art objects are well known in Germany [Conard 2009] and personal ornaments including small beads are found in France [White 2007]), the status of this raw material in the Spanish Aurignacian seems to have specific particularities which are difficult to explain with the data we have so far.

Although bone and ivory from different terrestrial and marine mammals (Pétillon 2008) and even shell (Jones, 1988) can be used to make weapons for hunting, the morpho-structural properties of antler make it the best osseous material for the manufacture of projectile points (Christensen and Tejero 2015). These properties were known and systematically exploited from very early on (the Early Aurignacian) by Spanish (and wider European) Aurignacians who worked antler of a single taxon of deer: Red deer (*Cervus elaphus*), although a few possible examples of Reindeer (*Rangifer tarandus*) have been reported (for instance at El Castillo: Tejero 2013). This choice was largely imposed by the faunal composition of Late Pleistocene Iberia biotopes, where the prevalence among deer species corresponded mainly to Red deer and Roe deer (*Capreolus capreolus*) (Castaños 1986, 2005). This association is very strongly reflected in the composition of Aurignacian faunal deposits, where Red deer is always the species best represented among cervids—even among the fauna of sites such as El Castillo (Cabrera 1984; Dari 2003; Liouville 2007). The use of Roe deer antler is not documented, probably owing to the fact that its use for technology manufacture is limited by its poor development and reduced cortical tissue thickness. Thus, the almost exclusive use of Red deer antler appears to be determined not so much by choice or cultural preference but by a double constraint: the preponderance of this taxon in the environment and the technical possibilities of the antler from this species.

Antler procurement modalities can be only partially determined. The presence of manufacturing waste from projectile points made from shed antler bases (e.g., El Castillo, Labeko Koba) indicate that at least part of the exploited antler came from collected shed antler. We do not have enough data to propose the generalization of this behavior in the sites mentioned above or in other sites with evidence of Aurignacian antler working. However, some indirect evidence seems to suggest that the collection of antler for the manufacture of hunting weapons was a common practice among Spanish Aurignacians. For instance, the absence of deer among the fauna hunted, as at Labeko (Altuna and Mariezkurrena 2000), or their capture principally in late winter and spring when the males of this species lack antlers, as observed at El Castillo (Pike-Tay et al. 1999). Moreover, the use of antler with a cortical thicknesses exceeding 6 mm in most of these sites can only have come from fully calcified and shed antlers or from antlers of adult males about to complete its annual cycle of growth (Averbouh 2000), indicating a selection from hunted animals which is not consistent with the archaeological record of the Spanish Aurignacian. However, in many cases no zooarchaeological studies have been undertaken to clarify this question.

Antler Weaponry Production

Getting the Blanks: A New Way to Transform the Osseous Raw Material

In relation to the anatomical parts of the antler worked, in most Spanish Aurignacian sites a preference for the A and B beams—that is, the main branches of the antlers—as blocks to be worked for projectile point production is found. This preference is dictated by two inherent factors of the projectile weapon: cortical osseous tissue thickness sufficient to ensure the strength of the weapon and the straightness of the blank which allows the future point to maintain its efficiency through a predictable trajectory under the laws of ballistics. To a lesser degree a minimum length is required. However, maintenance of projectile points which decreases its length from the theoretical original points (see below), demonstrates that this parameter is likely to be less important. These requirements are preferably provided (sometimes exclusively depending on the antler) from the antler beam. Generally, in tines and tine tips calcification is less and these sections are less straight (except in some cases of the eye tine of Red deer) than in the beams.

The technical steps between the supply of raw materials and the finalization of the point have rarely been discussed in the literature. However, obviously, obtaining blanks to fabricate the points—debitage—involves the transformation of the material to remove a fraction of it in order to be transformed into a projectile point. Today we know that the Aurignacians were obtaining blanks both to make split-based and simple-based points, by cross segmentation of the beams to obtain cylindrical blocks that would later be split. These ways to penetrate into the osseous matter were unknown before the Early Aurignacian and are different to those used later during the Gravettian (see Goutas 2016). The Gravettians were the 'inventors' (in our present state of knowledge) of the procedure known as 'double longitudinal grooving', although they also continued to occasionally use the splitting technique (Goutas 2004).

Bone exploitation modalities in the Early Aurignacian differ from those used for antler (Tartar 2009, 2012; Tejero 2013) and had already been employed by Neanderthals to procure blanks for their objects made in this raw material (awls, retouchers, etc.) (e.g., Vincent 1993; Armand and Delagnes 1998; d'Errico et al. 2003; Soressi et al. 2013). Bone fragments were obtained in most cases by a simple and expeditious fracturing by percussion that has no technical purpose but for food processing—that is, to fracture the bone to access the bone marrow. Therefore, the knowledge of how to work antler during the Early Aurignacian represents a working procedure for osseous raw materials completely new and best known today through the work of several researchers outlined below.

Knecht (1991, 1993) was one of the first researchers to talk about splitting as a procedure by which the Aurignacians produced blanks from antler, of more or less rectangular morphology, in order to make projectile points. Later, Liolios (1999) tested the experimental reproduction of this type of blank, though her work was more didactic than demonstrative. Recently, based on the analysis of the manufacturing waste and blanks of Spanish Aurignacian sites, the author in collaboration with M. Christensen and P. Bodu has reproduced the procedure of Red deer antler splitting to obtain blanks and compared the results to archaeological material (Tejero et al. 2012). This experimental program has allowed us to accurately characterize the way in which Aurignacians transformed antler raw material blocks in order to make their hunting weaponry.

The technical procedures for processing antler during the Aurignacian initially acted across the grain (segmentation: cutting through the fibers), later changing to a horizontal action (split: splitting and tearing the fibers). This choice seems determined by the physical and structural characteristics of antler, which we have seen, Aurignacians knew to perfection. Uncontrolled direct percussion (as was used repeatedly for bone) is an inappropriate technique for antler working owing to its relative degree of elasticity, and therefore, its capacity to absorb the blows. This feature is determined by the relationship between the organic fraction and mineral fraction of the raw material, which is characteristic of all osseous tissues but in the case of antler has a larger amount of the former, than that found in bone or ivory (in other words, antler is less mineralized) (Christensen 2004). Nevertheless, as shown by some recent studies focusing on the Badegoulian techno-typological tradition, it is possible to obtain antler flakes blanks by knapping, although an important section of the raw material is lost and the blanks morphology remains quite random (Averbouh and Pétillon 2011; Pétillon and Ducasse 2012; Borao et al. 2016). Regarding fracturing by splitting, this procedure exploits the natural disposition of the osseous fibers which are laid in longitudinal bundles, separating them in an expeditious way while allowing relative control of the width, and especially, very precise control of the length of all extracted blanks (Tejero et al. 2012). Indeed, this arrangement of the osseous fibers is reminiscent of vegetable materials, leading some to propose that the beginning of the osseous material exploitation was a simple transfer of wood working techniques to osseous (Liolios 1999, 2006). This is a very interesting hypothesis, but so far, lacks supporting evidence.

On the basis of evidence from Spanish sites, but also found in major French assemblages such as Isturitz and La Quina-Aval (Tejero 2014), the Aurignacians cross segmented antler with two clearly hierarchical objectives. The primary objective was the production of blocks from beam segments (A and B beams) which were subsequently split in order to obtain blanks with roughly rectangular or sub-triangular morphology ('*baguette*' type) (Fig. 5.2: 2). This objective may have been preceded by the preparation of the antler by eliminating unnecessary elements that hinder cross-sectioning (tine and tine tips), which in this case, would become the secondary aim clearly subsidiary to the previous one.

This procedure, which seeks to recover the useful (usually central) anatomical parts of the antler and eliminate other less useful sections, is supported by the waste documented in almost all studied sites, as well as the few objects other than projectile points which were made from this raw material. Recovered waste always consists of tines, tine tips, fragments of the junction of the beams A/B, and the antler bases (Fig. 5.2: 1), with occasional antler objects manufactured from these sectioned parts. Mostly, these other objects consist of tines used like intermediate or bevelled pieces (El Castillo, Hornos de la Peña, Labeko Koba, Covalejos, Conde, Cierro, etc).

The relationship between length and width of the blanks, both found in Spanish Aurignacian artefacts and experimental points (Tejero et al. 2012), disproves the claim of some authors that this procedure does not allow the formation of long and narrow blanks (Liolios 1999). On the contrary, a successful split fracture has various factors (start it straight, position with regard to the work plane, position of the intermediate piece on the blow, etc.) that, provided there is some anticipation of the artisan, allows the removal of long and relatively narrow blanks. Such blanks are indirectly evidenced by the existence of points over 210 mm (El Castillo Delta from Obermaier's excavations, Mallaetes XIII), but are documented directly in certain French sites (Isturitz and La Quina-Aval) where some *baguettes* exceed 200 mm in length with a width ranging between 20 and 30 mm (Tejero 2014).

No other technology has been documented alongside projectile points as objects made from antler *baguettes* in Spanish sites, reinforcing the idea that the exploitation of antler is linked exclusively to the manufacture of hunting equipment. Moreover, dimensional compatibility between blanks and points where both have been recovered indicates the adequacy of one to manufacture the other. Additionally, we have the intermediate technical elements (rough outs), as in the case of Conde A/B or Covalejos (B) (Fig. 5.2: 3), where the matching morphometric modules of successive stages (blank-rough out-projectile point) confirms this theory (Tejero 2013).

Transforming Blanks into Projectile Points

In the manufacturing phase, the overall volume and symmetry of the point are established. The edges are regularized, eliminating fracture planes which resulted from the splitting and removing of the blank from the cancellous fraction of the antler (though a fraction is still visible on almost all projectile points) (Fig. 5.2: 2). Here, only scraping was used and was utilized for both split-based and simple-based points.

The use of only one technique among several possibilities (abrading, polishing, incision, etc.), along with the absence of work that, from our perspective would be describe as 'aesthetic', has led some authors to speak of 'minimalism' in the making of these objects. This practice is documented in both the Aurignacian as well as in certain Gravettian assemblages (Goutas 2004). Differences in workmanship between various sites and within different levels of the same site is not found, except in the extent of surface worked. For both of the two Aurignacian point types, you can find worked pieces processed by scraping, while similar others have a rough (largely unworked) superior face. As we shall see, when referring specifically to each of the point types, we can propose a hypothesis to explain this difference for split-based points. However, this question, like many other unknowns about the technical aspects of the Aurignacian simple-based points, remains to be clarified for reasons including that the attention of researchers has been devoted largely to the tip of the Early Aurignacian points.

While both types of points share manufacturing methods, other aspects such as their morphometry or hafting system, are specific to each group. The sections below are devoted to the specific characteristics of split-based points and simple-based points respectively.

Split-Based Points

Although the Spanish corpus of split-based points is quantitatively limited in comparison to other areas of Europe (i.e., southwest France or the Swabian Jura [Knecht 1991; Liolios 1999; Tejero 2014]), these weapons are present in ten Spanish sites. These sites are found in both the Cantabrian and Mediterranean regions and represent a strong presence and wide distribution of this morphotype (Fig. 5.3).

Morphometric Design

Spanish Aurignacian split-based points feature an apparent diversity in the form of their base cleft. This difference can be seen on both completed tools as well as those discarded in manufacture. Other parameters, such as point dimensions, are more difficult to evaluate owing to point fragmentation and can only be considered as indicative in conjunction with other values. Careful analysis, however, shows that Spanish split-based points can be classify into two types of morphometric design: (1) points with an elliptical form; and (2) points with a bi-convex form (Fig. 5.3). A single specimen with a circular cross section has been identified, but in this case it was dictated by the choice of raw material (ivory) (White 1995; Christensen 1999). The production of an ivory blank was probably undertaken by simply taking advantage of a naturally exfoliated fracture, and thus, this specimen is different from all other examples made from antler. The exploitation of sub-fossil ivory is also documented in other European Aurignacian sites like Geissenklösterle (Liolios 1999).

Elliptical section points usually display medium or long lengths (around 60–110 mm). Edges are straight, converging towards the distal end, and the surface can be either completed scraped over or the superior face can be left rough. Points with the bi-convex design are generally smaller in size, with this value varying between a minimum of 30 mm and a maximum of 80 mm. Edges of these bi-convex points are often show a truncation (a rupture in its manufacture) from the end of the base or from the medial section. In almost all cases, the entire surface has been scrapped.

The interpretation of this dualism in Spanish Aurignacian point design is difficult owing to the small number of points available to study. It may simply be the result of a lack of morphometric standardization of this object, regional traditions and/or cultural variety, or even different use modalities of the hunting weapons with different morphology or varying size (different environments, different types of prey, etc.). None of these hypotheses can be verified from the available data. However, there is a technical feature common to both Spanish split-based points and those from other countries which may explain this circumstance (at least in part): the systematic resharpening and reshaping of broken tips during use.

Split-Based Point Maintenance

The relatively reduced availability of antler compared to others such as bone or lithic raw material, and the significant and complex technical investment in their manufacture, result in that points be systematically repaired. This behavior is not exclusive to the Early Aurignacian, also being documented within the Magdalenian (e.g., Pétillon 2006; Langley 2015), and it is foreseeable that new technical studies will identify

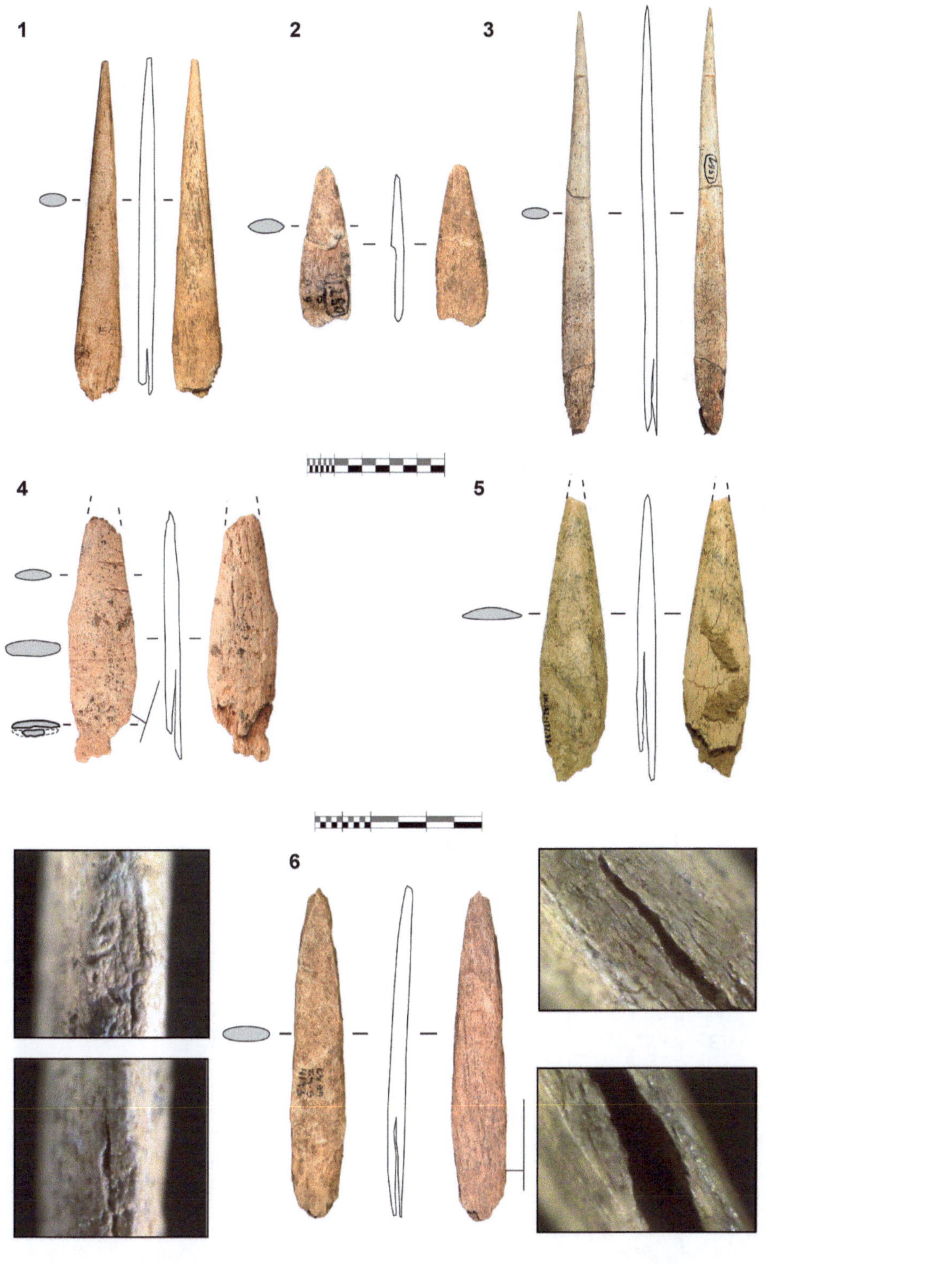

Fig. 5.3 Split-based points from the Spanish Aurignacian. (*1*, *3*) El Castillo. level Delta. (*2*) Morín 9. (*4*) La Viña XIII. (*5*) Labeko Koba V. (*6*) Covalejos B. *Bottom*: detail of lateral flattening of the base of the point (Magnification 15×, 20×)

evidence of this behavior in all techno-complexes of the European Upper Paleolithic.

In Western Europe, split-based points from different sites, such as those from Abri Poisson, La Quina-Aval, and Isturitz, display evidence for the recurrent recovery of broken projectile points (Tejero 2014) (Fig. 5.4). Liolios (1999) proposed a theoretical scheme for the resharpening and reshaping for these particular objects founded in the study of various French sites and of Geissenklösterle (Germany). In broad terms, this scheme corresponds to that observed for the

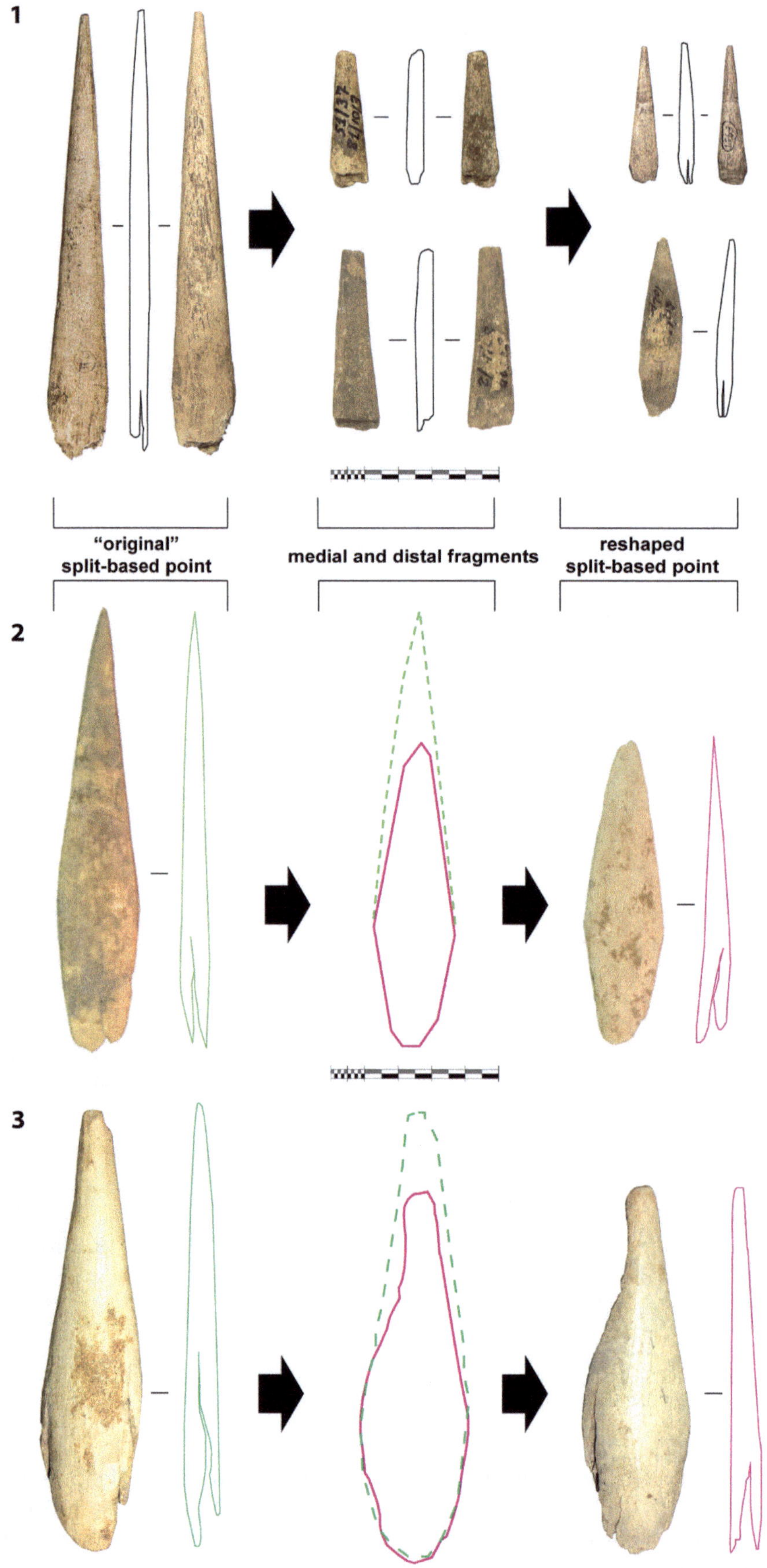

Fig. 5.4 Theoretical maintenance scheme for split-based points. (*1*) El Castillo. level Delta. (*2*) Abri Poisson (France) Aurignacian level. P. Girod excavations. (*3*) La Quina-Aval Aurignacian level

Spanish Aurignacian (Fig. 5.4). Following this theoretical schema, medial and distal fractures of the 'original' points were repaired to repoint the tip. This activity results in increasingly short projectile points until their small size makes them no longer functional. If we look at examples from El Castillo or Isturitz, exhausted points are around 50 mm in total length. According to experimental analyses, proximal fractures, in most cases, would result in the abandoning of the point, which cannot be repaired (Pétillon 2006).

This process explains the existence of the two morphometric types identified for this site and other Spanish sites—maybe even for other Europeans assemblages. The length of the El Castillo projectile points, as well as those from Isturitz cave and La Quina-Aval (Tejero 2014), is the most fluctuating value and likely corresponds to successive resharpening and reshaping cycles of broken points. In contrast, the width and thickness of the points remains quite constant throughout the corpus (including for whole and fragmented points). The width of the projectile base is theoretically determined by the need to attach the point to the shaft (presumably of wood which did not survive), and hence, can be considered the variable that limits the dimensions of the proximal part. Hence, the repair of broken points is done by scraping only to delineate a new distal tip without affecting the thickness of the body. This method is responsible for transforming an original elliptical point into a biconvex point, and also, for the truncation of edges with 'original' points having straight edges. From these data we can propose for most Aurignacian Spanish sites with split-based points, the existence of a single point design (first intention module *sensu* Liolios 1999) of elliptical cross-section. The resharpening of fractured projectile points would lead to a secondary point form that maintains both the width of the base and the overall thickness but becomes bi-convex in cross-section as the edges lose their original straight delineation.

Some Considerations on Hafting and Launching Modes

One of the most controversial issues regarding split-based points was the manufacture of the base, and consequently, everything that is related to its hafting system. On most Spanish split-based points, the base was made by simple indirect percussion according to the model proposed by Henri-Martin (1930). Recent work (Tartar and White 2013) has resumed the original hypothesis of Nuzhnyi (1998) who proposed an intermediate model between the proposals of Henri-Martin (1930) and Peyrony (1935). The author experimented with extracting a small material fraction of the base creating a so-called 'tongue-piece'. Thereafter, the crack initiated from the fracture and the base became cleft. Although this model has been tested experimentally and compared with the French assemblages of Abris Blanchard and Castanet, it is not possible to generalize and extend this approach to the Spanish points or other important French sites that we have studied including La Quina-Aval (Tejero 2014). Contrary to the impression given in Tartar and White (2013), 'tongued-pieces' are only present in one Spanish site, Covalejos (Tejero 2013), and only in the Early Aurignacian Level B (2) where we documented 3 examples, one being in a very poor state of preservation. The morphology of these three pieces is 'atypical' when compared to the French examples (Knecht 1991), and thus, their relationship with other split-based points is not yet demonstrated (Tejero 2013).

In contrast, split-based points of the Spanish Aurignacian share one technical feature which is likely related to the cleft base. The cleft is made after preparing a flattened section made by scraping on both sides (Fig. 5.3). This flattening is also documented in a number of French sites and is interpreted by Knecht as a way to control the length of the cleft (Knecht 1991:397). Indeed, if we compare the length of the flattened section with the clefts found in Spanish sites, we note that the clefts are always a few millimeters shorter than the flattened sections. This observation confirms that the flattened section is effective in stopping the spread of the slot, but cannot be said with certainty since there is a lack of experimental data on this issue.

The lack of experimental data also affects other important issues regarding these weapons: specifically their mode of use. For more recent periods of the Upper Paleolithic, different authors have explored the use of the bow or spearthrower to project points (see Langley et al. 2016). However, there exists no direct evidence for either of these launch objects in the Aurignacian (Cattelain 1994), nor does the few shooting experiments undertaken with Aurignacian points bring light to the issue. Knecht (1991, 1997) did, however, demonstrated the effectiveness of split-based and simple-based points as projectiles. Knecht hafted the points using a small wedge (*clavette* piece) in order to attempt to explain the existence of the 'tongued-pieces' as waste from the manufacture of these sockets, as well as determine if they were used in the hafting of the points. However, the researcher performed the shots with a modern device (crossbow), so we do not have data for their use with bows or spearthrowers (made perhaps as early as the Aurignacian on perishable materials such as wood), or perhaps, launched by hand. Furthermore, observations made by Knecht on the small amount of damage sustained by the points in use, even when penetrating the target by more than 20 cm, does not correspond with the archaeological evidence from at least some of the southwest European collections. Although the number of points available for the Spanish Aurignacian does not enable statistically significant data analysis, we have information from other sites in Western Europe. For instance, following the criteria for the identification of use fractures proposed by Pétillon (2006), we have documented nearly 50 distal and medial use fractures in a set of over 150 projectile points from Isturitz cave, along with ten proximal fractures (Tejero 2014). Ongoing research will allow us to compare these fractures with experimental examples, along with expanded traceological analysis to other sites.

Simple-Based Points

Morphometrical Features of a Heterogeneous Assemblage

Defined as elongated objects with a pointed distal tip, a variable cross section, and a simple hafting system, the simple-based point has been known by various names, including: pointed base points, biconical point, massive base point, or simply 'not a split-based point' (Hahn 1974; Leroy-Prost, 1974, 1975, 1979; Hahn 1988). I prefer to use the term 'massive base' or 'simple-based' point which is better suited to the reality and variability of the general morphology of these objects, which do not always exhibit a pointed proximal part or a biconical contour.

Some of these points are described as having '*à base raccourcie*' where its proximal part has a number of irregular steps (Mons 1988). The identification and characterization of the scraping technique known as 'scraping in diabolo' has been described and accurately characterized in several recent publications (Le Dosseur 2003; Chauvière and Rigaud 2005) although the first citations are older (Rigaud 1972). These artefacts are the result of a segmentation procedure which involves the progressive slimming down of the thickness of the element to be sectioned. To do this, localized, peripheral and unidirectional scraping is undertaken, pressing continuously on the raw material. The corpus best studied from a technical point of view (the Magdalenian levels from La Garenne. France), notes that in many cases these previously supposed projectile points are actually waste from manufacture of points whose blank was sectioned by scraping 'in diabolo' (Chauvière and Rigaud 2005). However, these same authors suggest the possibility that not all of these artefacts come from the same technical process (which may, in fact, come from an action of debitage, manufacture, or repair of a point), and must be considered on a case by case basis. Some of the evidence for the Spanish Aurignacian has been classified in various publications as '*à base raccourcie*' but in reality these artefacts are not a separate point morphology.

Chrono-culturally, all Spanish evidence for simple-based points are assigned to Evolved Aurignacian levels (Arbreda G, Mallaetes XIII, El Ruso IVB, Morin 5 inf and El Otero 6, 5, 4). In the characterization of the assemblages, perhaps the most remarkable feature is the morphological heterogeneity, both between sites and within site assemblages (Fig. 5.5). This heterogeneity may be the result of various factors, including: no standardization of piece production, the adequacy of each item to fulfil different functional hunting needs (perhaps different types of prey), or even, given that we cannot be sure that all pieces belong to the same occupation period, may correspond to the contribution of various groups with different technical traditions in successive occupations. Morphometrically the maximum lengths of intact points range between 224 and 93 mm. Width varies between 6 and 20 mm and thickness between 4 and 10 mm. Cross-sections are preferably elliptical, with a few sub-rectangular or biconvex examples. There is no correlation in these values that infer a specific use for simple-based points.

Spears Points Without Evidence for Resharpening and Reshaping

It is not possible to deduce from the analyzed pieces any evidence for resharpening, reshaping or recycling of simple-based points. The absence of such evidence may be owing to several factors, including: the limited sample size that has prevented examples with maintenance being included, greater availability of raw material that renders the reuse of the broken point unnecessary, and the relatively less investment in the manufacture of simple-based points as against split-based points. The last two possibilities, however, seem very improbable.

Some Considerations by Way of Summary and Conclusion

Despite the unique features of Spanish Aurignacian osseous projectile point assemblages—especially their limited number compared to other geographical areas—the study of hunting weapons still allows us to propose a series of reflections extending to the Western European Aurignacian.

Deer Antler Working: 'Complex' Transformation of an Osseous Raw Material

Throughout the entire Upper Paleolithic of Western Europe, prehistoric artisans chose deer antler to manufacture a large part of their hunting equipment, while bone was preferably

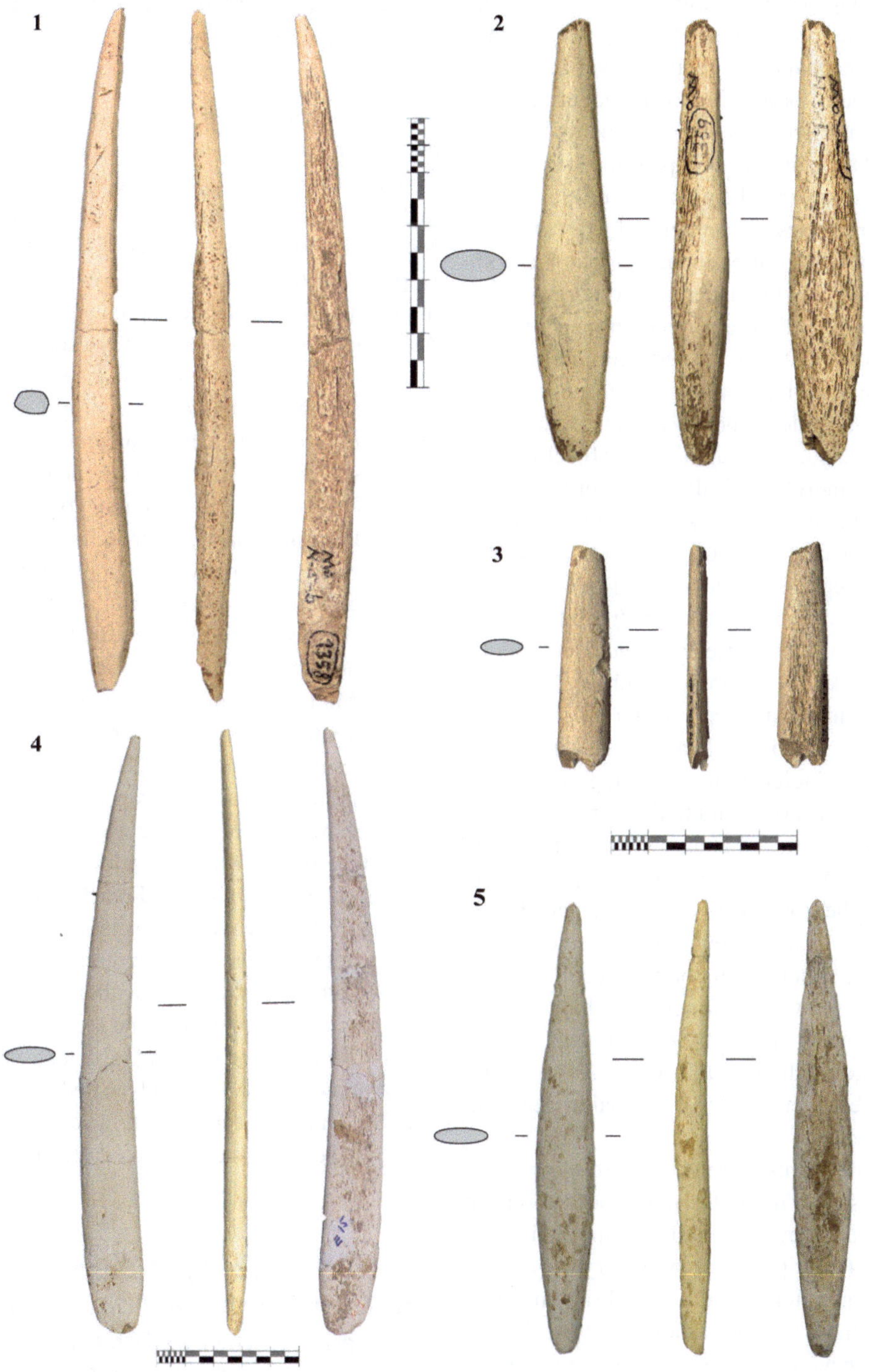

Fig. 5.5 Simple or massive base points from the Spanish Evolved Aurignacian. (*1*, *2*) Morín 5 Inf. (*3*) Arbreda G (*4*, *5*) Mallaetes XIII

used for the manufacture of domestic tools (Christensen and Tejero 2015). From the Early Aurignacian, however, both in Spain and in close by regions, the manufacture of hunting equipment is not limited by the choice of a particular raw material, but directly responsible for its exploitation. The parts of the antlers exploited (A and B beams), the working methods employed (transversal cross-sectioning and longitudinal splitting), the selection of antler pieces with high cortical values (above 6 mm thick), shows complex technical organization fundamentally different to the exploitation of bone.

When I talk about concepts like complexity or significant effort investment in the production of projectile points, I mean the mental and manual effort required over the entire *châine operatoire* of these tools. This effort begins at raw material procurement and requires the anticipation of the needs of the group, and the planning of expeditions to recover antler intact (when shed antler is being collected), as antler only maintains their properties and integrity for a few days after shedding (owing to different biotic and abiotic factors). Collection is a technical behavior closer to abiotic raw material supply (flint and/or other lithology) than to meat exploitation (or other resources related to hunted animals, such as skins or tendons). The knowledge necessary to choose the best methods for working antler to produce blanks and transform them into projectile points, imagining an object from a block of material that is morphologically different (although it is true that this is common to the work of lithic raw materials and not only exclusive of the Aurignacian but also the Mousterian when Neanderthals developed the lithic debitage method called levallois), hafting system configuration, and above all, the inclusion of these objects in the food supply chain of the group, make it a complex technical system requiring specific planning that is very different from that utilized in bone working. For bone, data from the sites with zooarchaeological studies indicate that bone fragments used or transformed into objects correspond to the remains of animals hunted and consumed (i.e., Labeko Koba: Altuna and Mariezkurrena 2000). Bone is fractured by direct percussion with minimal control, the goal of this fracturing not being technical, but alimentary. The bone fragments selected as tool blanks are transformed summarily or are used directly in order to make retouchers, awls, or smoothers which are the most representative objects in both Spanish and French Aurignacian assemblages (Schwab 2002; Tartar 2009; Tejero 2013; Schwab 2014; Tejero et al. 2016). No maintenance of bone objects is documented.

Conversely, projectile points were shown great care during manufacture and use. No doubt the main motivation for this treatment is determined by the functional imperatives of the projectile elements (strict planes of symmetry, regularity of their surfaces, hafting systems). But beyond raw material constraints, various authors propose a not strictly technical explanation for this behavior (Liolios 1999; Goutas 2004). The assignment of points to the hunting sphere in societies whose economic fundamentals revolve around hunting and consumption of different animal species, can certainly result in the tools gaining a special status. We must not forget that the viability of a human group includes the organization of an ensemble of complex systems (technical, economic, social, cultural), and a stable meat supply. Therefore, the objects used in their obtainment are all directly associated with the survival of the group.

Technological Data for an (In)homogeneous European Aurignacian

The traditional view of the Aurignacian as the first culture of the Upper Paleolithic and its association with *Homo sapiens* expansion across Europe, has led to the formulation of models that advocate the homogeneity of this techno-complex (i.e., Mellars 1989). Recent studies of lithic technology from early Upper Paleolithic sites in different geographical areas have improved our knowledge and changed our perceptions of the Aurignacian (i.e., Bon 2002; Teyssandier 2007). The study of osseous industries has also nuanced the perspective of the Aurignacian as a homogeneous entity. Liolios (1999) has demonstrated the existence in the apparently monotonous corpus of split-based points from some French and Central European sites, with what she calls different conceptions (morphometric designs). In our opinion, other technical aspects of these points also show heterogeneity if we look in detail at the European archaeological record from a technical and functional perspective. We have seen, for instance, that explanations for the manufacturing method for the point bases have been proposed (Henry-Martin, Peyrony) with a syncretic hypothesis of both most recently proposed (Nuzhnyi 1998; Tartar and White 2013). Perhaps the error of all these proposals reside in considering, implicitly, that all European Aurignacians always used the same way to prepare the bases of their projectile points. According to our own observations of the Spanish material, as well as at some important French sites (Isturitz, La Quina-Aval, Abri Poisson, etc.), base cleavage took place according to the site: either by direct percussion with prior preparation of a lateral flattening, by cleft without any preparation, or by extracting a fraction of material (generating a 'tongue piece'), along with posterior splitting (although with a somewhat different approach than that advocated by Nuznhny, Tartar and White), while in Isturitz not bifacial but unifacial sawing followed by bending is documented (Tejero 2014).

The Invention of Split-Based Points and Their Role in the Adaptation of First European H. sapiens During the Heinrich 4 Climatic Event: Research Perspectives on Aurignacian Osseous Weapons

Currently, many authors accept that climatic changes, especially those that occurred at a faster rate, have contributed significantly to the emergence of cultural innovations (Potts 1996; Ziegler et al. 2013). When the first modern humans arrived in Europe, about 40,000 years ago, they had to adapt to a new climate and the resulting new flora and fauna differed significantly from their African ecological environment. These *H. sapiens* European settlements, generally coincide chrono-culturally with the Early Aurignacian, and climatically with the Heinrich 4 event, which was a cold and dry period (Banks et al. 2013). While adaptation to these new environmental conditions was a key factor for their successful establishment in Europe, little is known about the aspects decisively affecting this adaptation process. Among others, the exploitation of antler as a raw material for making hunting weapons constitutes a key element of this process and is exclusive to Modern Human groups. However, we are still far from understanding the mechanisms of emergence and dissemination of this innovation and its importance in the human subsistence. For instance, what was the initial reason for antler exploitation? Antler projectile points were apparently never used by Neanderthals, so why their sudden appearance? We are currently working, with colleagues, on a project which aims at answering these questions by applying an interdisciplinary approach. Methods from archaeological and biomaterial science research fields will be combined to investigate material properties such as the splitting and fracturing behavior of different antler tissues, and to examine use wear through simulated hunting situations with experimentally prepared antler projectile points (Tejero 2014). These goals will provide insights into whether the first European *H. sapiens* distinctively *decided* which antler material to use for what purpose and tool, and yield explanations from a material science point of view for the behavior of antler tissue during processing and usage as a weapon.

To conclude, it seems that after decades of research, the first Modern Human osseous hunting weapons have not yet yielded all their secrets. To do this, we must not see them as an end, but as a means to approach one of the most fascinating aspects of prehistory: those related to hunting and subsistence of the Paleolithic hunter-gatherers.

Acknowledgments The author's research has been supported by a Beatriu de Pinós grant co-funded by the Catalan Government and European Community via the Marie Curie Actions of Seventh Framework Program for Research and Technological Development. This research is also part of the HAR2014-55131 research projects of the MICINN and the Quality Research Group of the Generalitat de Catalunya SGR2014-108. I'm very grateful to Marianne Christensen, Aline Averbouh, Nejma Goutas, Jean-Marc Pétillon, Boris Valentin, Michelle Langley and the two anonymous reviewers for providing useful comments about Upper Paleolithic hunting weapons. I also want to thank the Editor of this volume, Michelle Langley, for editing early drafts of this manuscript and for correcting the English, with any remaining errors being my own responsibility.

References

Altuna, J., & Mariezkurrena, K. (2000). Macromamíferos del yacimiento de Labeko Koba (Arrasate, País Vasco). In A. Arrizabalaga & J. Altuna (Eds.), *Labeko Koba (País Vasco). Hienas y humanos en los albores del Paleolítico Superior, Munibe N° 52* (pp. 107–181). San Sebastián: Sociedad de Ciencias Aranzadi.

Armand, D., & Delagnes, A. (1998). Les retouchoirs en os d'Artenac (couche 6c): Perspectives archéozoologiques, taphonomiques et expérimentales. In J.-P. Brugal et al. (Eds.), *Économie préhistorique: Les comportements de subsitance au Paléolithique. XVIIIè rencontres internationales d'archéologie et d'histoire d'Antibes. 23-25 Octobre 1997* (pp. 205–214). Sophia Antipolis: APDCA.

Averbouh, A. (2000). *Technologie de la matière osseuse travaillée et implications palethnologiques. L'exemple des chaines d'explotation du bois de cérvide chez les Magdaléniens des Pyrénées.* Ph.D. dissertation, Université Paris I Panthéon-Sorbonne.

Averbouh, A., & Pétillon, J.-M. (2011). Identification of "debitage by fracturation" on reindeer antler: Case study of the Badegoulian levels at the Cuzoul de Vers (Lot, France). In J. Baron & B. Kuffel-Diakoswka (Eds.), *Written in bones. Studies on technological and social contexts of past faunal skeletal remains* (pp. 41–51). Wroclaw: Institute of Archaeology, Wroclaw University.

Backwell, L. R., & d'Errico, F. (2016). Osseous projectile weaponry from early to late Middle Stone Age Africa. In M. C. Langley (Ed.), *Osseous projectile weaponry: Towards an understanding of Pleistocene cultural variability* (pp. 15–29). Dordrecht: Springer.

Banks, W. E., d'Errico, F., & Zilhão, J. (2013). Human–climate interaction during the Early Upper Paleolithic: Testing the hypothesis of an adaptive shift between the Proto-Aurignacian and the Early Aurignacian. *Journal of Human Evolution, 64*(1), 39–55.

Bernaldo De Quirós, F. (1982). *Los inicios del Paleolítico superior cantábrico.* Madrid: Ministerio de Cultura. Dirección General de Bellas Artes y Archivos.

Bertola, S., Broglio, A., Cristiani, E., de Estefani, M., Gurioli, F., Negrino, F., et al. (2013). La diffusione del primo Aurignaziano a sud dell'arco Alpino. *Preistoria Alpina, 47*, 17–30.

Bon, F. (2002). *L'Aurignacien entre mer et océan : réflexion sur l'unité des phases anciennes de l'Aurignacien dans le sud de la France.* Paris: Société Préhistorique Française.

Borao, M., Villaverde, V., & Aura, E. (2016). Debitage by fracturing in the osseous industry of Cova del Parpalló (Gandía-Valencia, Spain): A preliminary study. *Quaternary International, 403*(1), 118–131.

Cabrera, V. (1984). *El yacimiento de La Cueva de "El Castillo": Puente Viesgo, Santander.* Madrid: Instituto Español de Prehistoria, Consejo Superior de Investigaciones Científicas.

Caron, F., D'errico, F., Del Moral, P., Santos, F., & Zilhão, J. (2011). The reality of Neandertal symbolic behavior at the Grotte du Renne, Arcy-sur-Cure, France. *PLoS One, 6*(6), 21545.

Castaños, P. (1986). *Los macromamíferos del pleistoceno y holoceno de Vizcaya: Faunas asociadas a yacimientos arqueológicos.* Ph.D. dissertation, Universidad del País Vasco.

Castaños, P. (2005). Revisión actualizada de las faunas de macromamíferos del Wurm antiguo en la región Cantábrica. In R. Montes Barquin & J. A. Lasheras Corruchaga (Eds.), *Actas de la reunión científica: Nneandertales cantábricos, estado de la cuestión: Ccelebrada en el Museo de Altamira los días 20-22 de octubre de 2004, Monografías del Museo de Altamira N° 20* (pp. 201–207). Madrid: Ministerio de Cultura.

Cattelain, P. (1994). La chasse au paléolithique supérieur: arc ou propulseur ou les deux? *Archéo-Situla, 21–24*, 5–26.

Chauvière, F. X., & Rigaud, A. (2005). Les "sagaies" à "base raccourcie" ou les avatars de la typologie: Du technique au "non-fonctionnel" dans le Magdaléniene à navettes de la Garenne (Saint-Marcel, Indre). In V. Dujardin (Ed.), *Industrie osseuse et parures du Solutréen au Magdalénien en Europe, table ronde sur le Paléolithique supérieur récent, Angoulême (Charente, 28-30 mars 2003). Mémoire XXXIX de la Societé Préhistorique Française* (pp. 233–242). Paris: Societé Préhistorique Française.

Christensen, M. (1999). *Technologie de l'ivoire au Paléolithique supérieur: Caractérisation physico-chimique du matériau et analyse foctionnelle des outils de transformation. British Archaeological Reports International Series 751.* Oxford: Hadrian Books.

Christensen, M. (2004). Fiche caractères morphologiques, histologiques et mécaniques des matières dures d'origin animales. In D. Ramseyer (Ed.), *Matières et techniques, Industrie de l'os préhistorique Cahier XI* (pp. 17–27). Paris: Société Préhistorique Française.

Christensen, M., & Tejero, J.-M. (2015). La fabrication d'objets en matières dures animales. In M. Balasse, J.-P. Brugal, Y. Dauphin, C. Oberlin, E.-M. Geigl, & I. Reiche (Eds.), *Messages d'os. Archéométrie du squelette animal et humain* (pp. 73–90). Paris: Éditions des Archives Contemporaines.

Conard, N. J. (2009). A female figurine from the basal Aurignacian of Hohle Fels Cave in southwestern Germany. *Nature, 459*(7244), 248–252.

Dari, A. (2003). *Comportements de subsistance pendant la transition Paléolithique moyen-Paléolithique supérieur en Cantabrie à partir de l'étude archéozoologique des restes osseux des grands mammifères de la Grotte d'El Castillo (Puente Viesgo, Espagne)*. Ph.D. dissertation, Muséum National d'Histoire Naturelle.

d'Errico, F., Julien, M., Liolios, D., Vanhaeren, M., & Baffier, D. (2003). Many awls in our argument. Bone tool manufacture and use in the Châtelperronian and Aurignacian levels of the Grotte du Renne at Arcy-sur-Cure. In J. Zilhão & F. d'Errico (Eds.), *The chronology of the Aurignacian and of the transitional technocomplexes: Dating, stratigraphies, cultural implications: Proceeding of symposium 6. 1 of the XIV Congress of the UISPP (University of Liège, Belgium, September 2-6, 2001). Trabalhos de Arqueologia 33* (pp. 247–270). Oxford: Oxbow Books.

Goutas, N. (2004). *Caractérisation et évolution du Gravettien en France par l'approche techno-économique des industries en matières dures animales (étude de six gisements du Sud-ouest)*. Ph.D. dissertation, Paris I Panthéon-Sorbonne.

Goutas, N. (2016). Gravettian projectile points: Considerations about the evolution of osseous hunting weapons in France. In M. C. Langley (Ed.), *Osseous projectile weaponry: Towards an understanding of Pleistocene cultural variability* (pp. 89–107). Dordrecht: Springer.

Hahn, J. (1974). Analyse des sagaies du Paléolithique supérieur ancien en Europe. Méthodes et premiers résultats. In H. Camps-Fabrer (Ed.), *1èr. Colloque international sur l'industrie de l'os dans la préhistoire : Abbaye de Sénanque, 18-20 avril 1974* (pp. 119–128). Paris: Université de Provence.

Hahn, J. (1988). Fiche sagaie à base fendue. In H. Delporte, G. Hahn, L. Mons, D. Pinçon, & D. de Sonneville-Bordes (Eds.), *Fiches typologiques de l'industrie osseuse Préhistorique. Cahier I. Sagaies. Commission de nomenclature sur l'industrie de l'os préhistorique* (pp. 1–21). Aix-en-Provence: Université de Provence.

Hahn, J. (1995). Les ivoires en Allemagne: débitage, façonnage et utilisation au Paléolithique supérieur. In J. Hahn et al. (Eds.), *Actes de la Table ronde: Le travail et l'usage de l'ivoire au Paleolithique superieur. Ravello 29–31 mai 1992* (pp. 115–135). Rome: Istituto Poligrafico e Zecca dello Estato.

Henri-Martin, L. (1930). La station aurignacienne de La Quina. *Bulletin de la Societé Archéologique et Historique de la Charente, 50*, 5–84.

Jones, T. (1988). A Shell Projectile Point from the Big Sur Coast, California. *Journal of California and Great Basin Anthropology, 10*, 100–103.

Klein, R. (1995). Anatomy, behavior and Modern Human origins. *Journal of World Prehistory, 9*, 167–198.

Klein, R.-G. (2009). *The human career: Human biological and cultural origins*. Chicago: Chicago University Press.

Knecht, H. (1991). *Technological innovation and design during the Early Upper Paleolithic: A study of organic projectile*. Ph.D. dissertation, New York University.

Knecht, H. (1993). Splits and wedges: The techniques and technology of Early Aurignacian antler working. In H. Knecht, A. Pike-Tay, & R. White (Eds.), *Before Lascaux: The complex record of the Early Upper Paleolithic* (pp. 137–162). Boca Raton: CRC Press.

Knecht, H. (1997). Projectile points of bone, antler, and stone. Experimental explorations of manufacture and use. In H. Knecht (Ed.), *Projectile technology* (pp. 191–212). New York: Plenum Press.

Langley, M. C. (2015). Investigating maintenance and discard behaviours for osseous projectile points: A Middle to Late Magdalenian (c. 19,000–14,000 cal. BP) example. *Journal of Anthropological Archaeology, 40*, 340–360.

Langley, M. C., Pétillon, J.-M., & Christensen, M. (2016). Diversity and evolution of osseous hunting equipment during the Magdalenian (21,000–14,000 cal. BP). In M. C. Langley (Ed.), *Osseous projectile weaponry: Towards an understanding of Pleistocene cultural variability* (pp. 143–159). Dordrecht: Springer.

Le Dosseur, G. (2003). Sens et contre sens. Réflexions concernant l'orientation d'un geste technique observé sur des objets en matières osseuses du Levant. *Préhistoire Anthropologie Méditerranéennes, 12*, 115–127.

Leroy-Prost, C. (1974). Les pointes en matières osseuses de l'Aurignacien. Caractéristiques morphologiques et essais de définitions. *Bulletin de la Societé Préhistorique Française, 71*(2), 449–458.

Leroy-Prost, C. (1975). L'industrie osseuse aurignacienne. Essai régional de classification: Poitou, Charentes, Périgord. *Gallia Préhistoire, 18*(1), 65–156.

Leroy-Prost, C. (1979). L'industrie osseuse aurignacienne. Essai régional de classification: Poitou, Charentes, Périgord. *Gallia Préhistoire, 22*(1), 205–370.

Liolios, D. (1999). *Variabilité et caracteristiques du travail des matières osseuses au debut de l'Aurignacien: Approche technologique et economique*. Ph.D. dissertation, Paris X University.

Liolios, D. (2006). Reflections on the role of bone tools in the definition of the Early Aurignacian. In O. Bar-Yosef & J. Zilhão (Eds.), *Towards a definition of the Aurignacian: Proceedings of the symposium held in Lisbon, Portugal, June 25-30, 2002* (pp. 37–52). Lisbon: Trabalhos de Arqueologia 45.

Liouville, M. (2007). *Variabilité du cerf Elaphe (Cervus Elaphus Linné 1758) au cours du Pléistocène moyen et supérieur en Europe occidentale : Approches morphométrique, paléoécologique et cynégétique*. Ph.D. dissertation, Muséum National d'Histoire Naturelle.

Mellars, P. (1989). Major issues in the emergence of modern humans. *Current Anthropology, 30*(3), 349–385.

Mellars, P., & Stringer, C. (Eds.). (1989). *The human revolution: Behavioural and biological perspectives on the origins of modern humans*. Edinburgh: Edinburgh University Press.

Mons, L. (1988). Fiche "sagaie à base raccourcie". In H. Camps-Fabrer (Ed.), *Fiches typologiques de l'industrie osseuse préhistorique. Cahier 1: Sagaies* (pp. 1–9). Aix-en-Provence: Université de Provence.

Nuzhnyi, D. (1998). The preliminary results of experiments with Aurignacian split based points production, hafting and usage. *Préhistoire Européenne, 13*, 117–132.

Oakley, K. P., Andrews, P., Keeley, L. H., & Clark, J. D. (1977). A reappraisal of the Clacton spear-point. *Proceedings of the Prehistoric Society, 43*, 13–30.

Otte, M. (1995). L'ivoire paléolithique au Nord-Ouest européen. In J. Hahn et al. (Eds.), *Actes de la Table ronde: Le travail et l'usage de l'ivoire au Paleolithique superieur. Ravello 29–31 mai 1992* (pp. 103–113). Rome: Istituto Poligrafico e Zecca dello Estato.

Pétillon, J. M. (2006). *Des magdaleniens en armes. Technologie des armatures de projectile en bois de cervide du Magdalenien supérieur de la Grotte d'Isturitz (Pyrénnées-Atlantiques)*. Treignes: CEDARC.

Pétillon, J.-M. (2008). First evidence of a whale-bone industry in the western European Upper Paleolithic: Magdalenian artifacts from Isturitz (Pyrénées-Atlantiques, France). *Journal of Human Evolution, 54*(5), 720–726.

Pétillon, J.-M., & Ducasse, S. (2012). From flakes to grooves: A technical shift in antlerworking during the last glacial maximum in southwest France. *Journal of Human Evolution, 62*(4), 435–465.

Peyrony, D. (1933). Les industries "aurignaciennes" dans le bassin de la Vézère. *Bulletin de la Societé Préhistorique Française, 30*, 543–559.

Peyrony, D. (1934). La Ferrassie: Moustérien, Périgordien, Aurignacien. *Préhistoire, 3*, 1–92.

Peyrony, D. (1935). Le gisement Castanet, Vallon de Castelmerle, Commune de Sergeac (Dordogne). Aurignacien I et II. *Bulletin de la Societé Préhistorique Française, 42*, 418–443.

Pike-Tay, A., Cabrera, V., & Bernaldo De Quirós, F. (1999). Seasonal variations of the Middle-Upper Paleolithic transition at El Castillo, Cueva Morín and El Pendo (Cantabria, Spain). *Journal of Human Evolution, 36*, 283–317.

Potts, R. (1996). Evolution and climate variability. *Science, 273*, 922–923.

Rigaud, A. (1972). La technologie du burin appliquée au matériel osseux de la Garenne (Indre). *Bulletin de la Societé Préhistorique Française, 69*, 104–108.

Schwab, C. (2002). Fiche éclats daphysaires du Paléolithique Moyen et Supérieur: La grotte d'Isturitz (Pyrénées-Atlantiques). In M. Patou-Mathis (Ed.), *Retouchoirs, compresseurs, percuteurs: os à impressions et éraillures, Industrie de l'os préhistorique, Cahier X* (pp. 59–73). Paris: Société Préhistorique Française.

Schwab, C. (2014). *Les os à impressions de la Grotte d'Isturitz (Pyrénées Atlantiques)*. Ph.D. dissertation, Ecole du Louvre.

Soressi, M., Mcpherron, S. P., Lenoir, M., Dogandžić, T., Goldberg, P., Jacobs, Z., et al. (2013). Neandertals made the first specialized bone tools in Europe. *Proceedings of the National Academy of Sciences, 110*, 14186–14190.

Szmidt, C. C., Brou, L., & Jaccottey, L. (2010). Direct radiocarbon (AMS) dating of split-based points from the (Proto)Aurignacian of Trou de la Mère Clochette, Northeastern France. Implications for the characterization of the Aurignacian and the timing of technical innovations in Europe. *Journal of Archaeological Science, 37*(12), 3320–3337.

Tartar, E. (2009). *De l'os à l'outil - Caractérisation technique, économique et sociale de l'utilisation de l'os à l'aurignacien ancien. Étude de trois sites: l'Abri Castanet (secteurs nord et sud), Brassempouy (Grotte des Hyènes et Abri Dubalen) et Gatzarria*. Ph.D. dissertation, Université Paris I (Panthéon-Sorbonne).

Tartar, E. (2012). The recognition of a new type of bone tools in Early Aurignacian assemblages: Implications for understanding the appearance of osseous technology in Europe. *Journal of Archaeological Science, 39*(7), 2348–2360.

Tartar, E., & White, R. (2013). The manufacture of Aurignacian split-based points: An experimental challenge. *Journal of Archaeological Science, 40*(6), 2723–2745.

Tejero, J.-M. (2010). *La explotación de las materias duras animales en el Paleolítico superior inicial. Una aproximación tecno-económica a las producciones óseas Auriñacienses en la Península Ibérica*. Ph.D. dissertation, UNED University.

Tejero, J.-M. (2013). *La explotación de las materias óseas en el Auriñaciense. Caracterización tecnoeconómica de las producciones del Paleolítico superior inicial en la Península Ibérica. British Archaeological Reports International Series 2469*. Oxford: Archaeopress.

Tejero, J.-M. (2014). Towards complexity in osseous raw materials exploitation by the first anatomically modern humans in Europe: Aurignacian antler working. *Journal of Anthropological Archaeology, 36*, 72–92.

Tejero, J.-M., Christensen, M., & Bodu, P. (2012). Red deer antler technology and early modern humans in Southeast Europe: An experimental study. *Journal of Archaeological Science, 39*(2), 332–346.

Tejero, J.-M., & Grimaldi, S. (2015). Assessing bone and antler exploitation at Riparo Mochi (Balzi Rossi, Italy): Implications for the characterization of the Aurignacian in South-western Europe. *Journal of Archaeological Science, 61*, 59–77.

Tejero, J.-M., Arrizabalaga, A., & Villaluenga, A. (2016). The proto-Aurignacian and early Aurignacian retouchers of Labeko Koba (Basque Country, Spain). A techno-economic and chromo-cultural interpretation using lithic and faunal data. *Comptes Rendus Palevol, 15*, 994–1010.

Teyssandier, N. (2007). *En route vers l'ouest. Les débuts de l'aurignacien en Europe*. Oxford: Hadrian Books, Oxford.

Thieme, H. (1997). Lower Palaeolithic hunting spears from Germany. *Nature, 385*(6619), 807–810.

Vincent, A. (1993). *L'outillage osseux au Paléolithique moyen: une nouvelle approche*. Ph.D. dissertation, Université Paris X Nanterre.

White, R. (1995). Ivory personal ornaments or Aurignacian age: Technological, social and symbolic perspectives. In J. Hahn (Ed.), *Actes de la Table ronde: Le travail et l'usage de l'ivoire au Paleolithique superieur. Ravello 29-31 mai 1992* (pp. 29–62). Roma: Istituto Poligrafico e Zecca dello Stato.

White, R. (2007). Systems of personal ornamentation in the early Upper Palaeolithic: Methodological challenges and new observations. In P. Mellars, K. Boyle, O. Bar-Yosef, & C. Stringer (Eds.), *Rethinking the human revolution: New behavioural and biological perspectives on the origin and dispersal of Modern Humans* (pp. 287–302). Cambridge: McDonald Institute for Archaeological Research.

Yellen, J. E., Brooks, A. S., Cornelissen, E., Mehlman, M. J., & Stewart, K. (1995). A middle stone age worked bone industry from Katanda, Upper Semliki Valley, Zaire. *Science, 268*(5210), 553–556.

Ziegler, M., Simon, M. H., Hall, I. R., Barker, S., Stringer, C., & Zahn, R. (2013). Development of Middle Stone Age innovation linked to rapid climate change. *Nature Communications, 4*, 1905.

Zilhão, J., Angelucci, D. E., Badal-García, E., d'Errico, F., Daniel, F., Dayet, L., et al. (2010). Symbolic use of marine shells and mineral pigments by Iberian Neandertals. *Proceedings of the National Academy of Sciences, 107*(3), 1023–1028.

Chapter 6
Projectile Weaponry from the Aurignacian to the Gravettian of the Swabian Jura (Southwest Germany): Raw Materials, Manufacturing and Typology

Sibylle Wolf, Susanne C. Münzel, Krista Dotzel, Martina M. Barth, and Nicholas J. Conard

Abstract Here we describe the variability of projectile points made from bone, antler, and ivory recovered from cave sites in the Ach and Lone Valleys (Swabian Jura), focusing on Aurignacian and Gravettian assemblages. Based on the faunal provenience of the points, we recognize a distinctive change in raw material use from the Aurignacian to the Gravettian: during the Aurignacian antler was used for the small split-base points, bone for highly variable points, and ivory for the comparatively large and unstandardized points. During the Gravettian hardly any antler points have been found and bone points were manufactured from mammoth ribs. The raw materials tend to be associated with a specific type of point and *chaîne opératoire*.

Keywords Projectile point • Raw material preference • Early Upper Paleolithic • Massive-base point • Split-base point

S. Wolf (✉) • N.J. Conard
Institute of Pre- & Protohistory and Medieval Archaeology, Early Prehistory and Quaternary Ecology, University of Tübingen, Schloss Hohentübingen, Burgsteige 11, 72070 Tübingen, Germany

Senckenberg Center for Human Evolution and Palaeoenvironment, Tübingen, Germany
e-mail: sibylle.wolf@ifu.uni-tuebingen.de; nicholas.conard@uni-tuebingen.de

S.C. Münzel
Institute for Archaeological Sciences, University of Tübingen, Archaeozoology, Rümelinstr. 23, 72070 Tübingen, Germany
e-mail: susanne.muenzel@uni-tuebingen.de

K. Dotzel
Institute of Pre- & Protohistory and Medieval Archaeology, Early Prehistory and Quaternary Ecology, University of Tübingen, Schloss Hohentübingen, Burgsteige 11, 72070 Tübingen, Germany
e-mail: krista.dotzel@uconn.edu

M.M. Barth
Kultusministerium des Landes Sachsen-Anhalt, Turmschanzenstr. 32, 39114 Magdeburg, Germany
e-mail: Martina.Barth@mk.sachen-anhalt.de

Introduction

The aim of this chapter is to give an overview of those points made from bone, antler, and ivory dating to the Aurignacian and Gravettian assemblages of the Swabian Jura. This area includes the sites of Hohle Fels, Geißenklösterle, Sirgenstein, and Brillenhöhle, which are located in the Ach Valley between the towns of Blaubeuren and Schelklingen. The other cluster of cave sites of interest is located in the Lone Valley and includes Vogelherd, Hohlenstein-Stadel, and the Bockstein-complex (Fig. 6.1). Both valleys are branches of the Danube River. There seems to be a clear preference in raw material for Aurignacian and Gravettian people; while antler and ivory were the preferred raw materials during the Aurignacian, Gravettian points seems to be exclusively made of ribs, preferably mammoth ribs. These different raw material preferences had implications for the shape as well as for the functional properties of the points.

Research History of the Swabian Jura

The Swabian Jura has been the site of many archaeological and paleontological excavations since the mid-nineteenth century, and excavations are still ongoing today. Most of the investigated Paleolithic sites contain either Aurignacian, Gravettian, or both, techno-complexes within their deposit. In order to better understand the osseous technology to be described below, we provide a brief excavation history of the key sites of the Swabian Jura.

The first excavations in the renown Hohle Fels Cave near Schelklingen were conducted in 1870/71, and the University of Tübingen has conducted yearly excavations at this site almost every year since 1977 (Hahn 1989; Blumentritt and Hahn 1991; Conard et al. 2000). At this site, the archaeological

Michelle C. Langley (ed.), *Osseous Projectile Weaponry*, Vertebrate Paleobiology and Paleoanthropology,
DOI 10.1007/978-94-024-0899-7_6

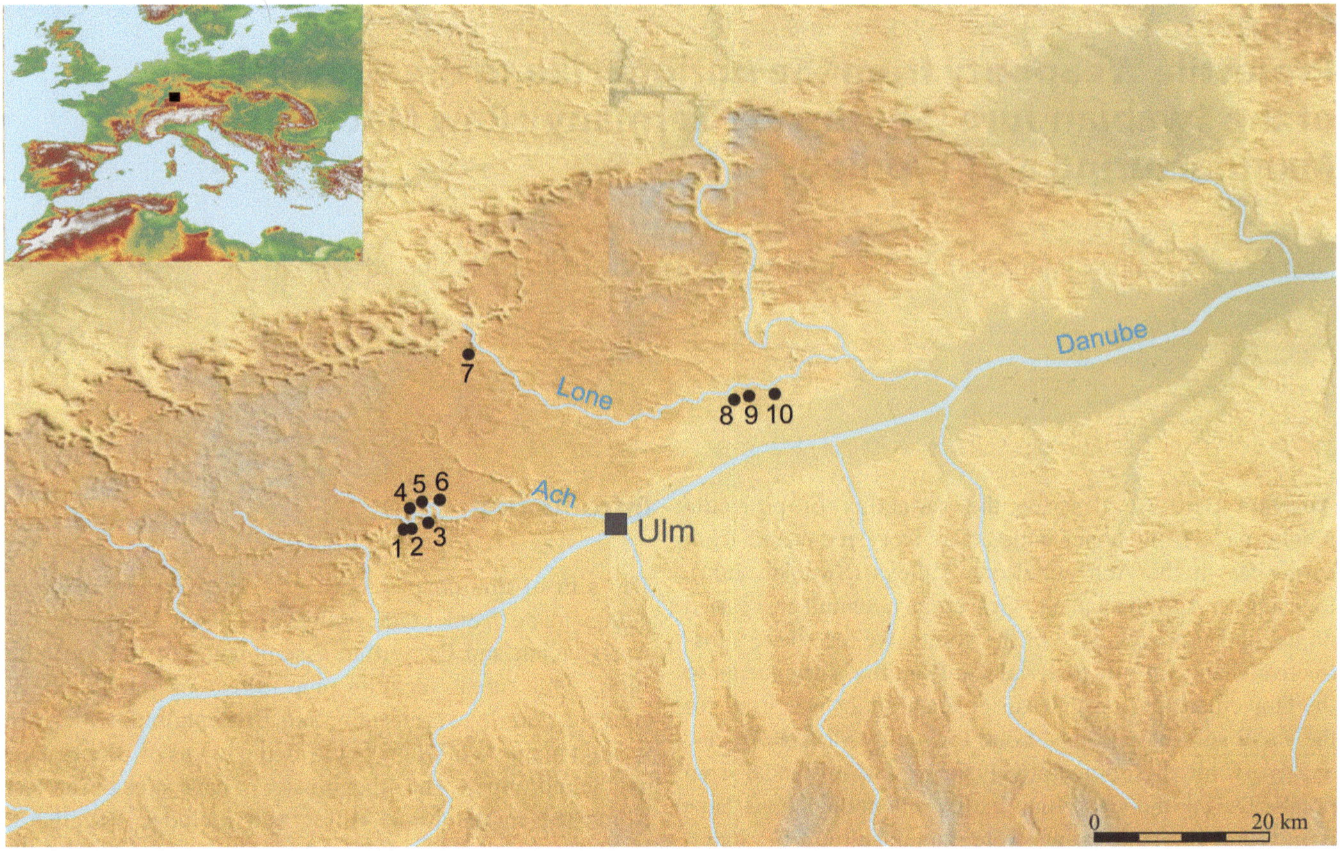

Fig. 6.1 Map of the caves of the eastern Swabian Jura: (1) Kogelstein; (2) Hohle Fels; (3) Geißenklösterle; (4) Sirgenstein; (5) Brillenhöhle; (6) Große Grotte; (7) Haldenstein Cave; (8) Bockstein; (9) Hohlenstein-Stadel; (10) Vogelherd. Map: University of Tübingen

horizons IIb to IIcf are Gravettian and date between 27,000 and 29,500 in uncalibrated calendar years. The Aurignacian layers, IId/e, III, IV, Va and Vb have provided dates between 29,500 and 35,700 years BP (Conard and Bolus 2003, 2006, 2008; Conard 2009).

Robert R. Schmidt excavated the Sirgenstein Cave, which lies in the valley between Hohle Fels Cave and Geißenklösterle Cave, in 1906 (Schmidt 1907, 1912). The Gravettian and Aurignacian layers here are designated II, III, IV and V and were occupied between 26,700 and 30,200 years BP (Conard and Bolus 2003, 2008). Joachim Hahn conducted excavations in the Geißenklösterle Cave between 1974 and 1991 (Hahn 1988). In 2001 and 2002 Nicholas J. Conard continued the work at this cave until he reached bedrock (Conard and Malina 2002, 2003). The Gravettian horizons Ip to Ic indicate an age between 24,400 and 32,900 years BP while the Aurignacian layers II and III date to between 29,300 and 39,000 years BP (Richter et al. 2000; Conard and Bolus 2003, 2008; Higham et al. 2012). Excavations at Brillenhöhle took place between 1955 and 1963 (Riek 1973). The Gravettian layers VII and VIII provide two old dates between 25,000 and 29,000 years BP (Riek 1973). The deeper layer, XIV, revealed only two Aurignacian points, which were directly dated to 30,400+240/-230 years BP and 32,470+270/-260 years BP respectively (Bolus and Conard 2006).

During his excavations in the Vogelherd Cave in 1931 Gustak Riek completely emptied the cave of sediments, dumping the backdirt onto the hill surrounding the cave (Riek 1934). The layers richest in finds were the Aurignacian layers IV and V, dating between 30,000 and 36,000 years BP. In contrast to these rich layers, Riek did not discover many Gravettian remains. Between 2005 and 2012 the Department of Prehistory and Quaternary Ecology of the University of Tübingen excavated the back dirt sediments of Riek's excavation. Because of the relatively rough excavation methods of the time of 1931, the new excavation was quite successful in finding an abundance of new artifacts, especially small finds (e.g., Conard et al. 2007, 2010). These artifacts, however, have no stratigraphic context and must be studied in tandem with finds from sites with well-documented stratigraphies.

Hohlenstein-Stadel, known for its famous lion-man (Schmid 1989; Kind et al. 2014), contains Aurignacian layers dated to between 31,500 and 35,000 years BP, but no significant Gravettian layers. The first significant archaeological investigations at Hohlenstein-Stadel took place between

1937 and 1939 by Robert Wetzel (1961). Between 2008 and 2013 Claus-Joachim Kind led excavations in front of and inside the cave (Kind and Beutelspacher 2010; Beutelspacher et al. 2011; Beutelspacher and Kind 2012; Kind et al. 2014).

Excavations at Bockstein Cave occurred on and off throughout the late nineteenth century through to the first half of the twentieth century (Schmidt 1912; Wetzel 1958; Wetzel and Bosinski 1969). The cave, as well as its entrance (Bockstein-Törle), has produced Aurignacian and Gravettian artifacts, however, the layers have proven difficult to distinguish from one another (Wetzel 1954; Krönneck 2012). The dates for the archaeological horizons IV to VI are between 20,400 (no AMS date) and 31,500 (AMS) years BP (Conard and Bolus 2003, 2008).

In 1972, Gerd Albrecht, Joachim Hahn, and Wolfgang Torke from the Institute of Prehistory and Quaternary Ecology of the University of Tübingen conducted the first and only systematic review and analysis of all organic projectile points from the Swabian Jura. They compared the Swabian points with other Aurignacian points from across Europe and conducted their analysis using innovative methods such as coding attributes and including statistical analysis (Albrecht et al. 1972). Since that time, however, many new projectile points have been excavated and no updated overview has been published. Here we update this work some 40 years later.

Materials and Methods

For the purposes of this chapter, we describe organic projectile points based on the criteria put forward by Albrecht et al. (1972; Fig. 6.2), and have thus measured the maximum length, width, and thickness of each point or point fragment. The main attribute of this artefact category is a pice from osseous material shaped into a pointed form. Projectile points are distinguishable from awls or other such pointed artifacts by their extensive shaping. They were whittled, scraped, or ground on all sides so that the artifact morphology is the result of carefully controlled manufacturing. In addition, these artifacts possess bases shaped in such a way to facilitate hafting.

During the Aurignacian and Gravettian different raw materials are documented for the production of projectile points. The people used bone, woolly mammoth ivory, and reindeer antler and each raw material possesses different properties that determine the manufacture and the function of the points (Albrecht 1977).

The identification of antler and ivory raw material is relatively simple, especially when compared to identifying the type and element of bone that was utilized as raw material for a point. Often only ribs can be identified, as these points exhibit a typical rib spongiosa (cancellous bone) on one side covered

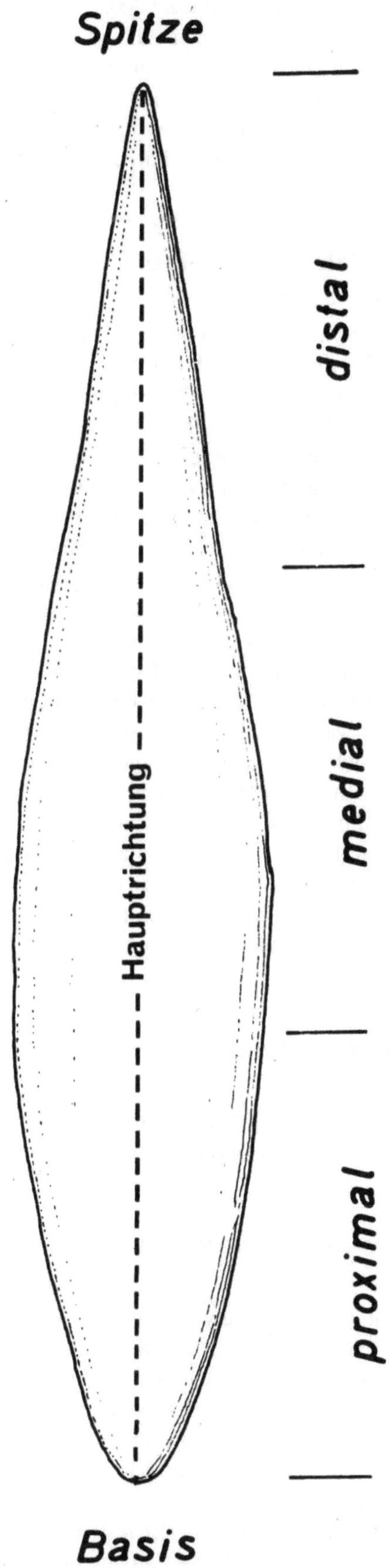

Fig. 6.2 Dimensions of a point. After Albrecht et al. (1972)

by a 'spongy' compacta (Münzel 2005). Another possibility is to use DNA to identify the animal and this method has been used to identify the raw material of the numerous Aurignacian bone points of Potočka zijalka, a high Alpine cave located in Slovenia. These latter points were probably made from cave bear long bones (Hofreiter and Pacher 2004).

Middle Paleolithic Points

Researchers have documented a handful of bone points ascribed to the Middle Paleolithic, though currently no ivory points have been identified for this period (Gaudzinski et al. 2005). In Germany, the first bone points appear during the Middle Paleolithic at the site of Salzgitter-Lebenstedt (Gaudzinski 1998). At this site, Neanderthals fashioned mammoth fibulae and ribs into pointed tools. At Vogelherd in the Swabian Jura, a similar tool, made of a split mammoth rib, has been documented from the late Middle Paleolithic layer VI. This tool is well preserved, with both the tip and the base whole. In addition, a massive-based bone point made from a horse-sized rib was excavated in 1931 (Bolus and Conard 2006; Fig. 6.4: 11). This point was recently directly dated to 31,310+240/-230 years BP, which, if correct, suggests it may instead originate from the Aurignacian. The Swabian site of 'Große Grotte', in the Ach valley, also produced a point from late Mousterian layers. This piece is a carefully worked antler point made from either reindeer or red deer (oral comm. Münzel 2013), and exhibits splintering at the tip, indicating it was well used (Wagner 1983).

Aurignacian Points

Aurignacian projectile points in the Swabian Jura all fit into one of two categories; massive-base points or split-base points.

Massive-Base Bone Points

These points take a variety of forms but generally have solid, rounded bases that were hafted by inserting them into a hollowed-out shaft. Most of the Aurignacian sites in the Swabian Jura have produced massive-base points, albeit not more than a few artifacts each. These points are highly variable in terms of shape and size. In particular, massive-base points are known from Sirgenstein, Hohle Fels, Geißenklösterle, Brillenhöhle, Bockstein-Törle, Hohlenstein-Stadel, and Vogelherd. These finds are described below.

In 1912, Robert R. Schmidt published a bone massive-base point recovered from Sirgenstein (Albrecht et al. 1972, Taf. 3, 24). Five fragments of antler points have been found at Hohle Fels, and one of these is likely a part of a split-base point (Fig. 6.4: 1). One bone massive-based point was also found here (Fig. 6.5: 4), and is a medial-proximal fragment made of mammoth/rhino rib. At Bockstein-Törle, excavations recovered two bone points with massive bases (Albrecht et al. 1972; Fig. 6.3: 1 and 3), while Hohlenstein-Stadel has revealed two bone massive-base points (Albrecht et al. 1972; Fig. 6.3: 2 and 4). Similarly, Brillenhöhle has produced two incomplete points from layer XIV (Riek 1973; Bolus and Conard 2006). One is probably a split-base point made of antler, while the other is a medial fragment of a bone massive-base point. Both have been recently dated revealed with the split-base point returning an age of 30,400+240/-230 years BP, and the massive-base point 32,470+270/-260 years BP (Bolus and Conard 2006).

Vogelherd has produced the greatest number of massive-base bone points from the Swabian Jura (n=6). These points come from layers IV and V, as well as from the recent back dirt excavations. The points from Vogelherd are highly variable (Fig. 6.4: 6–8). Three of the points are oval in cross-section (except for the narrowing tip which is sub-circular in section) and resemble split-base points in both size and shape. Two of the points are lozenge-shaped and were probably quite similar in size when complete. The last point is substantially different to the others (number 33/73_127). While the others have thicker oval or rectangular cross-sections, this point is quite flat, with a length and width much longer than the others. These massive-base points are all made of antler.

Interestingly, Geißenklösterle Cave produced no bone massive-base points despite its rich variety of other osseous artifacts. The only known point varieties from this cave are antler split-base points and ivory points with massive or double beveled bases.

Split-Base Points

Split-base points are found at many Aurignacian sites throughout Western and Central Europe (Albrecht et al. 1972; Knecht 1990), and take their name from the characteristic slit up the middle of their base. Aside from the split-base, these points can take a variety of shapes and sizes. Almost all split-base points are made from antler rather than bone, which is most likely owing to the specific biomechanical properties that antler possesses as a raw material. Antler is not as brittle as bone, with several researchers who have experimented with antler reporting that it is more pliable and easier to work than bone, especially when wet (Newcomer 1977; Bonnichsen 1979; Guthrie 1983; Tartar and White 2013). Given that many other forms of organic projectile points are made from bone instead of antler, it may be the case that antler is especially good for creating the characteristic split-base morphology.

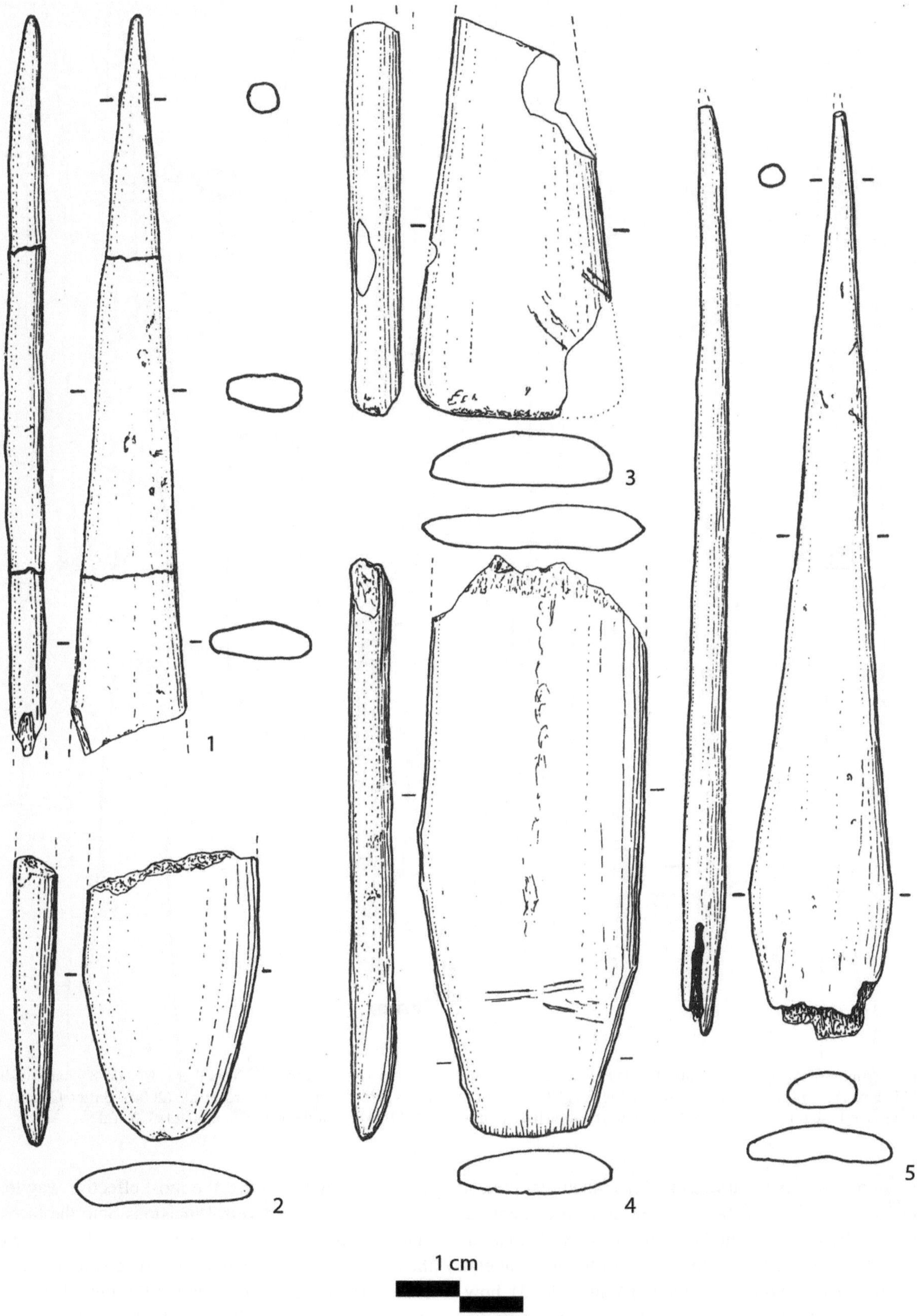

Fig. 6.3 Examples of Aurignacian points. Bockstein-Törle: (3) massive-base point, (1) point fragment; Hohlenstein-Stadel: (2, 4) massive-base points; Bockstein Cave: (5) split-base point. Drawings after Albrecht et al. (1972), Taf. 2

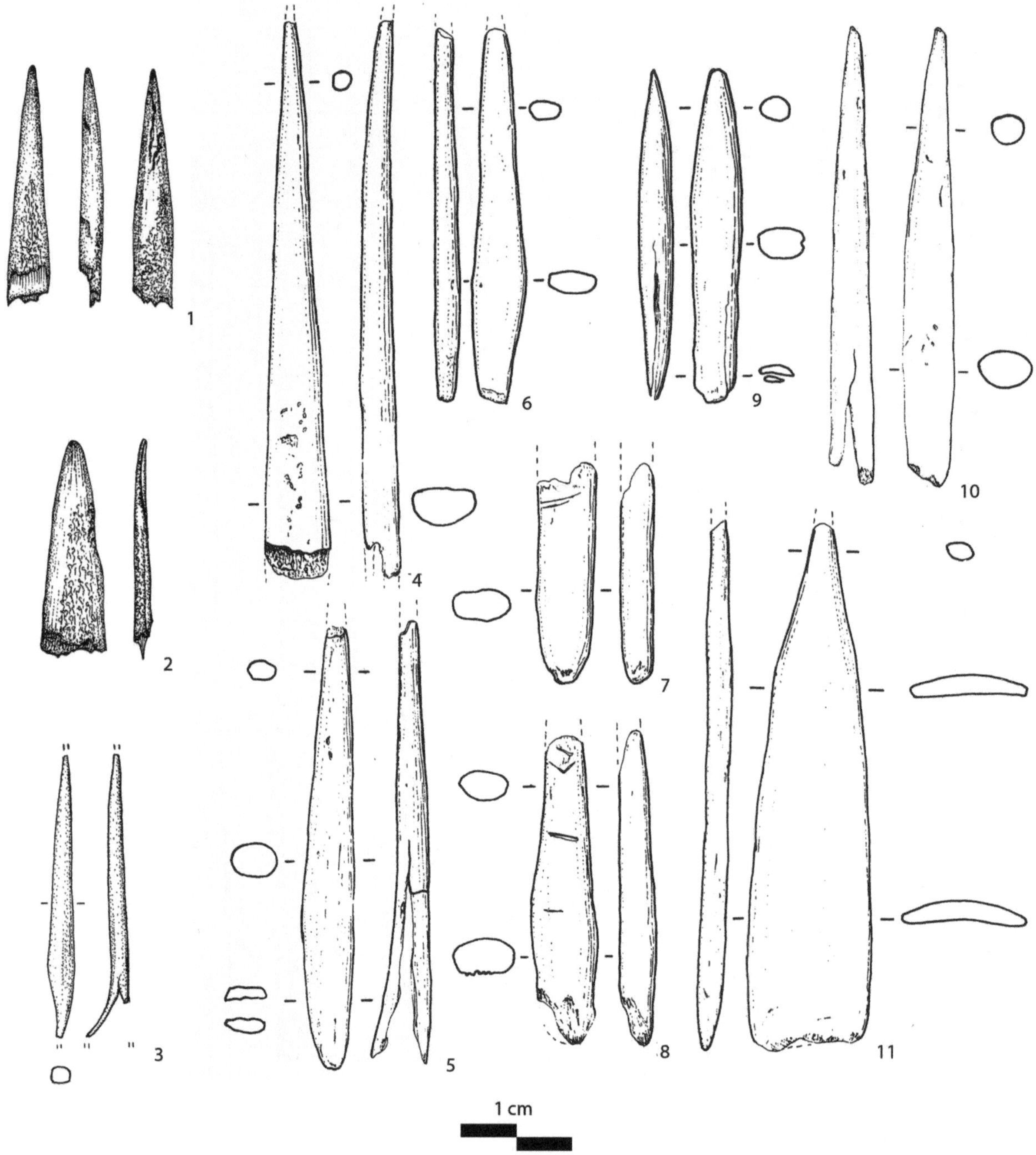

Fig. 6.4 Examples of Aurignacian points. Hohle Fels: (1–2) fragments from Hohle Fels IV, (3) split-base point from Hohle Fels Vb; Vogelherd VI: (11) massive-base point; Vogelherd V: (4, 5, 9, 10) split-base points, (6) massive-base point; Vogelherd IV: (7–8) massive-base points. Drawings 1, 2 after Conard et al. (2004), 3after Conard and Malina (2009), 4 – 11 after Albrecht et al. (1972), Taf. 4

The manner in which Aurignacian manufacturers created the split in their points has been somewhat of a contentious issue. Henri-Martin (1931) and later Knecht (1990) both argue that the split was created by simple cleavage to the basal end. Recent experimental work by Tartar and White (2013), however, found that splitting a point through simple cleavage was almost impossible. Instead they argue for a combination of Peyrony's (Peyrony 1935) and Henri-Martin/Knecht's method. They found that the most effective way to create the split was to cut transversal incisions onto the faces of a long blank where the desired base would be. They would then flex the blank on both sides until the force split the base (Peyrony 1935), which was then extended through cleavage. This created characteristic debitage in the form of a 'tongued piece'. This technique simultaneously created the 'tongued piece', the split, and removed material from inside of the wings of the

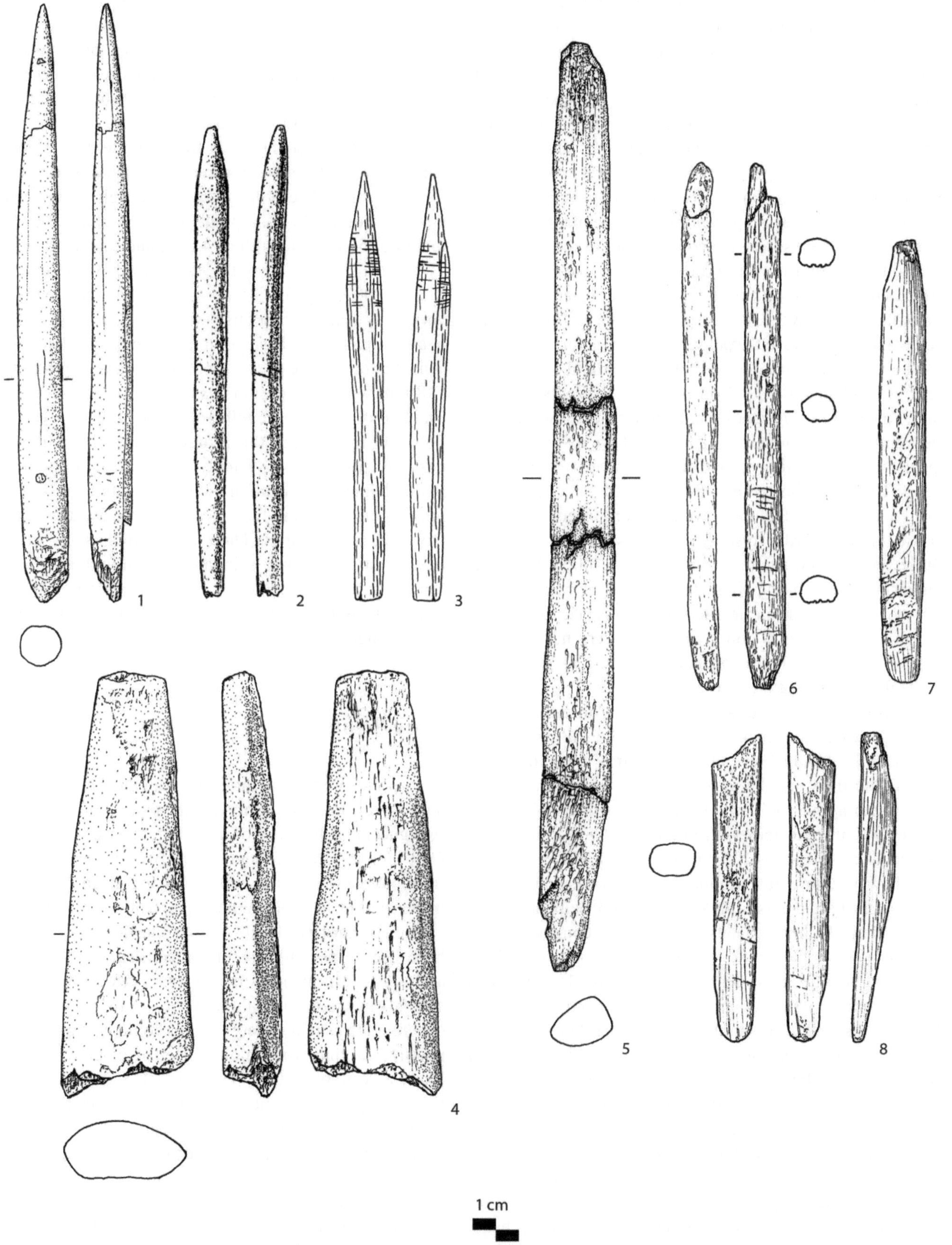

Fig. 6.5 Examples of Aurignacian points (1–4). Examples of Gravettian points (5–8). Hohle Fels AH IV (1, 2, 4), Hohle Fels AH Va (3), Hohle Fels AH IIb (5–6), Brillenhöhle AH VII (7–8). Ivory (1–3), mammoth/rhino rib (4–6), antler (7), unidentified bone (8). Massive-base points (5–6), double beveled base (7), single beveled base (8). Drawing 1 after Conard and Malina (2009), 2 after Conard and Malina (2006), 3 after Conard et al. (2003), 4 by R. Ehmann, 7 after Riek (1973), pl. 13, 10, 8 after Riek (1973), pl. 14, 7

point. While this argument is convincing, this construction method necessarily creates tongued pieces as debitage, which have not been observed in the Swabian Jura. Furthermore, Vogelherd has produced a handful of antler artifacts that appear to be point blanks roughly the size and shape of finished split-base points but which lack the split. If these artifacts are indeed split-base point blanks, this suggests that perhaps simple cleavage was, in fact, the preferred splitting method in the Swabian Jura. To construct a projectile point in the Swabian Jura then, it appears Aurignacian manufacturers' first extracted blanks from the compact part of the antler and shaped them into a roughly projectile point shape. Then the manufacturer would attempt to split the basal section of the blank through cleavage. If the split was successful, the final step would be to scrape the point blank into its final shape.

The split-base morphology almost certainly reflects a hafting mechanism. Based on her extensive experimental data, Knecht (1991) argues that a split-base allows hafting without the use of adhesive materials, if the distal end of a spear shaft was hollowed out into a U-shape to insert the base. To keep the points firmly fixed in the shafts, Aurignacian manufacturers would then insert a wedge inside the slit in order to splay the wings against the wood. Other researchers, such as Linda Owen (2005), however, suggest that split-base points were used as weaving or sewing tools rather than projectile points. Microscopic observations of split-base point tips, however, have shown impact fractures that are consistent with use as projectiles (Dotzel et al. in prep.; Tejero 2016).

Split-base points occur at several of the Aurignacian sites in the Swabian Jura, including Vogelherd, Geißenklösterle, Brillenhöhle, Hohle Fels, and Bockstein Cave. Vogelherd has produced by far the most split-base points out of the region with a total of 27 whole and fragmented points, followed by Geißenklösterle with 11 (Hahn 1988; Liolios 1999; Teyssandier and Liolios 2003; Dotzel 2011). The other three sites, however, have only produced one split-base point each.

The split-base points from Vogelherd and Geißenklösterle are a relatively homogenous group when compared with Aurignacian simple-based points. Unbroken points from these two sites range in length from 51 to 115 mm, with widths from 7 to 12 mm, and thicknesses from 4 to 7 mm. These points tend to be shorter and narrower than other varieties of organic projectiles, as well as split-base points from other regions (Albrecht et al. 1972). In terms of shape, the split-base points from these two sites also tend to be similar. Most of the points fall into one of two shape categories; 'curved' points and 'triangular' points. Triangular points are widest near their bases and feature straight lateral edges that taper evenly into a point with an overall shape that most closely resembles a triangle. Curved points, on the other hand, show lateral edges that are more rounded and gently slope toward the point. The maximum width of the latter type can occur anywhere along the shaft. Points from these sites also commonly feature cross-sections that resemble thick ovals or rectangles with rounded edges. While individual points from these sites vary in size and shape, makers usually adhered to a set range of patterns (Fig. 6.4: 4, 5, 9, 10).

The split-base points from Brillenhöhle, Hohle Fels, and Bockstein-Höhle, on the other hand, vary wildly both in form and size. The point from Hohle Fels is the smallest, nearly whole, split-base point in the Swabian Jura with a length of 51 mm, a width of 4 mm, and thickness of 3 mm. The piece derives from the deep layer Vb (Conard and Malina 2009; Fig. 6.4: 3), demonstrating that already in the very early Aurignacian this *fossile directeur* is present. The near complete split-base point from Bockstein-Höhle represents the other side of the spectrum with a length spanning 148 mm with a maximum basal width of 33 mm (Fig. 6.3: 5). In contrast to the Vogelherd, Geißenklösterle, and Hohle Fels split-base points, this artifact is quite flat, with a thickness of just 6 mm. Finally the split-base point from Brillenhöhle has a width of 23 mm and a thickness of 6 mm, making it a wide and flat basal fragment featuring straight, tapering, lateral edges. These three points show that split-base points were not standardized throughout the entire Swabian Aurignacian, even if the points from Vogelherd and Geißenklösterle were kept within a narrower range of morphologies and sizes.

Ivory Points

Ivory points were frequent throughout the Aurignacian times and were produced during all phases of the Swabian Aurignacian. The ivory of a mammoth tusk is composed of 60 % dentin, 30 % collagen and 10 % water (for detailed information see Locke 2008), making it an excellent raw material due to this unique composition which makes it extremely hard while also being elastic. Ivory appears to have been especially useful for constructing various tools and personal ornaments during this period (Wolf 2015). It was advantageous and attractive as a raw material because different forms in a variety of sizes could be carved from the massive dentine. The unique luster of ivory was also most likely a desirable trait (Conneller 2011), and in many cases the ivory points exhibit personalized characteristics, which demonstrate the expression of individuality.

To obtain ivory, people during the Aurignacian and Gravettian periods either hunted mammoths or collected tusks from the animals that perished in the landscape. So far, the evidence points more to systematic collection of tusks rather than hunting (Niven 2006; Wolf 2015). There are different methods for breaking down a tusk to create ivory projectile points. Manufacturers could: (1) etch a notch around the circumference of the tusk and then snap it; (2) split it length-wise into two halves; or (3) smash it using direct percussion (Khlopatchev and Girya 2010). These methods could also be used in combination. After the initial breaking down the ivory, manufacturers would use the groove and splinter technique to extract raw forms in the shape of long and slender rods. To obtain suitable blanks, Aurignacian and Gravettian people must have used the groove and splinter technique of blank extraction rather than try to flake the material as even a large flake is not regular enough to create a rod with the consistent thickness and length needed for a point longer than 200 mm. After the initial blank extraction, manufacturers would have chopped, scraped, ground, and smoothed the point until it reached its intended size and shape (Semenov 1957; Christensen 1999; Liolios 1999; Wolf 2015). Except for acquiring the ivory, all steps of the production sequence are documented in the collections of the Hohle Fels, Geißenklösterle and Vogelherd.

Ivory points were an important part of the Aurignacian toolkit. Altogether, 29 artifacts including five complete pieces are preserved in the archaeological record. These points show a great variety in size and shape but are all highly polished. In the Hohle Fels Cave, points (n = 11) have been excavated from all Aurignacian layers (Fig. 6.5: 1–3). These points show high variability in both form and size. Five pieces possess a massive base, one piece shows a double beveled base, and five points have bases which are indeterminate owing to preservation. The lengths of the completely preserved points vary between 93 and 238 mm. The widths vary between 6.5 and 40 mm, and the thickness between 6.2 and 14 mm. These points include a well-preserved 'Lautscher' or 'Mladeç' point (230 mm in length; Fig. 6.6), as well as a basal fragment of this same point type which would have been around the same size. The bases of these points bear an engraved cross-hatch pattern, likely to facilitate hafting. Two pencil-shaped pieces with a massive base and a round cross section have also been found. These pencil-shaped pieces have similar dimensions, except in length. One thick point even displays a curved groove on one side, which could be interpreted as a personal marking. The production sequence at Hohle Fels is well documented.

The excavations at Geißenklösterle produced points in the Aurignacian layers II and III (n = 5). Three pieces possess double beveled bases and two pieces have massive bases. They measure between 8 and 14 mm in width and between 6.5 and 11 mm in thickness (Hahn 1988).

At Vogelherd, three points came from layer V while the recent back dirt excavations produced an additional 13 items. At present about two thirds of the sediments from the excavations have been wet screened and sorted, so future work at Vogelherd may produce additional finds. So far, four pieces from Vogelherd have massive bases, four have double beveled bases, and the bases of five points remain undetermined. The length of the points with massive bases averages 4.6 mm and the width averages 0.9 mm. The artifacts from Vogelherd are generally consistent in size and shape and are relatively small. The manufacturers did carve points on site out of rods, though most of the ivory rods were used for the production of beads.

In summary, split-base points were made from antler and are quite numerous, while specimens made with a massive base are made either from bone or antler and are less numerous in comparison. Points made of ivory are again more frequent in their appearance in the Aurignacian record.

Gravettian Points

In the Swabian Jura, the Gravettian has been found only in the caves of the Ach Valley, including Hohle Fels, Geißenklösterle, Brillenhöhle and Sirgenstein. The Gravettian layers of all of these caves with the exception of Sirgenstein have produced a variety of tools and jewelry made of organic raw materials. The species that provided the majority of the raw material for organic tools were also the main game species and included mammoth, reindeer, and horse. More than 60 medial and 10 distal fragments derive from the Gravettian layers of these three cave sites. Raw material, similarities in shape, morphology, and cross-section as well as signs of impact-induced breakage suggest that these pieces, as well as some of the described basal fragments, should be interpreted as projectile points. The shape of the tips ranges from very pointed to rounded and blunted. Some of them show evidence of having reshaped tips through scraping (Barth 2007; Barth et al. 2009:14).

In contrast to the Aurignacian, Gravettian points were manufactured mainly from mammoth ribs and unidentified ribs of mammoth- to rhino-sized species (Münzel 2001, 2004, 2005), however, ribs of horses or of horse- to deer-sized animals and antler were also used. As very little on-site production of antler tools is recorded, we can assume that the few antler points found were brought as finished products into the caves of the Ach Valley (Barth 2007; Barth et al. 2009:16).

Production Sequence for Mammoth Ribs

Mammoth ribs used for projectile points were processed on-site in a standardized fashion. First they were notched along the edges on both sides to facilitate splitting (Münzel 2004:77, Figs. 5, 6). These split rib blanks could then be shaped into

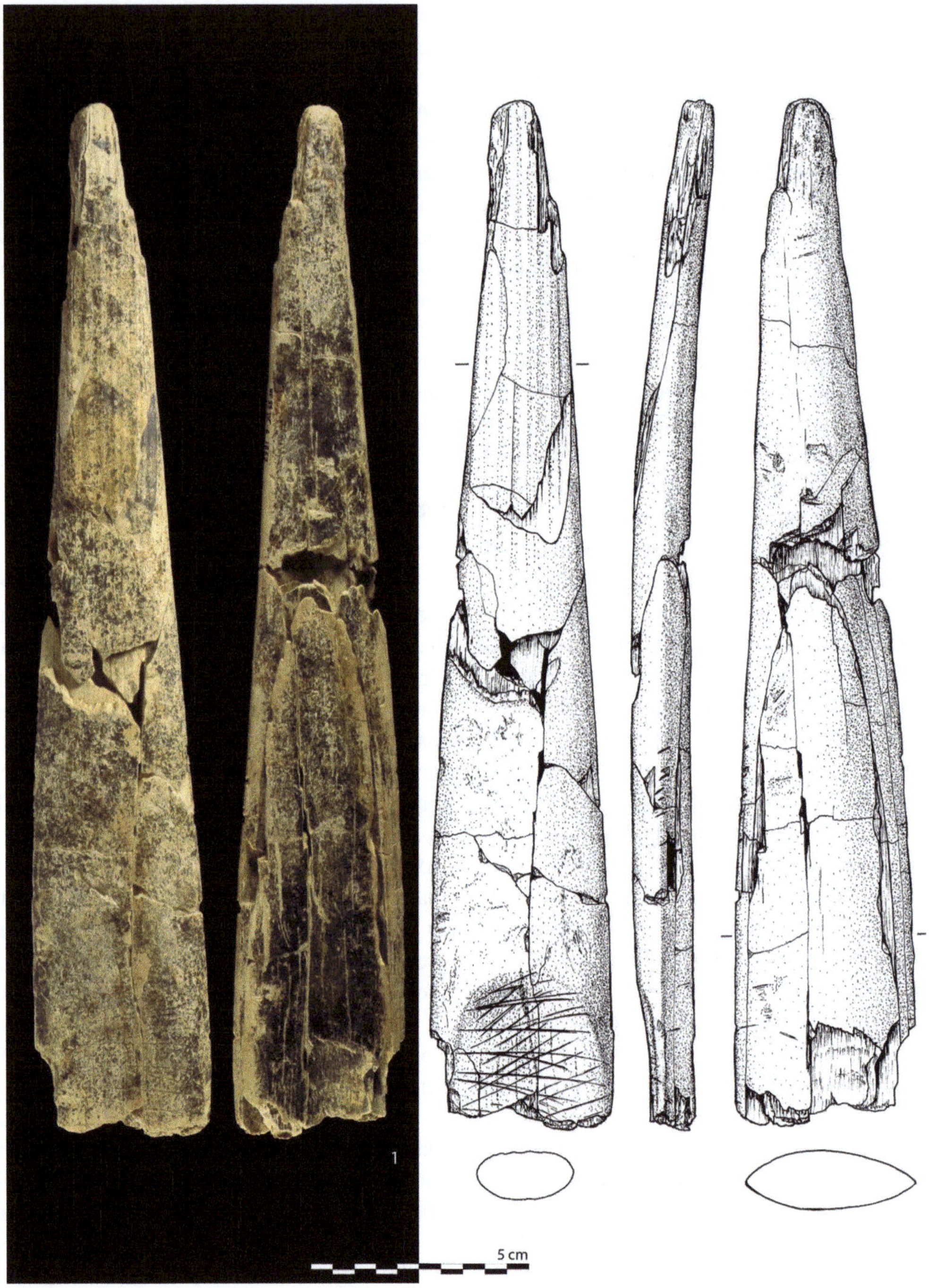

Fig. 6.6 Lautscher/Mladeç point from Hohle Fels AH IV. Photos by H. Jensen. Drawings after Conard and Malina 2007

different tools with several possible functions. They could be used as chisel-/wedge-like tools, used as burnishers or smoothing tools, or manufactured into projectile points. To manufacture the points, the split rib halves were ground along the edges and smoothed on both sides until they developed a typically circular, oval, or rectangular cross-section. At Geißenklösterle and Hohle Fels, all stages of this production sequence are well documented on-site (Barth 2007). Bone points from Brillenhöhle show the same manufacturing pattern (Riek 1973; Barth 2007). The length of the mammoth ribs as well as their straightness may have been an important prerequisite for the production of projectile points.

Among the complete and near complete preserved mammoth rib points (n=7), along with the clearly classifiable point fragments (n=23), four different point types could be identified (Barth 2007). These include: points with massive base, those with a single beveled base, with a double beveled base, and points *à base machonée.*

Points with a Massive Base

Altogether four nearly complete points with round bases come from the Gravettian layers in Brillenhöhle (n=2) and Hohle Fels Cave (n=2). They are made of mammoth ribs, except for one specimen of non-identifiable bone from Brillenhöhle. One point from Hohle Fels (145×11×8 mm) is cylindrical in shape (Fig. 6.5: 6). The cross-section is partly oval, partly rectangular. The base and the lower medial part are incised with a few irregular, parallel, transversal lines. The tip is splintered at one side. The other point is larger (201×15×12 mm) and broke into four fragments after being deposited (Fig. 6.5: 5). The cross-section changes from rectangular to oval at the terminal end. The base is slightly splintered and the tip is a little weathered. Compared to the points from Hohle Fels, the two specimens from Brillenhöhle are short and stocky (97×14×8 mm; 113×12×9 mm). Their shapes are cylindrical and slightly converging with round and oval cross-sections.

There are 20 basal fragments from Hohle Fels (n=3), Geißenklösterle (n=8) and Brillenhöhle (n=9). All bases from Geißenklösterle, eight from Brillenhöhle, and one from Hohle Fels are made of mammoth ribs. Reindeer antler served as raw material for one point only from Hohle Fels and another from Brillenhöhle. Most bases are slightly splintered, and two bases from Geißenklösterle and Brillenhöhle carry parallel, transversal incisions. One ivory basal point fragment was found in layer IIb in Hohle Fels (130×45 mm; Hiller 2003:18). This artifact has an irregular shape with the lower part of the base showing a scraped surface, while the pointed distal part is polished. In this case, the polish and further smoothing was likely carried out after the mounting or wrapping. So far, this artefact is the only ivory point known from the Gravettian of the Swabian Jura.

Points with a Single Beveled Base

Two points with single-beveled bases derive from layer VII of Brillenhöhle (Riek 1973: Fig. 13.9 & 14.7). One specimen is near complete. Its tip is tapered - suggesting that it was reworked after breaking—and broken. The other piece is a basal fragment with no further features (Fig. 6.5: 8). The bone used as raw material could not be further identified as the specimen was not available for reanalysis (Barth 2007:81).

Hohle Fels produced a basal fragment from layer IIc manufactured from mammoth rib. This piece is flat and slightly bent with a concave surface showing many parallel incisions, partly overlying each other. Unfortunately, it is too fragmented to clearly identify if it is, in fact, a point with a single beveled base.

Points with Double Beveled Base

Two examples of this point type were recovered from the caves of the Ach Valley. One near complete specimen from layer VII of Brillenhöhle is made of antler, probably reindeer (Fig. 6.5: 7). Its tip is broken and slightly drawn-in at one edge, perhaps indicating reworking of the tip. The double beveled base is roughened with chatter marks on the flat surfaces as well as on one edge.

The second double beveled base point is a basal fragment made from a mammoth rib recovered from layer IIcf at Hohle Fels Cave. Parallel and oblique incisions are present on both sides of the base.

Point à Base Machonée

In layer IIc of Hohle Fels Cave there is one small point (66×6×4 mm) produced from bone of an unidentified bear- to horse-sized animal. The tip is splintered and the base is tapered by *raclage en diabolo* (Barth 2007:43). At Gravettian sites in France, this technique was used as a technique of debitage, as well as a technique for repairing broken projectile points, so-called points *à base machonée* (Goutas 2004:146 & 573ff.). The specimen from the Hohle Fels is maybe an example of this type of manufacturing or maintenance activity (Table 6.1).

Discussion

Altogether 88 projectile points are known from the Aurignacian and 30 date to the Gravettian. These artifacts are common owing to the long research history in the Swabian Jura, and the detailed excavation methods utilized. While a gapless stratigraphic transition from the Aurignacian to the Gravettian is well documented in the caves of the Ach Valley (especially at Geißenklösterle and Hohle Fels), the Lone Valley produced scarcely any archaeological remains from the Gravettian (though rich in the Aurignacian).

The large mammal composition is broadly similar during the Aurignacian and Gravettian (Münzel and Conard 2004a, b). The caves of the Swabian Jura have revealed typical species of the Mammoth-steppe environment, such as mammoth, woolly rhinoceros, wild horse and reindeer. There is, however, a difference in the number of cervid species between the two time periods. During the Aurignacian four different cervids were present in the Ach Valley, namely giant deer, red deer, roe deer and reindeer. Each of these cervids requires different nutritional needs and represents

Table 6.1 Total number of points and fragments of points of the Swabian Aurignacian and Gravettian

		Ach Valley				Lone Valley				
	Point type/ Site	Hohle Fels	Geißen-klösterle	Sirgenstein	Brillenhöhle	Vogelherd	Hohlenstein	Bockstein Cave	Bockstein-Törle	Total
Aurignacian	Massive base (bone)	1		1	1		2		2	7
	Massive base (ivory)	11	5			13				29
	Massive base (antler)	5				6				11
	Split-based (antler)	1	11		1	27		1		41
	Total	18	16	1	2	46	2	1	2	88
Gravettian	Massive base (bone)	3	8		10					21
	Massive base (ivory)	1								1
	Massive base (antler)	1			1					2
	Single beveled	1 ?			2					3
	Double beveled	1			1					2
	à base machonée	1								1
	Total	8	8	–	14	–	–	–	–	30

different ecological niches. During the Gravettian period, however, only reindeer and red deer remained. This seems to indicate a climatic deterioration from the Aurignacian to the Gravettian in connection with the upcoming Last Glacial Maximum. This shift is also reflected in the avifauna from Geißenklösterle (Krönneck 2009). For carnivores such a shift is not visible, since their diet is based on the presence of game. Species such as cave and brown bear, hyena, lion, wolf, red and arctic fox are continuously present throughout both cultural periods.

Species which provided raw material for organic points, such as mammoth and reindeer, are present in both techno-complexes, but show a considerable bias towards specific elements. Concerning the sites in the Ach Valley, mammoth is mainly represented by ribs and ivory, with hardly any long bones, short bones or molars found. Similarly, reindeer is mainly represented by antler and metatarsi, which are elements important for tool making. Interestingly, a considerable change in the raw material preferences is seen from the Aurignacian to the Gravettian, even if there is no obvious shortage of one of the species (Münzel 2001, 2004). During the Aurignacian, reindeer antler and mammoth ivory were favored for point production. The manufacturers exclusively used antler to produce split-base points while ivory was used for a wide variety of point types.

Ivory points appear with the beginning of the Aurignacian and are present until the Gravettian. Except for the Lautscher point, which is characteristic for the Aurignacian, the ivory points of the Swabian Jura, in general, are not diagnostic for chronological purposes. This situation contrasts with the split-base point which appears from the very beginning of the Aurignacian and lasts until its end in the Swabian Jura (Bolus and Conard 2006). For the Swabian Aurignacian in general, the split-base point is used as a *fossil directeur*. Organic projectile points were abundant during the Swabian Aurignacian, and bone, antler, and ivory were used in ways well suited to the different qualities of each material. The Aurignacian people were intimately familiar with the properties and characteristics of the materials and knew how best to exploit them.

In comparison with the Aurignacian, almost all points from the Gravettian were manufactured from ribs. These ribs were from large mammals, such as mammoths, mammoth-to rhino-size animals, or horse-sized animals. These points made of mammoth ribs are a characteristic feature of the Gravettian layers at Geißenklösterle, Hohle Fels Cave, and Brillenhöhle (Barth 2007), and demonstrate a change from the utilization of antler and ivory to that of mammoth raw material within the Early Upper Paleolithic. According to Knecht (1991:235) the distribution of these "mammoth rib

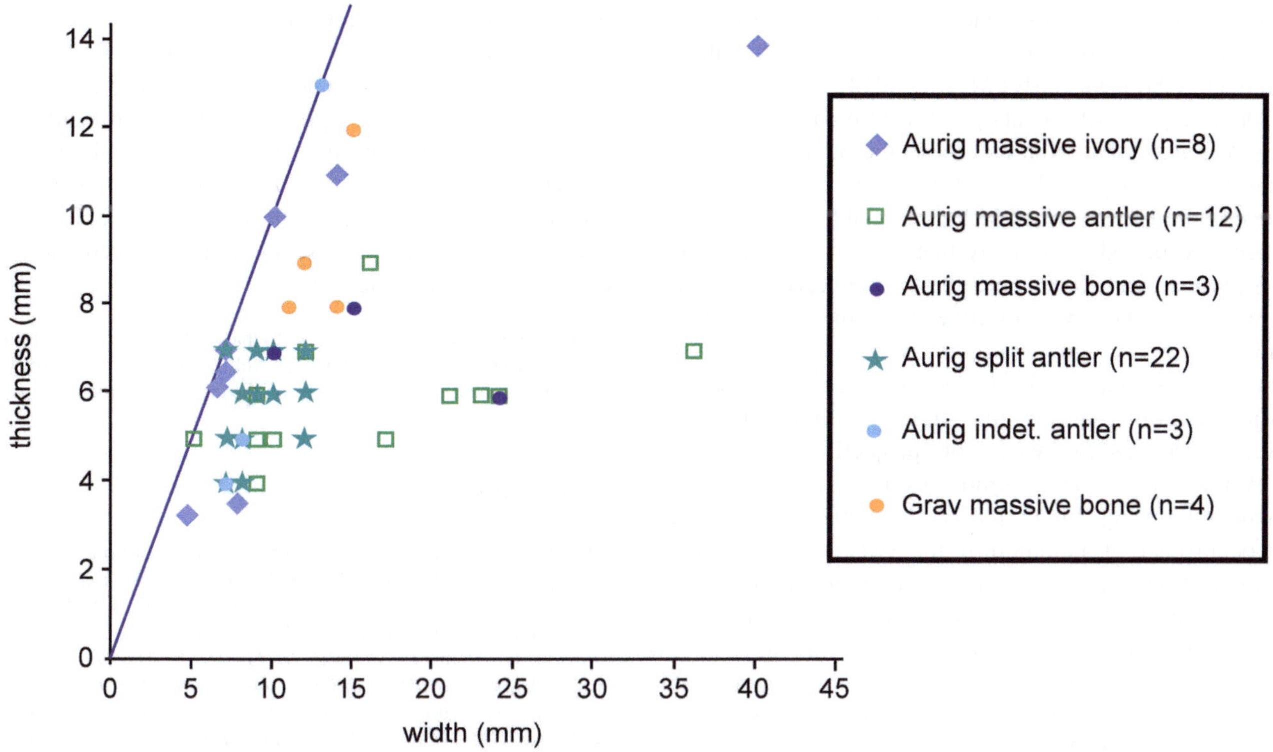

Fig. 6.7 Scatter plot of all complete or near complete point dimensions of the Swabian Aurignacian and Gravettian

points" is temporally and regionally limited to Gravettian sites in southern Germany. Mammoth ribs are of considerable size and the compact bone is thick enough to produce large projectile points. Shooting experiments demonstrate that bone points penetrate a carcass as deeply as do antler or ivory points (Knecht 1991:390), though their mechanical properties (hardness and brittleness) are less suitable for projectiles than antler and ivory.

Statistical analysis of the point dimensions found that split based points are best defined of all the Aurignacian and Gravettian osseous point types, owing to their tightly constrained dimensions (see Fig. 6.7 which only includes complete or almost complete specimens). This result, however, is not determined or dependent by the chosen raw material (antler), since thicker points were manufactured with massive bases from this same material in the Aurignacian. The ivory points from the Aurignacian have the largest dimensions (see the Lautscher point), especially in thickness, which is limited for antler but not for ivory. In the Gravettian there are not enough complete specimens to exactly define the group of "mammoth rib points" typical for southwest Germany (Knecht 1991). However, their width, thickness, and length are similar to those ivory points of the Aurignacian, and may replace them. The broader and flatter points with massive bases in the Aurignacian do not seem to have an analogous form in the Gravettian.

What happened during this transition from the Aurignacian to the Gravettian, and how do we explain this obvious change? Conard et al. (2004) postulated four different scenarios for the transition of the Aurignacian to the Gravettian in the Swabian Jura:

1. The local, gradual emergence of the new Gravettian material culture;
2. A fast development of the Gravettian *in situ*;
3. An extinction or migration of the Aurignacian people, followed by the arrival of the Gravettian people; or
4. A rapid adoption of the new artifact forms characteristic of the Gravettian from other regions with or without significant migration of people.

Based on the analysis of the lithic artifacts from Geißenklösterle and Brillenhöhle, Moreau argued for a regional development of the Gravettian out of the Aurignacian in the Swabian Jura (Conard and Moreau 2004; Moreau 2009, 2012). Bolus supports this hypothesis and states, based on the available lithic inventories, especially from

Geißenklösterle Cave, that the lithics indicate continuity or a slow transition of the Aurignacian forms to the forms of the Gravettian instead of a clear break between the two cultures (Moreau 2009; Bolus 2010; Moreau 2012).

Organic projectile points paint a different picture, however. We argue that, with respect to the organic artifacts, a clear break took place between the cultures. As mentioned above, there is no obvious lack of available animals during either time period. It is likely that a rapid cultural change took place around 30,000 uncalibrated radiocarbon years BP.

We cannot, however, totally exclude the possibility that limitations in raw material might have forced the Gravettian hunters to use mammoth ribs instead of antler or ivory for projectile points during that time (Barth et al. 2009). Rather than using ivory for projectiles, during the Gravettian it was used almost exclusively for personal ornaments during this period (Hiller 2003). Furthermore, the occurrence of mammoth in the Swabian Jura seems to diminish from the Aurignacian to the Gravettian and then again towards the Last Glacial Maximum (Barth et al. 2009), which may also help to explain this shift in raw materials. This is supported by Drucker's work with stable isotopes (^{13}C, ^{15}N). The typical ecological niche of mammoth with high $\delta^{15}N$ and low $\delta^{13}C$ values was gradually replaced during the Gravettian by horses in the Swabian Jura. This points to a deterioration of the living conditions for mammoth well before the Last Glacial Maximum (Drucker et al. 2015).

Furthermore, we know that at least two different systems of hunting weapons were present during both of these Upper Paleolithic periods: osseous points and lithic points. This is luckily reflected in a projectile point found embedded in the transversal process of a cave bear vertebra, recovered from the Gravettian layer IIc in Hohle Fels (Münzel et al. 2001; Münzel and Conard 2004b). This hunting lesion was caused by a triangular flint tip. With a length of 5 mm and a width and thickness of 2 mm, this would have been a remarkably small projectile with which to hunt a cave bear. The use of bow and arrow has not yet been documented in the Aurignacian or the Gravettian period. Because of this we assume that the weapon used in this case was a spear or a lance with a hafted flint tip, since osseous points with grooves or notches for inserting lithics are not known for this period. Furthermore, we know from experimental work with organic projectile points, that impacts of either lithic or osseous points are rarely distinguishable in bone (Letourneux and Pétillon 2008), and thus, leave little clearly identifiable damage on carcasses. This latter situation does not allow us to be able to clearly identify which prey was hunted with the osseous points.

To conclude, this chapter presented an overview of all osseous points from the Aurignacian and Gravettian of the Swabian Jura. It is obvious, especially at Hohle Fels, Geißenklösterle and Vogelherd, that these exceptionally rich sites allow a glimpse into the daily life of the first anatomically modern humans in Central Europe. The sites of the Ach Valley also provide a very good record of the transition from the Aurignacian to the Gravettian and the evolution of the Gravettian technology. Thus, even in this relatively small assemblage of projectile points from the Swabian Jura, a technological change in osseous weaponry technology and systems is well documented.

Acknowledgements We would like to thank Michelle Langley for inviting us to contribute to this volume, as well as two anonymous reviewers for their comments and suggestions on the draft of this manuscript.

References

Albrecht, G. (1977). Testing of materials as used for bone points of the Upper Palaeolithic. In H. Camps-Fabrer (Ed.), *Méthodologie appliquée à l'industrie de l'os préhistorique* (pp. 119–124). Paris: Colloques Internationaux du C.N.R.S. No. 568.

Albrecht, G., Hahn, J., & Torke, W. G. (1972). *Merkmalanalyse von Geschoßspitzen des mittleren Jungpleistozäns in Mittel- und Osteuropa*. Mainz, Germany: Verlag W. Kohlhammer GmbH.

Barth, M. (2007). *Familienbande? Die gravettienzeitlichen Knochen- und Geweihgeräte des Achtals (Schwäbische Alb)*. Rahden, Germany: Verlag Marie Leidorf GmbH.

Barth, M., Conard, N. J., & Münzel, S. (2009). Palaeolithic subsistence and organic technology in the Swabian Jura. In L. Fontana, F.-X. Chauvière, & A. Bridault (Eds.), *In search of total animal exploitation. Case studies from the Upper Palaeolithic and Mesolithic. Proceedings of the XVth UISPP Congress, Session C61, vol. 42, Lisbon, 4.-9- September 2006* (pp. 5–20). Oxford, England: BAR International Series.

Beutelspacher, T., Ebinger-Rist, N., & Kind, C. -J. (2011). Neue Funde aus der Stadelhöhle im Hohlenstein bei Asselfingen. In *Archäologische Ausgrabungen in Baden-Württemberg 2010* (pp. 65–70). Stuttgart, Germany: Landesamt für Denkmalpflege im Regierungspräsidium Stuttgart.

Beutelspacher, T., & Kind, C.-J. (2012). Auf der Suche nach Fragmenten des Löwenmenschen in der Stadelhöhle im Hohlenstein bei Asselfingen. In *Archäologische Ausgrabungen in Baden-Württemberg 2011* (pp. 66–70). Stuttgart, Germany: Landesamt für Denkmalpflege im Regierungspräsidium Stuttgart.

Blumentritt, R., & Hahn, J. (1991). Der Hohle Fels. In *Schelklinger Archäologischer Führer 1*. Schelklingen, Germany: Museumsgesellschaft Schelklingen, Heimatverein e.V.

Bolus, M. (2010). Continuity or Hiatus? The Swabian Aurignacian and the Transition to the Gravettian. In C. Neugebauer-Maresch & L.R. Owen (Eds.), *New aspects of the Central and Eastern European Upper Palaeolithic – Methods, chronology, technology and subsistence. Symposium by the Prehistoric Commission of the Austrian Academy of Sciences. Vienna, November 9–11, 2005* (pp. 139–150). Wien: Verlag der Österreichischen Akademie der Wissenschaften.

Bolus, M., & Conard, N. J. (2006). Zur Zeitstellung von Geschossspitzen aus organischen Materialien im späten Mittelpaläolithikum und Aurignacien. *Archäologisches Korrespondenzblatt, 36*, 1–15.

Bonnichsen, R. (1979). *Pleistocene bone technology in the Beringian refugium* (Mercury Series: Archaeology Survey of Canada Paper No. 89). Ottawa, ON, Canada: National Museum of Man.

Christensen, M. (1999). *Technologie de l'ivoire au Paléolithique supérieur* (BAR International Series 751). Oxford, England: Oxford University Press.

Conard, N. J. (2009). A female figurine from the basal Aurignacian of Hohle Fels Cave of southwestern Germany. *Nature, 459*, 248–252.

Conard, N. J., & Bolus, M. (2003). Radiocarbon dating the appearance of modern humans and timing of cultural innovations in Europe: New results and new challenges. *Journal of Human Evolution, 44*, 331–371.

Conard, N. J., & Bolus, M. (2006). The Swabian Aurignacian and its place in European prehistory. In O. Bar-Yosef & J. Zilhão (Eds.), *Towards a definition of the Aurignacian* (pp. 211–239). Trabalhos de Arqueologia 45. Lisbon, Portugal: Instituto Português de Arqueologia.

Conard, N. J., & Bolus, M. (2008). Radiocarbon dating the late Middle Palaeolithic and the Aurignacian of the Swabian Jura. *Journal of Human Evolution, 55*, 886–897.

Conard, N. J., Langguth, K., & Uerpmann, H. -P. (2000). Die Grabungen 1999 in den Gravettien-Schichten des "Hohle Fels" bei Schelklingen, Alb-Donau-Kreis. In *Archäologische Ausgrabungen in Baden-Württemberg 1999* (pp. 21–25). Stuttgart, Germany: Landesamt für Denkmalpflege im Regierungspräsidium Stuttgart.

Conard, N. J., Langguth, K., & Uerpmann, H.-P. (2003). Einmalige Funde aus dem Aurignacien und erste Belege für ein Mittelpaläolithikum im Hohle Fels bei Schelklingen, Alb-Donau-Kreis. In *Archäologische Ausgrabungen in Baden-Württemberg 2002* (pp. 21–28). Stuttgart: Landesamt für Denkmalpflege im Regierungspräsidium Stuttgart.

Conard, N. -J., Langguth, K., & Uerpmann, H. -P. (2004). Die Ausgrabungen 2003 in den Gravettien- und Aurignacien-Schichten des Hohle Fels bei Schelklingen, Alb-Donau-Kreis, und die kulturelle Entwicklung im frühen Jungpaläolithikum. In *Archäologische Ausgrabungen in Baden-Württemberg 2003* (pp. 17–22). Stuttgart, Germany: Landesamt für Denkmalpflege im Regierungspräsidium Stuttgart.

Conard, N. J., Lingnau, M., & Malina, M. (2007). Einmalige Funde durch die Nachgrabung am Vogelherd bei Niederstotzingen-Stetten ob Lonetal, Kreis Heidenheim. In *Archäologische Ausgrabungen in Baden-Württemberg 2006* (pp. 20–24). Stuttgart, Germany: Landesamt für Denkmalpflege im Regierungspräsidium Stuttgart.

Conard, N. J., & Malina, M. (2002). Neue Ausgrabungen in den untersten Schichten des Aurignacien und des Mittelpaläolithikums im Geißenklösterle bei Blaubeuren, Alb-Donau-Kreis. In *Archäologische Ausgrabungen in Baden-Württemberg 2001* (pp. 16–21). Stuttgart, Germany: Landesamt für Denkmalpflege im Regierungspräsidium Stuttgart.

Conard, N. J., & Malina, M. (2003). Abschließende Ausgrabungen im Geißenklösterle bei Blaubeuren, Alb-Donau-Kreis. In *Archäologische Ausgrabungen in Baden-Württemberg 2002* (pp. 17–21). Stuttgart, Germany: Landesamt für Denkmalpflege im Regierungspräsidium Stuttgart.

Conard, N. J. & Malina, M. (2006). Neue Ergebnisse zum Mittelpaläolithikum, zum Aurignacien und zu den letzten Neandertalern am Hohle Fels bei Schelklingen, Alb-Donau-Kreis. In *Archäologische Ausgrabungen in Baden-Württemberg 2005* (pp. 17–21). Stuttgart: Landesamt für Denkmalpflege im Regierungspräsidium Stuttgart.

Conard, N. J. & Malina, M. (2007). Die Ausgrabungen am Hohle Fels bei Schelklingen, Alb-Donau-Kreis. In *Archäologische Ausgrabungen in Baden-Württemberg 2006* (pp. 17–20). Stuttgart: Landesamt für Denkmalpflege im Regierungspräsidium Stuttgart.

Conard, N. J. & Malina, M. (2009). Spektakuläre Funde aus dem unteren Aurignacien vom Hohle Fels bei Schelklingen, Alb-Donau-Kreis. In *Archäologische Ausgrabungen in Baden-Württemberg 2008* (pp. 19–23). Stuttgart: Landesamt für Denkmalpflege im Regierungspräsidium Stuttgart.

Conard, N. J., Malina, M., & Zeidi Kulehparcheh, M. (2010). Neue Kunst und erste Einblicke in ungestörte Schichten am Vogelherd. In *Archäologische Ausgrabungen in Baden-Württemberg 2009* (pp. 57–61). Stuttgart, Germany: Landesamt für Denkmalpflege im Regierungspräsidium Stuttgart.

Conard, N. J., & Moreau, L. (2004). Current research of the Gravettian of the Swabian Jura. *Mitteilungen der Gesellschaft für Urgeschichte, 13*, 29–59.

Conneller, C. (2011). *An archaeology of materials. Substantial transformations in early prehistoric Europe*. New York: Routledge.

Dotzel, K. (2011). *Bone and antler technology at Vogelherd Cave: An investigation into aurignacian lives*. Master dissertation, Eberhard Karls Universität Tübingen.

Dotzel, K., Bolus, M., & Conard, N. J. (in prep.). Aurignacian Bone and Antler Projectiles from the Swabian Jura, Germany.

Drucker, D. G., Vercoutère, C., Chiotti, L., Nespoulet, R., Crépin, L., Conard, N. J., Münzel, S. C., Higham, T., van der Plicht, J., Lázničková-Galetová, M., Bocherens, H., Dobosi, V. T., Hannus, A., Johnson, E., Münzel, S. C., Scheer, A., & Villa, P. (2015). Tracking possible decline of woolly mammoth during the Gravettian in Dordogne (France) and the Ach Valley (Germany) using multi-isotope tracking (13C, 14C, 15N, 34S, 18O). *Quaternary International, 359–360*, 304–317.

Gaudzinski, S. (1998). Vorbericht über die taphonomischen Arbeiten zu Knochengeräten und zum faunistischen Material der mittelpaläolithischen Freilandfundstelle Salzgitter Lebenstedt. *Archäologisches Korrespondenzblatt, 28*, 323–337.

Gaudzinski, S., Turner, E., Anzidei, A. P., Àlverez-Fernández, E., Arroyo-Cabrales, J., Cinq-Mars, J., (2005). The use of Proboscidean remains in every-day Palaeolithic life. In G. Cavarretta, P. Gioia, M. Mussi, & M. R. Palombo (Eds.), *The world of elephants. Proceedings of the 1st International Congress, Consiglio Nazionale delle Ricerche, Rome 16–20 October* (pp. 179–194) Rom: CNR.

Goutas, N. (2004). *Caractérisation et évolution du Gravettien en France par l'approche techno-économique des industries en matièresdures animales (étude de six gisements du Sudouest)*. Thèse du doctorat, Université de Paris I, Panthéon-Sorbonne.

Guthrie, R.D. (1983). Osseous projectile points: biological considerations affecting raw material selection and design among Paleolithic and Paleoindian peoples. In J. Clutton-Brock and C. Grigson (Eds.), *Animals and archaeology 1: Hunters and their prey. British Archaeological Reports International Series 163* (pp. 273–294). Oxford: Archaeopress.

Hahn, J. (1988). *Die Geißenklösterle-Höhle im Achtal bei Blaubeuren I. Fundhorizontbildung und Besiedlung im Mittelpaläolithikum und im Aurignacien*. Stuttgart, Germany: Konrad Theiss Verlag.

Hahn, J. (1989). Neue Grabungen im Hohlen Felsen bei Schelklingen, Alb-Donau-Kreis. In *Archäologische Ausgrabungen in Baden-Württemberg 1988* (pp. 20–22). Stuttgart, Germany: Landesamt für Denkmalpflege im Regierungspräsidium Stuttgart.

Henri-Martin, L. (1931). *La station Aurignacienne de La Quina*. Angoulême, France: Impr. Ouvrière.

Higham, T., Basell, L., Jacobi, R., Wood, R., Bronk Ramsey, C., & Conard, N. J. (2012). Testing models for the beginnings of the Aurignacian and the advent of figurative art and music: The radiocarbon chronology of Geißenklösterle. *Journal of Human Evolution, 62*(6), 664–676.

Hiller, B. (2003). Die Nutzung von Elfenbein im Jungpaläolithikum des Hohle Fels bei Schelklingen. *Mitteilungen der Gesellschaft für Urgeschichte, 12*, 7–23.

Hofreiter, M., & Pacher, M. (2004). Using ancient DNA to elucidate raw material origin of bone points from Potočka zijalka (Slovenia): Preliminary results. In M. Pacher, V. Pohar & G. Rabeder (Eds.), *Potočka zijalka – Paleontologicla nad archaeological results of the campaigns 1997–2000* (pp. 201–210). Mitteilungen der Kommission für Quartärforschung der Österreichischen Akademie der Wissenschaften 13.

Khlopachev, G. A., & Girya, E. Y. (2010). *Secrets of ancient carvers of Eastern Europe and Siberia: Treatment techniques of ivory and reindeer antler in the Stone Age*. Naukaa, Russia: Sankt Petersburg.

Kind, C.-J., & Beutelspacher, T. (2010). Ausgrabungen 2009 im Stadel am Hohlenstein im Lonetal. In *Archäologische Ausgrabungen in Baden-Württemberg 2009* (pp. 62–69). Stuttgart, Germany: Landesamt für Denkmalpflege im Regierungspräsidium Stuttgart.

Kind, C.-J., Ebinger-Rist, N., Wolf, S., Beutelspacher, T., & Wehrberger, K. (2014). The Smile of the Lion Man. Recent Excavations in Stadel Cave (Baden-Württemberg, Southwest Germany) and the Restoration of the Famous Upper Palaeolithic Figurine. *Quartär, 61*, 129–145.

Knecht, H. (1990). *Technological innovations and design during the early Upper Paleolithic. A study of organic projectile technologies*. Ph.D. thesis, New York University.

Knecht, H. (1991). Projectile points of bone, antler and stone. Experimental explorations of manufacture and use. In H. Knecht (Ed.), *Projectile technology* (pp. 191–212). New York: Plenum Press.

Krönneck, P. (2009). Vögel – Bonusmaterial in der Archäozoologie. In: R. de Beauclair, S. C. Münzel, & H. Napierala (Eds.), *Knochen Pflastern ihren Weg. Festschrift für Margarethe und Hans-Peter Uerpmann*. BioArchaeologica 5. Rahden, Germany: Marie Leidorf.

Krönneck, P. (2012). *Die pleistozäne Makrofauna des Bocksteins (Lonetal – Schwäbische Alb). Ein neuer Ansatz zur Rekonstruktion der Paläoumwelt*. Tübingen, Germany: Universitätsbibliothek: https://tue.ibsbw.de/aDISWeb/app?service=direct/0/Home/$Direct Link&sp=S127.0.0.1:23022&sp=SAKSWB-IdNr367463628

Letourneux, C., & Pétillon, J. M. (2008). Hunting lesions caused by osseous projectile points: Experimental results and implications for the Upper Palaeolithic. *Journal of Archaeological Science, 35*, 2849–2862.

Liolios, D. (1999). *Variabilite et Characteristiques du travail de matières osseuses au debut de l'Aurignacien: Approche technologique et economique*. Thèse de doctorat, Université de Paris X.

Locke, M. (2008). Structure of Ivory. *Journal of Morphology, 269*, 423–450.

Moreau, L. (2009). *Geißenklösterle. Das Gravettien der Schwäbischen Alb im europäischen Kontext*. Tübingen, Germany: Kerns Verlag.

Moreau, L. (2012). Le Gravettien ancien d'Europe central revisité: Mise au point et perspectives. *L'Anthropologie, 116*, 609–638.

Münzel, S. C. (2001). The production of Upper Palaeolithic mammoth bone artifacts from southwestern Germany. In G. Cavarretta, P. Gioia, M. Mussi, & M.R. Palombo (Eds.), *The world of elephants. Proceedings of the 1st International Congress, Consiglio Nazionale delle Ricerche, Rome 16–20 October* (pp. 448–454). Rom: CNR.

Münzel, S. C. (2004). Subsistence patterns in the Gravettian of the Ach valley, a former tributary of the Danube in the Swabian Jura. In J. A. Svoboda & L. Sedláčková (Eds.), *The Gravettian along the Danube. Proceedings of the Mikulov conference, 20.–21. November 2002, Institute of Archaeology, AS CR, Brno* (pp. 71–85). Dolni Vestonice Studies 11.

Münzel, S. C. (2005). Mammoth remains in the Upper and Middle Paleolithic layers of Geißenklösterle Cave (Ach Valley, Swabian Jura, Southwestern Germany). Hunting season, acquisition of raw material and tool production at Geißenklösterle Cave. In D. Vialou, J. Renault-Miskovsky, & M. Patou-Mathis (Eds.), *Comportements des hommes du Paléolithique moyen et supérieur en Europe: Territoires et milieux* (pp. 39–49) Liège, Belgium: Eraul 111.

Münzel, S. C., & Conard, N. J. (2004a). Change and Continuity in Subsistence during the Middle and Upper Palaeolithic in the Ach Valley of Swabia (South-west Germany). *International Journal of Osteoarchaeology, 14*, 225–243.

Münzel, S. C., & Conard, N. J. (2004b). Cave Bear Hunting in the Hohle Fels, a Cave Site in the Ach Valley, Swabian Jura. *Revue de Paléobiologie, 23*(2), 877–885.

Münzel, S. C., Langguth, K., Conard, N. J., & Uerpmann, H. P. (2001). Höhlenbärenjagd auf der Schwäbischen Alb vor 30.000 Jahren. *Archäologisches Korrespondenzblatt, 31*(3), 317–328.

Newcomer, M. (1977). Experiments in Upper Palaeolithic Bone Work. In H. Camps-Fabrer (Ed.), *Methodologie appliqué et à l'industrie de l'os préhistorique* (pp. 293–302). Paris: CNRS.

Niven, L. (2006). *The Palaeolithic occupation of Vogelherd Cave. Implications for the subsistence behavior of late Neanderthals and early modern humans*. Tübingen, Germany: Kerns Verlag.

Owen, L. (2005). *Distorting the past. Gender and the division of labor in the European Upper Palaeolithic*. Tübingen, Germany: Kerns Verlag.

Peyrony, D. (1935). Le gisement de Castanet, vallon de Castelmerle, commune de Sergeac (Dordogne). *Bulletin de la Société Préhistorique Française, 32*, 418–443.

Richter, D., Waiblinger, J., Rink, W. K., & Wagner, G. A. (2000). Thermoluminescence, electron spin resonance and ^{14}C-dating of the late Middle and Early Upper Palaeolithic site of Geißenklösterle Cave in Southern Germany. *Journal of Archaeological Science, 27*, 71–89.

Riek, G. (1934). *Die Eiszeitjägerstation am Vogelherd im Lonetal*. Tübingen, Germany: Heine Verlag.

Riek, G. (1973). *Das Paläolithikum der Brillenhöhle bei Blaubeuren (Schwäbische Alb). Mit einem Beitrag von Wilhelm Gieseler and Alfred Czarnetzki. Teil 1*. Stuttgart, Germany: Verlag Müller & Gräff.

Schmid, E. (1989). Die altsteinzeitliche Elfenbeinstatuette aus der Höhle Stadel im Hohlenstein bei Asselfingen, Alb-Donau-Kreis. *Fundberichte aus Baden-Württemberg 14* (pp. 33–96). Stuttgart, Germany: Landesamt für Denkmalpflege im Regierungspräsidium.

Schmidt, R. R. (1907). Der Sirgenstein und die eiszeitlichen Kulturepochen Schwabens. *Fundberichte aus Schwaben, 15*, 2–7.

Schmidt, R. R. (1912). *Die diluviale Vorzeit Deutschlands. Mit Beiträgen von E. Koken und A. Schliz*. Stuttgart, Germany: E. Schweizerbartsche Verlagsbuchhandlung.

Semenov, S. A. (1957). *Prehistoric Technology. An experimental study of the oldest Tools and Artefacts from traces of Manufacture and Wear*. Russland 1957, 1964, 1. Englische Auflage, 1970, 2. Englische Auflage. Bath, England: Adams & Dart.

Tartar, E., & White, R. (2013). The manufacture of Aurignacian split-based points: An experimental challenge. *Journal of Archaeological Science, 40*(6), 2723–2745.

Tejero, J.-M. (2016). Spanish Aurignacian Projectile Points: An Example of the First European Palaeolithic Hunting Weapons in Osseous Materials. In M.C. Langley (Ed.), *Osseous projectile weaponry: Towards an understanding of Pleistocene cultural variability* (pp. 55–69). Dordrecht: Springer.

Teyssandier, N. & Liolios, D. (2003) Defining the earliest Aurignacian in the Swabian Alb: The relevance of the technological study of the Geissenklösterle (Baden-Württemberg, Germany) lithic and organcic production. In J. Zilhão & F. d'Errico (Eds.), *The chronology of the Aurignacian and the transitional technocomplexes. Dating, stratigraphies, cultural implications* (pp. 179–196). Lisbon, Portugal: Trabalhos de Arqueologia 33.

Wagner, E. (1983). *Das Mittelpaläolithikum der großen Grotte bei Blaubeuren (Alb-Donau-Kreis). Forschungen und berichte zur vor- und frühgeschichte Baden-Württembergs 16*. Stuttgart, Germany: Landesamt für Denkmalpflege im Regierungspräsidium Stuttgart.

Wetzel, R. (1954). Das Törle an der alten Bocksteinschmiede. *Mitteilungen des Vereins für Naturwissenschaft und Mathematik in Ulm, Heft, 24*, 3–20.

Wetzel, R. (1958). *Die Bocksteinschmiede mit dem Bocksteinloch, der Brandplatte und dem Abhang sowie der Bocksteingrotte. Ein Beitrag zur europäischen Urgeschichte des Lonetals und zur geschichtlichen Morphologie des Menschen. Mit eigener Bearbeitung besonderer Sachgebiete durch P. Filzer, U. Lehmann, R. Mosebach and E. Schmid. Teil I.* Stuttgart, Germany: Kohlhammer.

Wetzel, R. (1961). Der Hohlestein im Lonetal. *Mitteilungen des Vereins für Naturwissenschaften und Mathematik in Ulm (Donau), 26,* 21–75.

Wetzel, R., & Bosinski, G. (1969). *Die Bocksteinschmiede im Lonetal (Markung Rammingen, Kreis Ulm).* Veröffentlichungen des Staatlichen Amtes für Denkmalpflege Stuttgart, Reihe A, Heft 15.

Wolf, S. (2015). *Schmuckstücke – Die Elfenbeinbearbeitung im Schwäbischen Aurignacien.* Tübingen: Kerns Verlag, Germany.

Chapter 7
Gravettian Projectile Points: Considerations About the Evolution of Osseous Hunting Weapons in France

Nejma Goutas

Abstract The transition from the Aurignacian to the Gravettian witnessed important environmental, economic and social changes. These changes are especially evident in hunting weapons. Some Aurignacian points (split-based points) disappear from the archaeological record, others (simple-based points) remain, and new types (bevelled based points) appear, some of which will persist long after the Gravettian. Others become characteristic of the Gravettian, with some specific to certain phases ('Isturitz points', simple-based points with mesial incisions). In contrast with the Aurignacian, Gravettian projectile points become more and more refined and standardized. These changes are closely related to the introduction of a new method of blank extraction: the 'debitage by extraction with the groove and splinter technique' (DE with GST). Owing to the diversity of sedimentary, environmental, and cultural contexts in which Gravettian techno-complexes are found, this chapter will focus on osseous points discovered in France. The economic, environmental, and sociological factors involved in the transformation of this equipment during the Gravettian is outlined and discussed.

Keywords Gravettian • France • Osseous projectile point • Hafting system • Economical changes • Groove and splinter technique

N. Goutas (✉)
CNRS, ArScAn, UMR-7041, Équipe Ethnologie Préhistorique, 21, allée de l'Université, F-92023 Nanterre, Cedex, France

MAE - Maison René Ginouvès - Archéologie and Ethnologie, 21, allée de l'Université, F-92023 Nanterre, Cedex, France
e-mail: nejma.goutas@mae.cnrs.fr

Introduction

In this chapter, I will provide a synthesis on Gravettian osseous weaponry in France. This synthesis will revolve around several issues: what raw materials were used?, how were these weapons manufactured?, what morphotypes are known in the French Gravettian?, are all types historically reported really hunting weapons?, what do we know about the evolution of this equipment within the Gravettian?, are any points chrono-cultural markers of the Gravettian?, or even a specific phase of the French Gravettian?; and how do we interpret the changes which characterize the osseous Gravettian weapons? This work is based mainly on assemblages from seven French Gravettian sites (12 collections in total), relating to all phases of the Gravettian (Goutas 2004, 2009, 2008, 2013a), and complemented with published data. With the exception of Arcy-sur-Cure, located in Bourgogne (Renne and Trilobite caves), the studied sites are located in southwest France where most of the Gravettian osseous industries are found (Fig. 7.1).

Gravettian Osseous Projectile Points

The Gravettian is the second techno-complex of the Upper Paleolithic (30,000–20,000 uncal BP), succeeding the Aurignacian. It developed throughout Europe between the end of the Interpleniglacial (OIS 3) and the beginning of the Last Glacial Maximum (OIS 2), occurring in the context of general climate cooling in Europe, interspersed with warmer or more humid phases (Dansgaard-Oeschger interstadials 3 and 4) (Sanchez-Goñi and Harrison 2010; Blockeley et al. 2012).

In France, the oldest Gravettian industries date to around 29,000–28,000 uncal BP. According to the recent synthesis proposed by Pesesse (2013:79–80), we can distinguish seven

Michelle C. Langley (ed.), *Osseous Projectile Weaponry*, Vertebrate Paleobiology and Paleoanthropology,
DOI 10.1007/978-94-024-0899-7_7

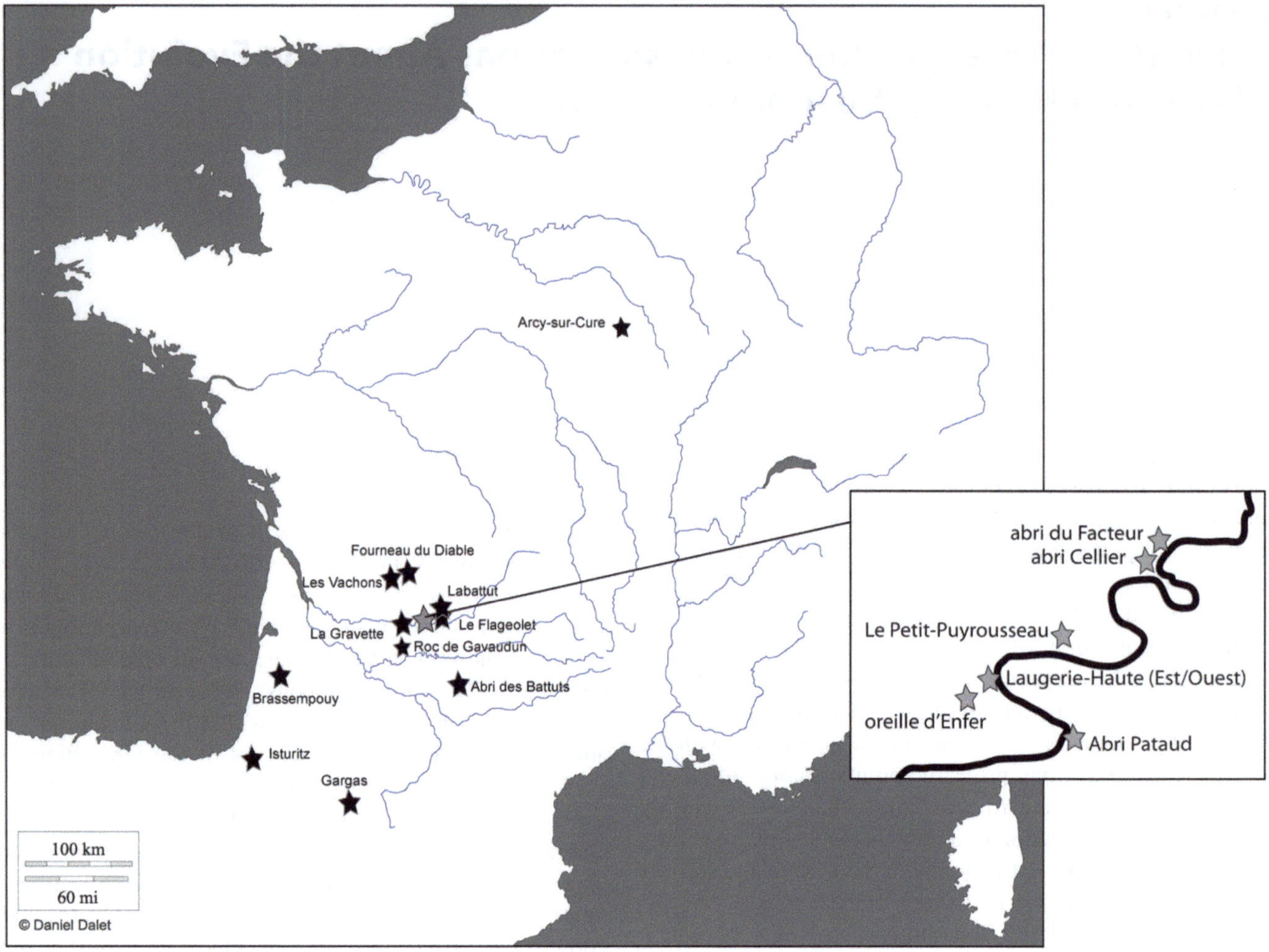

Fig. 7.1 Map of the main French Gravettian sites having delivered osseous industry (GEOATLAS, copyright 1999, Graphi-Ogre, modified map)

facies which can be grouped into four major chronological phases:

(1) An early phase (29–26 ka uncal BP): Including Bayacian, the Early Gravettian *sensu stricto*, and Gravettian with Font-Robert points or Fontirobertian.
(2) A middle phase (26–24 ka uncal BP): Gravettian with Noailles burins or Noaillian, and the Gravettian with Raysse burins or Rayssian.
(3) A recent phase: Also called 'Recent Gravettian'; and
(4) A final phase: Related to the Protomagdalenian industries or final Gravettian.

In France, Upper Paleolithic osseous weaponry is made mainly of cervid (mainly reindeer) antler (Fig. 7.2a). Unlike the Aurignacians (Liolios 1999; Tartar 2009; Tejero 2013), however, Gravettians did not use reindeer antler exclusively for hunting points, but also for other tools (e.g., *bâtons percés*, wedges etc.). Despite this fact, points made on bone are rare in France and are usually found restricted to a single or few items found in a limited number of sites (i.e., Isturitz, La Gravette, Laugerie-Haute, La Ferrassie). In the Gravettian economy, bone was instead mainly used for manufacturing domestic tools (Goutas 2004). The use of ivory is also rare in the Gravettian assemblages of France, as its use for projectile points, though it is more common than bone. In many sites, ivory points are found restricted to a few items, such as in the Noaillian levels of Isturitz cave, in the final Gravettian of Laugerie-Haute and the Labattut shelter (Alaux 1967–1968; Goutas 2004). Some sites, however, are distinguished by an unusual concentration of ivory points: either because of the quantity of points recovered, the characteristic of the points themselves, or even the particular context of their discovery.

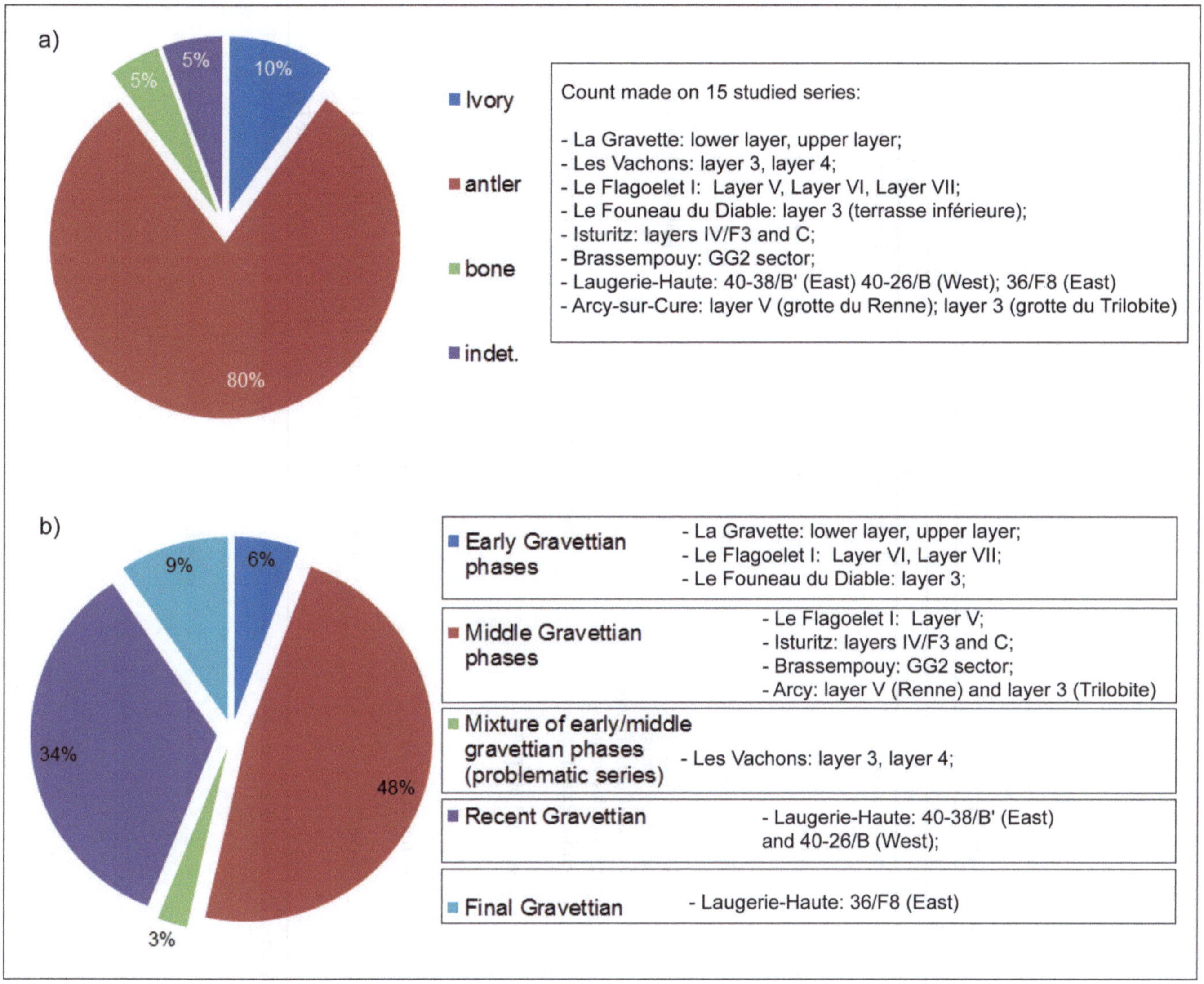

Fig. 7.2 Composition of Gravettian osseous point dataset (**a**) Raw materials used in French Gravettian sites; (**b**) Percentage of osseous points in the various main phases of the Gravettian

This first situation is the case for the recent Gravettian of Laugerie-Haute which has delivered twenty ivory points (Goutas 2004) associated with numerous lithic burins. Many of these points exhibit characteristic use wear of ivory working, suggesting shaping and perhaps *in situ* production of the implements (Christensen 1999). Arcy-sur-Cure (Rayssian facies) delivered seven massive section ivory points among which six are 'Isturitz points' although they also show specific technical and morphological features (see below) (Goutas 2013a). The last scenario is found at Brassempouy where a narrow space (GG2 sector) away from the main activity area and located about fifteen meters from the first Venus figurine, produced eight decorated ivory points associated with 102 lithic points. All of these projectiles points were not produced at this site, but were intentionally brought to and abandoned there, in some cases, after being used. The significance of this last assemblage is not yet clear, though its isolation, the investment in its manufacture, as well as the specificity of the used material, suggests an intentional and perhaps symbolic "exclusion" (Goutas and Simonet 2009).

Composition of the Osseous Weaponry Assemblages

One of the most distinctive characteristics of Gravettian points is the mix of technical and morphological criteria, as illustrated in Table 7.1. Some points are typologically at the junction of two different types, however, several distinct types exist.

Table 7.1 Chronological and typological distribution of osseous points during the Gravettian in France

Types/Sites	General chronological framework			
	Early gravettian	Middle gravettian	Recent gravettian	Final gravettian
Points with a single bevel made on one of the two main surfaces (superior or inferior face)	les Vachons/3[a], la Gravette	Isturitz, Pataud	Laugerie-Haute	Laugerie-Haute East, Pataud
Points with a single bevel made on one of the two sides (left or right)		Isturitz	Laugerie-Haute, Pataud	
Double bevelled points (bifacial)		Isturitz, les Vachons/ 4[c]	Laugerie-Haute	
"Isturitz points"	le Fourneau du Diable/ 3 (terrasse inférieure)[b]	Isturitz/III, Pataud, le Facteur, les Battuts, Arcy-sur-Cure (Trilobite), Labattut, les Vachons/ 4[c]	Laugerie-Haute	Laugerie-Haute East
Simple based points with a rounded or circular section	Fourneau du Diable[b], la Gravette, les Vachons/3[a]	Isturitz, Pataud	Laugerie-Haute, Pataud	Laugerie-Haute East
Simple based points with a flattened section	les Vachons/3[a]	Isturitz/C and F3/IV		
Simple based points with mesial incisions		Isturitz/III	Laugerie-Haute, Pataud	Laugerie-Haute East
Points with a mesial grooving	la Gravette, les Vachons/3[a], Pataud	Isturitz, les Battuts	Laugerie-Haute	Pataud
Points with a flattened mesial surface	Pataud	Isturitz/III	Laugerie-Haute, Pataud	Laugerie-Haute East, Pataud
Point with bilateral notches				Laugerie-Haute Est

[a]Problematic industry: Mixture of Early Gravettian, Noaillian and Aurignacian
[b]Risk of contamination with a Noaillian occupation
[c]Problematic industry: Mixture of Early Gravettian and Noaillian

Simple Based Points (Fig. 7.3a)

Also known as 'simple double-points' or 'biconoical points', the typo-technological data are not yet sufficiently advanced to highlight major differences in terms of shape and manufacture of this point type during the Gravettian, though it is noted that those which have been found in the Noaillian of Isturitz are wider and have a more flattened section than those known in the recent and final Gravettian of Laugerie-Haute (which have a rounded section).

The simple based points of Isturitz fall into two groups, depending on their size and on the morphology of their distal end:

1. Very narrow points whose distal end is relatively sharp and is formed by the convergence of the straight edges. On some points, the shaft has parallel edges which converge abruptly at the proximal end. The transition zone between these two sections is connected by a break in the profile, sometimes accentuated by a slight lateral offset of the proximal end. These points are on average 5–6 mm wide and 2–3 mm in thickness. Sections within this group are exclusively biconvex or flattened oval.
2. More massive section points. The distal end is usually smooth, with a 'spindle-shaped morphology'. There is not, at this section, a break between the shaft and the proximal end: the edges gradually converge towards the latter which gives an almost lozenge shape morphology.

A number of points form a *continuum* between these two morphological groups. Many simple based points exhibit use fractures caused by flexion (step, bevel, or splinter).

Single Bevelled Points (Fig. 7.3b)

According to Knecht (2000), single bevelled points are characteristic of Western Europe during the Gravettian. In France, these points are divided into two main sub-types (Fig. 7.3d–g):

(1) Points with a single bevel made on one of the two main surfaces (superior or inferior) of the point (*pointe à biseau simple facial*); and
(2) Points with a single bevel made on one of the two sides (left or right) of the point (*pointe à biseau simple latéral*).

According to Knecht, these points are primarily made on bone, however, my own observations have found that they are mainly of cervid antler (Goutas 2004). This difference in point of view is probably owing to the fact that Knecht took into account the large bone points found in the final Gravettian of Laugerie-Haute. For me, these points are intrusive and would come from overlying levels related to the 'Aurignacian V' (Proto-Solutrean). On the other hand, rare bone points which are thinner and smaller than these large examples, and which underwent a very different shaping process, seem to truly belong to the Gravettian. Knecht's assertion probably also relies on points discovered in the Upper Gravettian layer of Isturitz (layer III), which was excluded from this analysis. This level provides evidence of numerous contamination events from the Solutrean and Magdalenian layers, and the bone single bevelled points (with a circular section) which were found at this location are actually Solutrean (Fig. 7.4d–f). Their characteristics (raw material, morphology, size, techniques of shaping the bevel) are radically different from those identified in the underlying Gravettian levels. Additionally, there are identical points in the Solutrean of Isturitz (layer IIIa) and in the Aurignacian V of Laugerie-Haute (Goutas 2004:112–114).

Single bevelled points (some with striations on the bevel) are represented at several sites, but always limited in number. For example, Féaux reports a point with a "thin and exceptionally long bevel" from the upper layer of La Gravette (Sonneville-Bordes 1960:181). Layer 3 of Les Vachons, whose lithic industry attests to mixing between the Aurignacian and Early Gravettian (Pesesse 2008), perhaps with a Noaillian occupation, has also delivered a similar item (Goutas 2004). Items possibly indicative of the Early Gravettian phases, are thus, very rare.

During the Middle Gravettian, single bevelled points are known at Isturitz (F3/IV), where three examples exclusively made from reindeer antler and including a second intended point and one with an incised bevel are found. A fourth piece may also belong to this category. The Noaillian of Pataud (layer 4) also delivered three points (Bricker and David 1984), and Labattut shelter two (Alaux 1967–1968). Finally, for the same facies, Le Facteur shelter delivered a bone point with a slightly flattened convex bevel (Peyrony 1934). In contrast, in the Rayssian layer 5 of Le Flageolet I (Goutas 2004), these points are not known, but in the same context in Arcy-sur-Cure (Grotte du Trilobite/ layer 3), one example was discovered (Goutas 2013a).

For the recent Gravettian, they are known at Laugerie-Haute East and West (n = 13), where with the exception of one point made on ivory, only antler was used. Four of them have an incised bevel (Goutas 2004). One possible example could be present in layer 3 of Pataud (Bricker 1995), and in the final Gravettian, they are represented at Laugerie-Haute by seven examples, among which four are in antler and three on an indeterminate material. Two also have incised bevels (Goutas 2004). Points are also present in layer 2 of Pataud (Bricker 1995). Finally, in a less obvious contexts, de

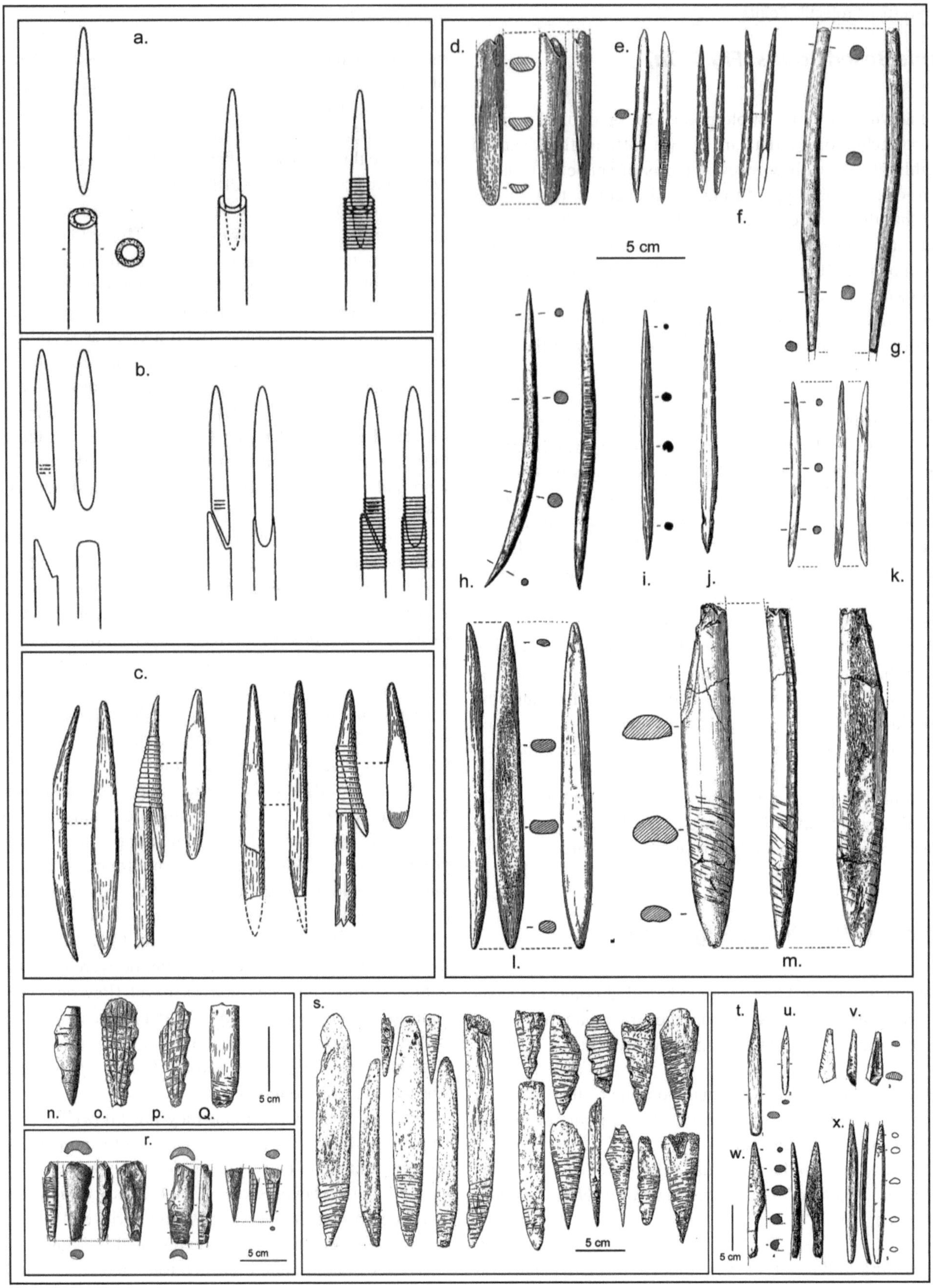

Fig. 7.3 Points and hafting systems identified in French Gravettian sites: (**a**) Socket-like hafting system with a single based point (after Knecht 1991b, fig. 4); (**b**) "Hafting by contact" system with a single bevelled point (after Knecht 1991b, fig. 4); (**c**) "Hafting by contact" system with a bipoint with a flattened mesial surface (Peyrony and Peyrony 1938, fig. 12: 23); (**d**) single beveled point from Pataud, final Gravettian (after Bricker 1995, fig. 20-c: 84); (**e**) single and incised beveled point from Laugerie-Haute (after Knecht 1991b, Fig. 1:121); (**f**) single beveled points from Laugerie-Haute, recent Gravettian (Peyrony and Peyrony 1938, fig. 6–6, 7: 15); (**g**) single lateral beveled

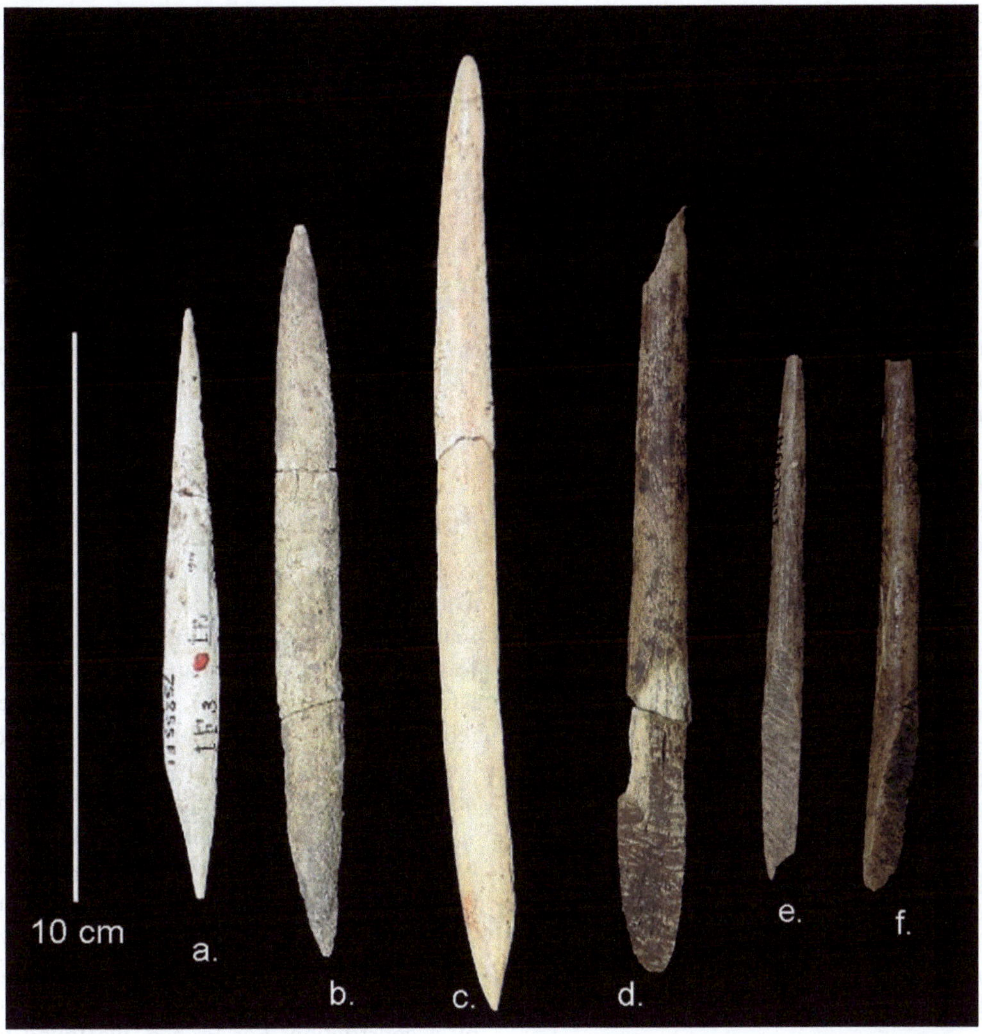

Fig. 7.4 Points from Isturitz Cave (Pyrénées-Atlantiques): (**a–c**) simple based points made on antler with a slight lateral offset of the proximal end (lower layers); (**d–f**) single and incised bevelled bone points (upper layer III). Musée d'Archéologie Nationale, Saint-Germain-en-Laye. photographs© Nejma Goutas

Sonneville-Bordes (1960) and Bricker (1995) reported an item with transverse striations in the 'Upper Périgordian' of Abri Pagès du Ruth (level D), while two probable single bevelled points were found in Laussel in a similar Gravettian level, according to Bouyssonie, the layers J and K of La Ferrassie (Sonneville-Bordes 1960).

In the assemblages I studied, there are both morphometric similarities and differences according to the chrono-cultural contexts in which these points were found. Whatever the context, however, single bevelled points have a medium or small size. For example:

- In Gravettian layer 3 at Les Vachons (predominantly Early Gravettian), the only single bevelled point found has a width of 16 mm and a thickness of 7 mm.
- In the Noaillian of Isturitz width varies between 7 and 13 mm, with a thickness ranging from 6 to 9 mm. In the Rayssien of Trilobite cave in Arcy-sur-Cure (layer 3), the

Fig. 7.3 (continued) point from Pataud, recent Gravettian (after Bricker and David 1984, Fig. 31-5:98); (**h**) bipoints with mesial incisions from Pataud, recent Gravettian (after Bricker and David 1984, Fig. 31-1:98); (**i**) point with mesial grooving from Pataud (after Bricker 1995, fig. 20-l: 84); (**j**) point with bilateral notches from Laugerie-Haute, final Gravettian (after Bordes 1958, fig. 22–15); (**k**) single based point from Pataud, recent Gravettian (after Bricker and David 1984, Fig. 31-6:98); (**l**) point with a flattened mesial surface from Pataud, final Gravettian (after Bricker 1995, fig. 20-b: 84); (**m**) "Isturitz point" from Pataud, Noaillian (after David 1985, fig. 45); (**n** to **q**) various fragments of "Isturitz points" from Gargas (**n**), Labattut (**o, p**), Téoulé (**q**) (after Saint-Périer and Saint-Périer 1952, fig. 68: 128); (**r**) "Isturitz points" from Pataud, Noaillian (David 1985, fig. 47); (**s**) "Isturitz points from Isturitz cave, Noaillian (Saint-Périer and Saint-Périer 1952, fig. 66, 67); (**t–x**) "Isturitz points from Le Facteur (**t, u**), Les Battuts (**v**), Roc de Combe (**w**), Roc de Gavaudun (**x**) (after Sonneville-Bordes 1972, fig. 2: 3)

only single based point discovered has a width of 7 mm and a thickness of 5 mm.

- In the recent Gravettian of Laugerie-Haute, points vary between 5 and 10 mm in width and between 4.5 and 9 mm in thickness;
- In the final Gravettian of this same site, width varies between 3 and 15 mm with thicknesses between 3 and 10 mm.

Only the recent Gravettian of Laugerie-Haute and layer 3 of Les Vachons have each delivered a complete artefact, in both cases with a length of 116 mm. An almost complete item from Laugerie-Haute, however, has an exceptional length of 385 mm, a width of 12 mm and a thickness of 10 mm.

For sections, both in the recent and final Gravettian, points with a single bevel are oval to elliptical (Laugerie-Haute, Les Vachons), whereas at Isturitz, in the Noaillian levels, planoconvex, biconvex and subtriangular sections are also seen.

On the bevel, in the middle Gravettian of Isturitz and the early Gravettian of Les Vachons, they are usually fairly short and have a flat surface. However, in the recent and final Gravettian of Laugerie-Haute, the bevel is usually very long, often curved, and ends with a sharpened extremity. On some, the bevel constitutes a third or even one half of the point's total length.

According to the data from Isturitz and Laugerie-Haute, it would thus appear that significant differences in the shape of single bevelled points may foreshadow an evolution of this type during the Gravettian. However, these data remain to be verified. Moreover, the situation is not so simple, as according to the illustrations published by David (1985), the Noaillian single bevelled points of Pataud differ from those of Isturitz. These latter artefacts present more similarities with those of the recent and final Gravettian of Laugerie-Haute, while the final Gravettian of Pataud has one point reminiscent of those of Isturitz (Fig. 3d). The single bevelled points from the Noaillian of Pataud are also very close to those found in Layer 3 of Les Vachons, and the Les Vachons point may then be intrusive. It could, in fact, come from layer 4 (upper), which is mainly referable to the Noaillian phase (Pesesse 2008; Simonet 2011).

Double Bevelled Points

Double bevelled points are rarer than single-bevelled points. An artefact was found in the early Gravettian of La Gravette (upper Layer, with Gravette points), and another comes from Les Vachons, layer 4, which is mainly related to a Noaillian phase (Pesesse 2008). At this latter site, the bifacial bevel is carved and the shaft carries a longitudinal groove on its upper surface. Another example is found in the Noaillian of Isturitz cave (layer F3/IV), while the recent Gravettian of Laugerie-Haute delivered three points of this type. Here, one point has a bifacial bevel with incisions, while another exhibits a bilateral bevel. At this site, their section is oval, elliptical or subquadrangular. The only near-complete point of this type measures 160 mm in total length (Goutas 2004). A few cases are also reported from La Roque Saint-Christophe (n = 1: Peyrony 1939), the Grand Abri de Laussel (Bricker 1995), and within layer E of Roc de Combe-Capelle (n = 2: Sonneville-Bordes 1960).

Points with a double bevel seem, however, missing from the final Gravettian, while a possible new type of point that we call '*point with bilateral notches*' ('*pointes à étranglement proximal*') seem to now appear.

Points with Bilateral Notches (Fig. 7.3j)

This particular type is present in the final Gravettian of Laugerie Haute, and is a thin double-point having in its proximal (or distal) section a small constriction created by two lateral notches (Fig. 7.3j). Three such artefacts have been recovered, two made from bone and one from reindeer antler; this last being of a slightly larger size. Their width varies between 4 and 8 mm, thickness 3–7 mm. The only complete specimen measures 143 mm in total length (Goutas 2004).

A similar type is reported by Kozlowski and Kozlowski (1977) for the recent phase of the Pavlovian (dated around 25–24,000 uncal BP) at the site of Moravany-Zakovska. This point corresponding to the subcategory 3.5 of the authors typology, and is sub-cylindrical with a conical base separated by a constriction. For France, we found this type only in Laugerie-Haute. Whether this point is a type specific to this site, or more generally to the final Gravettian we are not currently able to determine.

Points with Mesial Flattening: Gravettian or Solutrean Points?

These points exhibit mesial flattening on one side of the shaft measuring several centimeters, for which Peyrony proposed the following hafting system (Fig. 7.3c, i): "The flat section was intended to be applied against the bevel of a wooden shaft and the incisions would allow a strong ligature of both sides, thus forming a kind of prototype of a harpoon arrow." (Peyrony 1932:37; author's translation).

These points are (or seem to be) pointed at both ends and are represented at Pataud in almost the entire Gravettian sequence: early phase (n = 1), recent phase (n = 13) and final phase (n = 1) (layers 5, 3 and 2) – with the exception of the middle Gravettian (Bricker 1995). It is also represented in the Noaillian of Isturitz cave (layer III F3/IV), and in the recent (n=4) and final Gravettian (n=1) of Laugerie-Haute (Goutas 2004).

Where these points are present in a Gravettian context, it is interesting to note that there are always Solutrean or Proto-Solutrean layers also in the stratigraphy. At these sites, points with mesial flattening are present in both the Gravettian and Solutrean levels (Isturitz, Laugerie-Haute) or only in the Solutrean level as at Le Fourneau du Diable (Peyrony 1932; Goutas 2004). While we do not know if the Proto-Solutrean (layer 1) of Pataud also presented these points, they are reported for the Solutrean and Magdalenian of Cantabria (Knecht 1997; Pokines and Krupa 1997), and in the Solutrean of Les Rideaux cave (San Juan-Foucher 2005).

All these factors pose an unresolved question: are the points intrusive in the Gravettian? Or should we consider a sustainability of this type from early Gravettian to Solutrean? To answer this question, it is necessary to directly date these points and to resume their study—in terms of their production methods, raw material and morphology—and build in-depth intra-site and intra-period comparisons in order to identify any chrono-cultural specificities or indices of contamination. The systematic research of connections between fracture surfaces of points from different layers would also be useful in order to have a critical view of their stratigraphical distribution (Pétillon 2006).

In any case, these points are mostly made from reindeer antler and the flattened surface is always arranged on the lower side of the shaft. Only two of these artefacts (discovered in layer 3 of Pataud) were made from bone or ivory. Within this layer, which delivered the majority of these particular points, artefacts are short, broad, of sub-rectangular section, and have "*a flat facet sometimes incised transversely to the middle. The base is roughly cut or chopped into subconical form*" (Bricker 1995:103). According to the only representation that is given for this level (Bricker 1995:101, Fig. 27g), it seems that an artefact was found with a lateral flat side which continues until the end (distal or proximal?) of the point, though in reality, we cannot assert if it also affected the shaft, as the drawn item is only a pointed fragment. If this were the case, it would not be a point with a mesial flattened surface, but instead a single lateral bevelled point, which would explain the location of the flattened surface. However, the point from Pataud/layer 5 offers a mesial and facial flattening, with the latter being associated with a groove. Finally, a point from layer 2 of Pataud is also flattened in its medium part and has smoothed, sharpened ends (Fig. 7.3l; Bricker 1995).

Points with Mesial Incisions

These points are also double-ended points with an elliptical to oval section but which differ by the presence on their shaft of fine, transverse incisions that are more or less parallel. This type is present in the recent Gravettian of Laugerie-Haute (n=6) and Pataud (n=4). In the latter site (Bricker and David 1984), the bases are curve to the opposite side to the incisions. A point of this type is also present in the final Gravettian of Laugerie-Haute (Goutas 2004).

They are also found in the Gravettian with Noailles burins at Isturitz cave (layer III), begging one to wonder if the presence of this particular type reflects their continuation from the Noaillian phase to the recent Gravettian? Or do the points discovered in Isturitz reflect a palimpsest with a level that has a majority of Noailles age artefacts and a shorter occupation period related to a later phase of the Gravettian? Alternatively, it could be representative of a recent phase of the Noaillian where some elements of the lithic industry might be expected to be in transition (Clottes 1976; Simonet 2009).

Finally, these points with mesial incisions are reported by Peyrony and Peyrony (1938), in the Proto-Solutrean of Laugerie-Haute (Aurignacian V) where they may also be intrusive.

The 'Hafting by Contact' System: A Major Gravettian Innovation?

Biconical points, as well as the rare double bevelled points, have what is termed a 'male hafting system' (known since the Aurignacian), in which the point was inserted into a wooden shaft with a longitudinal socket (Stordeur 1987).

Bevelled based points, or those with a mesial flattened surface, inaugurated a new system: 'the hafting by contact' system (Fig. 7.3b, c; Knecht 1991a). The "projectile point and shaft are simply brought into contact against a relatively flat surface, the cohesion of the whole being secured by glue with possible ligature also used" (Pétillon 2006:18; this author's translation).

Single bevelled points, which used an oblique version of this hafting system, had the aim of reducing damage to the wooden shafts or foreshafts. This type of hafting is argued to be less destructive during violent impacts than one requiring the insertion of the point into a shaft or foreshaft (Pétillon 2006). Without invalidating or putting into question this assumption, it seems necessary to weigh the importance of single bevelled points in the Gravettian hunting systems of France. While this type of point is best represented in the later and final phases of the Gravettian at Laugerie-Haute and Pataud (Knecht 1991b:120), even here it is relatively poorly represented. In the collections of Laugerie-Haute, about twenty single bevelled points are found in the recent Gravettian and only four in the final Gravettian (Goutas 2004, 2009; cf. Knecht 1991a:470). This type of point is rare in all other phases of the Gravettian. For example, the Lower Noaillian at Pataud (layer 4) has produced only two examples (Vercoutère 2004:195). Furthermore, after a critical review of the stratigraphy and the osseous industry of Isturitz (upper and lower Gravettian layers), I found that some of the single bevelled points were actually intrusive and came from the overlying Solutrean and Magdalenian layers (Goutas 2004). Finally, the morphology of the most widespread Gravettian osseous points in France is not the single bevelled point, but rather the simple based point, which include various variations and which seem to suggest a diachronic evolution.

Chronological Evolution of Osseous Points During the French Gravettian

Ubiquitous Points and Points with a Strong Chrono-Cultural Association

Aurignacian split-based and the lozenge-based points disappear in the Gravettian, while simple based points persist. New types appear, among which some continue after the Gravettian (namely single bevelled points, double bevelled points, and points with a flattened mesial section or mesial grooving) (Fig. 7.3). Other types, however, are characteristic of the Gravettian and include the famous 'Isturitz point' and points with mesial incisions. Some point types are ubiquitous across the Gravettian, as is the case for simple based points, points with a facial bevel, and points with a flattened mesial surface.

On the other hand, some types are not (for the moment) attested for certain phases of the Gravettian. For example:

- Single bevelled points are known only for the middle (Isturitz) and the recent (Laugerie-Haute) Gravettian.
- Simple based points with mesial incisions are absent from the early Gravettian, with the single example from the final Gravettian (Laugerie-Haute) perhaps coming from the underlying level (recent Gravettian).
- The large simple based points with a flat section seem represented only in the Noaillian (Isturitz).
- 'Points with lateral notches' are known only in the final Gravettian (Laugerie-Haute).
- Bifacial bevel points are not attested to in the final Gravettian, whereas points with a mesial groove were not identified for the recent Gravettian

Revising the Status of 'Isturitz Points' as Fossile Directeurs for the Noaillian

For a long time, the 'Isturitz point' (stocky, with an incised end) was considered exclusive to the Noaillian (Fig. 7.3m–x), though they are now known for the Rayssien of Arcy-sur-Cure (Goutas 2013a). Layer V of Reindeer cave at Arcy-sur-Cure (a layer very similar to that of layer 3 at nearby Trilobite cave), however, has also delivered two Gravette points and 18 Microgravette points (Klarics 2003). According to Klaric, this association could foreshadow mixtures between a main occupation phase relating to the middle Gravettian (with Raysse burins), and other, shorter occupations dating to a more recent phase of the Gravettian. Currently there is no consensus on the interpretation of this association (Gravette points with Raysse burins), and it is difficult to argue that it is simply the result of mixed contexts, although this possibility must be taken into account. Only the discovery of a new series from reliable chrono-stratigraphic contexts will allow an answer to this question by showing either the repetition of this association, or stratigraphical disjunction of these two lithic *fossiles directeurs*. Nevertheless, no evidence for a Noaillian occupation having been observed in the lithic industry, and thus, the hypothesis of contamination with Noaillian industries seems rejected. In fact, the presence of Isturitz points in Arcy-sur-Cure confirms that these particular objects are not exclusive to Noaillian facies.

The Isturitz point is also associated with recent and final Gravettian contexts at Laugerie-Haute (Goutas 2004, 2013b) and Pataud (Flori 2013), along with more questionable contexts such as the early Gravettian of Le Fourneau-du-Diable, where the closeness of a Noaillian level does not exclude the possibility of contamination (Goutas 2008). For the Noaillian, these points are attested to at numerous sites and remain a strong cultural marker of this facies. The Noaillian levels of Isturitz yielded the richest corpus in France (n = 190), composing more than 70 % of the recorded examples. At sites other than Isturitz, and to a lesser extent, the Abri Pataud (n = 22, San Juan-Foucher and Vercoutère 2005), Isturitz points are generally poorly represented and usually fragmented (Goutas 2008).

Revising the Functional Status of Some Osseous Points: Hunting Points or Tools?

'Isturitz Points'

The term 'Isturitz point' (Sonneville-Bordes 1971, 1972) will be used here, rather than 'Isturitz spear' for two reasons. Firstly, the term 'spear' is a misnomer since it refers to the whole projectile weapon: from point through the shaft to the potential tail (Pétillon 1999). Second, even if we think that the term 'point' is not the most appropriate, it is nevertheless more neutral than 'spear', since it can also be applied to the active portion of a (non-projectile) tool and does not infer a sole hunting function (Goutas 2008).

The uniquely large sample of Isturitz points at Isturitz itself provides a statistically representative corpus that allows for a detailed analysis of these objects, which are so particular in many ways (see Goutas 2008). Since its discovery, the Isturitz point has generated several hypotheses related to its function (weapon or tool) and its mode of use (Goutas 2004; Vercoutère 2004; San Juan-Foucher and Vercoutère 2005), with the aim of determining if the striated end was the proximal or the distal part of the tool. A morphological, technological, and functional study reveals that these various hypotheses are not inevitably contradictory. One part served well and truly as hunting weapons, but others were indubitably domestic tools first and foremost. It is, as such, interesting to underline that the most massive points of Isturitz, those farthest from the morphological and technical characteristics expected for a real hunting weapon, are also those reinterpreted most frequently as tools. Among these items, some present bipolar wear resulting from use as an intermediate tool, while others display non-violent stigmata and are associated with a blunt active end suggesting a use in a gesture of friction. Thus, 'Isturitz points' while being very heterogeneous in their form and size, reflect several functional realities.

In addition, I have proposed a typology for this object which retains one main criterion of distinction: the localization of the striated end, which testifies, in my opinion, to different functions according to whether it is on the proximal (group A, n=23) or distal part (group B, n=36). The rest of the assemblage consists of fragmentary artefacts (approximately 77 %), and includes a number with a 'shortened base' (3 % by '*raclage en diabolo*' or sawing). Strangely, all the Gravettian sites which have produced such artefacts have a similar situation: the frequency of striated ends, the scarcity of the complete items, and the absence of complementary fragments to the striated ends. I believe this lack of non-striated ends ensues from two factors: (1) an abandonment of these ends outside the cave for reasons linked to the function of the points; and (2) an identification problem.

The functional reallocation of some of the 'Isturitz points' as domestic tools does not mean that the inhabitants of Isturitz did not use osseous projectile points. As we have seen, nearly 150 points with morpho-technical attributes of hunting weapons are in evidence at this site. These hunting points (bevelled and simple based points) distance themselves from the Isturitz point by:

(1) A more normalized production allowing for interchangeability of points into shafts;
(2) A thinner production. Some points have a width of only about 5–6 mm with a thickness not exceeding 3 mm; and
(3) A sharp active end.

Ultimately, the Isturitz point is not an entirely distinctive projectile point type, because the features which characterize it can be found on functionally very different items. It is not the Isturitz point in itself which constitutes a cultural marker of the middle phase (with Noailles burins), but instead it is a particular set of morphological attributes (and perhaps functionality) which the Gravettians applied to various categories of objects—not just projectile points. That is why, due to the diversity of the morphology, and probably the function of Isturitz points, I have suggested that the use of a more neutral term such as 'pieces with Isturitz type features' (Goutas 2008).

Bipoints

Although omnipresent in Gravettian sites, and across the Upper Paleolithic in general, bipoints constitute one of the object categories which are less well characterized. Functional hypotheses for these points are highly varied and include: straight fishhooks, a composite hunting weapon, or 'processing tools' for working plant materials (e.g., wood, bark, plants) or animal materials (leather, fur, tendons). Given the size diversity in this object category, however, it is likely that they were used for a variety of uses. If any of these bipoints turned out absolutely to be hunting weapons, it would greatly modify our conception of Gravettian weaponry, as they would significantly increase:

(1) The proportion of osseous points within assemblages; and
(2) The role of bone as the material of manufacturing this very specialized equipment.

Finally, it would highlight a phenomenon which would remain to be characterized for this period: that of 'microlithization' of osseous weapons, and its relationship with its equivalent in the field of lithic weapons.

Discussion: Chronological and Sociological Perspectives on the Archaeological Data

From Aurignacian to Gravettian: A Decrease in Osseous Points?

The Gravettian of Western Europe was, for a long time, considered as a phase of near abandonment of osseous points in favor of lithic points (Knecht 1991b; Cattelain 1995; Knecht 1997; O'Farell 2004). This idea, at least for France, has now been adjusted as the frequency of osseous projectile points are now found to be highly variable according to the assemblage and the studied facies. In some sites, points are very few (Le Flageolet I, layers VII and VI; La Gravette, lower and upper layers; Les Vachons, layer 3, etc.), and this is particularly the case for the early Gravettian at La Gravette/lower layer (n=5; Goutas 2004) and Pataud/ layer 5 (n=4+2 probable examples, Vercoutère 2004). It should be noted, however, that the osseous industry in general at these sites (with the exception of La Gravette, upper layer), are quantitatively low and often poorly preserved. In contrast, large and well preserved assemblages have a considerably larger sample of osseous points. For example, more than 150 points (excluding Isturitz points) or fragments of sharpened objects morphologically and technically compatible with this typo-functional category have been recovered from Noaillian contexts in Isturitz cave. Another hundred were found in the recent Gravettian of Laugerie-Haute, and nearly thirty in the final Gravettian of this same shelter (Goutas 2004, 2009). We must also consider if the apparent quantitative abundance of osseous points during the Aurignacian was not amplified by a particular research focus on the early Aurignacian, to the detriment of research on the end of the Aurignacian and the Gravettian.

Indeed, during the early Aurignacian, especially in the Perigord and the Pyrenees, osseous points are relatively abundant. In the 1990s, Knecht (1993:34) counted 341 split-based points in 16 French sites, however, more recent studies have greatly increased the size of the corpus of split-based points for the Aurignacian in Europe (about 700 points of this type are now known, including nearly 525 from 31 French sites [Tejero 2013, 2016]). These new data support previous hypotheses (Goutas 2009), on whether this quantitative explosion of weaponry is not a reflection of the Aurignacian standard strictly speaking, but rather corresponds more to a particular episode within this European cultural complex (the early Aurignacian), and to a regional context (southwest France) more specifically. In fact, the data are still too sparse for the very early phases (Proto-Aurignacian), as well as the recent Aurignacian (evolved Aurignacian) for setting up a chrono-cultural model, or for determining if these changes are diagnostic elements of major changes in behavior between the Aurignacian and Gravettian (Goutas 2009; Teyssandier et al. 2010; Tejero 2013).

To be able to support a real decrease of osseous weapons production and use between the Aurignacian to the Gravettian, we need detailed comparison of late Aurignacian production methods with those of the early Gravettian, taking into account differential preservation, site function, and the surrounding environment. To this end, the currently available data on the early Gravettian allows (with difficulty) a suitable comparison (Goutas 2009). For the recent phases of the Aurignacian, the only data available which does not come from a problematic or insufficiently documented stratigraphic context are those of the evolved Aurignacian from Pataud (layers 8, 7 upper and 6), however these data are still difficult to use as the osseous industry is few. Here, layer 8 delivered three osseous artefacts (not projectile points), layer 7 has 22 osseous pieces, including a fragment of an antler 'spear', and upper layer 7 is even poorer with only two pieces of osseous industry (including a lozenge based point) (Vercoutère 2004:111–132; Chiotti 2005). Finally, layer 6 delivered no examples of osseous industry at all (Gregoriani 1996). Thus, if we focus purely on the transition between the end of the Aurignacian and the beginning of the Gravettian it is difficult to assert that there was a decrease in osseous hunting weapons between these two periods.

Finally, there is no doubt that the proportion of Gravettian osseous points is currently underestimated. For France, an initial revision of this equipment (and in particular of the 'Isturitz point' type) has been provided (Goutas 2004, 2008, 2009), however, much work remains to be completed on the French corpus in order to have a precise typological and functional understanding of this technology within each site and each Gravettian phase. This situation leads us to raise another problem inherent to the Gravettian context, namely, the difficulty of creating a usable typology. This problem is the result of two reasons:

(1) The mix of technical features found on many of the points. For example, for the double-point, there is the classic one (those without a special feature), others with a flattened mesial surface, and again others with incisions engraved on their shaft. The first requires a socket-like hafting system (Fig. 7.3a), while the second requires a 'hafting by contact' system (Fig. 7.3c). The last type (those with mesial incisions) may have functioned as a type of barbed point (Peyrony 1936), but this idea is still to be confirmed (as mentioned above).
(2) The existence of a morphometric *continuum* between different point sub-types and perhaps between projectile points *sensu stricto*, as well as pointed objects which are mainly tool-making tools. The border between one type and the others still has to be defined. This case is especially true for Isturitz which provides the most important French corpus of Gravettian osseous points.

Mutations in Projectile Points Features

Let us now focus on the qualitative evolution of the points: first between the Aurignacian and Gravettian, and secondly during the Gravettian itself. In French sites, osseous projectile points are the artefact category most subject to change, however, the toolkit remains more or less stable throughout the Gravettian (Goutas 2004). In reality, this apparent stability hides subtle and complex developments. Thus, osseous tools still possess elements of chrono-cultural diagnoses which are insufficiently explored, especially when compared to the dramatic changes that accompany the evolution of hunting weapons throughout the Paleolithic.

In France, the Gravettian introduces a major conceptual breakthrough in the size and morphology of osseous points compared with those of the Aurignacian. If we had to describe broad trends, we could say that Gravettian points are generally much narrower and longer than those of the Aurignacian (Knecht 1991a; Liolios 1999; Goutas 2004). This is not to say that there are no fine Aurignacian points, only that they are not in the majority (Tejero, personal communication). In a number of cases, this narrowness (ratio of width/thickness) is the result of repair (having led to a decrease of their initial size and volume), while others are in fact the result of deliberate design (Liolios 1999; Tejero 2013; Tartar, personal communication). Which process resulted in each of these fine points, is not yet able to be determined.

Along with a general slimming down of points during the Gravettian, the shaft section changes to a round section, along with the mesial and proximal features outlined above. Some of these changes reflect the implementation of new hafting systems.

A Quantitative Development in Hunting Equipment During the Gravettian?

Most of the osseous points date to the Middle and Recent Gravettian, with few dating to the Early and Final Gravettian. For this last phase, of the four sites associated with it (Laugerie Haute, Pataud, Le Blot, Les Peyrugues), the last two have not delivered osseous points (Chauvière and Fontana 2005; Chauvière 2012). The assemblages reported for the Early Gravettian deliver very few points also (Fig. 7.2b).

This fact is probably amplified by the state of current research where more sites and more osseous industries are known for the Middle phase of the Gravettian, while Early Gravettian assemblages are often less well preserved than those of the later Gravettian phases. We cannot, however, exclude the possibility that these quantitative differences reflect a real change in techno-economic behavior, where osseous points increase between the early phases and the middle, recent and final phases (Goutas 2009). It must be emphasized, however, that Isturitz alone accounts for about 45 % of the Gravettian osseous points in France. This large cave would have been used during the Gravettian as a temporary aggregation site, where various Gravettian communities stemming from the North of Spain and from the southern part of the French Atlantic coast, would have congregated at particular times of the year to undertake economic and social practices (Lacarrière et al. 2011; Goutas and Lacarrière 2013; Normand et al. 2013). The very high frequency of osseous points (also lithic points) would therefore be connected to the execution of collective hunting at the site (essentially focused in the acquisition of bison during autumn). As previously stated, these "initial results lead us to balance the proposal made by Pike-Tay (1993) and Enloe (1993) from the study of Perigourdin sites in which Gravettians groups practiced 'opportunistic' hunting regardless of the aggregation of ungulates throughout the seasons" (Lacarrière et al. 2011:79).

Ultimately, even if it is true that in France Gravettian manufacturers mostly invested in stone for their weaponry, it nevertheless remains the case that in some sites and during certain periods, antler points also played a significant role in hunting activities (as shown by their number in the Noaillian at Isturitz cave, the recent and final Gravettian of Laugerie-Haute, and to a lesser extend in the Rayssien of Arcy-sur Cure) (Goutas 2004, 2009, 2013a). Besides, it is risky to oppose lithic against antler points purely on quantitative, functional and economical grounds as there are important differences in terms of conservation, use life, maintenance, repair, raw material supply, and complexity of the manufacturing system (Knecht 1991b; Cattelain 1995; Knecht 1997). Moreover, the use of wood as a substitute to antler also needs to be considered for the final Gravettian. This use of wood could reflects its use as a substitute to antler when there was limited access to the latter (Chauvière and Fontana 2005:144; Chauvière 2012).

Interpreting Changes in Gravettian Weaponry

To understanding the mechanisms involved in the observed changes in Gravettian weaponry, it is necessary to take into account the system of manufacture of this highly specialized equipment, the global economic system of the society and the environmental constraints. Having already dealt with these issues in detail in a previous work (Goutas 2009), we will not dwell on all of our arguments, but instead focus on some important ideas which reveal the existence of close links between the introduction of a new way to produce rods ('baguettes') and the evolution of Gravettian weaponry.

A Gravettian Innovation: Double Longitudinal Grooving (DLG)

Double longitudinal grooving (also called 'groove and splinter technique' by Clark and Thompson 1953: Goutas and Tejero 2016) is clearly attested from the Gravettian period. This process, however, does not substitute Aurignacian traditions in all places. These latter methods remain used in certain phases of the Gravettian in France (Goutas 2003, 2009), and in the Moravian site of Pavlov I (Klima 1987; Goutas 2013b), where variants of fracturation techniques were still used on antler to production large rods—in mixing a sectioning action, an indirect percussion method (‘*refend*’), and sometimes a short but deep groove (‘*rainurage/fendage*’).

DLG is quite simple to implement and has strong conceptual analogies with laminar knapping (Averbouh 2000). It allows the overcoming of the morphological and volumetric constraints of the osseous block, and the obtaining of perfectly predetermined blanks (lengthened, regular, narrow) which can be produced in series and favors greater standardization. In regional contexts, the place of DLG in techno-economic systems, as well as the explanatory factors for its use, vary considerably. Omnipresent in the French Gravettian, from the earliest phases its use is scattered throughout Central Europe: rare in the Swabian Jura, better represented in Austria (in more recent contexts) (Kamegg, Willendorf II/9), and little used in Moravia during the Pavlovian period (Otte 1981; Barth et al. 2009; Goutas 2013b). In Eastern Europe, its use remains anecdotal. To understand why this new method of debitage has known such success in Western Europe, but more of a scattered appearance elsewhere, requires the identification of the underlying motivations for this technique change (Goutas 2009).

New Hunting Needs, New Manufacture Methods for Osseous Points?

DLG is not more time efficient for equipment production, nor does it make them more effective (Goutas 2009). On the other hand, and according to Knecht (1991b), Gravettian points may possess better capacity for penetration than Aurignacian points owing to their tapered shape. The use of DLG is not related to this morphological change because (as we have seen) it allows the production of long, narrow and perfectly straight blanks. On the other hand, it seems more delicate to establish a direct link between the new shape and a better capacity for penetration. Indeed, Pétillon (2006:198) comments on the experimental work of Knecht, that “Aurignacian and Gravettian points were apparently tested [in Knecht’s experiments] with ‘equal conditions of shootings’—same target, same shafts, same propulsion system—but can we assert that it was the same in the Paleolithic?” (author’s translation).

In contrast, it appears that if DLG had been used for producing various items, hunting weapons were made exclusively using this debitage process. The invention of the DLG method is also concomitant of a change in the shape and fitting of osseous points (as discussed above). Environmental changes and changes in hunting strategies and techniques (O’Farell 1996, 2004) could have motivated Gravettians to seek new technical solutions which allowed them to produce blanks for projectile points that were both longer, finer, lighter and more standardized (Goutas 2009). They also did not allow breakage by indirect percussion, nor the so called ‘re-splitting procedure’ (‘*refend*’) used by their Aurignacian predecessors (Liolios 1999). However, recent experiments have helped us to improve our knowledge on this Aurignacian technique. It appears to allow control and predetermination in the blanks production (Tejero et al. 2012), however, the “rate of predetermination” is not comparable at all to that involved with the DLG which allows the reproduction of blanks, in series, exactly the same type of blank almost to the millimeter, without any limitation to length or width. That is why the term ‘total predetermination’ seems to me applicable only to DLG (Goutas 2004, 2009).

This change to greater standardization during the Gravettian could have been motivated by factors other than a change in manufacture method, though what modes of propulsion and tactics of game acquisition, which themselves, refer to issues of human group mobility (Cattelain 1994, 1995; Soriano 1995; Cattelain 1997; Pétillon 2006; Valentin 2006) and their demographic structure (Pelegrin 2000), still escape us. However, we note that if greater standardization of Gravettian points as against Aurignacian points cannot only be explained in terms of a change in debitage method (though shaping also plays a major role), DLG nevertheless greatly facilitates the change (Goutas 2009). In this context, the easier resharpening of Gravettian points, which can be “resharpened without modifying their general shape, while the repair of split based points—and to a lesser extent the lozenge point—requires repair on a larger scale” (after Knecht 1991b:135 in Pétillon 2006:198), as well as the possibility of better interchangeability of them in their hafts may have been a major advantage in the maintenance of this equipment (Knecht 1991a, b, 1997). This aspect was a considerable advantage if one considers it in terms of the mobility of human groups, hunting tactics implemented, and the rate of weapon loss which ensue from it (Cattelain 1995; Pelegrin 2000; O’Farell 2004). Valentin underlines that “the possibilities of maintenance to be finely estimated have to stand out both in [opportunities] which depend of the durability and the rate of loss of points, and in [facilities] which depend of the time and the working difficulty” (Valentin 2006:145, author’s translation). Finally, as with lithic weapons, the greater lightness and standardization of some osseous Gravettian points could reflect the use of a new mode of propulsion (the bow?) (Cattelain and Perpère 1993; Cattelain 1997).

If the direct link between the mutations that we have just covered (appearance of the DLG, morphological change, and diversification of osseous points) remains difficult to demonstrate, it is nevertheless very likely that these convergences are indicative of some larger processes. Similar observations were made for stone industry and have led some researchers (Pelegrin 2000; Valentin 2006) to consider the possibility of a subtle link between "the modification of the knapping techniques, and that other capital change, transformation of hunting weapon" (Valentin 2006:142, authors translation). Although this assumption is based on different contexts to those examined herein (Magdalenian and Azilian), and mainly on stone industry, it seems interesting to examine our osseous problems in this same way. Thus, and according to Pelegrin (2000), these changes (decrease in osseous weapons in favor of new lithic weapons and changes in the terms of debitage: knapping techniques, type of used hammer, etc.) could be an expression of significant changes in hunting techniques and strategies. As such, it is interesting to highlight that it is from the middle phase of the Gravettian that artefacts associated with the use of DLG (wastes, "baguette" blanks, finished objects) become more numerous and in parallel the number of osseous points becomes consistent.

A Co-Variation in the Changes Affecting Osseous and Lithic Points

Alongside the changes that characterize Gravettian osseous artifacts, we notice in the stone industry a development and a diversification of composite weapons, including "abrupt back points", which serve as the basis for the definition of the Gravettian (O'Farell 1996; Klaric 2003; Pesesse 2003; Foucher 2004; Guillermin 2004; Simonet 2010, etc.). There is also significant investment in knapping to produce standardized blades or bladelets. The Aurignacian concept of "*torsitude*" (twisted blanks) (Tixier 2005) reflecting the search for a convex sharp edge (D. Pesesse, personal communication) is replaced by the "Gravettian concept" of a straight sharp edge.

According to Simonet (2005), from a dimensional point of view an important *inter-* and *intra*site variability of Gravette points exists. This variability may reflect, depending on the context, a certain flexibility in manufacturing standards or different uses, however, most of these lithic points appeared to have served as projectile points (Cattelain and Perpère 1993; Soriano 1995; O'Farell 1996; O'Farell 2004; Simonet 2005).

An Evolution of Techniques and Hunting Strategies Between the Aurignacian and the Gravettian?

O'Farell underlined the existence of 'coincidences' or 'disturbing correlations' between the changes affecting stone weapons and environmental and faunal changes (O'Farell 1996, 2004). We shall evoke here only synthetically the conclusions of the author, which are based on an important argument which interests us for the rest of our discussion. O'Farell considers that the innovations observed in the Gravettian armament reflect an evolution of subsistence strategy (O'Farell 2004). Gravettian points (simple based points or bevelled base points) are a better adaptation to long-distance shooting, as well as greater multi-functionality, and a better capacity for maintenance than Aurignacian points. Relying particularly on ethnographic data, O'Farell questioned the possibility that the observed differences in hunting technology between the Aurignacian and Gravettian periods may reflect "the trend of the first to practice the shares of seasonal mass hunting of some species, while the second would spend more time in the acquisition of scattered animals" without it being necessary to invoke a 'specialization' of hunting (O'Farell 2004:135).

In summary, if as we suppose, there exists in France a subtle link between the emergence of the DLG method and the changes which occur in antler hunting equipment, too many unknowns (paleo-environmental, archaeozoological, technological) still limit us in our interpretations.

Conclusion.... and Beginnings for Other European Contexts

After all this background, it appears that French Gravettian sites have exhibited very specific mutations in the field of osseous hunting weapons. Nowhere else in Europe does such diversification of osseous weapons occur. Even the very rich collections of central (Pavlovian) and Eastern Europe ('Kostienki-Avdeevo culture') did not deliver a corpus as rich as those discovered in Isturitz and Laugerie-Haute. If evolutionary trends are beginning to emerge within the Gravettian hunting equipment of France, studies are still insufficient to determine precisely what these changes meant for the wider Gravettian.

In Central and Eastern Europe, the situation seems radically different (Goutas 2011, 2013b, c). On the one hand, osseous projectiles tend to be marginalized in favor of lithic points. On the other, the variety of types known in Western Europe is restricted to a more homogeneous production, dominated by the long ivory simple-based points. Presently across Europe, any of these points are sometimes decorated (Brassempouy, Predmost Ia, Avdeevo etc.). Moravian and Russian items are particularly large (dozens of centimeters in total length), whereas those of France are of a more modest size. If these dimensional differences are likely linked with differences in hunting strategies, they may also reflect a diversity of cultural conventions, or know-how in ivory working. Moravian and Russian human groups, societies which perfectly mastered ivory working, were able to develop technical solutions more adapted to this material resulting in the perceived differences.

The long ivory simple-based points may coexist alongside types which are sometimes exclusive of a region (e.g., Swabian Jura), or, of a site (e.g., Avdeevo). A feature of the Swabian Gravettian is both the extreme scarcity of ivory points and the use of a very particular type of point, realized on mammoth or rhino ribs (see Wolf et al. 2016). The latter seems very localized in time (early Gravettian) and space (southern Germany), and should be of high cultural value (Knecht 1991a; Barth et al. 2009). These points, oval in section, would be exclusively made from the spongy tissue and would present on some a side or facial bevel (Knecht 1993). In Russia, within the 'Kostienki-Avdeevo culture', particularly rare ivory points have been discovered in Avdeevo. These points are exceptionally lengthy (up to more than 500 mm) and curved, and are characterized by a sharpened, engraved base with two side spurs (Gvozdover 1995; Goutas 2011).

Ultimately, osseous points are often few within Gravettian assemblages with the notable exception of a handful of sites (i.e., Swabian Jura and Pavlov I in Moravia). All these quantitative and qualitative changes (significant decrease in osseous points in the central and eastern European Gravettian) and qualitative (introduction of new forms and new hafting systems) undoubtedly reflect new requirements in terms of hunting strategies. If the early invention of the bow can be considered as an explanation for these changes in the Western context, mutations of the hunting weapons in Central (Moravia) and Eastern (Russian plain) Europe seems intimately linked to the emergence of more intensive occupations of large camps and important changes in lifestyle (e.g., sites occupation over the long term) and in the exploitation of the available environmental resources.

Acknowledgements I warmly thank Michelle Langley for giving me the opportunity to collaborate to this collective book. I am also very grateful to my reviewers for their useful comments and to M. Langley for her essential reading of this paper.

References

Averbouh, A. (2000). *Technologie de la matière osseuse travaillée et implications palethnologiques. L'exemple des chaines d'exploitation du bois de cervidé chez les Magdaléniens des Pyrénées*. Ph.D. dissertation, Université Paris I Panthéon-Sorbonne.

Cleyet-Merle, J.-J., & Averbouh, A. (1995). Fiche Hameçons. In H. Camps-Fabrer (Ed.), *Éléments barbelés et apparentés* (pp. 83–89). Treignes, Belgium: CEDARC.

Alaux, J.-F. (1967–1968). Notes préliminaires sur l'abri périgordien des Battuts, commune de Penne (Tarn). *Fédération Tarnaise de Spéléologie-Archéologie, 5*, 37.

Baffier, D., & Julien, M. (1990). L'outillage en os des niveaux châtelperroniens d'Arcy-sur-Cure (Yonne). Mémoires du Musée de Préhistoire d'Île de France, actes du Colloque International de Nemours, 1988. In C. Farizy & J. Combier (Eds.), *Paléolithique Moyen Récent et Paléolithique Supérieur Ancien en Europe 3* (pp. 329–341). Nemours: Ed. APRAIF.

Barth, M., Conard, N., & Münzel, S. (2009). Palaeolithic subsistence and organic technology in the Swabian Jura. In L. Fontana, F.-X. Chauvière, & A. Bridault (Eds.), *In search of total animal exploitation. Case studies from the Upper Palaeolithic and Mesolithic. Proceedings of the XVth UISPP Congress, Session C61, 42, Lisbon, 4-9 September 2006. British Archaeological Reports. International Series 2040* (pp. 5–20). Oxford: John and Erica Hedges.

Blockeley, S. P. E., Lane, C. S., Hardiman, M., Rasmussen, S. O., Seierstad, I. K., Steffensen, J., et al. (2012). Synchronisation of palaeoenvironemental records over the last 60, 000 years, and an extended INTIMATE event stratigraphy to 48, 000 b2k. *Quaternary Science Reviews, 36*, 2–10.

Bordes, F. (1958). Nouvelles fouilles à Laugerie-Haute Est : premiers résultats. *L'Anthropologie, 62*(3, 4), 205–244.

Bricker, H. M. (1995). *Le Paléolithique supérieur de l'abri Pataud (Dordogne): Les fouilles de H.-L. Movius jr. Suivi d'un inventaire analytique des sites Aurignaciens et Périgordiens de Dordogne*. Paris: Maison des Sciences des Hommes, Document d'Archéologie Française 50.

Bricker, H. M., & David, N. C. (1984). *Excavation of the Abri Pataud, Les Eyzies (Dordogne): The Perigordian VI (Level 3) assemblage*. Cambridge, MA: Harvard University-Peabody Museum, Bulletin of American School of Prehistoric Research 34.

Cattelain, P., & Perpère, M. (1993). Tir expérimental de sagaies et de flèches emmanchées de pointes de la Gravette. *Archéo-Situla, 17–20*, 5–28.

Cattelain, P. (1994). La chasse au Paléolithique supérieur : Arc ou propulseur, ou les deux ? *Archéo-Situla, 21–24*, 5–26.

Cattelain, P. (1995). Armatures de projectile en pierre ou en matière dure animale : Un choix lié à l'environnement ? In O. Marcel (Ed.), *Nature et culture. Actes du colloque de Liège, 13-17 décembre 1993* (pp. 181–185). Liège, Belgium: Service de Préhistoire, Université de Liège (ERAUL 68).

Cattelain, P. (1997). Hunting during the Upper Paleolithic: Bow, spear-thrower or both? In H. Knecht (Ed.), *Projectile technology* (pp. 213–240). New York: Plenum Press.

Chauvière, F.-X., & Fontana, L. (2005). Modalités d'exploitation des rennes dans le Protomagdalénien du Blot (Haute-Loire, France). In V. Dujardin (Ed.), *Industrie osseuse et parures du Solutréen au Magdalénien en Europe. Table ronde sur le Paléolithique supérieur récent, Angoulême (Charente, 28-30 mars 2003)* (pp. 137–147). Mémoire de la Société Préhistorique Française 39.

Chauvière, F.-X. (2012). Objets d'ici et d'ailleurs : Les industries sur matières dures d'origine animale du Paléolithique supérieur du Blot (Cerzat, Haute-Loire). *Préhistoire du Sud-Ouest, 20*, 55–77.

Chiotti, L. (2005). *Les industries lithiques aurignaciennes de l'abri Pataud, Dordogne, France: Les fouilles de Hallam L. Movius Jr. British Archaeological Reports International Series 1240*. Oxford: Archaeopress.

Cleyet-Merle, J.-J. (1990). *La Préhistoire de la pêche*. Paris: Errance, Collection des Héspérides.

Clark, J. G. D., & Thompson, M. W. (1953). The groove and splinter technique of working antler in Upper Palaeolithic and Mesolithic Europe. *Proceedings of the Prehistoric Society, XIX*(2), 148–160.

Clottes, J. (1976). Les civilisations du Paléolithique supérieur dans les Pyrénées Françaises. *La Préhistoire Française, 1–2*, 1214–1231.

Christensen, M. (1999). *Technologie de l'ivoire au Paléolithique Supérieur: Caractérisation physico-chimique du matériau et analyse fonctionnelle des outils de transformation. British Archaeological Reports International Series 751*. Oxford: Hadrian Books.

David, N. C. (1985). *Excavation of the Abri Pataud, Les Eyzies (Dordogne): The Noaillian (Level 4) assemblage and the Noaillian*

Culture in Western Europe. Cambridge, MA: Harvard University, Peabody Museum, ASPRB 37.

Enloe, J. G. (1993). Subsistence organization in the early Upper Paleolithic: Reindeer Hunters of the Abri du Flageolet, couche V. In H. Knecht, A. Pike-Tay, & R. White (Eds.), *Before Lascaux: The complex record of the Early Upper Paleolithic* (pp. 101–115). Boca Raton, FL: CRC Press.

Flori, L. (2013). *Exploitation des matières dures d'origine animale au Gravettien récent. Exemple de la couche 3 de l'abri Pataud (Dordogne)*. Paris: Master 2 du Muséum National d'Histoire Naturelle.

Fontaine, A. (2000). Etude d'une ancienne collection du Gravettien, Site des Vachons (Voulgezac, Charente). *Bulletin de la Société Préhistorique Française, 97*(2), 191–198.

Fontaine, A. (2002). *Le site des Vachons (commune de Voulgézac-Charente). Etude typologique de la couche 4*. Paris: Mémoire de Master, Muséum d'Histoire Naturelle.

Foucher, P. (2004). *Les industries lithiques du complexe Gravettien-Solutréen dans les Pyrénées. Techno-typologie et circulation des matières siliceuses de part et d'autre de l'axe Pyrénées-Cantabres*. Thèse de doctorat, l'Université de Toulouse-Le Mirail.

Goutas, N. (1999). *Les foënes et les hameçons droits au Paléolithique supérieur en France*. Paris: Mémoire de Maîtrise, Université de Paris 1.

Goutas, N. (2003). Identification de deux procédés de débitage inédits du bois de Cervidés dans les niveaux Gravettiens de Laugerie-Haute Est et Ouest. *Paléo, 15*, 255–262.

Goutas, N. (2004). *Caractérisation et évolution du Gravettien en France par l'approche techno-économique des industries en matières dures animales (étude de six gisements du Sud-ouest)*. Doctorat de Préhistoire, l'Université de Paris 1 – Panthéon Sorbonne.

Goutas, N. (2008). Les pointes d'Isturitz sont-elles toutes des armes de chasse? *Gallia Préhistoire, 50*, 45–101.

Goutas, N. (2009). Réflexions sur une innovation technique Gravettienne importante: Le double rainurage. *Bulletin de la Société Préhistorique Française, 106*(3), 437–456.

Goutas, N. (2011). Новые данные о восточном граветте : взгляд на костяную индустрию. In G. Khlopatchev (Ed.), *Предметы вооружения и искусства из кости в древних культурах Северной Евразии (технологический и функциональный аспекты)* (pp. 27–47). Замятнинский сборник 2.

Goutas, N. (2013a). Nouvelles données sur l'industrie osseuse du Gravettien des grottes d'Arcy-sur-Cure (Yonne, France): Vers l'identification de nouveaux marqueurs techniques et culturels du Gravettien moyen à burins du Raysse. In P. Bodu, L. Chechmana, L. Klaric, L. Mevel, S. Soriano, & N. Teyssandier (Eds.), *Actes de la table ronde internationale: Le Paléolithique supérieur ancien de l'Europe du Nord-Ouest. Séance de la Société Préhistorique Française* (15-18 avril 2009, musée de Sens, Yonne) (pp. 89–115). Paris: Hors-Série Bulletin de la Société Préhistorique Française.

Goutas, N. (2013b). De Brassempouy à Kostienki : l'Exploitation technique des ressources animales dans l'Europe Gravettienne. In M. Otte (Ed.), *Les Gravettiens* (pp.105–160). Paris: Errance, Actes Sud.

Goutas, N. (2013c). New data on the osseous industry from Eastern Gravettian (Russia): Technological analyses and sociological perspectives, In F. Lang (Ed.), *The sound of bones. 8th Meeting of the Worked Bone Research Group, Salzburg, August 29 (September 3th 2011)* (pp. 133–154). Salzburg, Austria: Archaeoplus.

Goutas, N., & Simonet, A. (2009). Le secteur GG2 de la grotte du Pape à Brassempouy (Landes): Un dépôt intentionnel d'armes Gravettiennes? *Bulletin de la Société Préhistorique Française, 106*(2), 257–291.

Goutas, N., & Lacarrière J. (2013). L'exploitation des cervidés dans le Gravettien d'Isturitz. Une approche archéozoologique et technologique des ressources animales : De leur acquisition à leur utilisation. In C. de la Heras, J. A. Lasheras, A. Arrizabalaga, & M. de al Rasilla (Eds.), *Pensando el Gravetiense: Nuevos datos para la región cantábrica en su contexto peninsular y pirenaico. International colloquium, Musée d'Altamira, Cantabrie, Espagne (20–22 october 2011)* (pp. 565–592). Santillana del Mar, Spain: Associacíon de amigos de museo de Altamira.

Goutas, N., & Tejero, J.-M. (2016). Osseous technology as a reflection of chronological, economic and sociological aspects of Paleolithic hunter-gatherers: Examples from key Aurignacian and Gravettian sites in South West Europe. *Quaternary International, 403*, 79–89.

Gregoriani, S. (1996). *L'industrie osseuse et la parure des niveaux de l'abri Pataud (Les Eyzies-de-Tayac, Dordogne). Etude micromorphologique des surfaces*. Paris: Mémoire de Master 2, Muséum National d'Histoire Naturelle.

Guillermin, P. (2004). *Réflexions sur l'interprétation des industries Gravettiennes à partir de l'étude typo-technologique d'une occupation spécialisée: La couche E du gisement des Fieux (Miers, Lot)*. Mémoire de Master 2, Université de Toulouse-Le Mirail.

Gvozdover, M. (1995). *Art of the mammoth hunters. The finds from Avdeevo*. Oxford, England: Oxbow Monograph 49.

Kildéa, F., & Lang, L. (2011). Le Gravettien de la vallée du Cher: Le site de la Croix de Bagneux à Mareuil-sur-Cher. In N. Goutas, L. Klaric, D. Pesesse, & P. Guillermin (Eds.), *À la recherche des identités gravettiennes: actualités, questionnements et perspectives. Actes de la table ronde internationale du 6-8 octobre 2008, Aix-en-Provence* (pp. 273–289). Paris: Mémoire LII de la Société Préhistorique Française.

Klaric, L. (2003). L'unité technique des industries à burins du Raysse dans leur contexte diachronique. Réflexions sur la diversité culturelle au Gravettien à partir des données de la Picardie, d'Arcy-sur-Cure, de Brassempouy et du Cirque de la Patrie. Thèse de doctorat de Préhistoire, Université de Paris 1.

Klaric, L. (2010). Le Gravettien. In J. Clottes (Ed.), *La France préhistorique* (pp. 142–169). Paris: Éditions Gallimard.

Klima, B. (1987). Paleolitická Parohová industrie z Pavlova. *Památky archeologiké, LXXVIII*, 289–370.

Knecht, H. (1991a). *Technological innovation and design during the Early Upper Paleolithic: A study of organic projectile technologies*. PhD dissertation, New York University.

Knecht, H. (1991a). The role of innovation in changing Early Upper Paleolithic organic projectile technologies. *Technique et Culture, 17–18*, 115–144.

Knecht, H. (1993). Early Upper Paleolithic approaches to bone and antler projectile technology. In G. Larsen Peterkin, H. M. Bricker, & P. Mellars (Eds.), *Hunting and animal exploitation in the Later Paleolithic and Mesolithic of Eurasia* (pp. 33–47). Washington, DC: Archeological Papers of the American Anthropological Association 4.

Knecht, H. (1997). *Projectile technology*. New York: Plenum Press.

Knecht, H. (2000). Design strategy of early Upper Paleolithic bone and antler projectile technologies. In C. Bellier, P. Cattelain, & M. Otte (Eds.), *La chasse dans la Préhistoire. Hunting in Prehistory. Actes du Colloque International de Treignes 1990, Anthropologie et Prehistoire 111, ERAUL 51* (pp. 26–38). Liège: CEDARC.

Kozlowski, S. K., & Kozlowski, J. K. (1977). Pointes, sagaies et harpons du Paléolithique et du Mésolithique en Europe Centre-Est. In H. Camps-Fabrer (Ed.), *Méthodologie appliquée à l'industrie de l'os préhistorique, actes du 2ème colloque international sur l'industrie de l'os dans la préhistoire. Abbaye de Sénanque, 9–12 juin 1976* (pp. 205–227). Paris: Colloques Internationaux du CNRS 568.

Lacarrière J., Goutas, N., Normand, C., & Simonet, A. (2011). Vers une redéfinition des occupations Gravettiennes de la grotte d'Isturitz (Pyrénées-Atlantiques, France): Révision critique des collections "anciennes" par l'approche intégrée des données lithiques, fauniques et de l'industrie osseuse. In N. Goutas, L. Klaric, D. Pesesse, & P. Guillermin (Eds.), *À la recherche des identités Gravettiennes: Actualités, questionnements et perspectives. Actes de la table ronde internationale du 6–8 octobre 2008,*

Aix-en-Provence (pp. 67–83). Paris: Mémoire LII de la Société Préhistorique Française.

Langley, M. C., Pétillon, J.-M., & Christensen, M. (2016). Diversity and evolution of osseous hunting equipment during the Magdalenian (21,000–14,000 cal. BP). In M. C. Langley (Ed.), *Osseous projectile weaponry: Towards an understanding of Pleistocene cultural variability* (pp. 143–159). Dordrecht: Springer.

Liolios, D. (1999). *Variabilité et caractéristique du travail des matières osseuses au début de l'Aurignacien, Approche technologique et économique*. Thèse de doctorat, Université de Paris X

Normand C., Goutas, N., Lacarrière, J., & Simonet, A. (2013). El Gravetiense de la cueva de Isturitz : Nuevas investigaciones, nuevos datos. In C. de las Heras, J. A. Lasheras, A. Arrizabalaga, M. de la Rasilla (Eds), *Pensando el Gravetiense: nuevos datos para la región cantábrica en su contexto peninsular y pirenaico. International colloquium, Musée d'Altamira, Cantabrie, Espagne (20–22 October 2011)* (pp. 161–183). Santillana del Mar, Spain: Associacíon de amigos de museo de Altamira.

O'Farell, M. (1996). *Approche technologique et fonctionnelle des pointes de la Gravette, une analyse archéologique et expérimentale appliquée à la collection de Corbiac (Dordogne)*. Mémoire de Master 2, Université de Bordeaux I.

O'Farell, M. (2004). Les pointes de la Gravette de Corbiac (Dordogne) et considérations sur la chasse au Paléolithique supérieur ancien. In P. Bodu, & C. Clude (Eds.), *XXVe Congrès préhistorique de France, Nanterre 24-26 novembre 2000, Approches fonctionnelles en Préhistoire* (pp. 121–128). Paris: Société Préhistorique Française.

Otte, M. (1981). *Le Gravettien en Europe Centrale*. Dissertationes Archaeologicae Gandenses, Bruges, De Tempel.

Pesesse, D. (2008). *Les premières sociétés Gravettiennes. Analyse comparée des systèmes lithiques de la fin de l'Aurignacien aux débuts du Gravettien*. Mémoire de doctorat, Université Aix-Marseille.

Pesesse, D. (2013). Le Gravettien existe-t-il ? Le prisme du système technique lithique. In M. Otte (Ed.), *Les Gravettiens* (pp. 65–102). Paris: Errance, Actes Sud.

Pelegrin, J. (2000). Les techniques de débitage laminaire au Tardiglaciaire: Critères de diagnose et quelques réflexions. In B. Valentin, P. Bodu, & M. Christensen (Eds.), *l'Europe centrale et septentrionale au Tardiglaciaire. Table ronde de Nemours, 13-16 mai 1997* (pp. 73–86). Mémoires du musée de Préhistoire d'Île-de-France, 7.

Pétillon, J.-M. (1999). *Les pointes à base fourchue Magdaléniennes. Problèmes typologiques, techniques et fonctionnels*. Mémoire de Master 2, Université de Paris 1.

Pétillon, J.-M. (2006). *Des Magdaléniens en armes. Technologie des armatures de projectile en bois de cervidé du Magdalénien supérieur de la grotte d'Isturitz (Pyrénées-Atlantiques)*. Paris: CEDARC.

Peyrony, D. (1932). *Les Gisements préhistoriques de Bourdeilles (Dordogne)*. Archives de l'Institut de Paléontologie Humaine 10. Paris: Masson.

Peyrony, D. (1936). Le Périgordien et l'Aurignacien (Nouvelles observations). *Bulletin de la Société Préhistorique Française, 33*(11), 616–619.

Peyrony, D. (1939). Fouilles à la Roque-Saint-Christophe, commune de Peyzac-le-Moustier (Dordogne). *Bulletin de la Société Historique et Archéologique du Périgord, 3–4*, 248–269, 360–387.

Peyrony, D., & Peyrony, E. (1938). *Laugerie-Haute près des Eyzies (Dordogne)*. Archives de l'institut de paléontologie humaine 19. Paris: Masson.

Peyrony, E. (1934). Le gisement de La Forêt, commune de Tursac (Dordogne). XIe session du Congrès Préhistorique de France, Périgueux. *Bulletin de la Société Préhistorique Française, 31*, 424–430.

Pesesse, D. (2003). Approche du comportement technique au Gravettien: L'industrie lithique de l'unité OP10 de la Vigne Brun. Mémoire de D.E.A., Université de Provence, Aix-en-provence.

Pike-Tay, A. (1993). Hunting in the Upper Perigordian: A matter of strategy or expedience? In H. Knecht, A. Pike-Tay, & R. White (Eds.), *Before Lascaux: The complex record of the Early Upper Paleolithic* (pp. 85–99). Boca Raton, FL: CRC Press.

Pokines, J., & Krupa, M. (1997). Self-barbed antler spearpoints and evidence of fishing in the Late Upper Paleolithic of Cantabrian Spain. In H. Knecht (Ed.), *Projectile technology* (pp. 241–262). New York: Plenum Press.

San Juan-Foucher, C., & Vercoutère, C. (2005). Les "sagaies d'Isturitz" des niveaux gravettiens de Gargas (Hautes-Pyrénées) et de Pataud (Dordogne). Un exemple d'approche pluridisciplinaire et complémentaire de l'industrie osseuse. *Préhistoire Anthropologie Méditerranéennes, 12*, 75–94.

San Juan-Foucher, C. (2005). Aiguilles, sagaies et pendeloques: l'industrie solutréenne sur matière dure animale de l'abri des Harpons (Lespugue, Haute-Garonne). In V. Dujardin (Ed.), *Industrie osseuse et parures du Solutréen au Magdalénien en Europe, Actes du Colloque international d'Angoulême, 28–30 mars 2003* (pp. 161–176). Paris: Mémoire de la Société Préhistorique Française, 39.

Saint-Périer R., & Saint-Périer, S. (1952). *La grotte d'Isturitz III: Les Solutréens, les Aurignaciens et les Moustériens*. Archives de l'Institut de Paléontologie Humaine 25. Paris: Masson.

Sanchez-Goñi, M. F., & Harrison, S. P. (2010). Millennial-scale variability and vegetation changes during the last glacial: Concepts and terminology. *Quaternary Science Reviews, 29*(21–22), 2823–2827.

Simonet, A. (2005). Les armatures lithiques. Méthodes d'étude et enjeux : l'exemple des armatures gravettiennes d'Isturitz, Mémoire de DEA de l'Université de Toulouse II – Le Mirail.

Simonet, A. (2009). Les Gravettiens des Pyrénées. Des armes aux sociétés. *Munibe, 60*, 81–98.

Simonet, A. (2010). *Typologie des armatures lithiques Gravettiennes de la grotte d'Isturitz (Pyrénées-Atlantiques, France)*. British Archaeological Reports. *British Archaeological Reports International Series 2156*. Oxford: Archaeopress.

Simonet, A. (2011). La pointe des Vachons: Nouvelles approches d'un fossile directeur controversé du Gravettien à partir des exemplaires du niveau IV de la grotte d'Isturitz (Pyrénées-Atlantiques, France) et des niveaux 4 des abris 1 et 2 des Vachons (Charente, France). *Paleo, 22*, 271–298.

Sonneville-Bordes, D. (1960). *Le Paléolithique Supérieur en Périgord*. Bordeaux, France: Imprimere Delmas, CNRS.

Sonneville-Bordes, D. (1971). Un fossile directeur osseux du Périgordien à burins de Noailles. *Bulletin de la Société Préhistorique Française, 68*(2), 44–45.

Sonneville-Bordes, D. (1972). A propos des pointes osseuses à extrémité striée du Périgordien à burins de Noailles. *Bulletin de la Société Préhistorique Française, 69*(2), 37–38.

Soriano, S. (1995). *Analyse conceptuelle et fonctionnelle des pointes à dos microlithiques au Périgordien supérieur. Application à la série de Rabier (Lanquais, Dordogne)*. Mémoire de Maîtrise, Université de Paris X.

Stordeur D. (1987). Manches et emmanchements préhistoriques: quelques propositions préliminaires. In D. Stordeur (Ed.), *La Main et l'Outil. Manches et emmanchements préhistoriques. Table Ronde C.N.R.S. tenue à Lyon du 26 au 29 Novembre 1984* (pp. 11–34). Lyon, France: Travaux de la Maison de l'Orient.

Tartar, E. (2009). De l'os à l'outil. Caractérisation technique, économique et sociale de l'utilisation de l'os à l'Aurignacien ancien Étude de trois sites : l'Abri Castanet (secteurs nord et sud), Brassempouy (Grotte des Hyènes et Abri Dubalen) et Gatzarria, Thèse de doctorat, Université Paris I, Paris.

Tartar, E., & White, R. (2013). The manufacture of Aurignacian split-based points: An experimental challenge. *Journal of Archaeological Science, 40*(6), 2723–2745.

Tejero, J. M. (2013). *La explotación de las materias óseas en el Auriñaciense. Caracterización tecnoeconómica de las producciones del Paleolítico superior inicial en la Península Ibérica. British Archaeological Reports International Series 2469*. Oxford: Archaeopress.

Tejero, J.-M. (2016). Spanish Aurignacian projectile points: An example of the first European palaeolithic hunting weapons in osseous materials. In M. C. Langley (Ed.), *Osseous projectile weaponry: Towards an understanding of Pleistocene cultural variability* (pp. 55–69). Dordrecht: Springer.

Tejero, J. M., Christensen, M., & Bodu, P. (2012). Red Deer Antler technology and early Modern Humans in Southeast Europe: An experimental study. *Journal of Archaeological Science, 39*(2), 332–346.

Teyssandier, N., Bon, F., & Bordes, J.-G. (2010). Within projectile range: Some thoughts on the appearance of the Aurignacian in Europe. *Journal of Anthropological Research, 66*, 209–229.

Tixier, J. (2005). Torsitude à Ksar 'Aquil (Liban). In F. Lebrun Ricalens, J.-G. Bordes, & F. Bon (Eds.), *Productions lamellaires attribuées à l'Aurignacien, chaînes opératoires et perspectives technoculturelles. Actes du XIVe congrès de l'Union internationale des sciences pré et protohistoriques, Université de Liège, 2-8 Septembre 2001* (pp. 19–21). Luxembourg: Archéologiques 1, MNHA.

Valentin, B. (2006). *De l'Oise à la Vienne, en passant par le Jourdain. Jalons pour une Paléohistoire des derniers chasseurs*. Mémoire de Thèse d'Habilitation à Diriger des Recherches, Université de Paris 1.

Vercoutère, C. (2004). Exploitation de l'animal comme ressource de matières premières non alimentaires: Industrie osseuse et parure. Exemple de l'abri Pataud (Dordogne, France). Thèse du Muséum d'Histoire Naturelle.

Wolf, S., Münzel, S., Dotzel, K., Barth, M. M., & Conard, N. J. (2016). Projectile weaponry from the Aurignacian to the Gravettian of the Swabian Jura (Southwest Germany): Raw materials, manufacturing and typology. In M. C. Langley (Ed.), *Osseous projectile weaponry: Towards an understanding of Pleistocene cultural variability* (pp. 71–87). Dordrecht: Springer.

Chapter 8
Upper Paleolithic Bone and Antler Projectiles in the Spanish Mediterranean Region: The Magdalenian Period

Valentín Villaverde, J. Emili Aura Tortosa, María Borao, and Didac Roman

Abstract We have focused our study on the projectile points of Cova de Parpalló. The Magdalenian sequence in this archaeological site is one of the most complete in the Upper Paleolithic from the southwest of Europe. We have analyzed 334 pieces from an assemblage that includes well over 2000 finished objects, and consider these weapons as a representative sample from two well differentiated Magdalenian phases. The first period studied dates to the Badegoulian (layer 2.40–2.20 m), while the second is Upper Magdalenian (layer 0.80–1.00 m). The main aspects of this analysis are typological evaluation (hafting kinds or bases, sections, morphometry) and fracture patterns (position and kinds of fractures, distinguishing between use, post-depositional or recent fractures). These two assemblages are compared before the Upper Magdalenian of Parpalló is contextualized within the rest of the Upper Magdalenian from Mediterranean Iberian Peninsula. The assemblages from Cova de les Cendres and Cueva de Nerja lead us to establish a relationship between the studied sites and create a general vision of the geographic articulation of these cultures, including an evaluation of harpoons and gorges from these archaeological sites.

Keywords MIS2 • Iberia • Cova del Parpalló • Osseous projectile points • Badegoulian • Magdalenian

V. Villaverde (✉) • J.E. Aura Tortosa • M. Borao
Departament de Prehistòria i Arqueologia, Universitat de València, València, Spain
e-mail: Valentin.Villaverde@uv.es; Emilio.Aura@uv.es; M.Isabel.Borao@uv.es

D. Roman
SERP (Seminari d'Estudis i Recerques Prehistòriques). Secció de Prehistòria i Arqueologia, Universitat de Barcelona, València, Spain

Université de Toulouse-Le Mirail, TRACES-UMR 5608, Toulouse, France
e-mail: didacroman@ub.edu

Introduction

Iberian Peninsula is part of the Western Upper Paleolithic territory. Connections between Southern Europe and North Africa have been a recurring topic throughout the last 100 years, but lack of sufficient data makes it hard to assess those connections in depth.

There are many sites in the Spanish Mediterranean area with evidence for Magdalenian occupation, but few of them have yielded collections with abundant bone and antler weapons (Aura Tortosa 1995; Villaverde et al. 2012). Bora Gran d'en Carreras (Girona) and Cova del Parpalló (Gandia, Valencia) have the largest assemblages when compared with other sites. In the former, though, the number of pieces is only reaches almost 500 (Rueda Torres 1987), while there are well over 2000 pieces from Parpalló (Pericot 1942). Other sites with assemblages of about 100 pieces include Cova del Parco (Mangado et al. 2010), Cova Matutano (Olària 1999), Cova dels Blaus (Casabó 2012), Cova de les Cendres (Villaverde et al. 2012) and Cueva de Nerja (Aura Tortosa et al. 2012). Finally, other collections that have only a few pieces are Tossal de la Roca (Cacho and de la Torre 2005), Pirulejo (Cortés Sánchez 2008), Hoyo de la Mina (Ferrer Palma et al. 2006) and Volcán del Faro (Aparicio Pérez 2003).

Most of the papers published so far are typological studies. They hardly mention technological aspects of production, morphometric detail or assessment of fragmentation types and their causes. Only Tejero's (2005) work on the osseous industry at Cova del Parco includes a technological study. However, there is no doubt about the importance of Parpalló's weapon collection made of hard animal materials as it allows an evaluation of the evolution of the technology and typology of bone and antler points. Moreover, Parpalló is a benchmark for assessing the synchronization of Mediterranean Magdalenian evolution in relation to other southern European regional groups, particularly Cantabria and the south of France.

Michelle C. Langley (ed.), *Osseous Projectile Weaponry*, Vertebrate Paleobiology and Paleoanthropology,
DOI 10.1007/978-94-024-0899-7_8

The known importance of osseous implements, in fact, would not be the same without Bora Gran and Parpalló collections. This leads us to consider how appropriate it is to apply models which include marked regional differentiation by comparing the Spanish Mediterranean region to the rest of the Southwest European Magdalenian.

The study of Parpalló's collection is currently in progress. It is not possible to present quantitative data for all the Magdalenian layers, so in this work, we only deal with materials from two layers that correspond to two well differentiated phases of the period. We can count on large enough assemblages for both levels to offer insight into the variation perceived between two well differentiated periods from a sequential perspective.

One of these layers, the Upper Magdalenian, offers the possibility of comparing its data with data obtained from Cova de les Cendres (Villaverde et al. 2010; Borao 2012, 2013) and Cueva de Nerja (Aura Tortosa et al. 2002, 2010). This comparison can provide an assessment of the degree of uniformity and variation among osseous weaponry from the same period. This circumstance is particularly relevant when we observe the differences apparent in these two sites in relation to Parpalló concerning the role of harpoons and gorges (small bipoints).

Before we deal with the chronology and cultural frame of Parpalló's materials, it is important to point out that the site was excavated between the years 1929 and 1931. The excavation system was carried out by cutting layers of different potency, usually of 20 cm or higher in the different areas in which the cave was divided. The assemblages dealt with in the present study correspond to layer 0.80–1 m and layer 2.20–2.40 m. To these materials we have added some more from the Talud sector (the last area excavated but by natural layers) found at similar depth (Pericot 1942; Aura Tortosa 1995). The total number of pieces analyzed is 173 for the first layer and 161 for the second.

Chronological and Cultural Framework

Pericot (1942) established a close relationship between Parpalló's osseous industry evolution and the four earliest phases of the Magdalenian sequence of southwest France (Breuil 1912). He based his study on changes observed in the type of projectile and on decorative themes. Later studies have built on this work: Fullola (1979) and Aura Tortosa (1995) from the point of view of lithic industries, and Villaverde (1994) in the study of mobiliary art. As a result of these studies, which observed that lithic points were replaced at the end of the Late Glacial Maximum (LGM) by osseous points, the question of the identification of Badegoulian features in Early Magdalenian lithic industries has been raised.

Two questions have remained open in research carried out in recent years. The first one concerns Parpalló's relationship with other regions owing to the coincidences observed between the distribution of Badegoulian lithic industries and the "Solutrean territory". The second question is its chronology. Radiocarbon dating obtained in the late twentieth century raised an issue that has not yet been solved: synchronicity between Iberian Solutrean and French Badegoulian (Aura Tortosa 2007). Therefore, the duration of the Solutrean must be revised (Aura Tortosa and Jordá Pardo 2012). Recent dating of samples from Parpalló and Cendres—the only site with Early Magdalenian in the region (Villaverde et al. 2012)—suggests that the pace of developments in the Mediterranean region, the Ebro Valley and the Cantabrian region could be similar and close to that of southern France (Aura Tortosa et al. 2012).

Two layers from Parpalló that concern this topic have been selected for the present chapter. The first one (2.40–2.20 m) corresponds to the central core stratigraphic layers of Badegulian B Parpalló type (Aura Tortosa 1995). It presents a lithic assemblage of short and wide flakes and blades for implements that show continuous refitting. The most common implements are scrapers, notch-denticulate pieces and *raclettes*, and they have been associated to an important stage in the fabrication of osseous projectiles that were made earlier than the harpons (Aura Tortosa 2007). Owing to deficient recovering techniques, the microlaminar implements of this industry are poorly represented in this layer; but there are some types of carenated and dorsal front cores (Ducasse and Langlais 2007) that make it possible to identify microlaminar production of very small dimensions. There is no direct radiocarbon dating for this layer, but it could be in the vicinity of ca. 17,000–16,000 BP, before the earliest dates for the Early Magdalenian obtained from Cendres (Villaverde et al. 2012).

The second layer (1.00–0.80 m) coincides with the beginning of the Upper Magdalenian industries that has elements which clearly relates it to the rest of southwest Europe (Langlais 2010). Laminar and microlaminar productions are the basic component of lithic industries in this layer. There are high percentages of microlaminar implements (30–60 %) with important quantities of scrapers and burins. As in the first layer, we do not have direct dating here, but for layer 1.70–1.50 m it has been established (13,976 ± 300 BP (Birm 519), Bofinger and Davidson 1977). This time range is consistent with documentation from the region. From ca. 13,500 BP, these lithic types are associated with harpoons and to technological traits described as Upper Magdalenian (Aura Tortosa 1995; Romàn and Villaverde 2011).

Badegoulian Assemblage (Layer 2.40–2.20 m)

Projectile points prevail with 152 pieces in total, with three awls, one hammer and one pressure flaker added to these pieces.

For raw material, red deer antler (*Cervus elaphus*) clearly dominates over bone. Only nine pieces are made of bone, and three of these are fragments with no base, three are double points and one is a flat point. Altogether, bone pieces amount to 5.8 % of the total. Pieces from this layer present a high degree of fragmentation, often due, as evidenced by the observation of fractures, to postdepositional and modern causes. This fracturation is a factor that must be taken into account as some fragments might be joined and the total number of pieces will be reduced and change the degree of sequential integrity of the layers.

In antler, 18 pieces are complete or almost complete (12.94 %). There are 40 mesial fragments, 51 mesial fragments with part of their base and 13 base fragments. Most points in this level are very fragmented (91.45 % of pieces). Bases are more common than distal parts. The number of mesial distal fragments, when added together, is roughly the same as the number of proximal fragments.

It must be said that the distinction between mesial distal fragments, on the one hand, and mesial proximal on the other, is particularly complicated for bipoints. Therefore, lacking morphometric criteria so far, we prefer not to consider the implications that might result from this distribution. Nevertheless it must be pointed out that there is a high proportion of proximal parts that correspond to beveled points that, without doubt, had a base (Table 8.1).

In this section, we only take into account preserved fractured pieces, so any further assessment will have to wait until the three possible causes of fracture (use, postdepositional processes and recent breakage) are analyzed.

The number of fractures counted for the 152 points is 269. Out of these, 108 are use fractures (40.14 %), 96 are old postdepositional fractures (35.68 %), and in 65 cases, recent fracture where straight and fresh fracture planes can be observed are found (24.16 %).

The 108 use fractures have been classified according to their morphology. The following types have been identified: crushing, *languette* (bevel, *en marche* and *charnière*), splinter and split (Arndt and Newcomer 1986; Bertrand 1999; Pétillon 2000, 2002, 2005; Tejero 2005; Pétillon 2006; Tejero et al. 2013). Each type of fracture takes place in relation to the impact angle with the obstacle, the force and the haft (Pétillon 2000).

Taking into account the part of the piece analyzed, we have gathered the following data:

- Distal parts: three splinter fractures, 12 *languette* fractures and eight crushing fractures in the tip area; and in the proximal area, three splinter fractures and three *languette* fractures.
- Medial parts: one splinter fracture and nine *languette* fractures in the distal area; and 11 splinter fractures and five *languette* fractures on the proximal part.
- Proximal parts, or bases: 18 splinter fractures and 23 *languette* fractures; and three splinter fractures, four *languette*, 15 crushing fractures and one longitudinal split on the proximal part.

Crushing fractures are more common on the tip and proximal areas (Fig. 8.1: 1, 4). This makes sense for pieces that were used as projectiles. There is a high proportion of *languette* fractures on the distal and proximal parts of the pieces (Fig. 8.1: 2, 3, 5), and splinter ones are common on the area where the shaft meets the proximal part (Fig. 8.1: 7).

The disparity observed when comparing the three categories established (distal, mesial, and bases), suggests fractures associated to hunting episodes with different processes for the transportation of the prey. In any case, the high proportion of base fragments found indicates that weapons were probably repaired; broken parts of the base held to the haft being preserved and discarded at the site.

Where typology is concerned, we have found the following variations in the base form: 63 pieces have a simple bevel (41.4 %)—out of these 17 correspond to flat points—, 23 pieces have a pointed base, and 10 a simple base. From these data we can infer that the beveled hafting system prevailed. As for their size, 27 bevels are longer than a third of the total length of the piece; this represents 42.8 %. As for their longitudinal shape, 17 are concave, one convex and 37 straight. The bevel end is pointed in ten cases, simple in seven and truncated in one (Tables 8.2, 8.3 and 8.4).

The decorative patterns on the haft vary: three pieces have transversal line motifs, 13 present motifs made of longitudinal grooved lines, five have motifs distributed in series oriented crosswise, seven have cross motifs and one piece has a short and shallow groove.

The points, as a whole, present the following type of section: 84 circular, 12 oval, seven square, one triangular, 25 flat-oval, 10 flat-rectangular and one flat-convex. This shows that thicker shapes (round or angular) prevail (74.3 %) over flatter shapes, while round sections (87.1 %) also prevail over angular ones.

Ten points and an awl are either complete or can be reconstructed to their original size. The smallest pieces are two points with beveled bases which are approximately 4 cm long. There are three other beveled base points whose length is between 5 and 8 cm, and two more—one beveled and one bipoint—that are about 9 cm in length. The longest point measures 12.5 cm. All this indicates an important morphometric variation in beveled points (Fig. 8.2). Most pieces measure between 0.5 and 1 cm, though 18 are wider

Table 8.1 Preserved parts of the projectile points

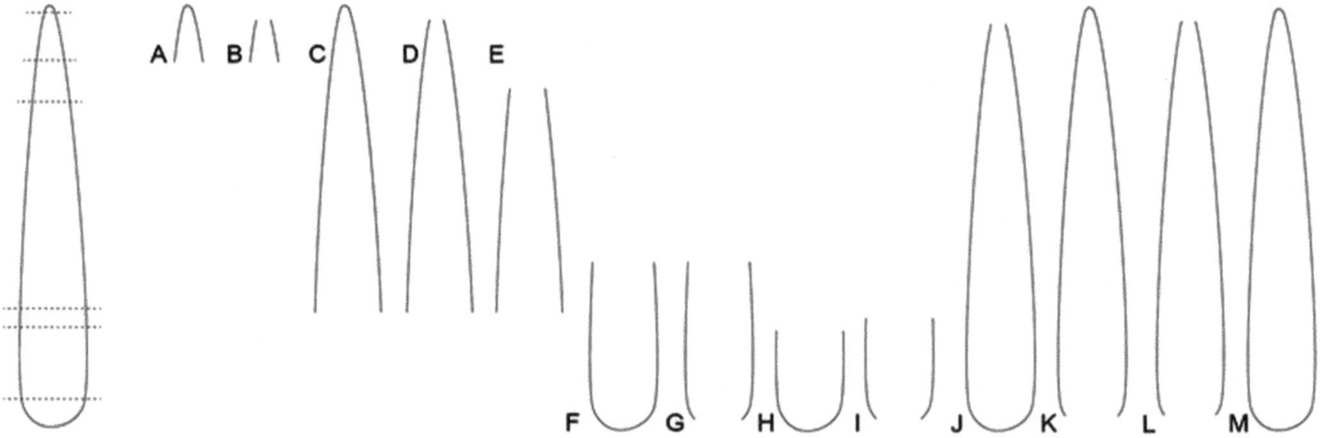

	Without base	Simple bevel	Pointed base	Simple base	Long point	Total
A	2					2
B	7				1	8
C	1					1
D	7		4		4	15
E	33	1	4			38
F		2	3		1	6
G		41	8	2		51
H		4		1		5
I		6	4	3		13
J		2		1		3
K				1		1
L		7		1		8
M				1		1
N						
	50	63	23	10	6	152

Parpalló (layer 2.20–2.40 m). Regarding piece fragmentation, the following possibilities have been considered: distal part that corresponds to the active part of the point (A); distal fragment with slight fracture of the tip (B); distal and mesial fragment (C); distal and mesial fragment with slight fracture of the tip (D); mesial fragment (E); mesial and base fragment (F); medial and base fragment with slight fracture at the tip (G); fragment of base (H); fragment of base with a slight fracture at the tip (I); piece with a slight distal fracture or at the point (J); piece with a slight fracture on the base (K); piece with slight distal and proximal fractures (L); complete piece (M); and finally, longitudinal fragment (N)

than this measurement (Fig. 8.3a). Finally, their thickness ranges between 0.40 and 0.79 cm (Fig. 8.3b).

Assessment of the bevel sizes is again conditioned by the few pieces that are complete; only eight. As a whole, size distribution is quite similar to that of the Upper Magdalenian layer, though in this Badegoulian layer there are no bevels smaller than 2 cm and there is a piece that is longer than 4 cm (Fig. 8.3c).

Upper Magdalenian Assemblage (0.80–1.00 m)

From a typological point of view, the assemblage is clearly dominated by projectile points (groups I, IV, VII and VIII of Barandiarán's [1967] typology), followed by a few awls, and needles or other materials which are scarce. The total number of points is 167. Most of them belong to group I: normal

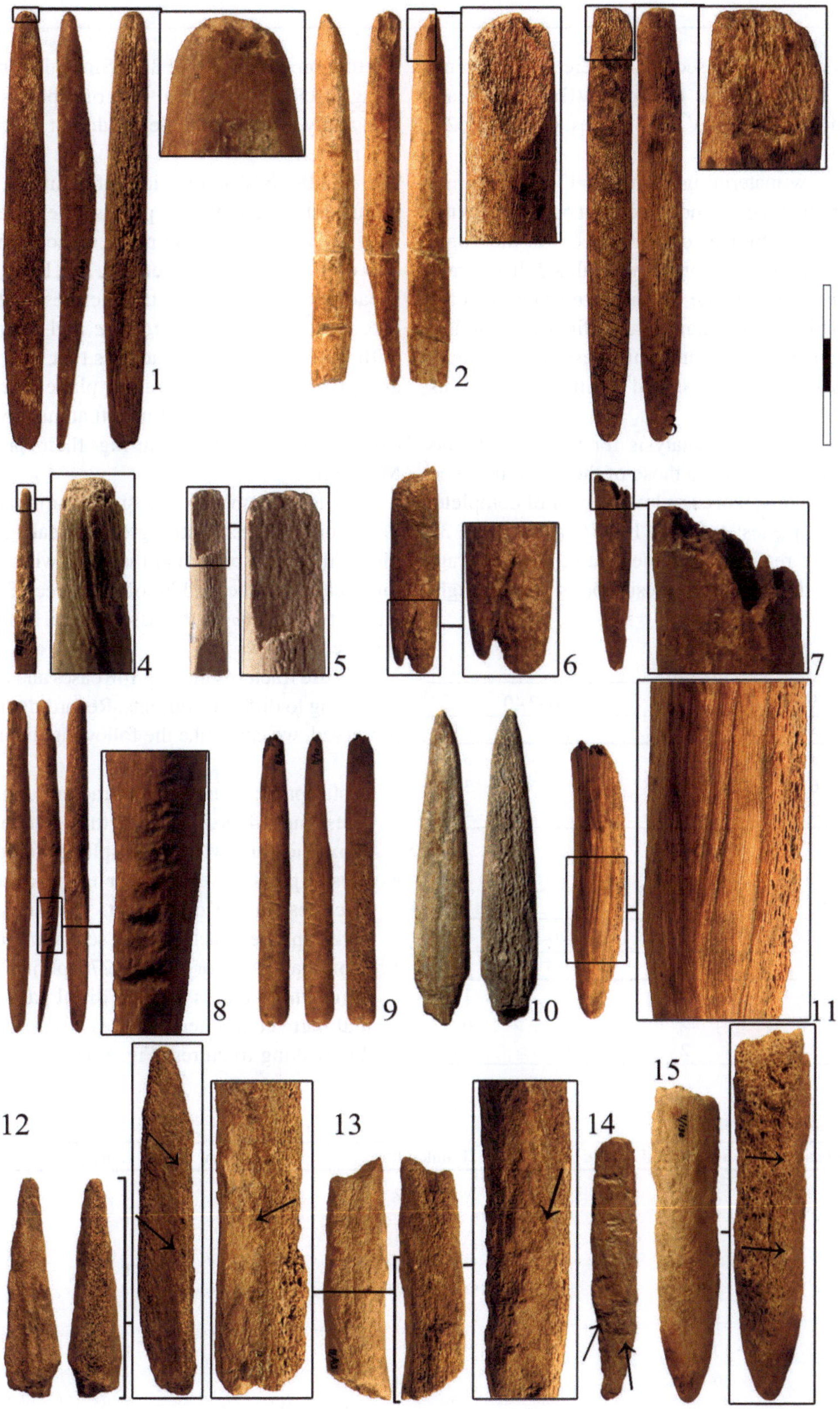

Fig. 8.1 Parpalló, layer 2.20–2.40 m: (*1*) Simple bevel point with a crushing fracture at the tip. (*2*) Simple bevel point with a *languette* fracture of bevel kind. (*3*) Double point with a *languette* fracture of *marche* kind. (*4*) Simple bevel point resharpened with a crushing fracture at the tip. (*5*) Simple bevel point with a *languette* fracture of *charnière* kind. (*6*) Simple bevel point with a split fracture at the base. (*7*) Simple bevel point with a saw teeth fracture at the mesial part. (*8*) Simple bevel point with traces of removal by direct percussion, maybe used for hafting. (*9*) Simple base point reworked from a simple bevel point. The bevel was preserved conforming the distal part. (*10*) Point *à base raccourcie*. (*11*) Long point which preserves double grooving traces and has been resharpened. (*12, 13*) Blanks obtained by fracturation. We can observe in the enlarged pictures the fracture planes. (*14*) Blank with flake morphology with traces of removal by direct percussion. (*15*) Simple base point which preserves fracture planes

dimension points with non-flat sections, whereas, long points belong to group IV, flat points to group VII and rods are included in group VIII. Group IV has 4 pieces, group VII, 25, and group VIII, 3.

Regarding the raw material used, out of 174 pieces only 11 (6.32 %) are made of bone, and if we limit ourselves to the study of weapons, the number of bone pieces is then only six: a distal fragment of a flat point; a mesial and distal fragment with a slight fracture at the tip with two grooves on the point; three pointed mesial fragments and, finally, a longitudinal pointed fragment. As for their thickness, all the pieces are rather small and could be classified as simple base points or bipoints.

The preservation/breakage analysis for these points has yielded results that are similar to those of the previous layer. As Fig. 8.5 shows clearly, there is a low number of complete, or almost complete, pieces; only 23 (13.77 %). There are 35 mesial pieces which retain part of their base, 34 base fragments and, finally, 11 distal and mesial parts with a slight fracture on the tip (Table 8.5). Since the limitations in identifying distal and proximal part of bipoints mentioned above also apply to this layer, we would not proceed any further on this matter.

The number of fractures identified in the 167 points is 311, taking into account that pieces often present fractures on their proximal and distal parts. Concerning the characteristics of the observed fractures, we have established three broad groups: pieces with fractures that were caused by use, with their usual *languette* and splinter types; pieces with post-depositional fractures that have no relation to use and present the usual straight plane fracture; and, finally, pieces with recent fractures that happened either in the excavation process or during their preservation in the Museum.

The total number of post-depositional and recent fractures numbers 162. The degree of breakage varies. There are pieces that have broken at the ends, whereas others are fractured closer to the middle of the piece. The recent fractures amount to 27.1 % of the assemblage, whereas use fractures represent 47.7 %.

The use fractures include 148 cases and have been classified according to different variants. Regarding the part of the piece preserved, we can make the following distinctions:

- Distal parts: four splinter fractures, nine *languette* fractures and 14 crushing fractures on the area of the tip. Proximal parts with seven splinter fractures.
- Mesial parts: seven splinter fractures and nine *languette* fractures on the distal part. And on the proximal part, seven splinter fractures and seven *languette* fractures.
- Proximal parts or bases: 27 splinter fractures and 14 *languette* fractures on their distal part. And on the proximal part six splinter fractures, 10 *languette* fractures and 27 crushing fractures (Fig. 8.4).

Table 8.2 Bevel proportion

Bevel	0.80–1.00		2.20–2.40	
	>1/3	<1/3	>1/3	<1/3
Pointed	3	16	7	3
Rounded	6	8	4	3
Truncated	–	–	1	–

Table 8.3 Bevel ends

	0.80–1.00		2.20–2.40	
	>1/3	<1/3	>1/3	<1/3
Straight	17	36	18	19
Concave	7	7	11	6
Convex		2	1	–

Table 8.4 Bevel longitudinal shape

	Without base	Simple and double bevel	Pointed base	Simple base	Long point	Rod	Total
A	1						1
B	14						14
C	1						1
D	6	3	1		1		11
E	33	3				1	37
F		1	3				4
G	1	17	9	5	1	2	35
H		6		1			7
I		27	2	4	1		34
J				1			1
K		1		1			2
L		13	3	2			18
M		1					1
N	1						1
	57	72	18	14	3	3	167

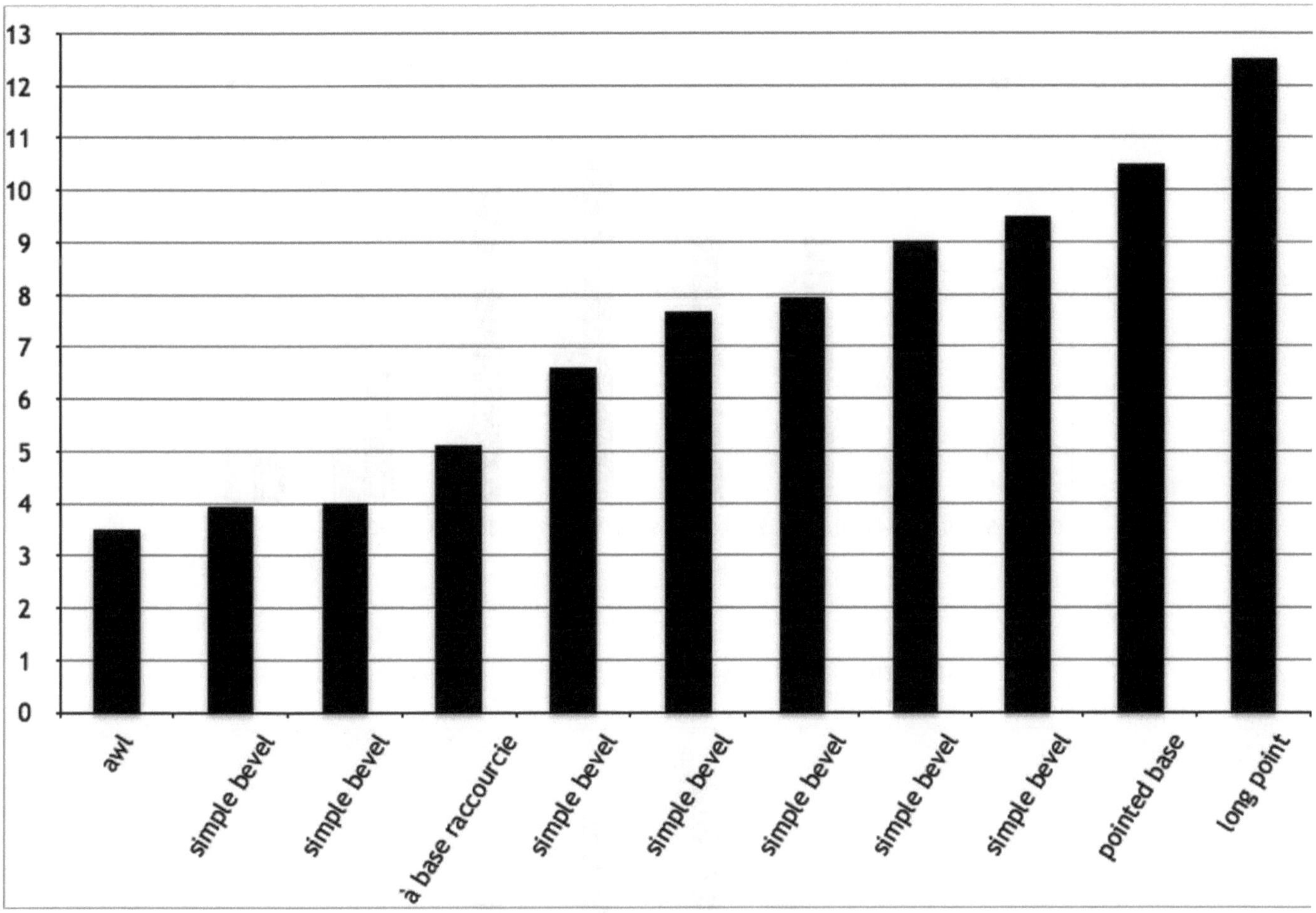

Fig. 8.2 Complete points length. Parpalló (2.20–2.40 layer)

This data is consistent with the use of the points as projectile tips. It shows frequent crushing of distal ends and bases, and splinter fractures on the base, whereas *languette* fractures affect equally distal, mesial and proximal parts. Frequency of fractures on the proximal and distal parts do not coincide, so they do not belong to the same broken weapon and instead arrived at the site during different hunting events.

The points found in this level are mostly single beveled points (70 cases, which represent 61.6 % of the pieces whose proximal part can be studied). There are only two double bevels (1.8 %) and 18 bipoint pieces.

In 24 cases bevels are longer than a third of the piece (23 are single bevel and one is double), this amounts to 34.8 % of all beveled bases. In this assemblage, bevels are longitudinally concave in seven cases and straight in 17. Their ends have only been preserved in nine cases—the rest are all broken: three are pointed and six rounded. Decorative themes are quite simple: grooved lines oriented crosswise (four pieces) or longitudinally (five pieces) that in one case have shallow convergent grooves and cross lines (Tables 8.2, 8.3 and 8.4).

Within the 45 cases of bevels that are shorter than a third of the piece, seven are concave, two are convex and 36 are straight. The end of the bevels can only be ascertained in 24 cases, 16 of which are pointed and eight rounded. Decorative patterns are longitudinal strokes in nine cases, longitudinal grooves are used in four pieces and crosswise transversal in series are also used in four cases, in one piece cross motifs are found.

The points found in this layer, regardless of their type of base, have different sections: 57 are circular, five are oval, 44 are square, 14 triangular, 12 flat-oval, 14 flat sub-rectangular and three are half-round. This data indicates certain balance between rounded and angular sections, regardless of whether the piece is thick or flat.

Morphometric assessment of the points is conditioned by the high degree of fragmentation in this collection. Only 24

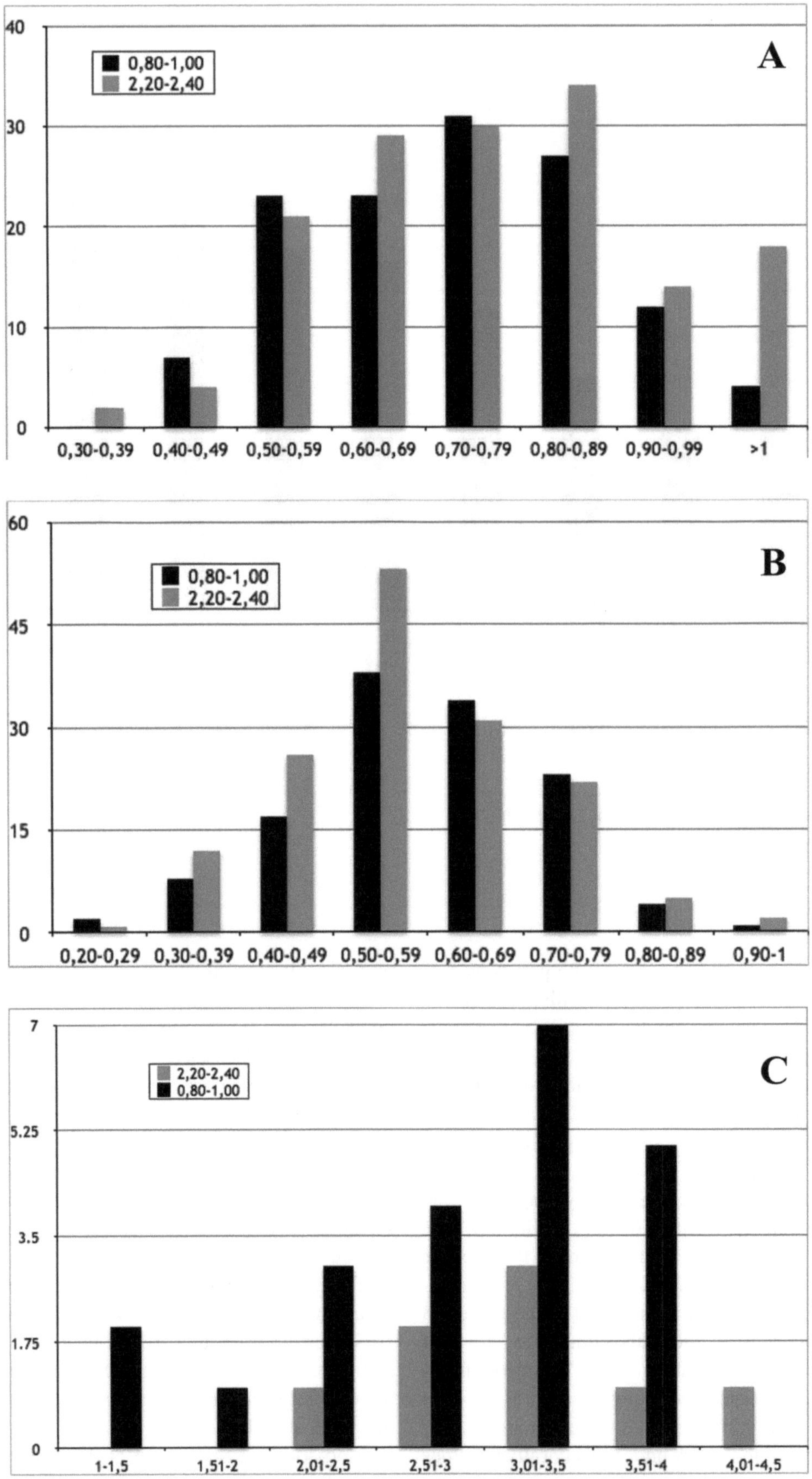

Fig. 8.3 Width (**a**), Thickness (**b**) and Bevel length (**c**) distributions of Parpalló points

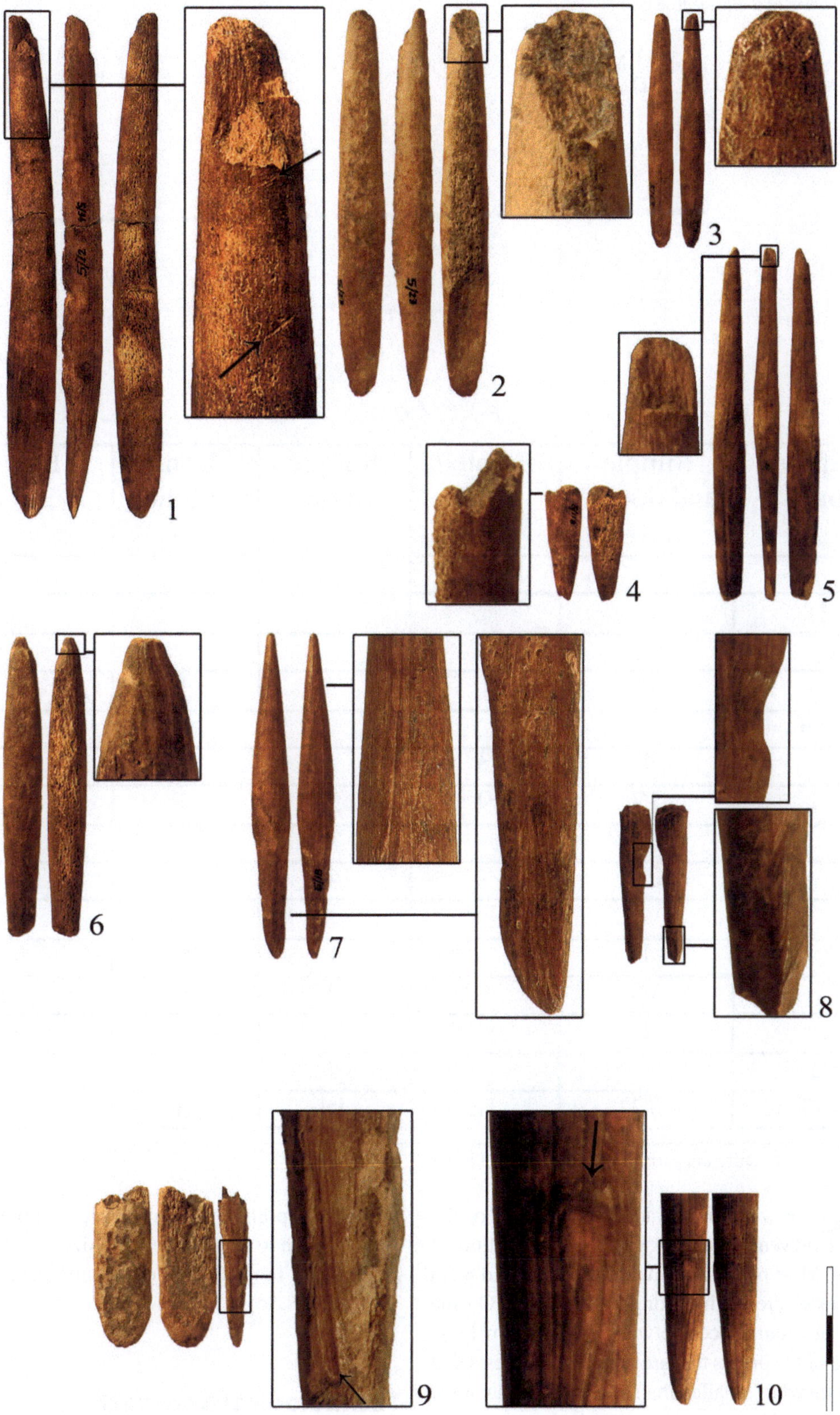

Fig. 8.4 Parpalló, layer 0.80–1.00 m: (*1*) Simple bevel point with use marks. (*2*) Simple bevel point with a *languette* fracture of bevel kind. (*3*) Double point with a *languette* fracture of *marche* kind. (*4*) Flat point with a saw teeth fracture at the mesial part. (*5*) Simple bevel point with a *languette* fracture of *marche* kind. (*6*) Double point resharpened at the tip. (*7*) Simple bevel point resharpened in the meso-distal part, and reworked in the proximal part by removing the original bevel and being replaced by a lateral bevel. (*8*) Flat point reworked with a hafting protuberance and resharpened at the base. (*9*) Simple bevel point which preserves double grooving traces. (*10*) Double point resharpened at the base, with a hafting protuberance

Table 8.5 Preserved parts of the projectile points

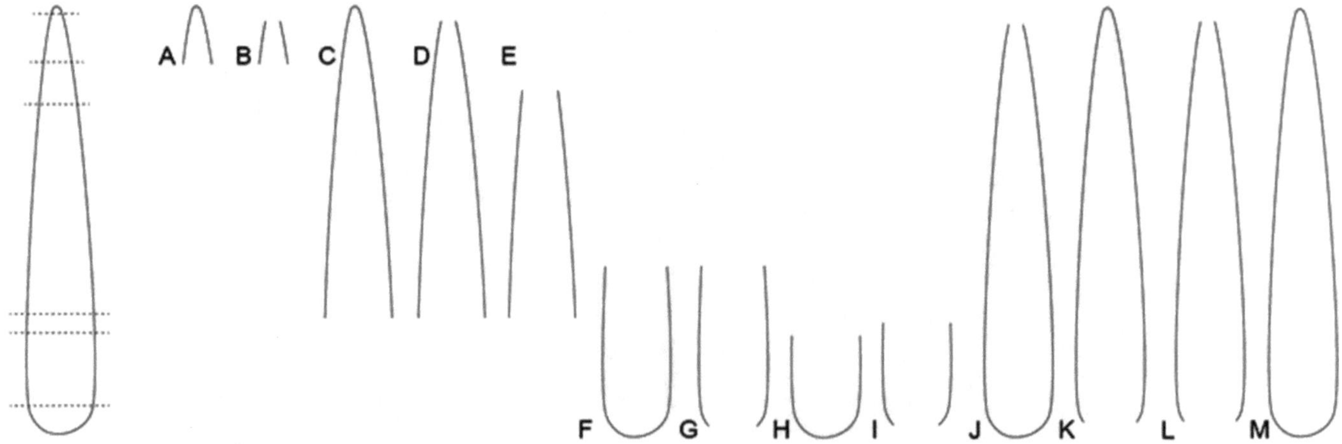

	Without base	Simple and double bevel	Pointed base	Simple base	Long point	Rod	Total
A	1						1
B	14						14
C	1						1
D	6	3	1		1		11
E	33	3				1	37
F		1	3				4
G	1	17	9	5	1	2	35
H		6		1			7
I		27	2	4	1		34
J				1			1
K		1		1			2
L		13	3	2			18
M		1					1
N	1						1
	57	72	18	14	3	3	167

Parpalló (layer 0.80–1.00 m). Different parts are described in Table 8.1

pieces are complete, or with slight fractures on the tip or the base that does not prevent an approximate reconstruction of their original sizes. Most pieces are small; between 5 and 8 cm long. Only two pieces are slightly longer than 9 cm. On the other hand, there are four pieces between 3 and 5 cm long. Therefore, the size of the pieces that are complete suggests that their format is somewhat smaller than those found in layer 2.20–2.40 m (Fig. 8.5). This is also confirmed when the width of the pieces is analyzed. Their width ranges from 0.50 to 0.89 cm, whereas for pieces from layer 2.20–2.40 m ranges from 0.50 to 1 cm. However, thickness does not present any variation, in both layers it is between 0.40 and 0.79 cm (Fig. 8.3a, b).

Only 22 points have preserved complete bevels, or complete enough to appreciate their size, and they reach between 2 and 4 cm long. In fact only three pieces are smaller than 2 cm (Fig. 8.3c).

Technological Approach to the Magdalenian of Parpalló

Here we show a preliminary technical study that will be developed in future works. We describe some technical elements identified by the method of manufacture and object typology.

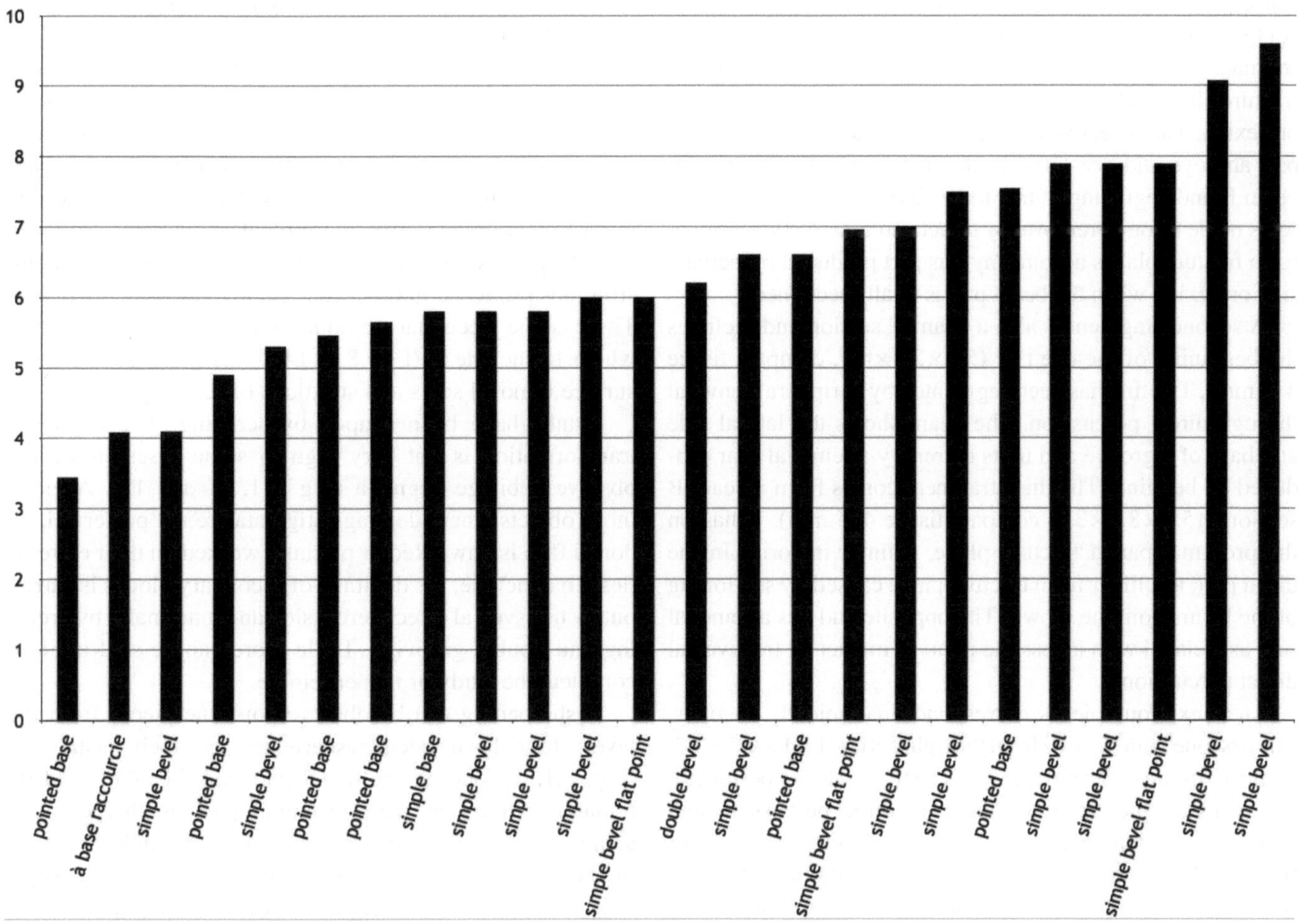

Fig. 8.5 Complete points length. Parpalló (0.80–1.00 layer)

Raw Material

As stated above, red deer antler (*Cervus elaphus*) is the raw material most used to manufacture projectile points at Parpalló. Both morphological and structural characteristics, as well as the faunal record (Davidson 1989) show that red deer is the only cervid present in this region. Most of deer remains represent adults and sub-adults males, but youngers, elders and females are also represented in low proportions.

The introduction of antler into the site comes from the hunting of the red deer. A preliminary examination of basal parts shows a huge predominance of antler obtained through hunting (12 in Badegoulian, 13 in Magdalenian) in contrast to gathered shed antler (one in Badegoulian, none in Magdalenian).

For bone, red deer (*Cervus elaphus*) and wild goat/ibex (*Capra pyrenaica*) are the best represented and in similar quantities. Other ungulates such as horses (*Equus ferus*) and auroch (*Bos primigenius*) are documented. Rabbit (*Orytolagus cunniculus*) is the taxon best represented for small sized animals (Pericot 1942:269). Finally, carnivores are represented by fox (*Vulpes vulpes*), lynx (*Lynx sp.*) and wild cat (*Felis sp.*).

Level of Preservation

The level of preservation in general is good and the technical study can be considered reliable. In spite of that, fire and losses of material are the main factors that affect the assemblage followed by concretions, vermiculations and finally tooth marks.

Technical Description

Layer 2.20–2.40 m

Next we describe some technological elements identified from this layer all of them of red deer antler (Cervus elaphus).

Waste products documented in this layer are scarce, with eight pieces identified for which anatomic origin is identifiable.

Three pieces are from beams. The first one comes from beam A (13.5×6.2×3.6, compact tissue >5 mm) and it has different stigmata of fracture techniques. The proximal part shows a fracture plane with splinter form produced by direct percussion or flexion. The objective is to section this part from the basilar part and eye and bez tines. In the distal part, the union with beam B and beginning of trez tine exhibits peripheral removal scars made through removal by direct percussion. Two splinter form fracture planes accompany this part produced by percussion or flexion when the beam part is finally sectioned.

A second fragment is also a beam A section and includes the beginning of the eye tine (5.4×2.4×4.1, compact tissue 4–5 mm). The tine has been segmented by peripheral removal through direct percussion. The beam shows the lateral side and base of a groove and in its extremity a removal scar produced by bending. The third fragment comes from a beam B section (15.3×8.1×3.2, compact tissue 4–5 mm). It has on the proximal part a fracture plane, splinter in form. In the distal part, a splinter form fracture plane caused by sectioning of the beam from the crown. The opposite end has a removal scar associated with a possible point of impact by transversal direct percussion.

The next four pieces correspond anatomically to tines. The first one is an eye tine from the right antler (15.4×4.5×3.8, compact tissue 4–5 mm). The proximal part shows peripheral removal scars produced by removal by direct percussion, and a splinter fracture plane produced by direct percussion or flexion. The outer face shows two longitudinal fracture planes that overlap. At the end of these fracture planes there are two bending removal scars. The distal part presents a saw teeth fracture plane. The second tine is a segment (7.8×3.2×2.8, compact tissue 4–5 mm), with the proximal part exhibiting peripheral removal scars made by removal by direct percussion and a splinter fracture plane formed by direct percussion or flexion. The inner face shows a fracture plane with some transversal impact points. The distal extremity presents a splinter fracture plane. The third piece is the beginning of a tine (5.5×2.4×4.9, compact tissue <4 mm) which proximal part has post-depositional fractures while the distal end shows a splinter fracture plane. The last one is a tine segment (8.4×2×2.1, compact tissue <4 mm) and presents two splinter fracture planes on each extremity.

The technological analysis of these waste products indicates an initial step where the primary block is sectioned into secondary blocks. Sectioning is carried out by different techniques like direct percussion, removal by direct percussion and flexion. Some of those pieces show a debitage by transversal fracturation techniques. Only one piece has the stigmata of a groove for blank production.

The waste products not identifiable anatomically consist of fragments of variable size (from 2 to 4 cm long and 0.5 to 1 cm wide), and which show fracture planes on their edges. We have identified three in this layer.

This debitage produces rectangular, elongated and convergent blanks. Of these pieces, 7 have fracture planes along their edges and a proximal splinter fracture plane. One is complete (5.6×1×0.7, compact tissue <4 mm), two are blanks of the rod type, with one having on both edges, longitudinal and convergent grooves. The other has the lateral and base of a groove only on one side, with the other side exhibiting a fracture plane. Two more blanks fragments are in the shaping process because they show on their edges scraping striations produced in order to regularize the fracture planes. There is one piece that we do not know in which category in which to include it (Fig. 8.1, 14). This piece shows on its surface removal scars and striations of scraping.

Blanks have been shaped by scraping. The degree of transformation is not very high in some cases so we can observe débitage stigmata (Fig. 8.1, 11 and 15). Also, on those objects where debitage stigmata are not preserved, the dorsal face is unworked or partially worked on their extremities. To conclude, the debitage of secondary blocks is carried out by transversal direct percussion and marginally by grooving and double grooving. Little more can be said until we complete the study of this ensemble.

Resharpening can be observed on nine pieces from this layer: four have been resharpened on their distal part (Fig. 8.1, 4), one on the distal part and the sides, and the remaining four, on their proximal part. Resharpening is achieved by means of strong scratches on the ends, or on part of them. It can expand towards the sides, and has as the objective of making damage surfaces more regular. On one of the proximal fractures, which is quite large, we have observed that the resharpening did not make it completely regular. On another fracture, a lateral lump remained that could have been left on purpose.

Reworked pieces are also found in this layer. The following five pieces are described as examples of this technique. The first one was originally a single beveled point broken at the meso-distal part; the fracture was rounded and a groove made. This piece has preserved the bevel that now shapes its distal part, and it has been turned into a simple base point (Fig. 8.1, 7). The second piece, also a single beveled point, fractured where the bevel begins, was modified by sharpening the piece there. It preserves its typological classification, but its morphology has been highly altered. Finally, there are three points with whittled (*raccourcie*) bases. They have all been made from meso-distal fragments of points, so their size has been reduced considerably. One of them, which is 6.45 cm, can be included in the parameter of complete pieces. But the other two are considerably smaller; 3.20 and 3.50 cm. This is interpreted as shaping debris and it comes from the calibration of the blank or maybe as removal points stuck in the carcass (Le Dosseur 2003; Chauvière and Rigaud 2005; Rigaud 2006). In this respect, it is worth remembering that in the Parpalló layer we are concerned with, there is an important variation in the size of beveled points, but none of them is smaller than 4 cm.

Chattermarks have also been found. These are the result of resharpening and reworking carried out by strong scraping that impacted against the bone fiber, producing these marks on the surface.

There are seven pieces that display some type of surface work associated with the haft; excluding bevel decoration. Four of these are made of small parallel incisions that are oblique and perpendicular to the longitudinal axis on the sides. One of the pieces has a bulk on a groove base, perhaps made when repairing the base after a fracture. Another piece has a groove on the base that was possibly used for fixing the point of a projectile. Finally, notches on the edge of the bevel probably produced by hafting were found on another piece.

Layer 0.80–1.00 m

Little can be said for this layer as the number of technological pieces is much reduced. We only have a small representation of waste products and blanks and preforms are totally absent. Some objects preserves their technological stigmata so we can say something, however, about the techniques employed in their debitage.

Waste products are consist of 14 tine tips of red deer antler, all of them with a splinter fracture plane in the proximal part. We hypothesize a preparation of the primary block by sectioning tine tips by direct percussion or flexion to explain these parts.

Some objects preserve debitage stigmata (Fig. 8.4, 9). They show two parallel or convergent grooves on their edges. The grooves preserved are composed by a lateral groove with longitudinal and parallel striations, and part of the base. It allows us to hypothesize a extraction of blanks by double grooving procedure.

Regarding the shaping of the implement surface, scraping has been used on all the pieces studied. It is worth mentioning at this point that the dorsal face of 21.15 % of the points have not been shaped and preserves the natural surface of the antler.

Many of the weapons that make up this layer's assemblage have been resharpened in order to repair damage or reworked when the fracture affected the piece in such a way as to make it impossible to repair. In the latter case, a different piece with a different function was made (Knecht 1997; Pétillon 2006). Resharpening affects the ends and edges of the pieces and reduces their volume (Knecht 1997), but maintains the morphology of the piece and its function (Pétillon 2006). Twenty-five points were resharpened by means of scraping in order to sharpen and to make the damage surface of the piece more regular. For most of them, resharpening takes place on the distal and proximal ends, even though sometimes it extends to the mesial parts and to the sides in order to make the outline more regular. The quantity of resharpening cases coincides with the number of fractures produced on those parts (Pétillon 2006).

Chattermarks are very common in all these resharpening and reworking cases. Six reworked pieces have been identified. Two of them are points *à base raccourcie* whose original type is impossible to determine. Another one is the distal and proximal part of a single beveled point that has been reversed and turned into a point *à base raccourcie*. On three single beveled points, made from older single beveled points, asymmetry produced by changes in the outline and crushing can be observed. Finally, there is a single beveled point with a lateral fracture of the bevel that reaches the beginning of this section; the piece was remade by turning the fracture into a lateral bevel.

Points *à base raccourcie* are also the result of recycling. Many authors coincide in saying that these are points stuck into the animal skeleton that could not be extracted and were cut off in order to recover the shaft; they would have been found in the sites for that reason (Plisson and Geneste 1989; Morel 1993; Bertrand 1999). This theory could be accepted in the case of single beveled points cut at the mesial part. But for two of the points found in this layer we would not know whether they were cut off on purpose or they are just the result of recycling (Chauvière and Rigaud 2005; Chauvière 2016) because they have the same distal finishing as other implements (Pétillon 2006).

There are different types of bases (bevel, double bevel, pointed base, simple base, *à base raccourcie*, and others with bulks and wavy shapes in its mesial-proximal and proximal outline of the artifacts). Apart from these incisions, grooves and bevels, scraped or cut by a knife are found in the proximal parts of the points. All these features are associated with specific functions and hafting techniques (Bergman 1987; Knecht 1997) and even make sure that the projectile does not lose force when it impacts a target (Arndt and Newcomer 1986). The efficiency of striations on the base to facilitate the hafting system, optimized by the use of adhesives, has been mentioned in the case of *pointes à base raccourcie* (Chauvière and Rigaud 2005).

Comparison Between the Assemblages from Parpalló

Both layers show similar patterns regarding the parts of the points documented and their type of fractures.

Studies indicate that distal and mesial-distal fractures are the most common (Arndt and Newcomer 1986; Bergman 1987; Bertrand 1999) because this is the area of the projectile that receives the brunt of impact. Crushing distal fractures are produced by a perpendicular impact of the projectile against the object, whereas for *languette* fractures, the impact angle is an acute one (Pétillon 2000, 2006). Splinter fractures have been documented in lateral impact cases (Odar 2012).

Although distal fragments are important, they are not the most abundant in the archaeological record. As J.M. Pétillon mentions (2006) when observing the results of his research, distal and mesial-distal fractures might have been caused in many cases by missing shots and it is logical to think that the distal parts found in sites arrived there stuck to the bodies of animals.

For mesial-proximal and proximal fragments the same type of fractures are identified, including splits. As in the case of the distal parts, these fractures are produced by the impact angle, the force of the blow, and the haft, that in this case has an important role in the resistance to the impact (Bertrand 1999). The high number of mesial-proximal and proximal pieces is owing to the fact that they have many possibilities in their arriving to the site. They could have been hafted to the shaft, extracted from the target for a possible reworking and, finally, inserted in the body of the animal (Pétillon 2006). In the two layers from Parpalló, the parts of the weapon found and their type of use fractures indicate that they arrived at the site within the bodies of the prey or hafted to shafts in order to be repaired.

We are only dealing with part of Parpalló's Upper Magdalenian and Badegoulian B industry. So we will limit ourselves to general aspects and leave the detailed assessment of both periods for a future study.

One of the first things that can be observed from a typological point of view is the absence of rods and double bevelled points in level 2.20–2.40 m. We should remember, in this respect, that in level 0.80–1.00 m a point with a double bevel base and three half-round rods were found, however.

Another contrasting fact between these two levels is found in the proportions of circular and angular section points which present different values both for spears with bevels >1/3 and those with bevels <1/3. So in layer 0.80–1.00 m all bevel points <1/3 (types 4.1–4.3 in Barandiarán's classification) have square or triangular sections (14) and the same is true for bevel points >1/3 (9). Whereas in layer 2.20–2.40 m, in both subtypes circular sections prevail (24) and there is a small amount of angular sections (3). These differences can be seen in the contrast between the distribution of the sections of points for each layer. In layer 0.80–1.00 m there is a certain balance between the number of rounded (62) and angular sections (58) of normal points, on the one hand, and that of flat points (12/14), on the other. But in layer 2.20–2.40 m, rounded sections (122) prevail both for normal and flat points. In this layer angular sections include 18 pieces.

The appearance of a certain type of single bevel in layer 2.20–2.40 m should be added to these differences. This type of piece has well defined stylistic features both in relation to its morphometric characteristics and the decoration of the shaft. It is a bevel point of small caliber (between 0.5 and 0.8 cm.), of a cylindrical body that can be straight, slightly curved and, less often, quadrangular. The bevel is predominantly concave and, on two or three of its sides, its shaft normally has patterns of broken lines or zig-zags made by juxtaposed angles. This type of piece has most frequently been found in this layer or in the one immediately above (2.00–2.20 m). It therefore introduces a well-defined stylistic feature in Badegoulian B.

Regarding the details observed on the bevels, where relative lengths are concerned, differences between both layers are not very pronounced. In both of them, lengths <1/3 are more common. But in layer 2.20–2.40 m the difference concerning lengths >1/3 is considerably reduced as they represent 42.9 % of beveled points.

Very few pieces have preserved the end of the bevel complete. So the analysis of that feature can only be but a tentative one, especially if we want to assess the differences that might exist regarding the proportion of that part of the piece in relation to the total length of the point. In bevel points >1/3, we can observe that a simple base prevails in layer 0.80–1.00 m, whereas in layer 2.20–2.40 m pointed ends are more frequent. Besides, among bevel >1/3 the pieces from this level, we can also find one with a truncated base. For bevel points <1/3, in layer 0.80–1.00 m, pointed ends are double the number of rounded ones; whereas in level 2.20–2.40 m there is a similar number of both types. Little can be said regarding the longitudinal shape of bevels, except that concave ends are more frequent in level 2.20–2.40 m (Fig. 8.5).

It is not easy to find differences in the proportions and variations of bipoints, but it is worth mentioning the appearance of a piece with a central crushing fracture in layer 2.20–2.40 m.

Finally, in order to conclude this comparison of typological features, it is important to insist on the variation observed between both levels regarding the role of flat points: they are far more common in layer 2.20–2.40 m (29.6 % of all points) than in layer 0.80–1.00 m (14.9 %). Besides this fact, it must be remembered that from the morphometric point of view, there are differences between both layers. Wide pieces are more common in layer 2.20–2.40 m, which shows the role flat sections have in this layer.

Osseous Equipment in the Mediterranean Upper Magdalenian

In this section we will limit ourselves to some comments regarding osseous industries from the Upper Magdalenian in the Spanish Mediterranean region (Fig. 8.7).

The most common types of osseous points in the region are beveled points of single or double bevels, bipoints and flat points. But other types have also been identified, such as *pointes à base raccourcie*, simple points and points with bulk or protuberance on the base. The Magdalenian osseous weapon inventory for the Spanish Mediterranean region also includes rods, short points and harpoons (Table 8.6).

In the last few decades, an increasing number of Upper Magdalenian occupations have been identified (Aura Tortosa 1995; Villaverde et al. 1998; Villaverde 2001; Casabó 2005). However, the number of sites where harpoons have been found is low in relation to the total number of sites identified: 68 pieces have been found in 13 sites, but they are distributed in a rather irregular way. Two sites claim the majority of the pieces: 21 in Bora Gran plus 19 in Cova de les Cendres. In Cendres, nine pieces from sector A level XI, four more from level IX, and six yet to be attributed to level IX and XI, should be added (Villaverde et al. 1999; Roman and Villaverde 2012).

The materials from Bora Gran present characteristics that have not been documented in the central-south sites of the Spanish Mediterranean region. Examples of these characteristics are: reindeer (*Rangifer tarandus*) remains, *sagaie* with a forked base, some decorative patterns of the Pyrenees style and the large size of some of the points. But apart from these, we have the discovery of bilaterally barbed harpoons that have made it possible to establish a relationship between the northeast of Spain and the south of France.

The core of Valencian sites where harpoons have been found (Cova Matutano, Blaus, Volcán del Faro, Parpalló, Foradada d'Oliva, Tossal de la Roca and Cova de les Cendres) is 350 km away from Bora Gran. Further south, some harpoons have also been found: Cueva de los Mejillones, Murcia (Martínez Andreu 1989) Cueva del Higuerón, Cueva Victoria, Hoyo de la Mina and Cueva de Nerja in Andalusia. But none of the harpoons found in these sites are of the bilateral type, nor present any of the characteristics that characterize those of Bora Gran mentioned above, which can therefore be considered as regionalization features.

Only two of the sites excavated in the last few decades offer significant osseous industry collections with elements that could complement the features described in Parpalló. These sites are Cova de les Cendres (Fig. 8.6a) and Cueva de Nerja (Fig. 8.6b). Both have hearths and evidence for manufacturing and repair of lithic and bone implements, and abundant remains of ornaments. Fauna in Cendres is characterized by the prevalence of red deer, but there are also small prey remains (rabbit and birds), all with abundant processing marks (Villaverde and Martínez Valle 1995; Martínez 1996; Real 2012). In Nerja, wild goat is the prevalent species and small prey is more diversified: rabbit, birds and abundant marine fauna (Pérez Ripoll and Raga 1998; Aura Tortosa et al. 2002, 2009; Jordá Pardo and Aura Tortosa 2009; Aura Tortosa et al. 2010; Álvarez-Fernández et al. 2014) (Fig. 8.7).

The Harpoons

Level XI harpoons from Cendres have been dated to between 15,017 and 16,288 cal. BP, and those from level IX to 14,743 cal. BP (Bergadà et al. 2013) In level XI, the technical transformation schema of antler weapons was studied and it is defined by a debitage by extraction (Borao 2012, 2013). Secondary blocks were exploited longitudinally by means of parallel or convergent double grooving on one or both ends. This procedure was completed by splitting and bending in order to extract a blank that had not been worked on previously. The blanks obtained were flat and rod type and their dimensions vary. Preforms have been worked at least on one of their faces by means of scraping. Finally, the finishing of the pieces was achieved by scraping.

On bone, we only have the remains of blanks rod type that testify the debitage by extraction of them by parallel or convergent double grooving.

The number of finished pieces recovered from level XI sector A is 66. Out of these, 54 are projectile points or harpoons that represents 81.8 %. The prevalent raw material is red deer antler, but the importance of bone should not be overlooked: ten points, one harpoon, one awl and nine needles are made of bone.

The main differences between Cendres and Parpalló's Upper Magdalenian layer are marked by the larger amount of points with a double beveled base and rods found in the former and in the importance of harpoons there for the characterization of the Upper Magdalenian.

The importance of harpoons for the region is documented by data obtained in studies of pieces from Cendres (Villaverde and Román 2006; Román and Villaverde 2011, 2012) and deserves further comment. A north–south gradation can be observed in relation to the raw material used for the manufacture of harpoons. For those made north of Cendres the prevalent raw material is red deer antler (27) over bone (5), whereas the reverse is the case south of the site, with ten pieces made from bone and four from antler. Data from Cendres in this respect is more balanced but antler (12) slightly prevails over bone (7).

Most harpoons are broken and the fragments preserved have less than five barbs, except for a 12 barb piece from Cova de les Cendres. Their average length is 95.5 mm. This clearly differentiates them from those from the French and Cantabrian areas, whose average lengths are 126 and 123 mm respectively.

As for their morphology, harpoons from the central and southern areas can be divided into two groups: large base pieces with a smaller number of barbs and pieces with a smaller base that have a larger number of barbs. Both types have short barbs that are close to the shaft and do not protrude beyond the width of the base. There are also a number

Table 8.6 Summary of Upper Magdalenian weaponry

	Parpalló: 2.40–2.20 m (a)	Bora Gran (b)	Cova del Parco II (c)	Cova Matutano (d)	Cova dels Blaus (e)	Volcán del Faro (f)	Parpalló: 1.00–0.80 m (aa)	Foradada d' Oliva (e)	Tossal de la Roca (g)	Cova de les Cendres XI (h)	Cueva de los Mejillones (i)	Cueva de Nerja (j)	Cueva del Higuerón (k)	Cueva Victoria (l)	C. Hoyo de la Mina (m)	El Pirulejo 4 (n)
Simple base spear points	6						8			1						
Point à base raccourcie	3	×					2		1							
Simple bevel spear points	46	×	3	3	1	×	60		1	6		1				
Double bevels spear points		××	5	5	1	×	2			7		7				
Pointed spear points	16	×					18									
Long spear points	6	×					4			1						
P. with two symmetrical protuberances												2				
Flat spear points	35		1	1			9									
Rods		××		2	3	×	2									5
Points (distal and mesial parts)	50	×××	11	26		×	58		15	54		29				15
Harpoons		21		7	1	1		1	1	9	3	5	2	2	2	
Gorges					×						40				×	
Total osseus assemblages	142	±500	31	28	>6	>1	167	>1	17	66	3	96	2	2	>2	26

(a) Ancient Magdalenian: this study; (aa) Upper Magdalenian: This study; (b) Rueda Torres (1987); (c) Tejero (2005); (d) Olària (1999); (e) Casabó (2012); (f) Aparicio (2003); (g) Cacho and de la Torre (2005); (h) Villaverde et al. (2012); (i) García del Toro (1985); (j) Aura Tortosa and Pérez (1998); (k) López and Cacho (1979); (l) Fortea (1986); (m) Ferrer et al. (2006); (n) Cortés Sánchez (2008)

×: present; ××: numerous; ×××: abundant

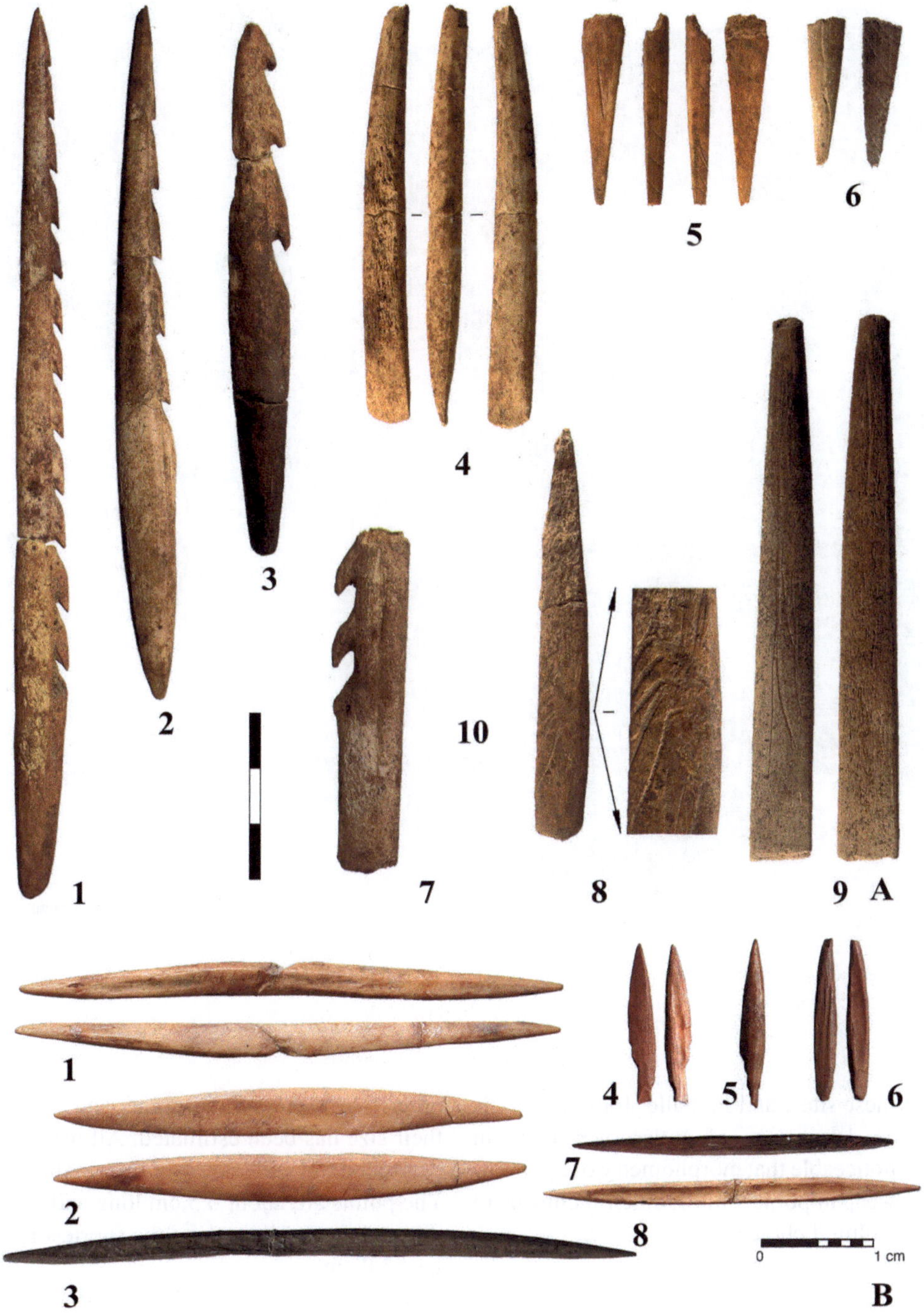

Fig. 8.6 (**a**) Cendres: *1–3 & 7* harpoons; *4–5* double bevel point; *6* rod; *8–9* simple bevel points. (**b**) Nerja: *1–8* gorges of diverse length

of pieces with their barbs hardly carved and whose function could differ from that of other harpoons. Finally, their bases are either cylindrical or slightly beveled, and some of them have double bevel.

It is not possible to establish differences in the function of the harpoons in Cendres collection. On complete pieces, size variation is the most significant feature, and for the fractured pieces there is a prevalence of distal or medial parts against proximal parts. However, this latter aspect may be a result of the difficulty of differentiating fractures bases of other weapons from harpoons, especially if the considerable length of this part of the harpoon is considered.

There is no reason to think that bone or antler raw material selection could be motived by morphometric or typological motives in harpoons elaboration of Cendres or other southern Upper Magdalenian sites. Red deer faunal remains

Fig. 8.7 Sites cited in the text (see Table 8.6)

are abundant in all these sites, and Parpalló, for example, is characterized by the domination of antler over bone in weapon points. It is noticeable that morphometry of harpoons is similar to others weapon points, all of which incites us to think that we face a cultural choice.

The Gorges

The Upper Magdalenian osseous industry from Nerja (NM16-14 y NV7-5) consists of 98 objects: 10 correspond to technological remains, 27 are objects made from antler and 71 made from bone. Some parts of middle size mammal skeletons (*Cervus elaphus*, *Capra pyrenaica*), small size mammals (*Oryctolagus cuniculus*) and sea birds (*Sula bassana*) can be recognized on the pieces.

Out of the 88 pieces that have been classified, nine are complete: one needle and eight short and thin points. For the other four pieces, three points and one rod are identifiable, their size has been estimated. All pieces are small, only one decorated rod with a broken line might have exceeded 10 cm. The points are about 6.5 cm long and the needle 7 cm. Short thin points are on average the shortest is 1.5 cm and the longest 3.8 cm.

Half of the pointed objects registered, correspond to short thin bipoints (40 objects). They often have scraping traces that can be related to a quick manufacturing process, or to a functional finish with the objective of improving them for a more secure assembly or suspension. Most sections are flat (oval or rectangular), but next to them in importance are circular and triangular sections. As for their thickness, more than 80 % of the pieces are >1.6 cm and 40 % show different grades of thermic alteration. A wide piece in process of manufacture has a spindle shape and two complete pieces that are smaller than 1.7 cm have a cut base, possibly as a result of being resharpened. Differentiation of thin points and needle fragments has not been completely solved yet because morphology, section and dimensions overlap. The uneven finish—abrasion and regular polish for needles, in contrast to striation and scraping for thin points, along with the scarce number of needles have been helpful factors for the classification process.

Thin and short bipoints have been classified as gorges or straight hooks (Aura Tortosa and Pérez 1998). Arguments in support of this interpretation are based on the occupation context and on archaeological and ethnologic comparisons. One of the arguments is that the pieces coincide with a moment of important exploitation of marine resources in Magdalenian occupations, but this idea is an indirect inference (Aura Tortosa et al. 2002, 2009; Jordá Pardo et al. 2011; Álvarez-Fernández et al. 2014).

There have been many archaeological references to similar objects for the Western Upper Paleolithic (Averbouh and Cleyet-Merle 1995), as is also the case for the Natufian (Campana 1989) and Iberomarusian and Capsian of North Africa (Camps-Fabrer 1967). Similar objects, but of a bigger size, have been described in La Vache's Magdalenian levels (Averbouh 2003: Fig. 262), in Mesolithic levels at Aizpea (Barandiarán 2002), at El Espertín (Bernaldo de Quirós and Neira 2007–2008), or in more recent contexts, highlighting their morphometric and functional diversity (Alday et al. 2011). To the association of these objects to leather working (as borers, pins, buttons, etc.), or as active barbed implements (Campana 1989), we add their use as hooks. This suggestion can be inferred from identical objects documented among native populations in northwest America (Reid 1910). There are few fractures in the thin points found in Nerja that could be related to impacts. Furthermore, due to their size and the ethnographic parallels mentioned above, it might be possible to consider that the pieces are gorges (passive projectiles) that were used at the end of tackles or in hook lines held by their middle part.

Conclusion

With the exception of Parpalló, bone and antler weapons are represented by very few pieces in the Upper Magdalenian levels of the Mediterranean. Their assemblages present similar features to those from Parpalló: single and double beveled points, rods and some harpoons. As for raw materials, the use of antler prevails for larger pieces, in contrast to objects made of bone that are smaller and rougher in their finish. The role of harpoons which we have assessed in relation to osseous weapons from Cendres, and the role of thin points from Nerja, are instance worth considering as specific cases. Likewise, we have insisted on the important differences between sites in the northeast of the Iberian Peninsula, especially Bora Gran, and the Southern sites. Technological data has made a first characterization of their processes possible, as has established relationships with regions that are further north (Pétillon and Averbouh 2012).

Regarding the quantification of weapons made from hard animal material, Parpalló's figures are particularly high, and not just in relation to other Mediterranean sites. This situation is made more apparent when we consider that the number of finished pieces from the whole Upper Magdalenian in Parpalló numbers over 1000. Owing to the importance of its osseous weapon collection, Parpalló is a crucial site in the western Paleolithic sequence of the Iberian Peninsula. The data we are managing so far makes up an initial representation of the development of different types of points, hafting and blank manufacturing techniques. With the help of this data, it will be possible to make progress in the areas of diachronic changes and regional relations.

Weapon point densities are difficult to evaluate in Mediterranean Magdalenian sites since excavations are old in some cases and no data on the volume of sediment is known (Bora Gran). However, if we consider the number of pieces, it is clear that some sites (Bora Gran, Parpalló, Cendres and Nerja, essentially) have a higher quantity of pieces, probably due to the intensity and duration of occupations. It is a significant fact that this trait is repeated in Parpalló along the entire Magdalenian sequence, which permits a diachronic evaluation of weapon points over a period that is not possible to evaluate in any other site of this region.

Osseous projectile weaponry shows important spatial adaptive strategies that are not apparent at the level of individual sites. This data is representative of the regional-scale foraging systems and adaptations to climate-driven changes in post-LGM environments. The increasing importance of microlithic elements and osseous artefacts may be in part a response to increasing population and the need to decrease the risk of hunting failure through investment in weapons technology (Barton et al. 2013).

An assessment of Parpalló's singularity would require models capable of explaining the reiterative occupation of the site and the concurrence of its symbolic elements. But going perhaps beyond the objective of the present paper, it is worth pointing out that the stylistic features and technological procedures observed in the Badegoulian and Magdalenian layers we have analyzed show supra-territorial relationships that clearly articulate this region of the Spanish Mediterranean within the French-Cantabrian Magdalenian.

Acknowledgments This research was supported by the Ministry of Science and Innovation (Spain) projects HAR2014-52761-R, HAR2013-46861-R and HAR2011-24878, and by the *Direcció General d'Universitat, Estudis Superiors i Ciencia* (*Genralitat Valenciana*) project PROMETEOII/2013/016. We are grateful to Dr. Helena Bonet for her help with the study of Parpalló materials in the *Museo de Prehistoria de Valencia*. We express our gratitude to Michelle Langley for her invitation to participate in this book and her editing labor.

References

Alday, A., Juez, L., Pérez-Romero, A., Adán, G., Santos, E., Galindo-Pellicena, et al. (2011). La industria ósea de El Portalón de Cueva Mayor (Sierra de Atapuerca, Burgos). Biapuntados, puntas de flecha y agujas, morfología y funcionalidad. *Munibe, 62*, 227–249.

Álvarez-Fernández, E., Carriol, R.-P., Jordá, J. F., Aura Tortosa, J. E., Avezuela, B., Badal, E., et al. (2014). Occurrence of whale barnacles in Nerja Cave (Málaga, southern Spain): Indirect evidence of whale consumption by humans in the Upper Magdalenian. *Quaternary International, 337*, 163–169.

Aparicio Pérez, J. (2003). *El paleomesolítico valenciano: Cova del Volcán del Faro: memoria de las excavaciones e inventario del material*. Valencia: RACV, Sección de Prehistoria y Arqueología.

Arndt, S., & Newcomer, M. (1986). Breakage patterns on prehistoric bone points. In D. A. Roe (Ed.), *Studies in the Upper Paleolithic of Britain and Northwest Europe. British Archaeological Reports International Series S296* (pp. 165–173). Oxford: Archaeopress.

Aura Tortosa, J. E. (1995). *El Magdaleniense Mediterráneo: la Cova del Parpalló (Gandía, Valencia)*. Valencia: Trabajos Varios, 91, Servicio de Investigación Prehistórica.

Aura Tortosa, J. E. (2007). Badegouliens et Magdaléniens du versant méditerranéen espagnol. *Bulletin de la Société Préhistorique Française, 104*, 809–824.

Aura Tortosa, J. E., & Pérez, C. I. (1998). ¿Micropuntas dobles o anzuelos? Una propuesta de estudio a partir de los materiales de la Cueva de Nerja (Málaga). In J. L. Sanchidrián & M. D. Simón (Eds.), *Las culturas del Pleistoceno superior en Andalucía* (pp. 339–348). Málaga: Patronato de la Cueva de Nerja.

Aura Tortosa, J. E., & Jordá Pardo, J. F. (2012). Solutrenses del Sur de Iberia en Transición. *Espacio, Tiempo y Forma Serie I, Nueva época Prehistoria y Arqueología, 5*, 149–169.

Aura Tortosa, J. E., Jordá Pardo, J. F., Pérez Ripoll, M., Rodrigo García, M. J., Badal García, E., & Guillem Calatayud, P. (2002). The far south: The Pleistocene-Holocene transition in Nerja Cave (Andalucía, Spain). *Quaternary International, 93–94*, 19–30.

Aura Tortosa, J. E., Jordá Pardo, J. F., Morales, J. V., Pérez, M., Villalba, M. P., & Alcover, J. A. (2009). Economic transitions in finis terra: The western Mediterranean of Iberia, 15–7 ka BP. *Before Farming, 2009*(2), 1–17.

Aura Tortosa, J. E., Jordá Pardo, J. F., Pérez Ripoll, M., Badal, E., Morales, J. V., Avezuela, B., et al. (2010). Treinta años de investigación sobre el Paleolítico superior de Andalucía: la Cueva de Nerja (Málaga, España). In X. Mangado (Ed.), *El Paleolítico superior peninsular. Novedades del siglo XXI* (pp. 149–172). Barcelona: SERP, Universitat de Barcelona.

Aura Tortosa, J. E., Tiffagom, M., Jordá Pardo, J. F., Duarte, E., Fernández de la Vega, J., Santamaría, D., et al. (2012). The Solutrean - Magdalenian Transition: A view from the southwest. *Quaternary International, 272–273*, 75–87.

Averbouh, A. (2003). Les petits éléments droits à double pointe. In J. Clottes & H. Delporte (Eds.), *La Grotte de la Vache (Ariège). Fouilles Romain Robert. I.- Les occupations du Magdalénien* (pp. 353–356). Paris: Éditions du CTHS.

Averbouh, A., & Cleyet-Merle, J. J. (1995). Fiches hameçons. In H. Camps-Fabrer (Ed.), *Fiches Typologiques de l'industrie osseuse préhistorique. Cahier VII. Éléments barbelés et apparentés* (pp. 83–99). Treignes: CEDARC.

Barandiarán, I. (1967). *El Paleomesolítico del Pirineo Occidental*. Zaragoza: Monografías Arqueológicas III.

Barandiarán, I. (2002). La industria ósea: elaboraciones sobre hueso, asta y concha. In A. Cava & I. Barandiarán (Eds.), *Cazadores-recolectores en el Pirineo navarro: sitio de Aizpea entre 8000 y 6000 años antes de ahora* (pp. 179–212). Pamplona: Diputación Foral de Navarra.

Barton, C. M., Villaverde, V., Zilhão, J., Aura, J. E., Garcia, O., & Badal, E. (2013). In glacial environments beyond glacial terrains: Human eco-dynamics in late Pleistocene Mediterranean Iberia. *Quaternary International, 318*, 53–68.

Bergadà, M., Villaverde, V., & Roman, D. (2013). Microstratigraphy of the Magdalenian at the Cendres Cave (Teulada-Moraira, Alacant, Spain): Formation and diagenesis processes. *Quaternary International, 315*, 56–75.

Bergman, C. A. (1987). Hafting and use of bone and antler points from Ksar Akil, Lebanon. In D. Stordeur (Ed.), *Le main et l'outil. Manches et emmanchements préhistoriques, Actes du colloque de Lyon (1984)* (pp. 117–126). Paris: Travaux de la Maison de l'Orient, n° 15, CNRS.

Bernaldo de Quirós, F., & Neira, A. (2007–08). Una pieza excepcional del Mesolítico cantábrico: el anzuelo de la cueva de "El Espertín" (Cuenabres, Burón, León). *Veleia, 24–25*(1), 571–579.

Bertrand, A. (1999). *Les armatures de sagaies Magdaléniennes en matière dure animal dans les Pyrénées. British Archaeological Reports International Series 773*. Oxford: Archaeopress.

Bofinger, E., & Davidson, I. (1977). Radiocarbon age and depth: A statistical treatment of two sequences of dates from Spain. *Journal of Archaeological Science, 4*, 231–243.

Borao, M. (2012). Estudio tecnológico y tipológico de los útiles fabricados sobre materias duras animales en el Magdaleniense superior de la Cova de les Cendres (Teulada-Moraira, Alicante). *Sagvntvm (P.L.A.V.), 44*, 17–37.

Borao, M. (2013). The technological study of the manufacturing process of the bone and antler industry in the Upper Magdalenian of Cova de les Cendres (Alicante, Spain). In F. Lang (Ed.), *The Sound of Bones. Proceedings of the 8th Meeting of the ICAZ Worked Bone Research Group in Salzburg 2011* (pp. 71–85). Salzburg: Schriften zur Archäologie und Archäometrie an der Paris Lodron-Universität Salzburg, Band 5. Archaeoplus.

Breuil, H. (1912). *Les subdivisions du Paléolithique supérieur et leur signification. Congrès International d'Anthropologie et d'Archéologie Préhistoriques*. Genève: Compte Rendu de la XIVe session.

Cacho, C., & de la Torre, I. (2005). Les harpons magdaléniens sur le versant méditerranéen espagnol. In V. Dujardin (Ed.), *Industrie osseuse et parures du Solutréen au Magdalénien en Europe. Actes de la table ronde sur le Paléolithique supérieur récent* (pp. 257–266). Paris: Société Préhistorique Française (Mémoire 39).

Campana, D. V. (1989). *Natufian and protoneolithic bone tools. The manufacture and use of bone implements in the Zagros and the Levant. British Archaeological Reports International Series 494*. Oxford: Archaeopress.

Camps-Fabrer, H. (1967). *Industrie osseuse Épipaléolithique et Néolithique du Maghreb et du Sahara. Types 31–54*. Algérie: Centre de Recherches Anthropologiques, Préhistoriques et Ethnographiques.

Casabó, J. (2005). *Paleolítico superior final y Epipaleolítico en la Comunidad valenciana*. Museo Arqueológico de Alicante-Marq: Serie Mayor, 3.

Casabó, J. (2012). Las industrias de la Cova dels Blaus (La Vall d'Uixó, Castelló). Aportación a la transición paleolítico-epipaleolítico en las comarcas septentrionales del País Valenciano. *MARQ, Arqueología y Museos, 05*, 19–51.

Chauvière, F.-X., & Rigaud, A. (2005). Les "sagaies" a "base raccourcie" ou les avatars de la typologie: du technique au "non-fonctionnel" dans le Magdalénien à navettes de la Garenne (Saint-Marcel, Indre). In V. Dujardin (Ed.), *Industrie osseuse et parures du Solutréen au Magdalénien en Europe. Table ronde sur le paléolithique supérieur récent. Angoulême (Charente), 28–30 mars 2003* (pp. 233–242). Paris: Société Préhistorique Française (Mémoires, 34).

Chauviére, F.-X. (2016). Osseous projectile points in the Magdalenian: 'True' points versus pointed waste-products. In M. C. Langley (Ed.), *Osseous projectile weaponry: Towards an understanding of Pleistocene cultural variability* (pp. 161–169). Dordrecht: Springer.

Cortés Sánchez, M. (2008). El Pirulejo (Priego de Córdoba). *Cazadores recolectores del Paleolítico superior en la sierra subbética*. Córdoba: Antiquitas 20.

Davidson, I. (1989). *La economía del final del Paleolítico en la España oriental* (p. 85). Valencia: Trabajos Varios del SIP.

Ducasse, S., & Langlais, M. (2007). Entre Badegoulien et Magdalénien, nos coeurs balancent... Approche critique des industries lithiques du Sud de la France et du Nord-Est espagnol entre 19000 et 16500 BP. *Bulletin de la Société Préhistorique Française, 104*, 771–785.

Ferrer Palma, J. E., Marqués Melero, I., Cortés Sánchez, M., Ramos Fernández, J., & Baldomero Navarro, A. (2006). Excavaciones en Cueva de Hoyo de la Mina (Málaga, Andalucía, España). Contrastación de una secuencia arqueológica clásica en el estudio del Tardiglaciar-Holoceno antiguo en el sur de la península Ibérica. In J. L. Sanchidrián Torti, A. M. Márquez Alcántara, & J. M. Fullola Pericot (Eds.), *IV Simposio de Prehistoria Cueva de Nerja. La Cuenca Mediterránea durante el Paleolítico Superior 38000–10000 años. Reunión de la VIII Comisión del Paleolítico Superior UISSP* (pp. 316–325). Málaga: Fundación Cueva de Nerja.

Fortea Pérez, F. J. (1986). El Paleolítico superior y Epipaleolítico en Andalucía. Estado de la cuestión cincuenta años después. *Actas del Congreso Homenaje a L. Siret* (pp. 67–78). Sevilla: Junta de Andalucía.

Fullola Pericot, J. M. (1979). *Las industrias líticas del Paleolítico Superior Ibérico*. Valencia: Trabajos Varios del SIP, 60, Servicio de Investigación Prehistórica.

Garcia del Toro, J. (1985). La Cueva de los Mejillones: nueva estación del Magdaleniense mediterráneo español con industria ósea. *Anales de Prehistoria y Arqueología, I*, 13–22.

Jordá Pardo, J. F., & Aura Tortosa, J. E. (2009). El límite Pleistoceno -Holoceno en el yacimiento arqueológico de la Cueva de Nerja (Málaga, España): Nuevas aportaciones cronoestratigráficas y paleoclimáticas. *Geogaceta, 46*, 95–98.

Jordá Pardo, J. F., Maestro, A., Aura Tortosa, J. E., Álvarez-Fernández, E., Avezuela, B., Badal, E., et al. (2011). Evolución paleogeográfica, paleoclimática y paleoambiental de la costa meridional de la Península Ibérica durante el Pleistoceno superior. El caso de la Cueva de Nerja (Málaga, Andalucía, España). *Boletín de la Real Sociedad Española de Historia Natural, 105*, 137–147.

Knecht, H. (1997). Projectile points of bone, antler and stone: Experimental explorations of manufacture and use. In H. Knecht (Ed.), *Projectile technology* (pp. 191–212). New York: Plenum press.

Langlais, M. (2010). *Les sociétés Magdaléniennes de l'Isthme Pyrénéen*. Paris: CTHS, collection Documents Préhistoriques, n° 26.

Le Dosseur, G. (2003). Sens et contre sens: réflexions concernant l'orientation d'un geste technique observé sur des objets en matière osseuse du Levant. *Préhistoire Anthropologie Méditerranéennes, 12*, 115–127.

López, P., & Cacho, C. (1979). La cueva del Higuerón (Málaga), Estudio de sus materiales. *Trabajos de Prehistoria, 36*, 11–81.

Mangado, X., Tejero, J. M., Fullola, J. M., Petit, M. A., García-Argüelles, P., García, M., et al. (2010). Nuevos territorios, nuevos grafismos: una visión del Paleolítico superior en Catalunya a inicios del siglo XXI. In X. Mangado (Ed.), *El Paleolítico superior peninsular. Novedades del siglo XXI* (pp. 63–83). Barcelona: SERP, Universitat de Barcelona.

Martínez Andreu, M. (1989). *El Magdaleniense superior en la costa de Murcia*. Murcia: Consejería de Cultura, Educación y Turismo.

Martínez, R. (1996). *Fauna del Pleistoceno superior en el País Valenciano: aspectos económicos, huellas de manipulación y valoración paleoambiental*. Ph.D. Dissertation, Universitat de València.

Morel, P. (1993). Impacts de projectiles sur le gibier: quelques éléments d'une approche expérimentale. In P. C. Anderson, S. Beyries, M. Otte, & H. Plisson (Eds.), *Traces et fonction: les gestes retrouvés, Actes du Colloque international de Liège, 8-10 décembre 1990* (pp. 55–57). Liège: ERAUL.

Odar, B. (2012). Archers at Potočka zijalka? *Arheološki vestnik, 63*, 433–456.

Olària, C. (1999). *Cova Matutano (Vilafamés, Castellón). Un modelo ocupacional del Magdaleniense superior-final en la vertiente mediterránea peninsular*. Castelló: Monografies de Prehistòria i Arqueologia Castellonenques.

Pérez Ripoll, M., & Raga, J. A. (1998). Los mamíferos marinos en la vida y en el arte de la Prehistoria de la Cueva de Nerja. In J. L. Sanchidrián & M. D. Simón (Eds.), *Las culturas del Pleistoceno superior en Andalucía* (pp. 251–275). Nerja: Patronato de la Cueva de Nerja.

Pericot, L. (1942). *La Cova del Parpalló (Gandía, Valencia)*. Madrid: C.S.I.C. Instituto Diego Velázquez.

Pétillon, J.-M. (2000). Les pointes à base fourchue magdaléniennes: Approche fonctionnelle. *Préhistoire Anthropologie Méditerranéennes, 9*, 29–55.

Pétillon, J.-M. (2002). Typologie et utilisation: l'exemple des pointes à base fourchue magdaléniennes. In M. Patou-Mathis, P. Cattelain, & D. Ramseyer (Eds.), *L'industrie osseuse pré- et protohistorique en Europe. Approches technologiques et fonctionnelles, Actes du colloque 1.6, XIVe Congrès de l'UISPP. Bulletin de la Société Préhistorique Française, 101*, 45–53.

Pétillon, J.-M. (2005). Tir expérimental de pointes à base fourchue en bois de renne. In V. Dujardin (Ed.), *Industrie osseuse et parures du Solutréen au Magdalénien en Europe, actes de la table ronde sur le Paléolithique supérieur récent, Angoulême (Charente) 2003* (pp. 243–256). Paris: Société Préhistorique Française (Mémoires, 39).

Pétillon, J.-M. (2006). *Des Magdaleniens en Armes. Technologie des armatures de projectile en bois de cervidé du Magdalénien supérieur de la Grotte d'Isturitz (Pyrénnés-Atlantiques)*. Treignes, Belgium: Artefacts 10, Editions du Cedarc.

Pétillon, J.-M., & Averbouh, A. (2012). Le travail du bois de renne dans les couches badegouliennes. In J. Clottes, J.-P. Giraud, & P. Chalard (Eds.), *Solutréen et Badegoulien au Cuzoul de Vers. Des chasseurs de rennes en Quercy*. Liège: ERAUL.

Plisson, H., & Geneste, J.-M. (1989). Analyse technologique des pointes à cran solutréennes du Placard (Charente), du Fourneau-du-Diable, du Pech-de-la-Boissière et de Combe-Saunière (Dordogne). *Paléo, 1*, 65–106.

Real, C. (2012). Aproximación metodológica y nuevos datos sobre los conjuntos arqueozoológicos del Magdaleniense superior de la Cova de les Cendres. *Archivo de Prehistoria Levantina, XXIX*, 99–120.

Reid, C. H. (1910). *British Museum handbook to the ethnographical collections*. Oxford: Oxford University Press.

Rigaud, A. (2006). Étude technologique des baguettes demi-rondes de Labastide (Hautes-Pyrénées). *Archéologie des Pyrénées Occidentales et des Landes, 25*, 229–246.

Romàn, D., & Villaverde, V. (2011). Los arpones del Magdaleniense superior mediterráneo. Valoración tipológica y cronoestratigráfica a partir de nuevas piezas halladas en la Cova de les Cendres (Teulada-Moraira, País Valenciano). *Zephyrus, LXVII-1*, 27–34.

Román, D., & Villaverde, V. (2012). The Magdalenian harpoons from the Iberian Mediterranean, base don pieces from Cova de les Cendres (Teulada-Moraira, Valencian región). *Quaternary International, 272–273*, 33–41.

Rueda Torres, J. M. (1987). La industria òssia del Paleolític Superior de Serinyà: Reclau Viver i Bora Gran d'En Carreres. *Cypsela, VI*, 229–236.

Tejero, J. M. (2005). *El treball de l'os a la Prehistòria. Anàlisi morfotècnica de la indústria sobre matèries dures animals de la cova del Parco (Alòs de Balaguer, Lleida)*. Barcelona: Societat Catalana d'Arqueologia.

Tejero, J.-M., Avezuela, B., Maicas, R., & Cacho, C. (2013). Osseous material industry from the Magdalenian levels of La Peña de Estebanvela (Segovia). A preliminary technological study. In F. Lang (Ed.), *The sound of bones. Proceedings of the 8th meeting of the ICAZ worked bone research group in Salzburg 2011* (pp. 237–249). Salzburg: Schriften zur Archäologie und

Archäometrie an der Paris Lodron-Universität Salzburg, Band 5. Archaeoplus.

Villaverde, V. (1994). *Arte paleolítico de la Cova del Parpalló. Estudio de la colección de las plaquetas y cantos grabados y pintados*. Valencia: Museo de Prehistoria.

Villaverde, V. (2001). El Paleolítico superior: El tiempo de los cromañones. Periodización y características. In V. Villaverde (Ed.), *De neandertales a cromañones. Los inicios del poblamiento humano en las tierras valencianas* (pp. 177–218). Valencia: Universidad de València.

Villaverde, V., & Martínez Valle, R. (1995). Características culturales y económicas del final del Paleolítico superior en el Mediterráneo español. In V. Villaverd (Ed.), *Los últimos cazadores. Transformaciones culturales y económicas durante el Tardiglaciar y el inicio del Holoceno en el ámbito mediterráneo* (pp. 79–117). Alicante: Colección Patrimonio, 22.

Villaverde, V., Martínez, R., Badal, E., Guillem, P., García, R., & Menargues, J. (1999). El Paleolítico superior de la Cova de les Cendres (Teulada, Moraira). Datos proporcionados por el sondeo efectuado en los cuadros A/B 17. *Archivo de Prehistoria Levantina, XXIII*, 6–65.

Villaverde, V., Aura Tortosa, J. E., & Barton, C. M. (1998). The Upper Paleolithic in Mediterranean Spain: A review of current evidence. *Journal of World Prehistory, 12*(2), 121–198.

Villaverde, V., & Román, D. (2006). Los arpones del Magdaleniense superior de la Cova de les Cendres y su valoración en el context del Magdaleniense mediterráneo. *Munibe, 57*(2), 2017–2225.

Villaverde, V., Roman, D., Martínez, R., Badal, E., Bergadà, M. M., Guillem, P. M., et al. (2010). El Paleolítico superior en el País Valenciano. Novedades y perspectivas. In X. Mangado (Ed.), *El Paleolítico superior peninsular. Novedades del siglo XXI* (pp. 85–113). Barcelona: SERP, vol. XXI, Universitat de Barcelona.

Villaverde, V., Roman, D., Pérez Ripoll, M., Bergadà, M. M., & Real, C. (2012). The end of the Upper Magdalenian in the Mediterranean Basin of the Iberian Peninsula. *Quaternary International, 272–273*, 33–41.

Chapter 9
A Review of the Osseous Projectile Points from the Upper Paleolithic of Portugal

Marina Almeida Évora

Abstract In this chapter, I provide an overview of Portuguese Upper Paleolithic projectile points made from mammal bone and red deer antler. These artefacts were recovered from nine archaeological sites, located in two geographic regions: Estremadura and Algarve. The majority of the Upper Paleolithic osseous assemblages from Portugal come from old excavations, and as early studies of the points are rare or only preliminary in nature, our understanding of this industry in Portugal is poorly understood. Consequently, this chapter will address morphologic and functional variability of the Portuguese technology, and focus on several aspects including fracture and stigmata patterns remaining from their manufacture and use. As a preliminary conclusion, it appears that these osseous projectile points share features with similar others from the Southern Iberian region.

Keywords Gravettian • Solutrean • Magdalenian • Fracture types • Morphology

Introduction

This chapter reviews Portuguese Upper Paleolithic osseous projectile artefacts and the archaeological sites from which they were recovered. The information presented here brings together data from various Portuguese archaeological journals and congress proceedings, much of it never before available in English (e.g., Aubry et al. 1992; Aubry and Moura 1993, 1994; Cardoso and Gomes 1994; Moura and Aubry 1995; Zilhão 1995; Aubry et al. 1997; Chauviére 2002; Almeida et al. 2004; Bicho et al. 2004).

Most of these assemblages come from old excavations which were often subjected to artefact selection first in the field, and then later, in the museum where they were curated, they must, therefore, be interpreted with caution. A few assemblages such as Vale Boi, Algarve and Buraca Grande, Estremadura come from more recent excavations and can therefore be considered more representative, though with potential sample size problems, of each geographic areas. As for Vale Boi, all sediments were sieved and all artefacts collected, the faunal assemblage is large, well preserved allowing the identification of osseous utensils in the various stages of production.

The Portuguese projectile points described herein were examined first with the naked eye, then with the use of a binocular microscope. All surface alterations and basic morphometric data were recorded.

The Archaeological Sites and the Sample

The archaeological sites that preserved organic projectile points consist of rockshelter and cave sites located mainly in Estremadura (Buraca Grande, Abrigo do Lagar Velho, Lapa dos Coelhos, Gruta do Caldeirão, Casa da Moura, Gruta da Furninha, Lapa da Rainha, Gruta das Salemas). The only site outside of this region is Vale Boi Rockshelter, located in Southwestern Algarve (Fig. 9.1). Table 9.1 provides an overview of the projectile point data described below.

Buraca Grande

The archaeological site of Buraca Grande is a rockshelter located in Serra de Sicó, near Pombal (Leiria). It was discovered in 1990 by T. Aubry and H. Moura and its stratigraphic sequence is characterized by a lower level with no absolute dates, an overlying level with Gravettian, Proto-Solutrean

M.A. Évora (✉)
ICArEHB - Interdisciplinary Center for Archaeology and Evolution of Human Behaviour, FCHS da Universidade do Algarve, Campus de Gambelas, 8005-139 Faro, Portugal
e-mail: marevora@gmail.com

Michelle C. Langley (ed.), *Osseous Projectile Weaponry*, Vertebrate Paleobiology and Paleoanthropology,
DOI 10.1007/978-94-024-0899-7_9

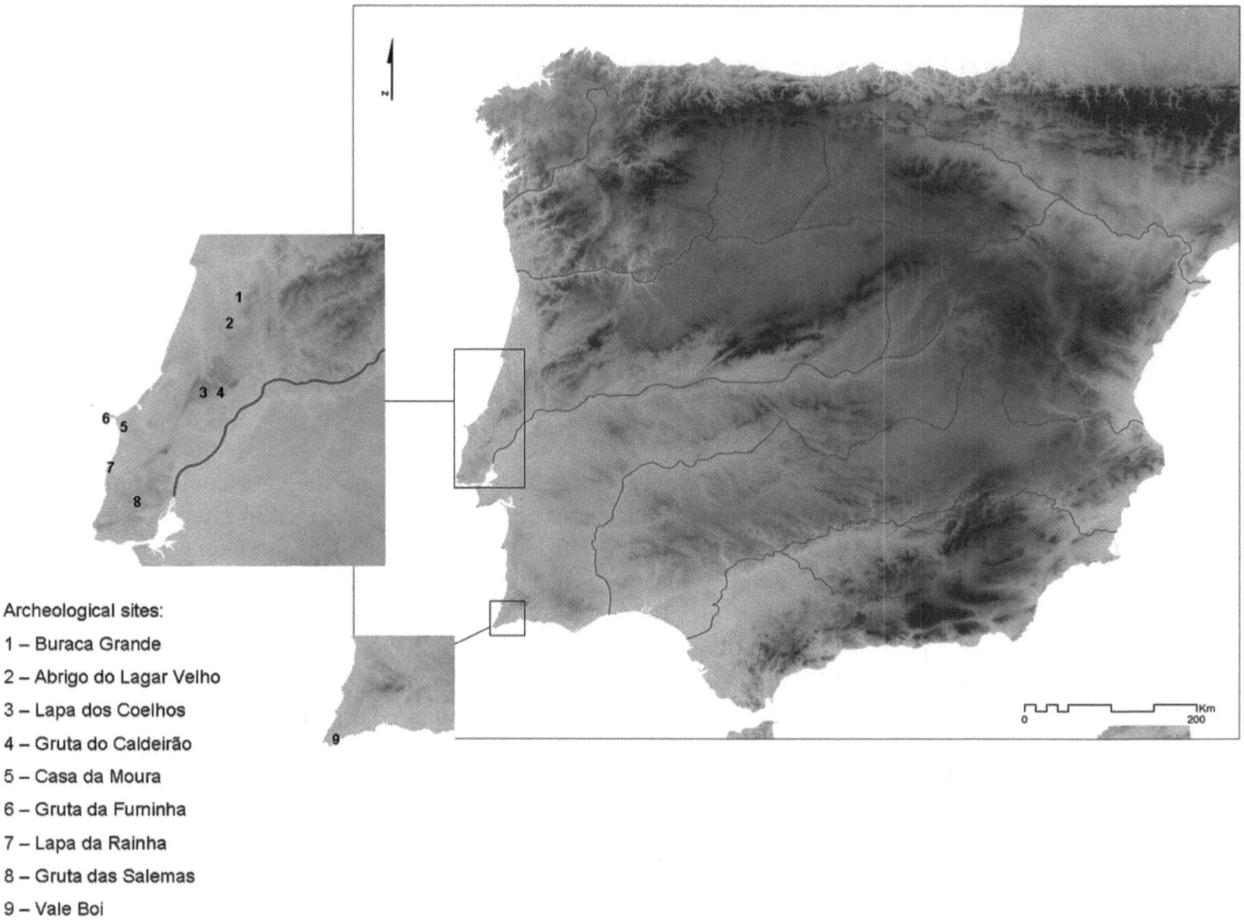

Fig. 9.1 Archaeological sites mentioned in text

and Solutrean artefacts, and at the top of the sequence, another problematic level with mixed artefacts dating from the Upper Paleolithic to modern times (Aubry et al. 1992). The rockshelter is composed of two chambers, but it was from the rearmost chamber (square K17) that a *baguette demi-ronde* fragment was recovered in a mixed sediment, and the resulting date of 13,050 ± 100 years BP provided the chronological placement of the archaeology (Aubry et al. 1997).

Also recovered from this site were eight osseous projectile points. Of these, three are of the simple base type with a convergent morphology, two have plane-convex sections and one an elliptical mesial section. Total length of these points ranges between 52 and 105 mm. One of these artefacts dates to the Gravettian, while the other two are Magdalenian. All are made from red deer antler. There is also a Gravettian single bevel point with a convergent morphology, a plane-convex mesial section and a total length of 67 mm (Fig. 9.2a). A Magdalenian *baguette demi-ronde* made from red deer antler (69 mm total length) was also found. This last artefact has a plane-convex mesial section and several diagonal striations on its inferior face. The superior face is decorated with small concavities, placed in pairs (side by side) along the length of the artefact (Fig. 9.2b). Other fragments were also recovered at Buraca Grande: mesial fragments made from red deer antler and mammal bone with fusiform morphology and diversified mesial sections ranging from 31 to 50 mm in total length and 7 to 10 mm in thickness (Table 9.1).

Abrigo do Lagar Velho

The archaeological site of Lagar Velho rockshelter is located in the Lapedo Valley, near Leiria, on the base of a limestone outcrop facing north. The site was subjected to earthmoving by the land owner with 2–3 m of sediments removed from the rockshelter (Zilhão and Almeida 2002). The archaeological materials recovered from sector TP (Hanging Remnant deposit) include two osseous projectiles. These came from levels TP06 and C6 (terminal Gravettian) dating to 22,000 ± 180 years BP (Angelucci 2002).

Table 9.1 Inventory of Portuguese Upper Paleolithic projectile points: *ind* indeterminate, *G* Gravettian, *PS* Proto-Solutrean, *S* Solutrean, *M* Magdalenian

Archaeological Site	Chronology	Level/Layer	Raw material	Artefact	Type	Morphology	Distal section	Mesial section	Proximal section	Total length (mm)	Total width (mm)	Total thickness (mm)	Deposited in
Buraca Grande	M	K17A/12	Antler	Whole	Projectile—baguette demi-ronde	Lozange	Elliptic	Plano-convex	Elliptic	69	12	6	DRCC- Leiria
Buraca Grande	M	C8C/C9A	Antler	Mesial	Projectile?—indeterminate	Fusiform	–	Circular	–	41	7	7	DRCC- Leiria
Buraca Grande	G	K20.357.C9B	Mammal bone	Distal/mesial	Projectile—indeterminate	Fusiform	Fusiform	Elliptic	Elliptic	50	11	9	DRCC- Leiria
Buraca Grande	M	K17 B/6 C9A.C14 (C8c)	Antler	Mesial	Projectile—indeterminate	Indeterminate	–	Circular	–	31	10	10	DRCC- Leiria
Buraca Grande	G	K19.15	Antler?	Whole	Projectile—simple bevel	Lozange	Circular	Plano-convex	Plano-convex	67	8	6	DRCC- Leiria
Buraca Grande	M	K17. C8C-C9A.103	Antler	Whole	Projectile—simple base	Lozange	Plano-convex	Plano-convex	Plano-convex	105	12	6	DRCC- Leiria
Buraca Grande	M	N19.C9A.15	Indeterminate	Proximal/mesial	Projectile—simple base	Lanceolate	–	Plano-convex	Plano-convex	52	12	6	DRCC- Leiria
Buraca Grande	G	L17A/4 toca	Antler	Whole	Projectile—simple base	Lanceolate	Elliptic	Elliptic	Elliptic	52	10	5	IPA Pombal
Abrigo Lagar Velho	G	C6, parte W, bl abatim.	Mammal bone	Distal/mesial?	Projectile—indeterminate	Fusiform	Elliptic	Elliptic	–	34	6	5	DGPC-Lisboa
Abrigo Lagar Velho	G	Camada TP08	Antler?	Distal/mesial	Projectile—indeterminate	Fusiform	Elliptic	Elliptic	–	83	6	5	DGPC-Lisboa
Lapa dos Coelhos	M	C4.NA17.G3.QSE	Mammal bone	Whole	Harpoon—simple base	Lanceolate	Elliptic	Elliptic	Elliptic	44	4	3	DGPC-Lisboa
Lapa dos Coelhos	M	C4.NA12.F3.QNW	Mammal bone	Whole	Harpoon—bipointed	Fusiform	Circular	Elliptic	Elliptic	39	3	2	DGPC-Lisboa
Lapa dos Coelhos	G	NA37.CM8	Mammal bone	Distal/mesial	Projectile—indeterminate	Fusiform	Elliptic	Elliptic	–	42	11	8	DGPC-Lisboa
Gruta do Caldeirão	M	Eb.O13.117.E7	Antler	Whole	Projectile—simple base	Lanceolate	Elliptic	Elliptic	Elliptic	59	13	6	MNA
Gruta do Caldeirão	M	Eb.O11.Sc242.E7	Mammal bone	Proximal	Harpoon?	Fusiform	–	–	Plano-convex	40	8	6	MNA
Gruta do Caldeirão	S-M	Fa.P13.Sc290	Indeterminate	Distal	Projectile—bipointed?	Fusiform	Elliptic	Assymetric	–	33	7	6	MNA
Casa da Moura	G	Unknown	Mammal bone	Whole	Projectile—simple bevel	Lanceolate	Elliptic	Trapezoidal	Trapezoidal	145	13	11	MG
Casa da Moura	G	Unknown	Antler	Proximal/mesial	Projectile—simple base	Lanceolate	–	Elliptic	Elliptic	80	17	12	MG
Casa da Moura	G	f=1.60	Mammal bone	Proximal	Projectile—simple bevel	Indeterminate	–	Quadrangular	Quandrangular	44	9	6	MG
Casa da Moura	S-M	–	Mammal bone	Distal/mesial	Projectile	Indeterminate	Circular	Circular	–	54	11	9	MG
Gruta da Furninha	Unknown	Unknown	Antler	Whole	Projectile—simple base	Lanceolate	Oval	Oval	Oval	835	127	67	MG
Lapa da Rainha	M	Corte II sec 7 a 10	Mammal bone	Whole	Projectile—trapue	Lozange	Plano-convex	Plano-convex	Plano-convex	66	14	4	UA
Gruta das Salemas	G	Unknown	Mammal bone	Whole	Projectile—simple base	Lanceolate	Circular	Elliptic	Elliptic	179	14	11	MG
Gruta das Salemas	S	t-c	Mammal bone	Whole	Projectile—bipointed	Fusiform	Elliptic	Elliptic	Elliptic	140	8	6	MG
Vale Boi	G	G24.18	Antler	Whole	Projectile—bipointed	Fusiform	Elliptic	Circular	Elliptic	138	8	8	UALG
Vale Boi	G	H23.13	Mammal bone	Whole	Projectile—bipointed	Fusiform	Circular	Circular	Circular	61	5	5	UALG
Vale Boi	G	H23.13	Mammal bone	Whole	Projectile—bipointed	Lanceolate	Elliptic	Circular	Circular	82	10	8	UALG

(continued)

Table 9.1 (continued)

Archaeological Site	Chronology	Level/Layer	Raw material	Artefact	Type	Morphology	Distal section	Mesial section	Proximal section	Total length (mm)	Total width (mm)	Total thickness (mm)	Deposited in
Vale Boi	G	H24.26	Indeterminate	Distal/mesial	Projectile—indeterminate	Fusiform	Circular	Circular	Elliptic	37	5	4	UALG
Vale Boi	S	J15.B2	Antler	Whole	Projectile—bipointed	Fusiform	Circular	Circular	Circular	896	80	64	UALG
Vale Boi	G	J18.4.9	Antler	Mesial	Projectile—indeterminate	Indeterminate	–	Oval	–	137	81	56	UALG
Vale Boi	G	J18.4.6	Antler	Proximal	Projectile—simple bevel	Fusiform	–	oval	–	135	77	49	UALG
Vale Boi	G	G25.21	Mammal bone	Distal	Projectile—pointed	Indeterminate	Circular	–	–	10	10	4	UALG
Vale Boi	S	G24.9	Mammal bone	Mesial	Projectile—indeterminate	Fusiform	–	Plano-convex	–	28	6	6	UALG
Vale Boi	G	G24.17	Antler	Mesial	Projectile—indeterminate	Fusiform	Oval	Oval	Oval	333	91	69	UALG
Vale Boi	S	J20.19	Antler	Distal	Projectile—indeterminate	Indeterminate	Oval	Oval	–	175	88	67	UALG
Vale Boi	S	J20.16	Antler	Distal	Projectile—indeterminate	Indeterminate	Oval	Oval	–	128	56	47	UALG
Vale Boi	G	H24.9	Antler	Mesial	Projectile—indeterminate	Fusiform	Plano-convex	Plano-convex	–	218	62	52	UALG
Vale Boi	G	G24.20	Mammal bone	Mesial	Projectile—indeterminate	Fusiform	–	Plano-convex	–	185	63	50	UALG
Vale Boi	S	H24.4	Antler	Distal	Projectile—indeterminate	Indeterminate	Plano-convex	–	–	17	71	65	UALG
Vale Boi	G	H24.21	Mammal bone	Mesial	Projectile—indeterminate	Fusiform	–	Elliptic	–	19	4	3	UALG
Vale Boi	G	H24.28	Antler	Mesial/proximal	Projectile—indeterminate	Fusiform	Elliptic	Elliptic	–	58	9	7	UALG
Vale Boi	G	H24.26	Indeterminate	Distal/mesial	Projectile—indeterminate	Fusiform	Oval	Oval	–	513	85	73	UALG
Vale Boi	PS	Z26.12	Indeterminate	Mesial	Projectile—indeterminate	Fusiform	–	Circular	–	108	56	56	UALG
Vale Boi	PS	Z25.5	Mammal bone	Distal	Projectile—indeterminate	Indeterminate	–	Oval	–	117	49	28	UALG
Vale Boi	G	O28.6	Antler	Mesial	Projectile—indeterminate	Fusiform	–	Oval	–	31	79	84	UALG
Vale Boi	G	O28.11	Antler	Mesial	Projectile—indeterminate	Fusiform	–	Oval	–	113	51	39	UALG
Vale Boi	S	H15.5	Indeterminate	Distal/mesial	Projectile—indeterminate	Indeterminate	–	Oval	–	163	57	49	UALG
Vale Boi	G	L20.4.6	Antler	Mesial	Projectile—indeterminate	Indeterminate	–	Plano-convex	–	86	70	60	UALG
Vale Boi	S	J16.C1	Antler	Mesial	Projectile—indeterminate	Indeterminate	–	Oval	–	137	85	66	UALG
Vale Boi	S	I17.B6	Antler	Proximal	Projectile—simple bevel	Fusiform	–	Oval	–	258	11	83	UALG
Vale Boi	S	H17.C1	Antler	Distal	Projectile—indeterminate	Indeterminate	–	Oval	–	191	89	52	UALG

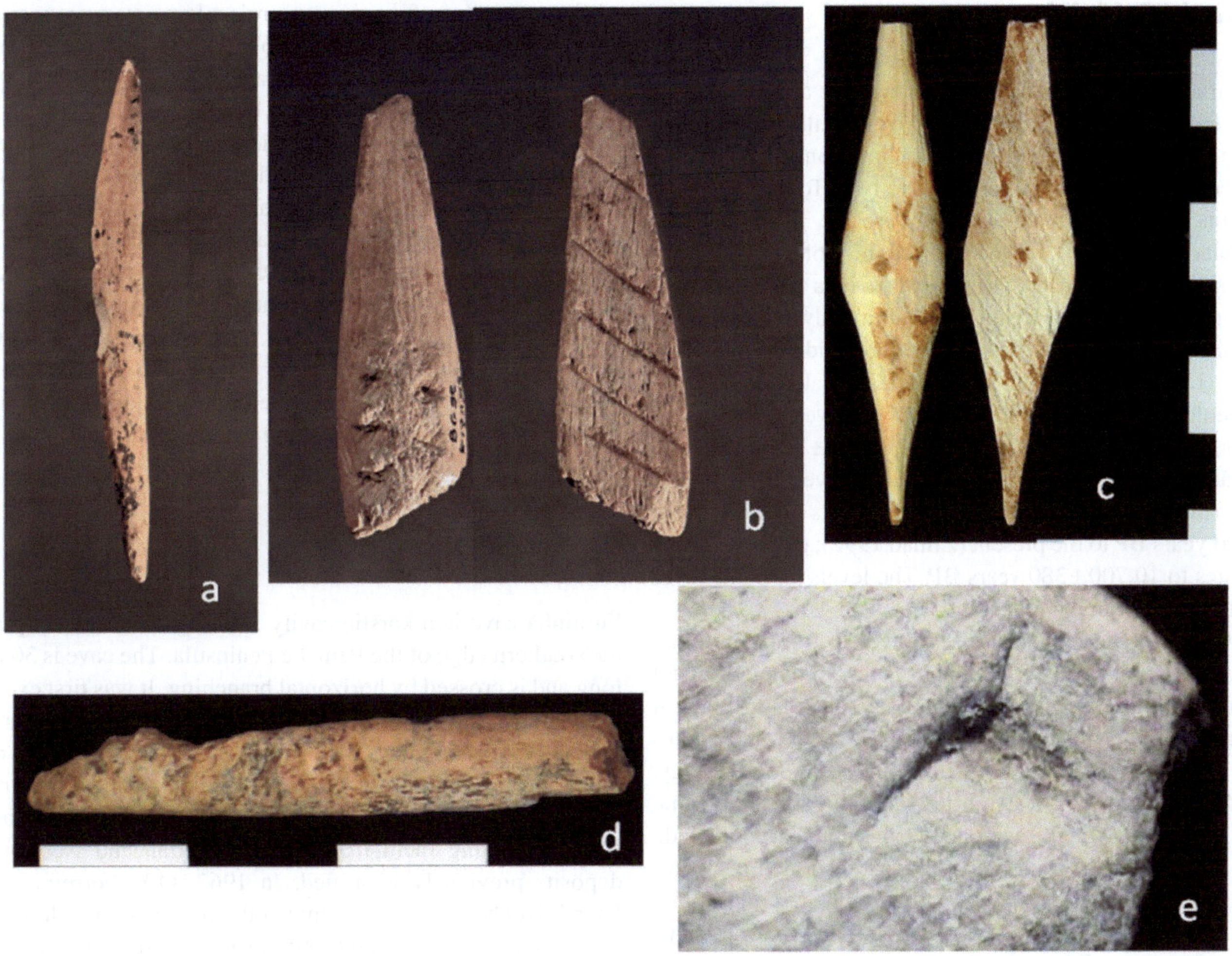

Fig. 9.2 Buraca Grande: (**a**) Gravettian single bevel, (**b**) Magdalenian baguette demi-ronde (photographed by Jaime Abrunhosa); Lapa da Rainha: (**c**) Baguette demi-ronde; Gruta do Caldeirão: (**d**) Magdalenian fragment of an harpoon (?); Gruta da Furninha: (**e**) Incised mark in the proximal end of the simple base point (20× magnification)

One of the recovered projectiles is made from mammal bone, with fusiform morphology and elliptical mesial section (34 mm total length, 5 mm thickness). Since it is a distal/mesial fragment its typology cannot be determined. The second point is also a distal/mesial fragment, possibly made from red deer antler. It also has a fusiform morphology and an elliptical mesial section (83 mm total length, 5 mm thickness).

Lapa dos Coelhos

The Lapa dos Coelhos archaeological site is located near the Almonda spring in Torres Novas. Several archaeological excavations have uncovered an Upper Paleolithic sequence with three human occupations (Almeida et al. 2004). The site has eight stratigraphic layers covering the time span from the Upper Paleolithic to historical periods. Layer 3 has an AMS date of 11,660 ± 60 years BP, while Layer 4 a ^{14}C date of 12,240 ± 60 years BP, corresponding to the Upper Magdalenian period. The artefacts have suffered little post-depositional movement, with several lithic artefacts associated with fish vertebrae and bones, found together with two organic artefacts interpreted as fishhooks (Almeida et al. 2004). Both artefacts are short tools made from mammal bone, one of them has a lanceolate morphology and an elliptic mesial section, the other tool has a fusiform morphology also with elliptic mesial section, and present a longitudinal groove, parallel to the long axis of the tool, that extends from the fracture on the proximal end until the mesial area (see Table 9.1).

Two osseous projectile points were recovered from Layer 4. They are both made on mammal bone; one is a simple base type with a lanceolate morphology and an elliptical mesial section; the second is bipointed with fusiform morphology, also with an elliptical mesial section. Another point was recovered from Layer 8, and is made from mammal bone. It has a convergent morphology and an elliptical mesial section.

Gruta do Caldeirão

Caldeirão cave was first excavated in 1979 and then again from 1982 to 1988 by a team led by J. Zilhão. The cave entrance faces south and is located in limestone hilly country crossed by the Nabão River, near the city of Tomar. The cave contains sediments resulting from erosive processes that took place during the Pleistocene until the Holocene, and has human occupations dating to several periods (Zilhão 1995). The stratigraphic sequence corresponds mainly to Pleistocene deposits: in the base of the sequence are Middle Paleolithic layers (L-Q), rich in hyena remains (Zilhão 1995); next the archaeological layers Fa-K, the top of layer K dating to 28,000 years BP and layer Fc to 18,840±200 years BP (Zilhão 1995). On top of these layers, are layers A, B, C and Ec, with an accumulation of sediments dating from 18,000 years BP to the present (Zilhão 1992); the top of layer Eb dates to 10,700±380 years BP. The levels atop this layer date to the Neolithic (Zilhão 1995).

Two projectile points were recovered from Caldeirão cave. From Solutrean/Magdalenian layer Fa, came a distal fragment with a fusiform morphology and asymmetrical mesial section. From the Magdalenian layer Eb, came a complete antler projectile point with convergent morphology, simple base and elliptical mesial section. From this same layer came a proximal fragment of a possible harpoon made from mammal bone with plano-convex section (Fig. 9.2d). This proximal fragment has parallels with harpoons illustrated in Julien (1999:135, see Figs. 1:1–2 and 6). These harpoons share a small lateral bulge in the proximal end, and as stated by Julien (1999), this is an attribute found in some Spanish Magdalenian unilateral harpoons.

Gruta da Casa da Moura

Casa da Moura cave is located in a limestone outcrop on the Cesaredas plateau, near Óbidos. The cave entrance is a 4 m deep well with access to a wide room, divided in two parts by a substantial block. In 1865 and 1866 N. Delgado performed the first excavations at the site, and recovered artefacts from near the entrance well. Breuil defined an Upper Paleolithic human occupation at the site in 1918 (Zilhão 1995), and in 1987, L. Straus carried out further archaeological work in the cave confirming the stratigraphy proposed by Delgado, along with the fact that the cave entrance was open before the Solutrean, then being occupied by wolfs (Straus et al. 1988). The cave has a date of 25,090±220 years BP obtained from a wolf mandible recovered from the base of the stratigraphic sequence, though above the travertine (Straus et al. 1988; Zilhão 1995). According to Zilhão (1995), the osseous artefacts came from Delgado's 'Inferior Deposit' can be attributed to the Final Gravettian and Upper Solutrean.

From this cave, four osseous projectile points were recovered. A complete single bevel point (Fig. 9.3a) made from mammal bone has a slight lanceolate morphology and a trapezoidal mesial section. The simple bevel has oblique striations that are extended over the entire width and length of the bevel. This point also has a groove located in the center of the inferior face, touching the bevel. There is also a proximal fragment of a simple bevel, also made from mammal bone, though with a quadrangular proximal section. Additionally, a mesial-proximal fragment of a simple base point (Fig. 9.3b) was recovered. This artifact is made from antler, and exhibits a lanceolate morphology and elliptical mesial section. Finally, there is a distal-mesial fragment with an indeterminate morphology (but circular mesial section) made from mammal bone.

Gruta da Furninha

Furninha cave is a karstic cavity forming a littoral scarp at the southern edge of the Peniche Peninsula. The cave is 30 m long and is crossed by horizontal branching. It was first excavated under the direction of N. Delgado during 1879 and 1880. H. Breuil in 1918, mentioned the presence of Paleolithic artefacts in several stratigraphic levels, providing the first recognition of Upper Paleolithic artefacts in this cave, separating them from the Mousterian and Neolithic deposits previously identified. In 1962, O.V. Ferreira and later J. Roche in 1974 recognized the presence of Solutrean stemmed points. J. Zilhão (1995) then worked on the materials from Furninha cave, concluding that the human occupation of the cave was ephemeral and that the stemmed points were from Neolithic or Chalcolithic cultural periods. Bicho and Cardoso (2010) refuted Zilhão's conclusions. According to these last authors, there are two main sedimentary complexes in Furninha with both being excavated in totality by Delgado: (1) the top deposit corresponding to Neolithic burials; (2) the lower deposit dated to MIS 4, 3 and 2 and has almost 9 m of depth. This deposit included faunal remains and many Mousterian lithic artefacts, as well as Solutrean points similar to those found in Vale Boi and in the Spanish Levantine region (Bicho and Cardoso 2010), confirming the earlier interpretation of Ferreira and Roche.

Amongst the artefacts that were recovered from Furninha cave, Bicho and Cardoso (2010) found a complete simple base point, with lanceolate morphology. It has an oval mesial section. Its stratigraphic provenience is unknown, but contrary to what is stated by these authors, this osseous projectile has parallels with a Gravettian bone point from Vale Boi (Évora 2008), and *not* Solutrean bone points. The proximal end has a vertical fracture and on its inferior face a depressed area in which two triangular marks are incised into the bone are visible. These marks are probably owing to hafting techniques (Fig. 9.2e).

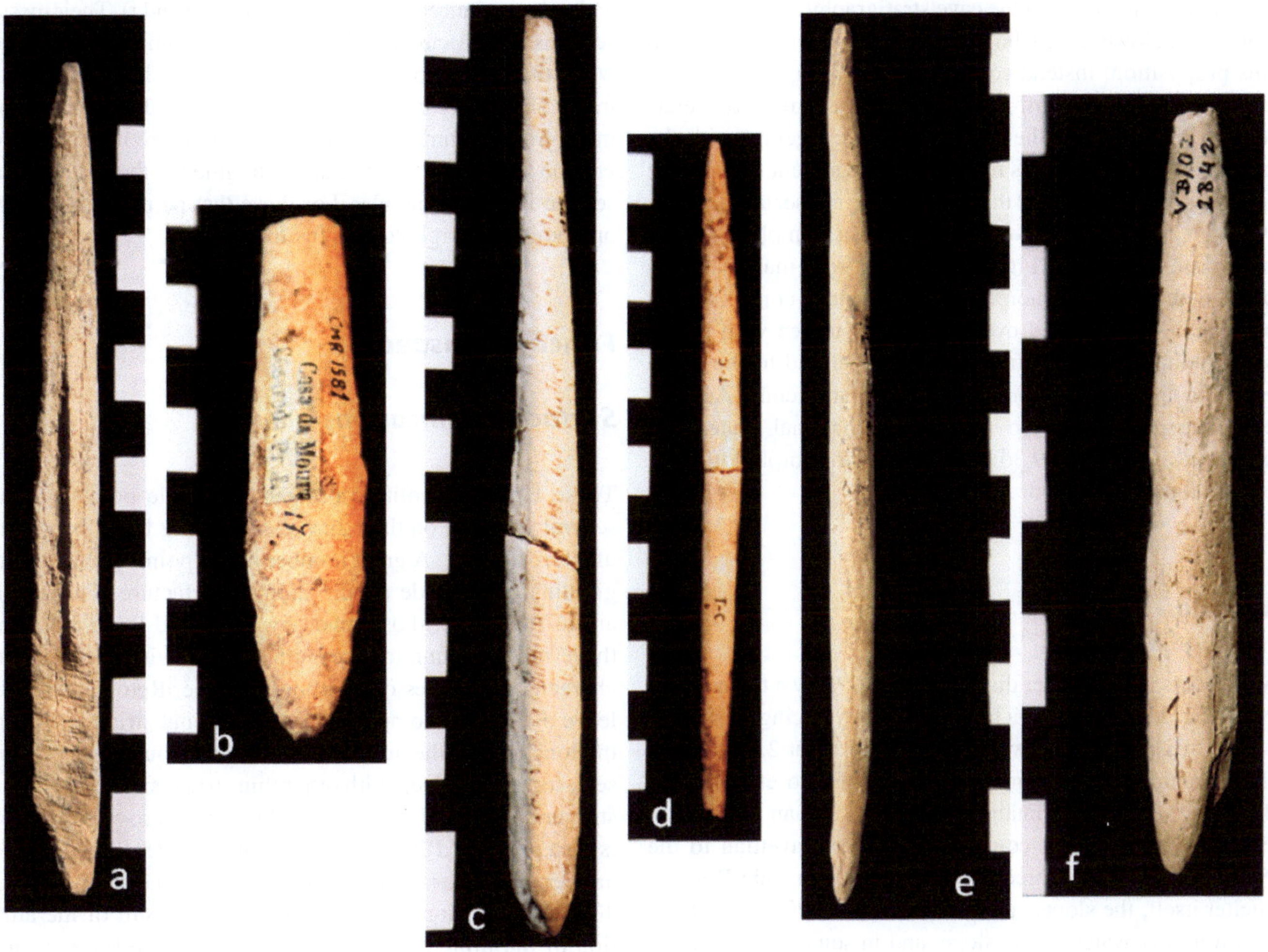

Fig. 9.3 Casa da Moura: (**a**) Single bevel point, (**b**) Simple base point fragment; Gruta das Salemas: (**c**) Simple base point, (**d**) Bipointed point; Vale Boi: (**e**) Gravettian bipointed point, (**f**) Gravettian bipointed point

Lapa da Rainha

This cave is located on the left bank of the Alcabrichel river valley near Torres Vedras. F. Almeida excavated the cave during 1968 and 1969, and a longitudinal profile (II) allows an understanding of the entire stratigraphic sequence, which contains seven layers and two human occupations (layers 4 and 3). In 1987, A. Marks tested the cave and was able to conclude, based on the presence of a Solutrean point, that the earliest human occupation dated to the Solutrean (Cardoso 1993; Marks et al. 1994; Zilhão 1995). The cave was used mainly by carnivores and only as a sporadic shelter by humans (Zilhão 1995).

From this site a whole *baguette demi-ronde* with lozenge morphology and plane-convex mesial section was recovered (Fig. 9.2c). This point has no decoration on its upper surface, but has several diagonal incisions on the inferior face that extend from one edge to the other. The distal end has a perpendicular fracture. This *baguette demi-ronde* has parallels with other Middle Magdalenian artefacts recovered from Mas d'Azil, France (Feruglio and Buisson 1999). The presence of this artefact at Lapa da Rainha may indicate a Magdalenian occupation, probably sporadic, that may have not have been recognized or documented during the 1968–1969 excavation work, owing to the mixed state of the sediments.

Gruta das Salemas

Salemas cave is located on the top of the slope of the Lousã river valley, near Loures. It was discovered by L.A. Castro who worked there in 1959, and latter O.V. Ferreira and J.C. França totally excavated the cave in 1959–1960 (Zilhão 1995). The cave has Neolithic burials, along with Solutrean, Gravettian, and Middle Paleolithic artefacts (Zilhão 1995). The bone tools said to be recovered from level III have an unsecured chronology, but Zilhão (1995) attributes the bone industry to the Aurignacian or Gravettian, the latter being the best represented level in the cave stratigraphy, along with the Solutrean. However, Bicho (2000, 2005) does not agree with this proposition, instead considering the Aurignacian occupation of the cave doubtful, arguing that the Dufour bladelets identified came from the Gravettian layer together with the backed bladelets. For this reason, the assignment of the bone industry to Gravettian or the Solutrean is not secure.

From Salemas cave was recovered a complete simple base projectile point (Fig. 9.3c), made from mammal long bone, with lanceolate morphology and a plane-convex mesial section. It is decorated over the entire surface with small, short oblique incisions from the proximal end to the distal end. The distal end exhibits a perpendicular fracture. Another complete projectile, also made from mammal bone, is a bipointed point (Fig. 9.3d), with fusiform morphology and elliptical mesial section.

Vale Boi

Vale Boi is a rockshelter discovered in 1998 by a team led by N. Bicho. It is located in a limestone valley, facing west, near Vila do Bispo, in Southwestern Algarve, about 2.5 km from the Atlantic Ocean (Manne et al. 2012; Bicho et al. 2013). The cave sediments contain evidence for human occupation covering every techno-complex from the Gravettian to the Neolithic. There are three excavation areas in Vale Boi: the shelter itself, the slope, and the terrace. In 2000, the first test pits were excavated in the slope, and in square G25, human occupation levels with Magdalenian, Proto-Solutrean, Gravettian and possibly Mousterian associations were recorded (Bicho et al. 2004). These were found without sterile layers between them. Three AMS dates exist for this sequence: c. 24,500 years BP, c. 17,600 years BP and c. 18,500 years BP (Bicho et al. 2004). Additionally, for square Z27, there is an age determination of c. 22,500 years BP. The deposits held *in situ* artefacts, including body ornaments, bone tools, portable art, well preserved terrestrial and marine fauna, and lithic assemblages. The rockshelter area corresponds to a shelter that collapsed at the end of the Last Glacial Maximum (LGM). The chronology covers the Solutrean and Magdalenian cultural periods. The terrace has the longest sequence at the site, with the Neolithic being represented by ceramics, lithics, and wild and domestic animals. A single human tooth dates from the Mesolithic. It also has Solutrean and Proto-Solutrean occupations. Bellow this layer (layer 4), there is a Gravettian occupation dating to c. 25,000 years BP, which is represented by lithic tools, adornments, marine and terrestrial fauna, and portable art. Layer 5 and 6 date to c. 28,000 years BP.

From Vale Boi were recovered four whole projectiles and 23 fragments of projectiles (see Table 9.1). The complete artefacts include bipointed points, three with fusiform morphology and only one lanceolate example (Fig. 9.3e and f). Their mesial sections are all circular. Two are made from mammal bone, while the other two are made from red deer antler. The fragments are mainly made from antler (n=14), though a few are made from mammal bone (n=5), and four are indeterminate. For the Gravettian and Solutrean fragments, the main mesial section is oval or circular. For those that permit analysis, the predominant morphology is fusiform.

Functional aspects

Surface Modifications

These Upper Paleolithic osseous projectile points present several *stigmata* on their surfaces resulting from their manufacture and use. A great number of the points preserve longitudinal stria made during their manufacture. These stria are of two types: those made by retouched lithic tools and those made by unretouched lithic tools, with each leaving characteristic traces on the bone surface. Retouched edges leave a *stigmata* composed of longitudinal stria, parallel to the long axis of the artefact, sometimes grouped together in sets, sometimes deep with other thin stria inside them. These stria are present all over the surface. A micro-wave pattern is also present (d'Errico and Giacobini 1985). These waves are perpendicular to the longitudinal stria and can be seen at 40× magnification. These marks are the result of the attrition of the lithic edge when passing over the bone surface (d'Errico and Giacobini 1985; Évora 2008). The unretouched tool, on the other hand, leaves a different type of stria. These stria are also longitudinal and parallel to the long axis of the artefact, but are thin, not too deep, and are usually not grouped in sets (d'Errico and Giacobini 1985; Évora 2008). Additionally, some stria are probably the result of using abrasives, such as sand or a stone with coarse grains in the final part of the manufacturing process, or even perhaps, left as a result of the resharpening of the distal end. This scenario appears to be the case for a Gravettian point from Lapa dos Coelhos.

Some projectile points also show near their distal end, short striations with an oblique and transversal orientation in relation to the long axis of the artefact that could be the result of use. In the case of the Furninha projectile point, a slight concavity can be seen close to the proximal end and may have resulted from the hafting of the point causing a compression of the bone in this specific location.

The Gravettian bone point from Gruta das Salemas has a faceted surface and presents short horizontal and oblique incisions along the entire surface. These incisions have a V section and some are deeper than others. Inside some of the incisions, we can see fine longitudinal striations. These incisions were made after the longitudinal striations which resulted from the manufacture of the bone point. The *baguette demi-ronde* from Buraca Grande also has decorations on its superior surface as mention above. Similarly, the Gravettian point from Cova da Moura has its surface faceted and several oblique lines incised along its single bevel. All projectiles with bevel bases have oblique incisions located on this section and are part of the hafting techniques. This same method is seen on the inferior surface of the *baguette demi-ronde* from Lapa da Rainha.

Fracture Types

In general four types of impact fractures are represented in the assemblages: oblique, *languette*, perpendicular and splinter, though some points present on their distal end three or four negative scars resulting from direct impact, as is the case for a Casa da Moura Gravettian projectile point (for example).

For the Gravettian assemblage, the predominant fractures for the distal end are perpendicular and oblique; for the mesial section: *languette*; and for the proximal end, oblique. For the Solutrean, there are fewer artefacts and only two oblique and one vertical fractures for the distal end were recorded, as was one splinter fracture for the proximal end. For the Magdalenian phase, the predominance is *languette* fractures for the distal end, oblique and vertical types on mesial sections, and the oblique type on the proximal end. These types of fractures indicate that the osseous points have all been used as they are characteristic of direct impact against a hard surface (such as bone) during hunting. In particular, oblique and *languette* fractures result from flexion, voluntary or accidentally, in a specific area of the projectile point that was not attached or hafted to the spear (Bertrand 1999; Pétillon 2006).

Discussion and Conclusions

Presented above are the osseous projectile points that have thus far been recovered from Upper Paleolithic contexts in Portugal. There are certain limitations which are mainly owing to:

1. Sample size: only a few complete points are known;
2. Preservation: although most faunal assemblages are well preserved as they were recovered from rockshelters;
3. Taphonomic modifications: only a few points have well preserved surfaces. This situation is mainly owing to bone fragmentation, carbonate coating on surfaces, rodent teeth marks, manganese oxide stains, trampling, cracks, osseous dissolution, and varnish. The varnish that was used in the laboratory to mark the museum inventory numbers remains a problem, making it very difficult or even impossible to observe and record manufacturing traces left on the surfaces; and
4. Almost all of the archaeological sites were excavated in the nineteenth or early twentieth centuries, and thus, not all osseous fragments were recovered in the field. Additionally, materials were sorted again in museums and more material may have been lost.

Each of these factors limit the identification of the manufacturing process of the Portuguese osseous tool tradition. Furthermore, it was only after 1990 that archaeologists began to pay more attention to this type of material culture, providing the first technological analyses.

In summary, seven Portuguese archaeological sites have Gravettian occupations, three Solutrean, and four Magdalenian occupations. The typology of the artefacts shows a predominance for the simple based and bipointed types of projectile points with a fusiform or convergent morphology. Relating to raw material choices, there is not a clear distinction between antler or mammal bone during the Gravettian (in terms of point morphology) as both were used for point manufacture during this period. A similar determination cannot be made for the Solutrean, owing to the small quantity of complete points preserved. In the Magdalenian, however, there seems to be a preference for antler to manufacture simple based points with a convergent morphology, which may be owing to the hafting techniques used as well as to the kinds of fish and game hunted. Since the points are mostly fragmented, it is difficult to make informed inferences about these choices, however, the fact that there are more distal and mesial fragments than complete elements, may indicate that Upper Paleolithic hunter-gatherers transported and butchered captured game at the habitation sites or butchering sites rather than at the killing sites. This suggestion can be made as broken point fragments remain inside the carcass until it was butchered, the fragments then being retrieved and discarded at the sites where they were preserved and later discovered.

Interestingly, only two points are decorated: a Gravettian point from Gruta das Salemas and a Magdalenian *baguette demi-ronde* from Buraca Grande. Could this lack of decorated points be interpreted as a stylistic preference within these groups of hunter-gatherers? The absence of decoration on osseous projectile points could indeed be a stylistic mark (LeMoine 1999), differentiating human groups living in Estremadura from those in Algarve. As both decorated points came from Estremadura, none from Vale Boi (Algarve), and there are several decorated stone plaquettes at this Algarve site (demonstrating the use of decoration on other media), this suggestion seems possible.

Most Portuguese Upper Paleolithic archaeological sites are located in Estremadura, perhaps resulting from the fact that this area was intensively surveyed since the nineteenth

Century, and has received much attention from archaeologists since that time. The survey for Upper Paleolithic archaeological sites in Algarve only began in 1998 with a project named "*A Ocupação Humana Paleolítica do Algarve*" (Paleolithic Human Occupation in Algarve), led by N. Bicho. This difference in the number of sites and osseous projectiles between the regions could thus be explained by the intensity of archaeological surveys and the number of sites found and excavated so far. Despite this fact, however, Vale Boi remains the archaeological site with the largest sample of Upper Paleolithic projectile points and other osseous tools in Portugal.

The Portuguese Upper Paleolithic bone industry as a whole share features with Southern Iberia. Here, some of the osseous projectiles points were recovered from rockshelter sites located near the coast where hunter-gatherers exploited coastal resources, while others are from rockshelter sites located inland from where they could exploit land and fluvial resources as well (see Villaverde et al. 2016). Besides hunting ungulates, some points appear to have been used for fishing (Évora 2013). Marine resources were exploited from the Gravettian and it is during this period that we recorded more osseous points entering the archaeological record. Examples include two projectiles from Lapa dos Coelhos that were recovered in association with fish remains. Also a number of artefacts classified as fishhooks: one from the Gravettian deposit in Vale Boi (Portugal) and another one from the Magdalenian of Nerja (Spain) which share similar morphology (Aura and Pérez 1998; Bicho et al. 2004; Évora 2008, 2013). The opposite correlation occurs during the Solutrean, when a regression of the coastline takes place as a consequence of the Last Glacial Maximum. The sites located near the shoreline from this period are presently most probably under water. Then, during the Magdalenian, the coastline reached near today's limits (Haws et al. 2011).

Another similarity to Southern Iberian archaeological sites is the fact that during the Gravettian there is a high frequency of projectile points, as opposed to the Solutrean and, to a lesser extent the Magdalenian. This observation cannot be solely attributed to a change in climate, as during the Upper Paleolithic, major climatic changes did not affect Southern Iberia like it did other regions to the north (Salgueiro et al. 2010). This fact is demonstrated in faunal assemblages previously published (Yravedra 2001a; Manne 2010), which show that the animal resources hunter-gatherers exploited continued to be the same, although not in the same frequencies. Thus, mammal bone and antler were always available as a raw material (as shed antler or hunted red deer) during the whole of the Upper Paleolithic (Évora 2013). The lower frequency of organic projectiles points and other categories of osseous artefacts during the Solutrean, and even during the Magdalenian phases, could instead be owing to a change in raw material choices for manufacturing points for hunting and fishing.

Furthermore, the osseous industry has been shown to have been adapted to all environments that hunter-gatherers exploited in Southern Iberia as shown by archaeological sites located in coastal areas like Vale Boi, Furninha, Rainha, Moura (Portugal), Nerja, Mejillones, Cendres (Spain), and other sites located inland and in fluvial areas, like Buraca Grande, Lagar Velho, Coelhos, Caldeirão, Salemas (Portugal), Pirulejo, Ambrosio and Parpalló (Spain) which are close to major rivers (Évora 2013). These locations allowed the exploitation of different kinds of habitats and their diverse resources (fish, shellfish, birds and mammals) (Villaverde et al. 1998; Yravedra 2001a, b; Davis 2002; Asquerino and Riquelme 2005; Bicho et al. 2006; Manne 2010; Villaverde et al. 2012).

In conclusion, more research is necessary to understand the technological processes of manufacturing osseous tools during the Upper Paleolithic in Portugal. In particular, it is necessary to review faunal assemblages in order to identify bone and antler fragments with debitage and manufacturing marks. This task is what we expect to accomplish in the future in order to enrich our picture of Upper Paleolithic hunter-gatherers living in these territories.

Acknowledgments I wish to thank all the people and Institutions who gave me the opportunity to study the osseous projectile points from the Upper Paleolithic archaeological sites in Portugal: N. Bicho, J. L. Cardoso, T. Aubry, F. Almeida, L. Raposo (Museu Nacional de Arqueologia, Lisbon), M. Ramalho (Museu Geológico, Lisbon) and J. Zilhão. I thank Nuno Bicho for his comments and for reviewing the manuscript. I also thank Célia Gonçalves for providing the map of Iberian Peninsula. I would like to thank Thierry Aubry and Jaime Abrunhosa for providing the images of a projectile point from Buraca Grande. I also thank the two anonymous reviewers for their comments and suggestions to the manuscript. All images were photographed by M.A. Évora except when mention otherwise.

References

Almeida, F., Angelucci, D., Gameiro, C., Correia, J., & Pereira, T. (2004). Novos dados para o Paleolítico Superior Final da Estremadura Portuguesa: resultados preliminares dos trabalhos arqueológicos de 1997–2002 na Lapa dos Coelhos (Casais Martanes, Torres Novas). *Promontória, 2*(2), 157–192.

Angelucci, D. E. (2002). The Geoarcheological Context. Portrait of the Artist as a Child. The Gravettian human skeleton from the Abrigo do Lagar Velho and its Archeological Context. *Trabalhos de Arqueologia, 22*, 58–91.

Asquerino Fernández, M. D., & Riquelme Cantal, J. A. (2005). Aproximación al aprovechamiento faunístico de mamíferos en el yacimento de Paleolítico Superior de "El Pirulejo". In *La Cuenca Mediterránea durante el Paleolítico Superior. IV Simposio de Prehistoria, Cueva de Nerja* (pp. 256–271). Nerja.

Aubry, T., Fontugne, M., & Moura, M. H. (1997). Les occupations de la Grotte de Buraca Grande depuis le Paleolithique Supérieur et les apports de la séquence holocène a l'étude de la transition Mésolithique/Néolithique au Portugal. *Bulletin de La Société Préhistorique Française, 94*(2), 182–190.

Aubry, T., & Moura, M.H. (1993). *Nouvelles donnes sur les occupations paleolithiques de la region de Rendinha (Serra de Sicó,*

Portugal). Coimbra, Portugal: Separata da 3ª Reunião do Quaternário Ibérico 439–449.

Aubry, T., & Moura, M. H. (1994). *Paleolítico da Serra de Sicó. Trabalhos de Antropologia e Etnologia* (Vol. XXXIV, Fasc. 3–4). Porto, Portugal: Actas do 1° Congresso de Arqueologia Peninsular, SPAE 43–60.

Aubry, T., Moura, M. H., & Zilhão, J. (1992). Dados preliminares sobre a organização estratigráfica da Buraca Grande do Vale do Poio Novo (Redinha), Memórias e Noticias. *Publicações do Museu e Laboratório Mineralógico e Geológico, 114*, 119–132.

Aura Tortosa, J. E., & Pérez Herrero, C. (1998). Micropuntas dobles o anzuelos? Una propuesta de estudio a partir de los materiales de la Cueva de Nerja (Málaga). In J. L. Sanchidrián Torti & M. D. Simón Vallejo (Eds.), *Las Culturas del Pleistoceno Superior en Andalucía* (pp. 339–348). Nerja, Spain: Patronato de la Cueva de Nerja.

Bertrand, A. (1999). *Les armatures de sagaies Magdaléniennes en matière dure animal dans les Pyrénées. British Archaeological Reports International Series 773*. Oxford: Archaeopress.

Bicho, N. F. (2000). *Revisão Crítica dos conhecimentos actuais do Paleolítico Superior Português. Proccedings of the 3° Peninsular Archaeology Congress* (Vol. II, pp. 425–442). Porto, Portugal: ADECAP

Bicho, N. F. (2005). The extinction of Neanderthals and the emergence of the Upper Paleolithic in Portugal. *Promontória, 3*, 174–228.

Bicho, N., & Cardoso, J. L. (2010). *Paleolithic occupations and lithic assemblages from Furninha Cave, Peniche (Portugal)* (pp. 17–38). LXVI: Zephyrus.

Bicho, N., Haws, J., & Hockett, B. (2006). Two sides of the same coin—rocks, bones and site function of Picareiro Cave, central Portugal. *Journal of Anthropological Archaeology, 25*, 485–499.

Bicho, N., Manne, T., Marreiros, J., Cascalheira, J., Pereira, T., Tátá, F., et al. (2013). The ecodynamics of the first modern humans in Southwestern Iberia: The case of Vale Boi, Portugal. *Quaternary International, 318*, 102–116.

Bicho, N., Stiner, M., & Lindly, J. (2004). Shell ornaments, bone tools and long distance connections in the Upper Paleolithic of Southern Portugal. In M. Otte (Ed.), *La Spiritualité* (pp. 71–80). Liège, Belgium: Université de Liège, ERAUL 106.

Cardoso, J. L. (1993). *Contribuição para o conhecimento dos grandes mamíferos do Plistocénico Superior de Portugal*. Oeiras, Portugal: Câmara Municipal de Oeiras.

Cardoso, J. L., & Gomes, M. V. (1994). Zagaias do Paleolítico Superior de Portugal. *Portvgalia, Nova Série, XV*, 07–31.

Chauviére, F. X. (2002). Industries et parures sur matières dures animales du Paléolithique Supérieur de la grotte de Caldeirão (Tomar, Portugal). *Revista Portuguesa de Arqueologia, 5*, 05–28.

d'Errico, F., & Giacobini, G. (1985). Approche methodologique de l'analyse de l'outillage osseaux, un exemple d'étude. *L'Anthropologie, 89*, 457–472.

Davis, S. J. M. (2002). The mammals and birds from the Gruta do Caldeirão, Portugal. *Revista Portuguesa de Arqueologia, 5*, 29–98.

Évora, M. A. (2008). Artefactos em haste e em osso do Paleolítico Superior Português. *Promontória, 6*, 9–50.

Évora, M. A. (2013). Osseous industry and exploitation of animal resources in Southern Iberia during the Upper Palaeolithic. *Quaternary International, 318*, 35–45.

Feruglio, V., & Buisson, D. (1999). Accolements de pièces à section demi-ronde. In M. Julien, A. Averbouh, D. Ramseyer, C. Bellier, D. Buisson, P. Cattelain, et al. (Eds.), *Préhistoire d'Os. Recueil d'etudes sur l'industrie osseuse préhistorique* (pp. 143–149). Aix-en-Provence, France: Publications de l'Université de Provence.

Haws, J., Funk, C. L., Benedetti, M. M., Bicho, N. F., Daniels, J. M., Minckley, T. A., et al. (2011). Paleolithic Landscapes and Seascapes of the West Coast of Portugal. In N. Bicho, J. A. Haws, & L. G. Davis (Eds.), *Trekking the shore. Changing coastlines and the antiquity of coastal settlement* (pp. 203–246). Interdisciplinary Contributions to Archaeology. New York: Springer.

Julien, M. (1999). Une tendence créatrice au Magdalénian: À propos de stries d'adhérence sur quelques harpons. In M. Julien, A. Averbouh, D. Ramseyer, C. Bellier, D. Buisson, P. Cattelain, M. Pathou-Mathis, & N. Provenzano (Eds.), *Préhistoire d'os. Recueil d'etudes sur l'industrie osseuse préhistorique* (pp. 133–142). Aix-en-Provence, France: Publications de l'Université de Provence.

LeMoine, G. (1999). Skeletal technology in context: an optimistic overview. In A. M. Choyke & L. Bartosiewicz (Eds.), *Crafting bone: Skeletal technologies through time and space, Proceeding of the 2nd meeting of the (ICAZ) worked bone research group* (BAR International Series 937). Oxford, England: J. and E. Hedges. 1–7.

Manne, T. (2010). *Upper Paleolithic foraging decisions and early economic intensification at Vale Boi, Southwestern Portugal*. Unpublished Ph.D. dissertation, University of Arizona.

Manne, T., Cascalheira, J., Évora, M., Marreiros, J., & Bicho, N. (2012). Intensive subsistence practices at Vale Boi, an Upper Paleolithic site in southwestern Portugal. *Quaternary International, 264*, 83–99.

Marks, A., Bicho, N., Zilhão, J., & Ferring, C. R. (1994). Upper pleistocene prehistory in Portuguese Estremadura: Results of preliminary research. *Journal of Field Archaeology, 21*, 53–68.

Moura, M. H., & Aubry, T. (1995). *A Pré-História Recente da Serra de Sicó. Trabalhos de Antropologia e Etnologia, Actas do 1° Congresso de Arqueologia Peninsular* (Vol. XXXV, Fasc. 3). Porto, Portugal: SPAE. 113–132.

Pétillon, J.-M. (2006). *Des Magdaleniens en armes. Technologie des armatures de projectile en bois de cervidés du Magdalénien Superieur de la Grotte D'Isturitz (Pyrénées-Atlantiques)*. Treignes, Belgium: Editions du CEDARC.

Salgueiro, E., Voelker, A. H. L., Abreu, L., Abrantes, F., Meggers, H., & Wefer, G. (2010). Temperature and productivity changes off the western Iberian margin during the last 150 ky. *Quaternary Science Reviews, 29*, 680–695.

Straus, L. G., Altuna, J., Jackes, M., & Kunst, M. (1988). New excavations in Casa da Moura (Serra D'El Rei, Peniche) and at the Abrigos de Bocas (Rio Maior), Portugal. *Arqueologia, 18*, 65–95.

Yravedra Sainz de los Terreros, J. (2001a). *Zooarquelogía de la Península Ibérica. Implicaciones tafonómicas y paleoecológicas en el debate de los homínidos del Pleistoceno Medio-Superior. British Archaeological Reports International Series 979*. Oxford: Hadrian Books Ltd.

Yravedra Sainz de los Terreros, J. (2001b). *La subsistencia en el Paleolítico Superior inicial en la región mediterránea de la Península Ibérica*. Alcoi, Spain: Recerques del Museo d'Alcoi, Núm. 10. 7–16.

Villaverde Bonilla, V., Román, D., Pérez Ripoll, M., Mercè Bergadà, M., & Real, C. (2012). The end of the Upper Palaeolithic in the Mediterranean Basin of the Iberian Peninsula. *Quaternary International, 272–273*, 17–32.

Villaverde, V., Aura, J. E., Borao, M., & Roman, D. (2016). Upper palaeolithic bone and Antler projectiles in the Spanish Mediterranean Region: The Magdalenian period. In M. C. Langley (Ed.), *Osseous projectile weaponry: Towards an understanding of Pleistocene cultural variability* (pp. 109–130). Dordrecht: Springer.

Villaverde, V., Aura, J. E., & Barton, C. M. (1998). The Upper Paleolithic in Mediterranean Spain: a review of current evidence. *Journal of World Prehistory, 12*, 121–198.

Zilhão, J. (1992). *A Gruta do Caldeirão. O Neolítico Antigo*. Trabalhos de Arqueologia 6. Lisboa, Portugal: IPPAR.

Zilhão, J. (1995). *O Paleolítico Superior da Estremadura Portuguesa*. Unpublished Ph.D. dissertation, Universidade de Lisboa.

Zilhão, J., & Almeida, F. (2002). The Archeological Framework. Portrait of the Artist as a Child. The Gravettian human skeleton from the Abrigo do Lagar Velho and its Archeological Context. *Trabalhos de Arqueologia, 22*, 29–57.

Chapter 10
Diversity and Evolution of Osseous Hunting Equipment During the Magdalenian (21,000–14,000 cal BP)

Michelle C. Langley, Jean-Marc Pétillon, and Marianne Christensen

Abstract The Magdalenian is largely defined by its diverse array of sophisticated osseous projectile weaponry. Found throughout Western Europe, Magdalenian assemblages include antler points hafted using a variety of technological designs, unilaterally and bilaterally barbed points, self-barbed points, bivalve points made of two half round rods (*baguettes demi-rondes*), and composite antler/lithic projectile points, not to mention foreshafts and spearthrower elements. As perhaps the richest assemblage of osseous projectile weaponry manufactured during the Pleistocene worldwide, a thorough understanding of this technology is essential for building a cohesive understanding of not only the technological choices made by hunter-gatherers in Western Europe during this period, but the wider economic and social role that osseous projectile technology can play in a Pleistocene age culture. This chapter aims to outline the diversity of Magdalenian osseous projectile weaponry as well as its evolution throughout the period 21,000–14,000 cal BP. How these implements were designed, manufactured, used, and maintained is described.

Keywords Western Europe • Antler • Whale bone • Spearthrower • Hunting kit • Hafting

M.C. Langley (✉)
Institute of Archaeology, University of Oxford, Oxford, UK

Archaeology and Natural History, College of Asia and the Pacific, Australian National University, Canberra, ACT, Australia
e-mail: michelle.langley@anu.edu.au

J.-M. Pétillon
CNRS – UMR 5608, Laboratoire TRACES, Université Toulouse Jean-Jaurès, Toulouse, France
e-mail: petillon@univ-tlse2.fr

M. Christensen
Institut d'Art et d'Archéologie, Université de Paris I Pantheon-Sorbonne, UMR 7041 – ArScAn, Équipe Ethnologie Préhistorique, Paris, France
e-mail: marianne.christensen@univ-paris1.fr

Introduction

Southwestern France and Northern Spain is often described as the 'core area' of development for the late Upper Paleolithic techno-complex known as the Magdalenian (e.g., Sacchi 2003). Indeed, this region's rich archaeological record of stratified cave and rockshelter sites with excellent preservation qualities, enables us to reconstruct the continuous evolution of Magdalenian osseous technology from 21,000 to 14,000 cal BP (or 17,500–12,000 BP; all dates were calibrated with the IntCal13 dataset). While we will here focus on the evidence recovered from this core area (Fig. 10.1), it should be noted that Magdalenian deposits have been identified as far north as Belgium and Germany, to Switzerland in the east, and to the coast of Southern Spain (see Villaverde et al. 2016).

Studies of Magdalenian technology have usually been conducted on a regional or local basis. This approach has resulted in the establishment of several conflicting chronological frameworks for the same geographically restricted archaeological entity. Recent research, however, has integrated lithic and osseous assemblages with faunal data, along with a growing radiometric framework, allowing the identification of three successive phases: Lower, Middle and Upper Magdalenian (Fig. 10.1) (Pétillon 2006, 2009; Langlais 2010, 2011; Pétillon et al. 2011). Fortunately, radiocarbon and climatic data for this period are rather precise (compared to earlier periods), enabling us to fit the Magdalenian within a well-defined environmental framework. While the Lower Magdalenian corresponds to the end of the Last Glacial Maximum (LGM), the Middle Magdalenian is contemporary with the Heinrich-1 Event (H1), and the Upper Magdalenian witnesses the last phase of the H1 followed by the warm up of the GIS-1 (Bølling). Characteristics of the osseous toolkit found for each of these three phases are described below.

Michelle C. Langley (ed.), *Osseous Projectile Weaponry*, Vertebrate Paleobiology and Paleoanthropology,
DOI 10.1007/978-94-024-0899-7_10

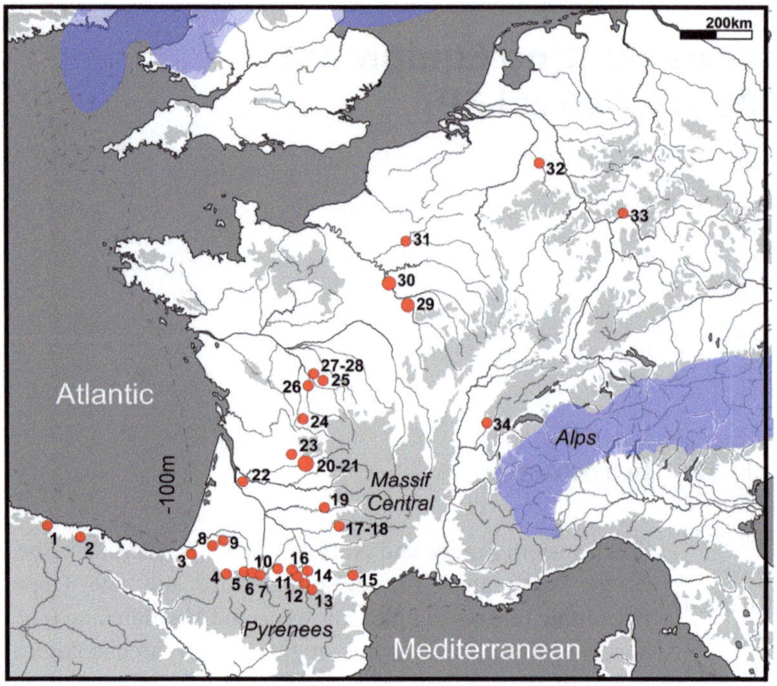

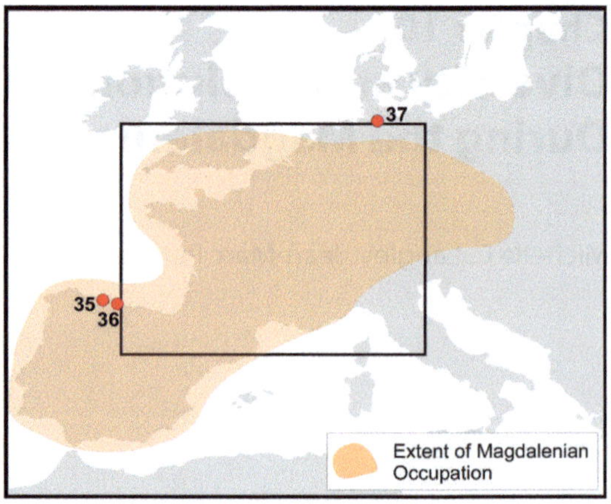

THREE PHASE CHRONOLOGY	
Lower Magdalenian	21,000 - 19,000 cal. BP
Middle Magdalenian	19,000 -16,500 cal. BP
Upper Magdalenian	16,500 -14,000 cal. BP

Fig. 10.1 Sites mentioned in text: (1) Cueto de la Mina; (2) Altamira; (3) Isturitz; (4) Saint-Michel d'Arudy; (5) Les Espélugues, Lourdes; (6) Aurensan; (7) Labastide, Gourdan; (8) Duruthy; (9) Brassempouy; (10) Les Scilles, Les Harpons; (11) Enlène; (12) Massat; (13) La Vache, Bédeilhac; (14) Mas d'Azil; (15) Gazel; (16) Montconfort; (17) Bruniquel; (18) Courbet, Gandil; (19) Petit-Cloup-Barrat; (20) Lascaux; (21) Cap Blanc, Laugerie-Basse, La Madeleine; (22) Saint-Germain-La-Rivière; (23) Combe-Saunière I; (24) Grotte de l'Abbe; (25) La Garenne; (26) Grotte du Bois-Ragot; (27) Taillis-des-Coteaux; (28) Roc-aux-Sociers; (29) Pincevent; (30) Etiolles; (31) Verberie; (32) Andernach-Martinsberg, Gönnesdorf; (33) Mannheim; (34) Combe-Buisson; (35) La Paloma, Las Caldas; (36) Tito Bustillo; (37) Stellmoor

The Magdalenian Hunting Kit: Characteristics and Changes

The Lower Magdalenian: c. 21,000–19,000 cal BP (17,500–15,500 BP)

The Lower Magdalenian is amply documented in the Iberian Peninsula, though we will concentrate on recent data from southwest France, a region where the Lower Magdalenian osseous industry has been the subject of little research until after the year 2000. Osseous hunting equipment belonging to the first Magdalenian groups in this area are currently documented at Gandil (Langlais et al. 2007), Scilles (Pétillon et al. 2008; Langlais et al. 2010), Saint-Germain-La-Rivière (Pétillon and Ducasse 2012; Pétillon in Langlais et al. 2015), Taillis-des-Coteaux (Primault et al. 2007) and Petit-Cloup-Barrat (Chauvière in Ducasse et al. 2011). It is very likely that the collection from Lascaux (Leroy-Prost 2008) also dates to this period.

Projectile points were manufactured exclusively from the compact tissue of reindeer antler; the blanks being extracted from the beam using the well-known groove-and-splinter technique (Clark and Thompson 1953; Semenov 1964). Typologically, the points are not very diversified. Almost all of them have a 'massive' base, that is, conical or spatulate in shape with no bevel, and likely hafted via a socket at the top of the projectile shaft. This hafting system was proven to be efficient in experiments, and supporting evidence for this configuration is found on one of the Lascaux points which exhibits traces of hafting on the proximal section (Leroy-Prost 2008). The second hafting system identified for Lower Magdalenian points is represented at Gandil rockshelter, where two self-barbed points (curved points with mesial flattening) were recovered. These weapon tips were probably hafted diagonally so that the proximal part protruded as a barb (Fig. 10.2) (Pokines and Krupa 1997).

Dimensions of Lower Magdalenian points are highly variable. At Scilles and Lascaux, several specimens reach 11–15 mm in width, 9–11 mm in thickness and 230–450 mm in total length, implying the use of very large antlers (from large adult male reindeer) as raw material. Conversely, at Saint-Germain-la-Rivière, Gandil, and Lascaux, many points have much smaller dimensions, c. 70–150 mm in length, 6–9 mm in width and 5–8 mm in thickness. It is not yet possible to tell if these differences are functional (related to different projectile types) or if they represent chronological variations within the Lower

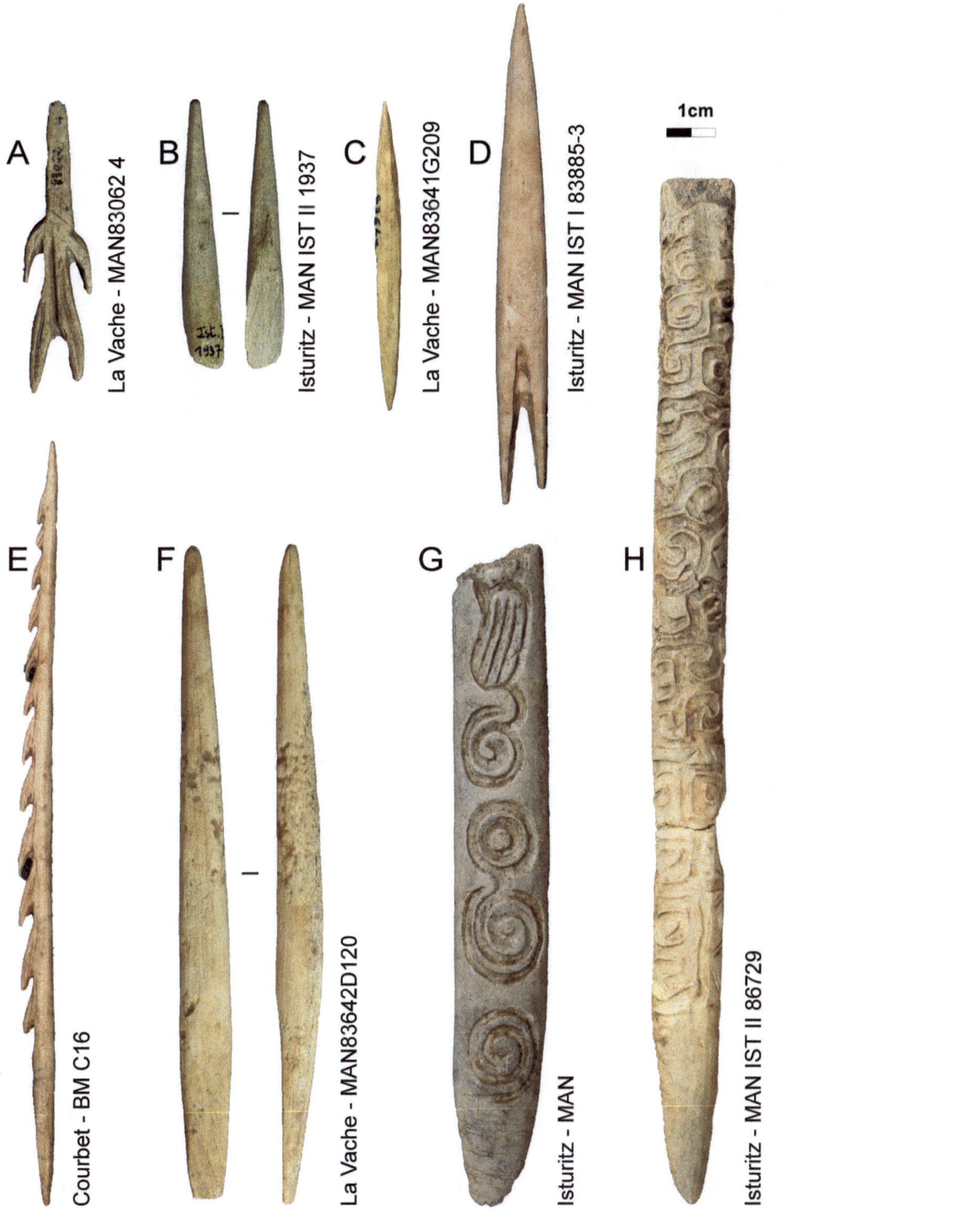

Fig. 10.2 Examples of Magdalenian antler projectile points. (**a**) *Foëne* or 'eel spear'; (**b**) Lussac-Angles point; (**c**) Dart; (**d**) Fork-based point; (**e**) Unilaterally barbed point; (**f**) Double-Bevel based point; (**g** and **h**) *Baguette demi-rondes* or Half Round Rods (HRRs). Photos by M.C. Langley, with permission of the *Musée d'Archaéologie Nationale*

Magdalenian. Another possibility is to consider them as adaptations to different local and/or seasonal availabilities in reindeer antler. This last hypothesis would be in accordance with the identified technical flexibility of the lithic system (Langlais et al. 2012).

A number of points (c.17 % at Scilles and 40 % at Saint-Germain-la-Rivière) have longitudinal grooves which were probably used to hold microliths. Recent projectile experiments have included replicas of these Lower Magdalenian composite weapon tips, leading to the suggestion that the

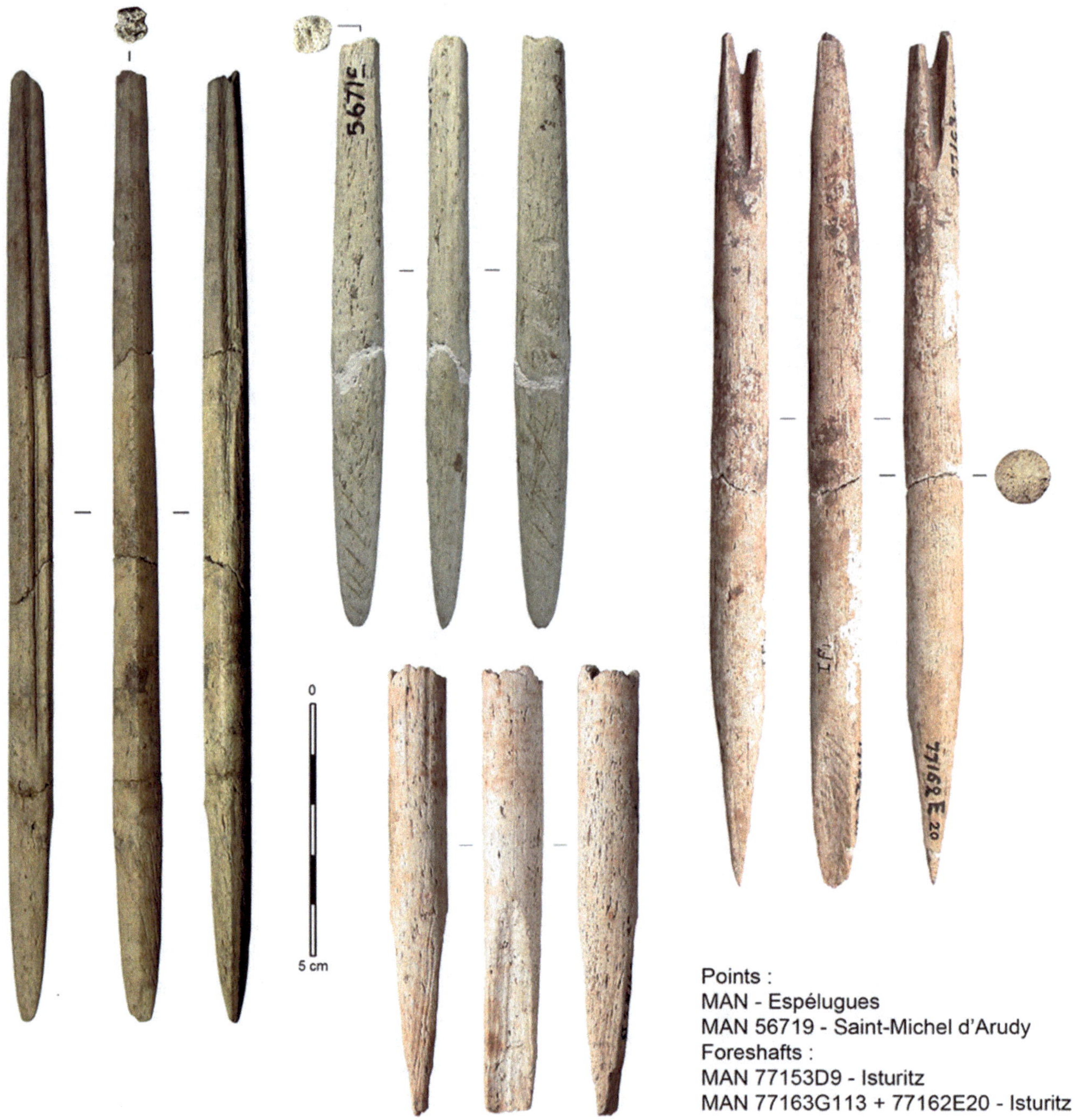

Fig. 10.3 Examples of whale bone projectile technologies. Photos courtesy of J.M. Pétilllon, with permission of the *Musée d'Archaéologie Nationale*

bladelets were arranged head-to-tail, simultaneously creating a sinuous cutting edge and negating the curved profile of the individual bladelet (Pétillon et al. 2011). This method of bladelet mounting may also have been implemented on points without engraved longitudinal grooves (instead simply stuck to the side), and has been shown experimentally to allow equally effective penetration of the lithic insets into the target (see Figs. 10.3 and 10.4 for examples). This type of composite antler/lithic projectile point is the most common weapon tip recovered from Lower Magdalenian contexts in southwest France, and though both the antler points and the bladelets show some variation, their basic design remains the same.

The Lower Magdalenian hunting kit as it is in southwest France therefore includes at least three different classes of projectile tips: composite antler/lithic points, distally-hafted lithic points (especially shouldered points), and the self-barbed antler point.

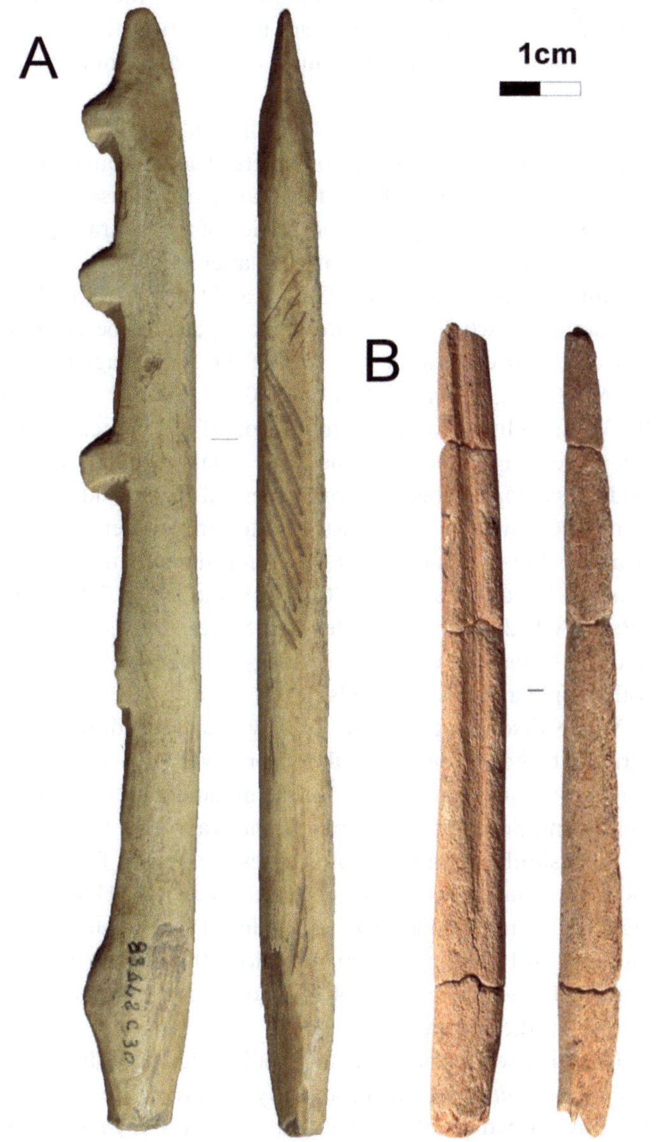

Fig. 10.4 Examples of composite antler/lithic projectile points: (*Left*) A unilaterally barbed point with striations on its mesial section to assist in the adherence of bladelets from La Vache; and (*Right*) A point with mesial groove for the insertion of bladelets from Saint-Germain-la-Riviere. Photos: M.C. Langley and J.-M. Pétillon, with permission of the *Musée d'Archaéologie Nationale*

The Middle Magdalenian: c. 19,000–16,500 cal BP (15,500–13,500 BP)

The beginning of the Middle Magdalenian witnesses a sharp rise in the intensity of bone and antler working, with collections of worked osseous items from these contexts being significantly richer than any of the preceding Upper Paleolithic cultures. This development is particularly reflected in the increased amount of osseous hunting equipment present within the Magdalenian archaeological record.

As in the Lower Magdalenian, osseous equipment in this middle phase is dominated by points made of reindeer antler, and whose blanks were extracted using the groove-and-splinter technique. However, this technique is now used in a way that allows for higher productivity: multiple longitudinal grooving, that allows for several parallel blanks to be extracted from a single large antler beam. The use of large module antler might be linked to the desire to produce projectile points of overall larger dimensions. Indeed, at Saint-Germain-la-Rivière (one of the few sites in southwest France that contains a succession of Lower to Middle Magdalenian levels *and* provides enough osseous projectile points for a statistical metric study), the beginning of the Middle Magdalenian is marked by an increase in the average width and thickness of points (Pétillon in Langlais et al. 2015). These changed requirements concerning the dimensions of the projectile points might have prompted the more intensive exploitation of the largest antler available, i.e., mature antler from large adult male reindeer, usually collected after shedding.

The economic importance of these antlers for the manufacture of hunting equipment suggests that Middle Magdalenian groups invested significant effort into their collection and processing, and that the production of antler equipment was likely integrated within an annual (if not multi-annual) economic cycle (Averbouh 2000, 2005).

These size requirements may also have been the impetus for the first exploitation of cetacean bone for the production of projectile technology (Fig. 10.3). Cetacean bone implements were first reported by one of the authors (JMP) in 2008 from the Middle to Late Magdalenian layers of Isturitz (Pyrénées-Atlantiques) (Pétillon 2008a). The use of this type of bone was found to be restricted to the manufacture of projectile weaponry (points and foreshafts), though several implements had been recycled into wedges. The raw material was likely scavenged from whales stranded on the Atlantic shore that was then 50–60 km walking distance west of the cave. It appears that only finished points were brought to the site as no manufacturing waste has been identified.

Although cetacean bone points represent only a small proportion of the weapon kit, they know a wide diffusion, attesting to direct or indirect long-range contacts between sites. Further examination of Pyrenean assemblages found cetacean bone implements in assemblages from 10 additional sites scattered from the Atlantic side to the Ariège in the central Pyrenees (Pétillon 2013).

During the Middle Magdalenian, two other technical solutions are used to manufacture large points. In the current state of our knowledge, it seems that these solutions are not known in the early phases of the Middle Magdalenian (not before c. 17,700 cal BP or 14,500 BP) and are mostly found in the Pyrenean region. The first solution is the use of 'half-round rods' (*baguettes demi-rondes)*, or HRRs (Fig. 10.2g, h). These implements are characterized by their unique cross-section: a convex side opposite to a flat one bearing oblique incisions to aid attachment. Mounted in pairs, the flat sides of two half-round rods were fastened together to form a 'bivalve' point with a bi-convex, rounded cross-section (as demonstrated by several *in situ* archaeological finds, as well as refittings between complementary fragments: see Feruglio and Buisson 1999). Thus two half-round rods with a 12 mm width and a 6 mm thickness could be paired to form a point with a 12 mm circular diameter. Experiments by Rigaud (2006) demonstrated that the cohesion of the two halves is sufficient for the use of these bivalve points as projectile tips, provided that a proper adhesive is used. Perhaps the most advantageous aspect of this bivalve technique is that it allows for the construction of large points when there is a scarcity of antler with a thick compact tissue layer. Half-round rods are usually found in a fragmentary state and it is difficult to reconstruct the original length of the points, but the largest complete specimens can reach lengths above 200 mm (230 mm at Harpons Cave in Lespugue; 260 mm at Montconfort; 262 mm at Grotte de l'Abbe; 280 mm at Labastide; 300 mm at Gazel; 340 mm at La Vache; 360 mm at Isturitz).

The Middle Magdalenian toolkit also includes massive base points and self-barbed points as found in the Lower Magdalenian. Here again, the frequent presence of longitudinal grooves on some of the points strongly hints at the use of laterally-hafted flint bladelets. The main novelty for this later period is the appearance and wide diffusion of simple-bevelled and double-bevelled points (Fig. 10.2f). In the early phases of the Middle Magdalenian (c. 19,000–17,700 cal. BP or 15,500–14,500 BP), one particular sub-type of single-bevelled point, termed a 'Lussac-Angles point' (Fig. 10.2b), shows an especially wide diffusion encompassing a large part of the Middle Magdalenian cultural area—from Tito Bustillo (Spain) in the west to Gazel (France) in the east, to Roc-aux-Sorciers (France) in the north (cf. Fig. 10.1). Double-bevelled points, on the other hand, are mostly known from the early phases of the Middle Magdalenian in the northern half of France (the so-called *Magdalénien à navettes*: Allain et al. 1985; Houmard and Jacquot 2009).

The Upper Magdalenian: 16,500–14,000 cal BP (13,500–12,000 BP)

The Upper Magdalenian toolkit inherits most of its technological and economical traits from the Middle Magdalenian. The intensity of bone and antler working remains the same and the richness of the osseous industry does not decrease from the previous phase. Large reindeer antler remains the material of choice for projectile points, and, as was the case during the Middle Magdalenian, was worked using the longitudinal grooving technique. Again, the exploitation of this raw material appears to have been integrated within an annual or multi-annual cycle (Averbouh 2000, 2005; Pétillon 2006).

Cetacean bone continues to be exploited, though it is now exclusively used to produce foreshafts. The combination of antler point and whale bone foreshaft perhaps uses to the best advantage the respective properties of the two raw materials: the large dimensions of the latter and the superior toughness of the former (MacGregor and Currey 1983). This production of whale bone foreshafts appears to be chronologically restricted to the first half of the Late Magdalenian c. 16,500–15,300 cal BP or 13,500–12,800 BP (Pétillon 2013). In addition to the examples found in Pyrenean sites, recent re-examination of the Late Magdalenian site of Andernach-Martinsberg (German Central Rhineland), more than 1000 km to the northeast, resulted in the identification of a possible foreshaft which had been manufactured from cetacean bone dated to 15,970–15,500 cal. BP (13,110±50 BP; OxA-V-2218-40) (Langley and Street 2013). This find again indicates the existence of long-distance movement of both people and raw materials.

While discussing the use of raw materials during this period, it should be noted that ivory was also exploited, though was quite rarely used for the manufacture of hunting equipment. Some projectile points and fragments with cross sections similar to those of projectiles are known from sites such as Marsoulas, Mas d'Azil, Laugerie Haute-Est, Verberie, and Gönnersdorf (Kandel 1995; Otte 1995; Christensen 1999; Fradet 2004; Pétillon 2008b). The rare use of ivory for weaponry is probably connected to the relative scarcity of the raw material.

Bone is used to manufacture small projectile points such as *éléments bipointe*—if these really are projectile points—and tiny barbed points. Julien and Orliac (2003) report the existence of around 40 specimens of these latter artefacts, whose dimensions are exceptionally diminutive. For example, those from the Grotte du Bois-Ragot (Vienne) do not exceed 45 mm in total length (Christensen and Chollet 2005), with the unilaterally barbed La Vache examples not exceeding 69 mm in total length and the sole bilaterally barbed example 77 mm in total length and a width of only 5.5 mm (Langley 2013). These artefacts may have been used as a barbed arrow for use on small prey such as birds (as in Weniger's [2000] functional typology), or they may be examples of artisan virtuosity—a product made by an individual demonstrating their manufacturing ingenuity and skill. Another possibility is that these small tools were children's toys. In all, we are currently unable to determine which of these interpretations is most likely.

Implements with longitudinal grooves persist in assemblages both in the Pyrenees and the Paris Basin, suggesting that the manufacture of composite lithic/antler projectile tips continue to be utilized during the Upper Magdalenian. In fact, one of the only composite lithic/antler points to survive to discovery in Western Europe dates to this period. The implement was recovered from Pincevent and has two microlith components glued onto each side of the antler point (Leroi-Gourhan 1983). Half-Round Rods, on the other hand, while present during the early phases of the Upper Magdalenian, appear to be discontinued during the terminal phases (ca. 15,300–14,000 cal BP or 12,800–12,000 BP).

Major changes in osseous hunting equipment from the Middle to the Upper Magdalenian are twofold. The first shift is a change in hafting systems. Unbarbed points with 'massive' or single bevel bases seem to disappear or to be extremely rare, while double-bevelled points and forked-base points become increasingly frequent (Fig. 10.2d, f). The double-bevelled base form is documented in all regions occupied by Upper Magdalenian sites and throughout the whole period. In fact, all of the (quite rare) osseous hunting equipment recovered from northern Upper Magdalenian sites (15,250–13,900 cal BP—after Debout et al. 2012) are double-bevelled points (48 pieces from 6 sites in the Paris Basin). Interestingly, the Paris Basin points are significantly longer when compared to contemporary Pyrenean points (140 mm versus 100 mm) (Bertrand 1999; Christensen 2008; Pétillon 2008b), and may indicate differences in both raw material availability and cultural attitudes towards exhausted implements (Langley 2015).

Fork-based points, on the other hand, are geographically limited to a smaller area stretching from La Paloma and Las Caldas in Asturias through Isturitz to La Vache in the central Pyrenees (cf. Fig. 10.1). Very few specimens were found outside this narrow, 600 km long east-west zone. Chronologically, forked bases are restricted to the first phase of the Upper Magdalenian (ca. 16,500–15,300 cal BP or 13,500–12,800 BP) before disappearing completely. Projectile experiments organized at CEDARC in 2003–2004 were intended to determine if this highly differentiated 'success' had any technological reason, i.e., if the double bevel was in any way technically superior to the forked base. Results demonstrated, however, that the two types are equally efficient, and that, if the forked bases indeed break more often, this defect is compensated by a better cohesion of the hafting: the detachment or dis-alignment of the point is more frequent with the double-bevelled points that have a tendency to slip in their hafting socket (Pétillon 2006). Thus the preference for the double-bevelled point and the discontinuation of the fork-based points outside certain geographical and chronological limits appears to be a cultural choice that is not constrained by technological specifications.

While self-barbed points disappear at this time, multi-barbed antler points with either one or two rows of barbs, and usually with a conical section base designed to be inserted in a socket (see examples in Figs. 10.2 and 10.4) are developed and constitute the second major projectile development from the Middle to the Upper Magdalenian. Barbed points are documented for the whole Upper Magdalenian and in all regions, however, certain traits appear specific to particular areas (see Lefebvre 2011). Barbed points with a perforated base are not encountered outside the northern coast of Spain (González Sainz 1989) and they appear specific to the final phase of the Magdalenian (Lefebvre 2011). Another type of barbed point, those with a single row of angular barbs and a tanged base (see Villaverde et al. 2016), is specific to the Mediterranean coast (Combier 1967). These points appear more or less synchronically between 14,500 and 14,000 cal. BP (Lefebvre 2011). In most regions, points with one row of barbs (unilaterally barbed) always outnumber those with two rows of barbs (bilaterally), but this situation is extremely accentuated in Upper Magdalenian sites from the northern coast of Spain where unilaterally barbed points represent more than 80 % of the total (Julien 1982). On the Mediterranean coast of Spain, with few exceptions, only unilaterally barbed points are known (Torres 1987; Cacho Quesada and Torre Sainz 2005; Villaverde and Roman 2005–2006). However, the situation is reversed in western France north of the Pyrenees (Périgord and Quercy) where bilaterally barbed points dominate assemblages by more than 80 % (Julien 1982).

These uni- and bilaterally barbed points had a significant place for more than 1000 years between 16,000–14,500 cal BP (cf. *infra*), representing approximately 10–40 % of all osseous points recovered from Upper Magdalenian assemblages in southwestern France and northeastern Spain (Pétillon 2009). Although they are usually termed 'harpoons' in the French literature, it is not clear whether they were used in the true sense of a harpoon: as detachable points connected to the shaft by a line, or as fixed (immobile) points securely fastened to the shaft (Pétillon 2009).

In sum, the evolution of the osseous hunting equipment in the Upper Magdalenian is marked by two opposed phenomena: the wide diffusion of two technical ideas (the double-bevelled point and the barbed point) and the development of regional specificities (the forked-base point, several types—or combinations of types—of barbed points).

More on the Production, Maintenance, and Recycling of Magdalenian Projectile Points

The previous sections outlined the composition of the osseous projectile toolkit at different points throughout the Magdalenian, briefly mentioning temporal changes in the manufacturing techniques of the weaponry discussed. Here these techniques will be described in more detail, along with a discussion of how this weaponry was repaired and recycled.

Production Methods

As stated above, the generic 'groove-and-splinter' technique was employed throughout the Magdalenian to produce rod-like blanks for the manufacture of projectile points. The term 'groove-and-splinter' describes the method employed—'to hollow' out the grooves on both sides of the future blank [*'l'accent étant mis sur la technique utilisée pour creuser les sillons de part et d'autre de la baguette'*] (Averbouh et al. 2010:65). However the technical procedure for the removal of blanks from the beam varies according to the position of the two (or more) grooves incised into the antler. Depending on the choices made during the process of working, it was sometimes necessary to use another technique in order to totally circumscribe the blank before proceeding to detachment. These technical choices changed between the chrono-cultural periods and even within a single phase. Their identification is essential to characterizing groups and their techno-economical behaviors.

During the Lower Magdalenian, the groove-and-splinter method is present but there are few complete waste products to identify which particular procedure was followed. Those waste products identified mostly attest to the extraction of one blank from an antler beam and sometimes another on the beztine, but it is impossible to know if the grooves were incised parallel or convergent to each other. Occasionally two blanks were extracted on the same beam, while the extraction of multiple blanks from a single beam is seldom identified (but found at Cap Blanc, Saint-Germain-La-Rivière [Pétillon and Ducasse 2012]). Multiple longitudinal grooving becomes more frequent during the Middle Magdalenian after 19,000 cal BP and was used during the Upper Magdalenian, at least in the Pyrenees area (Averbouh 2000).

During the Middle Magdalenian, multiple methods of blank extraction were used and frequently exploited the complete peripheral section of the antler beam, resulting with up to 5 parallel blanks being extracted from the same section of antler (Averbouh 2000). This process was especially applied to the largest antlers with the thickest compact tissue (*compacta* thicknesses of 6–10 mm); and can be seen as a way to maximize the productivity of this category of antler.

On the other hand, waste products from La Garenne (Indre) indicate that single blanks were also removed above the beztine, and sometimes continued onto the lateral face to produce several blanks together (Rigaud 2004). At the Pyrenean site of Enlène, it was found that two or more blanks were extracted on the front and back face of the beam leaving a characteristic triangular waste product (Averbouh 2000).

In Pyrenean and Aquitaine sites during the Upper Magdalenian, various methods are represented. Two specific blank morphologies having been recognized from the waste products:

1. 'Jagged base': (*'base dentelée'* in Averbouh 2000) the whole periphery of the antler beam was used to produce several narrow blanks with a mostly quadrangular section (La Vache: Averbouh 2000; Isturitz: Pétillon 2006; Saint Michel d'Arudy: Pujol 2009; La Madeleine: Treuillot 2011, etc.); and
2. 'Arched base': (*'base en arceau'* in Pétillon, 2006; *'base en bascule'* after Averbouh, pers. comm.) the antler beam is detached from the base with two opposite grooves on the lateral and medial sides that overlap the first third of the beztine and converge on the posterior side of the beam (Isturitz: Pétillon 2006; St Michel d'Arudy: Pujol 2009; Gourdan: Sgard 1999; Laugerie-Basse: Houmard 2004; Espelugues: Omnès 1980, etc.). The aim of this method is apparently to produce a secondary block made from the antler's beam, which will then be partitioned into several blanks. 'Arched bases' are also known from the Middle Magdalenian.

Besides these two very specifics ways of debitage by extraction, Pyrenean Magdalenians also produced single blanks from other parts the beam or tines. During the same period, but in the Paris basin (Etiolles, Pincevent, and Verberie), waste products indicate that one long single broad blank, generally with a plano-convex cross-section, was produced by extraction along the whole length of the anterior face or latero-facial (as in the case of Etiolles) of the antler beam. The grooves are either convergent (Etiolles) or parallel (Pincevent, Verberie). Waste products from this kind of extraction correspond to a so-called 'bow-shaped' matrix. The rod shaped blank was probably cut into several short blank according to the size required (Christensen and Averbouh 2005; Averbouh 2006, 2010; Averbouh et al. 2010).

At Grotte du Bois-Ragot (Vienne), one single blank was extracted on the back face of the antler beam from an adult animal (10 mm of *compacta*), with grooves on the lateral sides so that the blank would be wide and have a plano-convexe section. This method was probably used in order to get blanks for the production of bilaterally barbed points, which need additional space for barb production. Wide blanks were also extracted from the lateral sides of the beam, and more narrow blanks were produced through multiple extractions around the beam. The objectives were sometimes mixed so within one debitage by extraction it was possible to get wide and narrow blanks for different purposes (unbarbed points, barbed points, etc.) (Christensen unpublished data).

Overall, Magdalenian blank production is characterized by this systematic use of the 'groove-and-splinter' technique in order to get standardized blanks and serial products. This technique distinguishes Magdalenian groups from those of previous periods (Aurignacian, Gravettian, Solutrean, and Badegoulian) who used other methods to produce blanks such as fracturation by direct percussion, or by indirect percussion ('splitting'), or mixed procedures involving partial grooving and fracturation techniques (e.g., Liolios 1999; Goutas 2009; Averbouh and Pétillon 2011; Pétillon and Ducasse 2012; Tejero 2013).

Evidence for the Repair of Magdalenian Antler Projectile Points

Experiments with various bone and antler projectile points have demonstrated that careful maintenance of osseous weaponry is essential to prevent catastrophic failure of the weapon in use (e.g., Tyzzer 1936; Arndt and Newcomer 1986; Stodiek 1990; Pokines 1993, 1998; Stodiek 2000). Evidence for the maintenance of Magdalenian points is commonly observed in assemblages—if requiring careful examination to identify. This evidence comes in the form of striations, facets, uneven surfaces, and partially erased decorations indicating the resharpening of a distal point; the abrasion of broken barb scars on barbed points; and repaired or completely remade bases—all in an effort to extend the use life of the individual implement.

Rejuvenation stigmata have been identified on uni- and bilaterally barbed points (Julien 1982; Julien and Orliac 2003; Christensen and Chollet 2005; Langley 2015), fork-based points (Pétillon 2006), and single- and double-bevelled based points (Stodiek 1993; Langley 2015). For fork- and bevel-based points, it is assumed that a use, damage accumulation, rejuvenation, and re-use cycle occurred multiple times until the length of the point became too short to continue in use—this minimum length being determined by either functional efficiency and/or cultural ideas regarding form.

Examination of Middle—Late Magdalenian barbed and bevel based points from 24 sites located throughout France and southern Germany has shown that the distal extremity (the tip) was carefully rejuvenated using facets of various widths which were often then scraped or ground to produce a smooth surface (Figs. 10.5 and 10.6) (Langley 2015). Bevel based points, in addition to the repeated rejuvenation of the distal extremity, also had their bases repaired and remade when these sections fractured. Fracture at or just above the base is a common fracture pattern for bevel-based weapons (as was found in experiments mentioned above), and it appears new bases (both carefully and hastily made) were worked onto surviving distal-mesial fragments to continue the use life of an individual weapon tip (Fig. 10.6c).

Barbed points, because they have a number of morphological landmarks (barbs, barbs-to-base shaft section, basal bulb) and are almost universally decorated (engraved, and the La Vache assemblage provides examples that were also painted with red and black pigments [see Buisson et al. 1989]), allow the researcher to more easily identify evidence for maintenance than their unbarbed counterparts. Both uni- and bilaterally barbed varieties are argued to have been designed with a piercing, conical section tip which becomes spatulate in form once rejuvenated (Julien 1982; Langley 2014, 2015). Barbs may be intentionally removed in the process of rejuvenating the point, or having broken off the shaft in use and leaving a jagged scar were carefully ground down or scraped away until completely or near completely removed. Similarly, the bulb protuberances at the base can be ground away if broken, or the entire base reworked into a bevel or conical shape if fractured in use (Langley 2015). These repairs were usually undertaken carefully, and where hasty repair have been executed, it appears to have taken place at the end of the implements use life as these poorly executed repairs are usually identified on implements reduced in total length and which exhibit previous maintenance which was carefully executed (see Fig. 10.5 for an example of the phenomenon). It is usually this last round of repair which is most visible to the analyst.

Recycling of Projectile Weaponry During the Magdalenian

Magdalenian antler points were also commonly recycled into other functional tools. For example, both barbed and unbarbed points were reused as wedges at La Vache, Isturitz, La Madeleine, Laugerie-Basse, Grotte du Bois-Ragot, and Courbet (Pétillon 2006; Cholet-Kritter 2009; Treuillot 2011; Langley 2013) (Figs. 10.5d and 10.6d). Where the point did not already have a bevel or spatulate extremity which could function as the chisel end of the wedge, a bevel was crudely fashioned for its reuse in this function.

Barbed points were also reworked in order to be recycled. For example, sections of either or both the inferior and superior faces of an implement may be scraped and abraded to be reused as 'spatulas' as occurred at La Vache, Laugerie-Basse, and Courbet (see Fig. 10.5 for examples). Other tools that barbed and unbarbed points were recycled into include awls (La Madeleine, Aurensan), burnishers (Isturitz), and foreshafts for projectile weapons (Isturitz, Laugerie-Basse). Similarly, several mesial-proximal fragments of bevel-based points and HRRs at Isturitz and La Madeleine had a single, neat hole drilled through their shaft presumably to be worn as ornamentation (Saint-Périer 1936; Langley 2013).

Implements were also sometimes reworked into another type of projectile point. For example, several unilaterally barbed points recovered from Isturitz were manufactured out of fragments of HRRs. This recycling was evident as each of

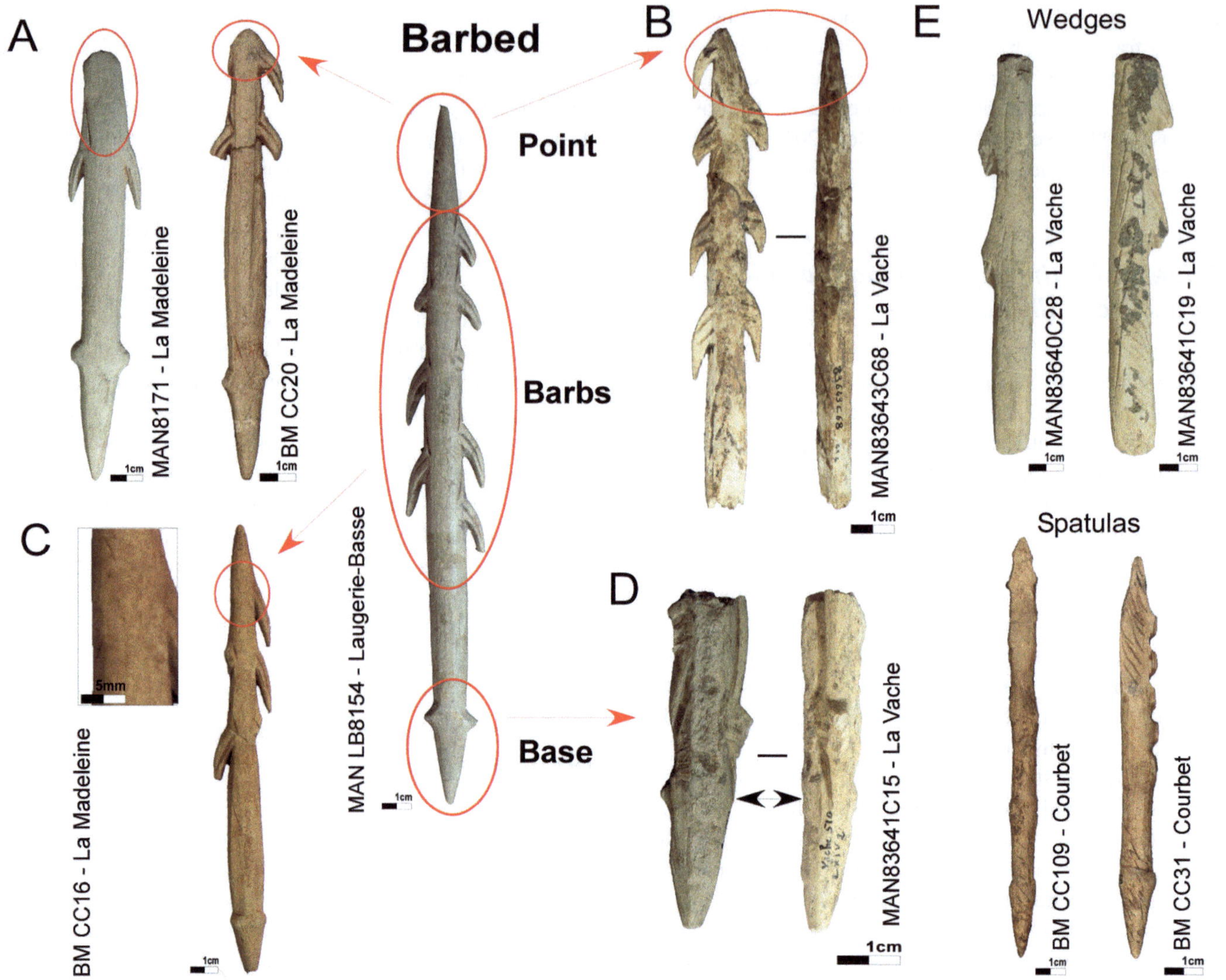

Fig. 10.5 Examples of rejuvenated and recycled barbed points. Photos courtesy of M.C. Langley, with permission of the *Musée d'Archaéologie Nationale*

these unilaterally barbed points had the thin, oval section and the regular oblique striations on one face characteristic of HRRs. In another case from this same site, a unilaterally barbed point was found to have been reworked into a bilaterally barbed point, and a single bilaterally barbed point reworked into a double bevel based point. A bilaterally barbed point recovered from La Vache was reworked into a double beveled based point, while double bevel based points from La Madeleine and Aurensan were discarded while being reworked into barbed (both uni- and bilaterally barbed) weapon tips (Langley 2013).

While this is only a brief overview of the types of evidence for the repair and recycling of osseous projectile points evident during the Magdalenian, it nevertheless outlines the range of maintainence that were regularly undertaken by implement caretakers (who may also have been their owners) and the diversity of tools into which projectile points were recycled.

Hafting and Launching Modes: Direct and Indirect Data

No indisputable evidence for the use of the bow exists for the Magdalenian. The curved wooden artefact from Mannheim (Germany), dated 18,060–17,650 cal BP (14,680 ± 70 BP; Rosendahl et al. 2006), has been tentatively identified as a bow fragment but this attribution cannot be unquestionably confirmed. To date, the oldest definite direct evidence for the use of the bow is still the large collection of Ahrensburgian pine arrow shafts from Stellmoor, Germany (Rust 1943), two millennia younger than the end of the Upper Magdalenian (Fischer and Tauber 1986). Other Mesolithic finds of bow and arrows from Germany, Denmark, Sweden, and Russia are all dated to later millennia (see reviews in Junkmanns 2001; Cattelain 2006). Furthermore, while rock art produced

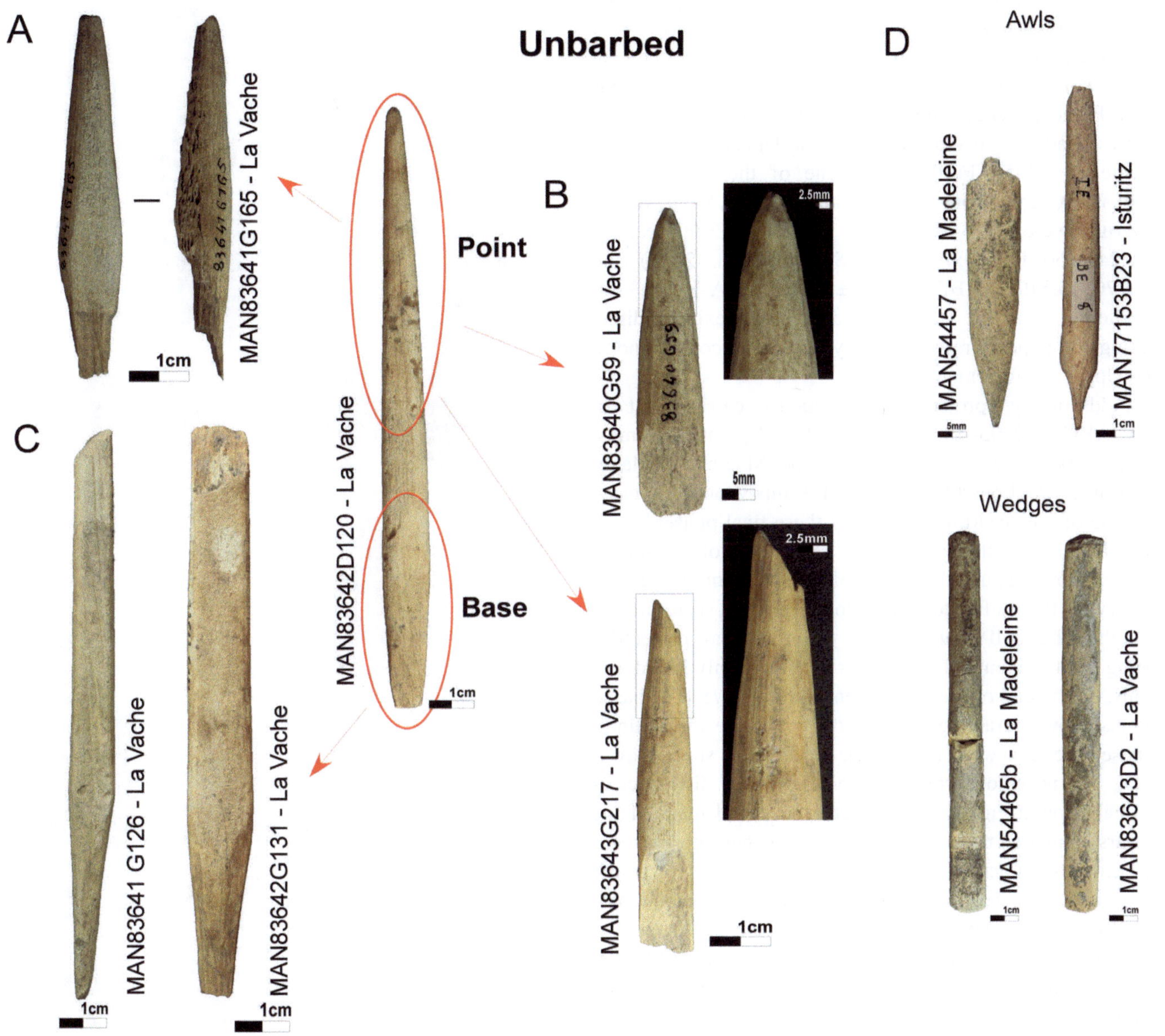

Fig. 10.6 Examples of rejuvenated and recycled unbarbed points. Photos courtesy of M.C. Langley, with permission of the *Musée d'Archaéologie Nationale*

by Magdalenian peoples include possible images of spears and spearthrowers (see Baffier 1990 for a summary of these images), no depictions of bows have been identified. However, it does not mean that the bow was not used in earlier periods: the possible existence of the bow in the Gravettian has repeatedly been raised (e.g., Cattelain 1997).

The use of the spearthrower, on the other hand, is documented by over 100 spearthrower distal parts ('hooks'), generally made of reindeer antler, found in one Solutrean and 32 Magdalenian sites located throughout Western Europe, but mostly within France (Stodiek 1993; Cattelain 2005). The chronology of spearthrowers throughout the Magdalenian, however, is still debated (Cattelain 2005). With the exception of the single spearthrower hook from Combe-Saunière I, whose attribution to the Solutrean might be questioned, the oldest specimens are dated to c.19,000–18,300 cal BP (15,500–15,000 BP), which corresponds to the very beginning of the Middle Magdalenian in southwest France (Cattelain 2004; Gonzalez Morales and Straus 2009). Most of the remaining spearthrower hooks, including the famous decorated specimens, were recovered either from ancient excavations with no reliable stratigraphy or from Middle Magdalenian layers. With a few exceptions, the 24 available radiocarbon dates (Cattelain 2005) show a regular chronological distribution between c.17,500 and 15,300 cal BP (14,300–12,800 BP), that is, during the Middle Magdalenian.

For the Upper Magdalenian, the situation is quite confused. It is generally considered that there are very few, if any, spearthrower hooks in this period. However, recent reassessment of the evidence suggest that several specimens from France, Switzerland, and Germany each adorned with horses heads might indeed date to the Upper Magdalenian (Cattelain 2005). Furthermore, one of the spearthrower hooks recovered from Isturitz and one from Saint-Michel d'Arudy were dated respectively to 14,480–13,960 cal BP (12,245±60 BP, OxA-19837; Szmidt et al. 2009) and 16,070–15,510 cal BP (13,155±75 BP, OxA-X-2523-44; Pétillon et al. 2015). These dates are the only existing direct datings of a spearthrower hook by AMS, and correspond to the Upper Magdalenian.

Additionally, projectile experiments carried out at CEDARC in 2003–2004 with fork-based points, the most frequent type of antler point in the Upper Magdalenian layer at Isturitz (Pétillon 2006), provided results which supported the use of the spearthrower during these final phases of the Magdalenian. In this experiment, 78 replicas of fork-based points were manufactured out of reindeer antler. Half of them were then hafted to spears propelled with a spearthrower and the other half to arrows shot with a bow, all projectiles being used in identical conditions against freshly killed animals. At the end of the experiment, 14 of the 78 fork-based points showed fractures on one or two of the fork's tines. These fractures were always the result of a spearthrower shot, and never occurred with the bow. This difference is probably owing to the much greater size and mass of the spears compared to the arrows, as well as their more irregular trajectory. Each of these parameters obviously place the point under greater bending forces upon impact, sometimes resulting in the snapping of the forked base (Fig. 10.7).

Proximal fractures are very frequent on the fork-based points from Isturitz: out of 419 specimens, 95 show fracture damage at the fork. The majority of these fractures (68 %) have close equivalents in the experimental sample. These similarities between the archaeological and experimental samples are determining enough to conclude that the Isturitz fork-based points were probably used to tip spears projected with a spearthrower, rather than arrows shot with a bow.

Thus, it is generally assumed that the end of the Magdalenian saw the replacement of the spearthrower by the bow among Western European hunter-gatherers, and recent research does not contradict this hypothesis.

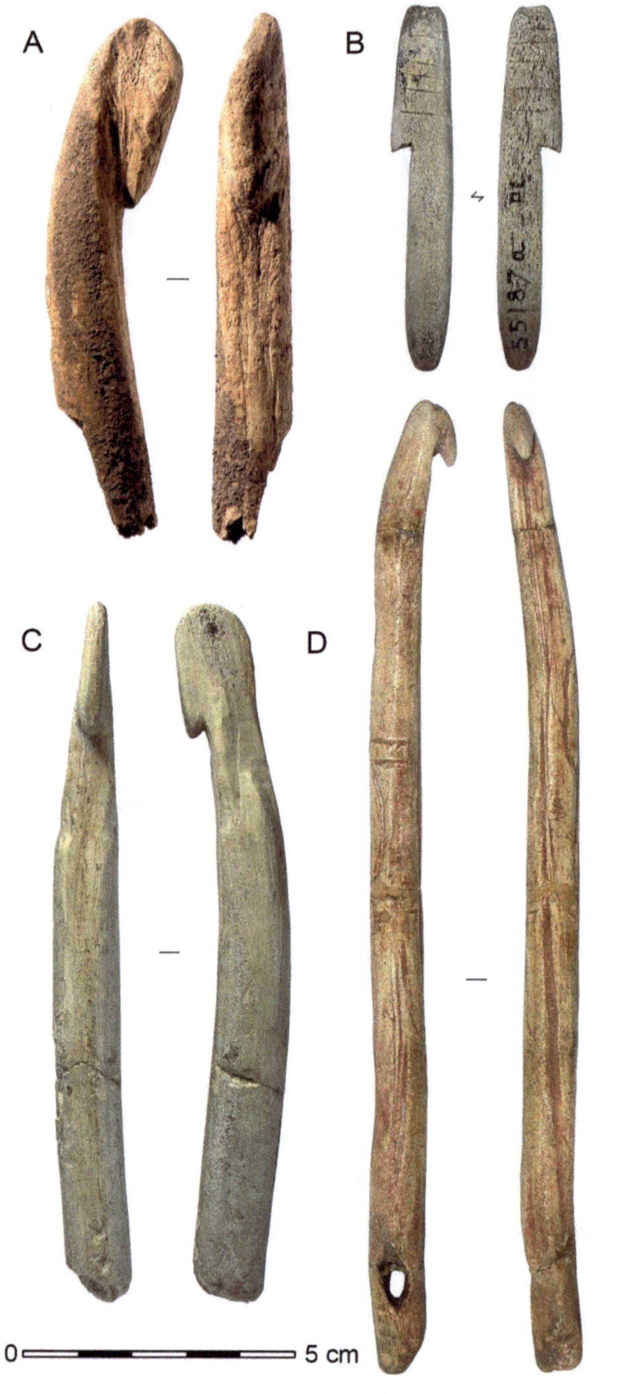

Fig. 10.7 Antler spearthrower hooks. (**a**) Distal fragment dated to 14,550–13,880 cal. BP (12,245±60 BP). (**b**) Complete specimen with engraved lines and proximal bevel (for hafting). (**c**) Distal fragment. (**d**) Complete specimen with (unfinished?) engraving below the hook. (**d**) complete specimen with proximal perforation (for hafting) and engraved decoration (all pictures by J.-M. Pétillon except A, by Pierre Cattelain)

Barbed or Unbarbed?

The great diversity in antler projectile weaponry manufactured and employed is suggestive of a toolkit which enabled the exploitation of a great range of fauna at any one time during the Magdalenian. Furthermore, environmental changes which occurred during the Magdalenian had an impact on the fauna available for exploitation during different periods. Reindeer, Horse, Bison, and Saiga antelope were the main prey of the Lower and Middle Magdalenian hunters, but during the Upper Magdalenian these species progressively left the area and were replaced by ungulates less dependent on cold/dry conditions and open landscapes (by Red Deer, especially). Thus, we are led to ask the question: which weapon was used for what fauna? Barbed or unbarbed?

This question is quite difficult to answer without direct association between faunal remains and projectile points. Osseous points irremediably wedged in bones are extremely rare within Paleolithic contexts, with to our knowledge, only three cases currently known: one Aurignacian from the French cave Combe-Buisson, where a small distal part of a projectile point was found in a bone, but the exact taxon is not known. The second is from the Epigravettian Siberian site of Kokorevo I where an osseous projectile point is wedged in a scapula from a Bison (Cordier 1990; Letourneux and Pétillon 2008). The final example was found at the UK site of Poulton-le-Fylde where one unilaterally barbed point was found amongst the ribs of a deer and another "found lying across the lateral condylar surface of the left metatarsus" (Jacobi et al. 2009:15).

This situation does not enable us to get closer to the exact type(s) of game against which barbed or unbarbed points were used. Ethnographic information can highlight possible connections between type of projectile points and targeted game (see Weniger 2000 for a possible correlation of Paleolithic/Mesolithic weaponry and ethnographical weapons; also Christensen et al. 2016).

The relationship between a particular kind of game and a specific weapon type in an archaeological context may indicate which was used for what game. For example, the appearance and development of Upper Magdalenian barbed points might be linked generally to the intensification of the capture of small animals (birds, fish, lagomorphs, etc.) at the beginning of the Upper Magdalenian (Le Gall 1992; Laroulandie 2003; Cochard 2004; Costamagno et al. 2008, 2009; Pétillon 2009), though others argue for their use in fishing activities (e.g., Weniger 2000).

Evidence for Social Interaction from Weaponry

Highly decorated weaponry is a feature of the Magdalenian toolkit and has been used in the past to investigate interaction between different sites and regions. Conkey (1980), through the analysis of engraved decoration on both barbed and unbarbed projectile points from 27 sites located throughout Cantabrian Spain, suggested that Altamira and possibly Cueto de la Mina were possible aggregation sites where various regionally dispersed peoples came together at regular intervals. Similarly, Bahn (1982) has argued on the basis of the distribution of portable artworks (including decorated weaponry) that Mas d'Azil and Isturitz were 'supersites', that is, large aggregation sites, while Enlène, Espélugues, Saint Michel d'Arudy, Massat, and Gourdan were proposed as possible locations for smaller aggregations.

Particular weaponry which was used in the proposal of these interaction networks include the HRRs with 'spiral' decoration which is so distinctive that some researchers have suggested that they may be the work of a single craftsperson (see two examples in Fig. 10.2g, h) (e.g., Saint-Périer 1920; Bahn 1982). The largest number of these distinctively decorated implements have been recovered from Isturitz, with five examples identified at Espalungue in Arudy, ten at Espélugues in Lourdes, and others at Les Harpons in Lespugue and Bourrouilla in Arancou (Passemard 1920; Saint-Périer 1929, 1947; Bahn 1982; Fritz and Roussot 1999). The simple fact that the largest set of these weapon tips were recovered from Isturitz is said to be suggestive that this site is the origin of these weapons, with the points being distributed either through direct or in-direct trade to the remaining sites throughout the Pyrenees (Bahn 1982).

Similarly, Bahn (1979, 1982) argues that the series of famous fawn and bird spearthrowers recovered from Mas d'Azil, Bédeilhac, and Saint Michel d'Arudy, with possible fragmentary pieces from Saint Michel d'Arudy, Isturitz, Enlène, Mas d'Azil, and Bruniquel (Cartailhac 1903), suggest not only that this particular design was popular among the Magdalenian peoples, but that if the implements themselves were not manufactured at a particular site/s and traded to surrounding groups, then the *design idea* moved around the landscape and was copied by various craftspeople (Cattelain 2005).

Researchers have also used the 'proliferation' in parietal and portable artworks as well as the extravagant decoration of weaponry to suggest that Magdalenian populations were reacting to an increase in social tensions as a result of changing environmental conditions (Geist 1978; Bahn 1982; Jochim 1983). If this hypothesis is correct, then projectile points and their launch systems during this period may not only have been effective hunting weaponry, but also functioned in the social arena.

Summary and Conclusion

The range of osseous projectile weaponry is so diverse and prolific during the Magdalenian of Western Europe that providing an outline of its components and chronological changes in a single chapter is a difficult task. Magdalenian craftspeople were expert antler workers and used their accumulated knowledge to manufacture complete hunting toolkits which were suitable to exploit a wide range of fauna. Some of these weapons were beautifully decorated with images of the most significant species (mostly horse, bison, deer) and were carefully maintained throughout their use life. While Magdalenian weaponry has been studied for the past 150 or more years, it continues to provide new insights into the technological, subsistence and social aspects of Magdalenian life.

Acknowledgements We would like to thank the University of Oxford, Université de Paris I, and the Australian National University for supporting this research. We would also like to thank the Musée d'archéologie Nationale (Saint-Germain-en-Laye), The British Museum (London), and the Musée de Préhistoire (Lussac-Les-Châteaux) for permission to study their collections and include photographs of their curated artefacts. Parts of this chapter are adapted from a previous publication with M. Langlais in a book edited by G. Khlopachev (Pétillon and Langlais 2011): both are thanked for prompting the initial work.

References

Allain, J., Rigaud, A., Kozłowski, J. K., & Desbrosse, R. (1985). Le Magdalénien à navettes. *Gallia Préhistoire, 28*, 37–124.

Arndt, S., & Newcomer, M. H. (1986). Breakage patterns on prehistoric bone points: An experimental study. In D. A. Roe (Ed.), *Studies in the Upper Palaeolithic of Britain and Northwestern Europe. British Archaeological Reports International Series 269* (pp. 165–173). Oxford: Oxbow.

Averbouh, A. (2000). *Technologie de la matière osseuse travaillée et implications palethnologiques: l'exemple des chaînes d'exploitation du bois de cervidé chez les Magdaléniens des Pyrénées.* Ph.D. dissertation, Université de Paris I Panthéon-Sorbonne.

Averbouh, A. (2005). Collecte du bois de renne et territoire d'exploitation chez les groupes magdaléniens des Pyrénées ariégeoises. In D. Vialou, J. Renault-Miskovsky, & M. Patou-Mathis (Eds.), *Comportements des Hommes du Paléolithique Moyen et Supérieur en Europe: Territoires et milieu* (pp. 59–70). Liège, Belgium: Université de Liège – Service de Préhistoire.

Averbouh, A. (2006). L'industrie en matières osseuses de l'unité Y127. In P. Bodu, M. Julien, B. Valentin, & G. Debout (Eds.), *Un dernier hiver à Pincevent. Les Magdaléniens du niveau IV0 (Pincevent, La Grande Paroisse, Seine-et-Marne), Gallia Préhistoire* (Vol. 48, pp. 144–146). Paris: Éditions du CNRS.

Averbouh, A. (2010). Utilisation et transformation des matières osseuses au Buisson Campin (Verberie, Oise). In E. Zubrow, F. Audouze, & J. G. Enloe (Eds.), *The Magdalenian household: Unraveling domesticity* (pp. 77–90). New York: State University of New York Press.

Averbouh, A., Christensen, M., & Letourneux, C. (2010). Altérations taphonomiques et industrie osseuse: Quelle approche pour quel objectif? Le cas de la dissolution et de son action sur les vestiges en matière osseuse travaillée. In C. Thiébaut, M.-P. Courront, & A. Averbouh (Eds.), *Mise en commun des approches en taphonomie, Actes du workshop 16 du XVème Congrès de l'UISPP, Lisbonne, 2006* (pp. 65–74). Les Eyzies-de-Tayac, France: SAMRA (Supplément à *Paléo* 3).

Averbouh, A., & Pétillon, J.-M. (2011). Identification of 'debitage by fracturation' on reindeer antler: Case study of the Badegoulian levels at the Cuzoul de Vers (Lot, France). In J. Baron & B. Kufel-Diakowska (Eds.), *Written in bones. Studies on technological and social contexts of past faunal skeletal remains* (pp. 41–52). Wroclaw, Poland: University of Wroclaw.

Baffier, D. (1990). Lecture technologique des représentations paléolithiques liées à la chasse et au gibier. *Paléo, 2*, 177–190.

Bahn, P. G. (1979). *The French Pyrenees: An economic prehistory.* Cambridge, England: University of Cambridge.

Bahn, P. G. (1982). Inter-site and inter-regional links during the Upper Palaeolithic: The Pyrenean Evidence. *Oxford Journal of Archaeology, 1*(3), 247–268.

Bertrand, A. (1999). *Les armatures de sagaies Magdaléniennes en matière dure animale dans les Pyrénées. British Archaeological Reports International Series 773.* Oxford: Hadrian Books.

Buisson, D., Menu M., Pinçon, G. & Walter, P. (1989). Les objets colores du Paléolithique supérieur: Cas de la grotte de La Vache (Ariège). *Bulletin de la société Préhistorique Française, 86*, 183–191.

Cacho Quesada, C., & Torre Sainz, I. (2005). Les harpons magdaléniens sur le versant méditerranéen espagnol. In V. Dujardin (Ed.), *Industrie osseuse et parures du Solutréen au Magdalénien en Europe. Actes de la table ronde sur le Paléolithique supérieur récent, Angoulême (Charente), 28-30 mars 2003* (pp. 257–266). Paris: Société Préhistorique Française (Mémoire 39).

Cartailhac, É. (1903). Les stations de Bruniquel sur les bords de l'Aveyron. *L'Anthropologie, 14*, 129–150.

Cattelain, P. (1997). Hunting during the Upper Paleolithic: Bow, spear-thrower, or both? In H. Knecht (Ed.), *Projectile technology* (pp. 213–240). New York: Plenum Press.

Cattelain, P. (2004). Un propulseur inédit de la grotte du Placard (Vihonneur, Charente, France). *Notae Praehistoricae, 24*, 61–67.

Cattelain, P. (2005). Propulseurs magdaléniens: marqueurs culturels régionaux. In V. Dujardin (Ed.), *Industrie osseuse et parures du Solutréen au Magdalénien en Europe. Actes de la table ronde sur le Paléolithique supérieur récent, Angoulême (Charente), 28–30 mars 2003* (pp. 301–317). Paris: Société Préhistorique Française (Mémoire 39).

Cattelain, P. (2006). Apparition et évolution de l'arc et des pointes de flèches dans la Préhistoire européenne (Paléo-, Méso-, Néolithique). In P. Bellintani & F. Cavulli (Eds.), *Catene operative dell'arco Preistorico: Incontro di archeologia sperimentale* (pp. 45–66). Trento, Italy: Soprintendenza Per i Beni Archeologici.

Cholet-Kritter, P. (2009). *Les Pointes de Projectiles à Biseau Simple au Magdalénien Moyen dans la Grotte d'Isturitz, Pyrénées-Atlantiques.* Approche Typologique et Fonctionnelle. Mémoire de Master 1, Université de Paris I.

Christensen, M. (1999). *Technologie de l'ivoire au Paléolithique supérieur. Caractérisation physico-chimique du matériau et analyse fonctionnelle des outils de transformation. British Archaeological Reports International Series 751.* Oxford: Hadrian Books.

Christensen, M. (2008). Le travail des matières osseuses, une activité qui se précise. In M. Olive, M. Christensen, N. Pigeot, & Y. Taborin (Eds.), *Les fouilles d'Etiolles (2007–2008). Rapport de 2e année de fouille triennale au Ministère de la Culture* (pp. 83–89). Unpublished report.

Christensen, M., Legoupil, D., & Pétillon, J.-M. (2016). Hunter-gatherers of the old and new world: Morphological and functional comparisons of osseous projectile points. In M. C. Langley (Ed.), *Osseous projectile weaponry: Towards an understanding of Pleistocene cultural variability* (pp. 237–252). Dordrecht: Springer.

Christensen, M., & Averbouh, A. (2005). Les vestiges en bois de cervidé des occupations Magdaléniennes d'Étiolles. In M. Olive, M. Christensen, N. Pigeot, & Y. Taborin (Eds.), *Les fouilles d'Etiolles 2005* (pp. 23–27). Unpublished report.

Christensen, M., & Chollet, A. (2005). L'industrie sur bois de cervidé et os des niveaux magdaléniens et aziliens de Bois-Ragot: étude préliminaire. In A. Chollet & V. Dujardin (Eds.), *La Grotte du Bois Ragot à Gouex (Vienne), Magdalénien et Azilien, essais sur les Hommes et leur environnement* (pp. 223–260). Paris: Société Préhistorique Française (Mémoire 38).

Clark, G. J. D., & Thompson, M. W. (1953). The groove and splinter technique of working antler in Upper Paleolithic and Mesolithic Europe. *Proceedings of the Prehistoric Society, 19*(1), 148–160.

Cochard, D. (2004). *Les Léporidés dans la subsistance des paléolithiques du sud de la France*. Ph.D. dissertation, Université de Bordeaux I.

Combier, J. (1967). *Le Paléolithique de l'Ardèche dans son cadre Paléoclimatique* (p. 4). Bordeaux, France: Publications de l'Institut de Préhistoire de l'Université de Bordeaux.

Conkey, M. W. (1980). The identification of prehistoric hunter-gatherer aggregation sites: The case of Altamira. *Current Anthropology, 21*, 609–630.

Cordier, G. (1990). Blessures préhistoriques animales et humaines avec armes ou projectiles conservés. *Bulletin de la Société Préhistorique Française, 87*, 462–480.

Costamagno, S., Cochard, D., Ferrié, J.-G., Laroulandie, V., Barbaza, M., Cazals, N., et al. (2008). Nouveaux milieux, nouveaux gibiers, nouveaux chasseurs? Évolution des pratiques cynégétiques dans les Pyrénées du Tardiglaciaire au début du Postglaciaire. *Bulletin de la Société Préhistorique Française, 105*, 17–27.

Costamagno, S., Laroulandie, V., Langlais, M., & Cochard, D. (2009). Exploitation du monde animal sur le versant nord des Pyrénées durant le Tardiglaciaire. In J.-M. Fullola, N. Valdeyron, & M. Langlais (Eds.), *Els Pirineus i les àrees circumdants durant el Tardiglacial. Mutacions i filiacions tecnoculturals evolucion paleoambiental, Actes du XIVe Colloque international d'Archéologie de Puigcerdà, Hommage à G. Laplace, Novembre 2006* (pp. 185–209). Puigcerdà, Spain: Institut d'Estudis Ceretans.

Debout, G., Olive, M., Bignon, O., Bodu, P., Chehmana, L., & Valentin, B. (2012). The Magdalenian in the Paris Basin: New results. *Quaternary International, 272–273*, 176–190.

Ducasse, S., Castel, J.-C., Chauvière, F.-X., Langlais, M., Camus, H., Morala, A., et al. (2011). Le Quercy au coeur du Dernier Maximum glaciaire. La couche 4 du Petit Cloup Barrat et la question de la transition Badegoulo-Magdalénienne. *Paléo, 22*, 101–154.

Feruglio V., & Buisson, D. (1999). Accolements de pièces à section demi-ronde. In M. Julien, A. Averbouh, & D. Ramseyer (Eds.), *Préhistoire d'os* (pp. 143–149). Aix-en-Provence, France: Publications de l'Université de Provence.

Fradet, C. (2004). *Le travail de l'Ivoire au Paléolithique Supérieur: l'Exemple du Magdalénien*. Toulouse, France: Université de Toulouse II Le Mirail.

Fritz, C. & Roussot, A. (1999). L'art mobilier. In C. Chauchat (Ed.), *L'Habitat Magdalénien de la Grotte du Bourrouilla à Arancou (Pyrénées-Atlantiques), Gallia Préhistoire* (Vol. 41, pp. 54–97). Paris: Éditions du CNRS.

Fischer, A., & Tauber, H. (1986). New C-14 datings of Late Paleolithic cultures from northwestern Europe. *Journal of Danish Archaeology, 5*, 7–13.

Geist, V. (1978). *Life strategies, human evolution, environmental design. Toward a biological theory of health*. New York: Springer.

Gonzalez Morales, M. R., & Straus, L. G. (2009). Extraordinary Early Magdalenian finds from El Mirón Cave, Cantabria (Spain). *Antiquity, 83*, 267–281.

González Sainz, C. (1989). *El Magdaleniense Superior-Final de la región Cantábrica*. Santander, England: Universidad de Cantabria.

Goutas, N. (2009). Réflexions sur une innovation technique Gravettienne importante: le double rainurage. *Bulletin de la Société Préhistorique Française, 106*(3), 437–456.

Houmard, C. (2004). *L'Exploitation du bois de cervidé à Laugerie-Basse: Analyse typo-technologique du matériel Magdalénien de l'Abri Classique*. Paris: Mémoire de DEA, Université de Paris I.

Houmard, C., & Jacquot, E. (2009). Des têtes de projectile composites à "La Garenne". In J. Despriée, S. Tymula, & A. Rigaud (Eds.), *Données récentes sur le Magdalénien de "La Garenne" (Saint-Marcel, Indre) et la place du Magdalénien "à Navettes" en Europe* (pp. 137–150). Argenton-sur-Creuse: Archéologie du val de Creuse en Berry, Bulletin de l'ASSAAM Special Issue.

Jacobi, R.M., Higham, T.F.G., & Lord, T.C. (2009). Improving the chronology of the human occupation of Britain during the Late Glacial. In M. Street, N. Barton, & T. Terberger (Eds.), Humans, environment and chronology of the Late Glacial of the North European Plain (pp. 7–25). Mainz: Verlag RGZM.

Jochim, M. A. (1983). Palaeolithic cave art in ecological perspective. In G. N. Bailey (Ed.), *Hunter-Gatherer economy in prehistory: A European perspective* (pp. 212–219). Cambridge, England: Cambridge University Press.

Julien, M. (1982). *Les harpons Magdaléniens*. Paris: Éditions du CNRS (Suppléments à *Gallia Préhistoire* 17).

Julien, M., & Orliac, M. (2003). Les harpons et les éléments barbelés. In J. Clottes, & H. Delporte (Eds.), *La Grotte de La Vache, Tome I: Les occupations du Magdalénien* (pp. 221–274). Paris: Éditions de la Réunion des Musées Nationaux/Éditions du Comité des Travaux Historiques et Scientifiques.

Junkmanns, J. (2001). *Arc et flèche. Fabrication et utilisation au Néolithique*. Bienne, Switzerland: Musée Schwab.

Kandel, D. (1995). Objets en ivoire du Musée des Antiquités Nationales. In J. Hahn, M. Menu, Y. Taborin, & F. Widemann (Eds.), *Le travail et l'usage de l'ivoire au Paléolithique supérieur. Actes de la Table-ronde de Ravello, 29–31 mai 1992* (pp. 95–102). Italy: CUEBC.

Langlais, M. (2010). *Les sociétés Magdaléniennes de l'Isthme Pyrénéen*. Paris: CTHS (Documents Préhistoriques, 26).

Langlais, M. (2011). Process of change in Magdalenian societies in the Pyrenean isthmus (20-16 ky cal BP). *Antiquity, 85*, 715–728.

Langlais, M., Ladier, E., Chalard, P., Jarry, M., & Lacrampe-Cuyaubère, F. (2007). Aux origines du Magdalénien quercinois: Les industries de la séquence inférieure de l'abri Gandil (Bruniquel, Tarn-et-Garonne). *Paléo, 19*, 341–366.

Langlais, M., Pétillon, J.-M., Beaune, S. A., Chauvière, F.-X., Letourneux, C., Szmidt, C., et al. (2010). Une occupation de la fin du Dernier Maximum glaciaire dans les Pyrénées: Le Magdalénien inférieur de la grotte des Scilles (Lespugue, Haute-Garonne). *Bulletin de la Société Préhistorique Française, 107*, 5–51.

Langlais, M., Costamango, S., Laroulandie, V., Pétillon, J.-M., Discamps, E., Mallye, J.-B., et al. (2012). The evolution of Magdalenian societies in South-West France between 18,000 and 14,000 cal BP: Changing environments, changing tool kits. *Quaternary International, 272–273*, 138–149.

Langlais, M., Laroulandie, V., Costamagno, S., Pétillon, J.-M., Mallye, J.-B., Lacrampe-Cuyaubère, F., et al. (2015). Premiers temps du Magdalénien en Gironde: réévaluation des fouilles Trécolle à Saint-Germain-la-Rivière (France). *Bulletin de la Société Préhistorique Française, 112*, 5–58.

Langley, M. C. (2013). *Investigating maintenance and discard patterns of middle-late Magdalenian antler projectile points*. Ph.D. dissertation, University of Oxford.

Langley, M. C. (2014). Magdalenian antler projectile point design: Determining original form for uni- and bilaterally barbed points. *Journal of Archaeological Science, 44*, 104–116.

Langley, M. C. (2015). Investigating maintenance and discard behaviours for osseous projectile points: A Middle to Late Magdalenian (c. 19,000–14,000 cal. BP example. *Journal of Anthropological Archaeology, 40*, 340–360.

Langley, M. C., & Street, M. (2013). Long range inland-coastal networks during the Late Magdalenian: Evidence for individual acquisition of marine resources at Andernach-Martinsberg, German Central Rhineland. *Journal of Human Evolution, 64*, 457–465.

Laroulandie, V. (2003). Exploitation des oiseaux au Magdalénien en France: état des lieux. In S. Costamango & V. Laroulandie (Eds.), *Mode de vie au Magdalénien: Apports de l'archéozoologie, Actes du XIVe Congrès UISPP, Liège 2001. British Archaeological Reports International Series 1144* (pp. 129–138). Oxford: Hadrian Books.

Lefebvre, A. (2011). *Les Pointes barbelées magdaléniennes. Étude typologique, géographique et chronologique*. Mémoire de Master 2 Dissertation, Université de Toulouse le Mirail.

Le Gall, O. (1992). Les Magdaléniens et l'ichtyofaune dulçaquicole. In J.-P. Rigaud, H. Laville, & B. Vandermeerch (Eds.), *Le peuplement Magdalénien. Paléogéographie physique et humaine* (pp. 277–285). Paris: Éditions du CTHS.

Leroi-Gourhan, A. (1983). Une tête de sagaie à armatures de lamelles de silex de Pincevent. *Bulletin de la Société Préhistorique Française, 80*, 154–156.

Leroy-Prost, C. (2008). L'industrie sur matières dures animales. In A. Glory, B. Delluc, & G. Delluc (Eds.), *Les recherches à Lascaux (1952–1963)* (pp. 119–166). Paris: Éditions du CNRS (Suppléments à *Gallia Préhistoire,* 39).

Letourneux, C., & Pétillon, J.-M. (2008). Hunting lesions caused by osseous projectile points: Experimental results and archaeological implications. *Journal of Archaeological Science, 35*, 2849–2862.

Liolios, D. (1999). *Variabilité et Caractéristiques du Travail des Matières Osseuses au Début de l'Aurignacien: Approche Technologique et Économique*. Ph.D. dissertation, Université de Paris X.

Macgregor, A. G., & Currey, J. D. (1983). Mechanical properties as conditioning factors in the bone and antler industry of the 3rd to the 13th century AD. *Journal of Archaeological Science, 10*, 71–77.

Omnès, J. (1980). *Le gisement préhistorique des Espélugues à Lourdes, Hautes-Pyrénées: Essai d'inventaire des fouilles anciennes*. Tarbes, France: Centre Aturien de Recherches sous Terre.

Otte, M. (1995). L'ivoire paléolithique au Nord-Ouest européen. In J. Hahn, M. Menu, Y. Taborin, & F. Widemman (Eds.), *Le travail et l'usage de l'ivoire au Paléolithique supérieur. Actes de la Table-ronde de Ravello, 29–31 mai 1992* (pp. 103–114). Italy: CUEBC.

Passemard, E. (1920). Les spirales de la caverne d'Isturitz. *Bulletin de la Société Préhistorique Française, 17*, 150–154.

Pétillon, J.-M. (2006). *Des Magdaléniens en armes. Technologie des armatures de projectiles en bois de cervidé du Magdalénien Supérieur de la Grotte d'Isturitz (Pyrénées-Atlantiques)*. Treignes, Belgium: Éditions du Cedarc.

Pétillon, J.-M. (2008a). First evidence of a whale-bone industry in the western European Upper Paleolithic: Magdalenian artifacts from Isturitz (Pyrénées-Atlantiques, France). *Journal of Human Evolution, 54*, 720–726.

Pétillon, J.-M. (2008b). Spécificités des armatures osseuses magdaléniennes du Bassin parisien. In B. Valentin (Ed.), *Habitats et peuplements tardiglaciaires du Bassin parisien, Projet collectif de recherche, Bilan des activités de 2006 à 2008* (pp. 25–30). http://hdl.handle.net/2332/1479. Accessed 1 June 2013.

Pétillon, J.-M. (2009). What are these barbs for? Preliminary reflections on the function of the Upper Magdalenian barbed weapon tips. *Palethnologie, 1*, 69–102.

Pétillon, J.-M. (2013). Circulation of whale-bone artifacts in the northern Pyrenees during the Late Upper Paleolithic. *Journal of Human Evolution, 65*, 525–543.

Pétillon, J.-M., & Ducasse, S. (2012). From flakes to grooves: A technical shift in antler working during the last glacial maximum in southwest France. *Journal of Human Evolution, 62*, 435–465.

Pétillon, J.-M., & Langlais, M. (2011). Эволюция костяного и каменного оружия эпохи финального Палеолита; Мадленские сообщества (20,000–14,000 cal BP) в Юго-Западной Европе. In G. A. Khlopachev (Ed.), *Предметы вооружения и искусства из кости в древних културах Северной Евразии (технологический и функциональный аспекты)* (pp. 69–95). Sankt-Peterburg, Russia: Nauka (Zamjatninskij sbornik).

Pétillon, J.-M., Langlais, M., Beaune, S. A., Beukens, R., Chauvière, F.-X., David, F., et al. (2008). Le Magdalénien de la grotte des Scilles (Lespugue, Haute-Garonne). Premiers résultats de l'étude pluridisciplinaire de la collection Saint-Périer. *Antiquités Nationales, 39*, 1–15.

Pétillon, J.-M., Bignon, O., Bodu, P., Cattelain, P., Debout, G., Langlais, M., et al. (2011). Hard core and cutting edge: Experimental manufacture and use of Magdalenian composite projectile tips. *Journal of Archaeological Science, 38*, 1266–1283.

Pétillon J.-M., Langlais M., Kuntz D., Normand C., Barshay-Szmidt C., Costamagno S., et al. (2015). The human occupation of the northwestern Pyrenees in the Late Glacial: New data from the Arudy basin, lower Ossau valley. *Quaternary International, 364*, 126–143.

Pokines, J. T. (1993). *Antler points from El Juyo (Santander, Spain): Form, manufacture and parameters of use*. Masters thesis, University of Chicago.

Pokines, J. T. (1998). Experimental replication and use of Cantabrian Lower Magdalenian antler projectile points. *Journal of Archaeological Science, 25*, 875–886.

Pokines, J. T., & Krupa, M. (1997). Self-barbed antler spearpoints and evidence of fishing in the late Upper Paleolithic of Cantabrian Spain. In H. Knecht (Ed.), *Projectile technology* (pp. 241–262). New York: Plenum Press.

Primault, J., Brou, L., Langlais, M., & Guérin, S. (2007). Le Magdalénien inférieur à microlamelles à dos de la grotte du Taillis des Coteaux à Antigny (Vienne, France). *Bulletin de la Société Préhistorique Française, 104*, 5–30.

Pujol, É. (2009). *L'exploitation des matières osseuses chez les Magdaléniens de Saint-Michel d'Arudy (Pyrénées-Atlantiques). Caractérisation des schémas de production par l'approche technologique du matériel osseux*. Mémoire de Master 2, Université de Paris I.

Rigaud, A. (2004). Fiche débitage du bois de renne au Magdalénien. L'exemple de La Garenne (Indre, France). In D. Ramseyer (Ed.), *Matières et techniques, industrie de l'os préhistorique* (pp. 79–87). Paris: Société Préhistorique Française.

Rigaud, A. (2006). Étude technologique des baguettes demi-rondes de Labastide (Hautes- Pyrénées). *Archéologie des Pyrénées occidentales et des Landes, 25*, 229–246.

Rosendahl, G., Beinhauer, K. W., Löscher, M., Kreipl, K., Walter, R., & Rosendahl, W. (2006). Le plus vieil arc du monde? Une pièce intéressante en provenance de Mannheim, Allemagne. *L'Anthropologie, 110*, 371–382.

Rust, A. (1943). *Die alt- und Mittelsteinzeitlichen funde von Stellmoor*. Neumunster, Germany: Karl-Wachholtz Verlag.

Sacchi, D. (2003). *Le Magdalénien: Apogée de l'art Quaternaire*. Paris: La Maison des Roches.

Saint-Périer, R. de. (1920). À propos des gravures en spirales de la grotte d'Isturitz. *Bulletin de la Société Préhistorique Française, 17*, 154.

Saint-Périer, R. de. (1929). Les baguettes sculptées dans l'art paléolithique. *L'Anthropologie, 39*, 43–63.

Saint-Périer, R. de. (1936). *La Grotte d'Isturitz II: le Magdalénien de la Grande Salle*. Paris: Masson (Archives de l'Institut de Paléontologie Humaine, Mémoire 17).

Saint-Périer, R. de. (1947). Les derniers objets magdaléniens d'Isturitz. *L'Anthropologie, 51*, 393–415.

Semenov, S. A. (1964). *Prehistoric technology*. London: Cory, Adams and Mackay.

Sgard, L. (1999). *Technologie du bois de renne dans les niveaux magdaléniens à Isturitz (Pyrénées-Atlantiques): le problème des modes de débitage*. Paris: Mémoire de DEA, Université de Paris I.

Stodiek, U. (1990). Jungpaläolithische Speerschleudern und Speere: Ein Rekonstruktionsversuch. *Experimentelle Archäologie in Deutschland. Archäologische Mitteilungen aus Nordwestdeutschland, 4*, 287–297.

Stodiek, U. (1993). *Zur Technologie der jungpaläolithischen Speerschleuder. Eine Studie auf der Basis archäologischer, ethnologischer und experimenteller Erkenntnisse*. Tübingen, Germany: Archaeologia Venatoria (Tubinger Monographien zur Urgeschichte, 9).

Stodiek, U. (2000). Preliminary results of an experimental investigation of Magdalenian antler points. In C. Bellier, P. Cattelain, & M. Otte

(Eds.), *La chasse dans la préhistoire/Hunting in prehistory* (pp. 70–78). Bruxelles, Belgium: Société Royale Belge d'Anthropologie et de Préhistoire/Service de Préhistoire de l'Université de Liège/Centre d'Études et de Documentation Archéologiques.

Szmidt, C., Pétillon, J.-M., Cattelain, P., Normand, C., & Schwab, C. (2009). Premières dates radiocarbone pour le Magdalénien d'Isturitz (Pyrénées-Atlantiques). *Bulletin de la Société Préhistorique Française, 106*, 583–601.

Tejero, J.-M. (2013). *La explotación de las materias óseas en el Auriñaciense. Caracterización tecnoeconómica de las producciones del Paleolítico superior inicial en la Península Ibérica. British Archaeological Reports International Series 2469.* Oxford: Archaeopress.

Torres, J. M. R. I. (1987). La indústria óssia del Palaolític Superior de Serinyá: Reclau Viver I Bora Gran d'En Carreras. *Cypsela, VI*, 229–236.

Treuillot, J. (2011). *Le Magdalénien de la Madeleine: Une exploitation intensive des matière dures animales? Technologie des outillages en matières dures animales du Magdalénien Indéterminé de l'Abri de la Madeleine.* Paris: Mémoire de Master 2, Université de Paris I.

Tyzzer, E. E. (1936). The "simple bone point" of the shell-heaps of the northeastern Algonkian area and its probable significance. *American Antiquity, 1*(4), 261–279.

Villaverde, V., & Roman, D. (2005–2006). Los arpones del Magdaleniense superior de la Cova de les Cendres y su valoración en el contexto del Magdaleniense mediterráneo. *Munibe, 57*(2), 207–225.

Villaverde, V., Aura, J. E., Borao, M., & Roman, D. (2016). Upper palaeolithic bone and Antler projectiles in the Spanish Mediterranean Region: The Magdalenian period. In M. C. Langley (Ed.), *Osseous projectile weaponry: Towards an understanding of Pleistocene cultural variability* (pp. 109–130). Dordrecht: Springer.

Weniger, G.-C. (2000). Magdalenian barbed points: harpoons, spears and arrowheads. In C. Bellier, P. Cattelain, & M. Otte (Eds.), *La chasse dans la préhistoire/Hunting in prehistory* (pp. 79–87). Bruxelles, Belgium: Société Royale Belge d'Anthropologie et de Préhistoire/Service de Préhistoire de l'Université de Liège/Centre d'Études et de Documentation Archéologiques.

Chapter 11
Osseous Projectile Points in the Magdalenian: 'True' Points Versus Pointed Waste-Products

François-Xavier Chauvière

Abstract Magdalenian osseous projectile points have the highest morpho-technical diversity in terms of size, technical design and hafting systems of all European Upper Paleolithic systems. A few years ago, some of these points, based on morphological and metrical criteria, were designated as 'points with a shortened base' (or "with a shrunken base", "nibbled", "stemmed", "unworked", or "flaked and broken"), and interpreted as a distinct projectile point type. Recent technological and functional analyses of these "points with a shortened base", however, have enabled us to replace them within their *chaîne opératoire* and reclassify them as pointed waste-products. The only common feature of these artefacts is their specific segmentation, which occurs in different phases of the production sequence (blank extraction, shaping and recycling), and corresponds to objectives attained through the use of various techniques, tools and actions. This same stigmata is observed on other artefact types (*baguettes demi-rondes*), not all of which are hunting implements (pierced batons and needles). In this chapter, the variability of these pointed waste-products is described and the real quantitative proportions of 'true' projectile points in Magdalenian assemblages and the European Upper Paleolithic is outlined.

Keywords European Upper Paleolithic • Technological analysis of bone industries • *Chaîne opératoire* • Manufacturing waste

F.-X. Chauvière (✉)
Office du Patrimoine et de l'Archéologie de Neuchâtel, Section Archéologie, Laténium, Espace Paul Vouga, 2068 Hauterive, Switzerland
e-mail: francois-xavier.chauviere@ne.ch

Introduction

In European archaeological assemblages attributed to the Magdalenian *sensu lato*, pointed artefacts made from bone, ivory and cervid antler are among the most numerous objects manufactured from osseous materials. Some are thought to have had a precise function, which is sometimes indicated by their typological designation ('awl', 'pin', 'spear point', 'forked-based point', 'split-based point', 'harpoon', etc.) (Leroy-Prost 1975, 1979; Julien 1982; Delporte et al. 1988; Camps-Fabrer et al. 1990; Pétillon 2006). Beyond these classifications, which are not fully agreed upon, we can distinguish two types of functions: one employing points to perforate soft materials (such as an awl), and the other as projectile points. Technological analysis has shown a nearly perfect match between the mechanical properties of the materials and the choices of prehistoric peoples regarding what raw material to use to make which tools (e.g., Knecht 1997; Liolios 1999; Averbouh 2000; Knecht 2000; Legrand 2002–2003; Christensen 2004; d'Errico et al. 2004; Goutas 2004; Pétillon 2006). The dichotomy between the use of bone (mostly for domestic tools) and the use of antler (mostly for projectile weapons), is an effective means for identifying the mode of use for a recovered object.

Projectile points played an essential role in the hunting strategies of Upper Paleolithic human groups in Europe, and their presence at a location can attest to the carrying out of various subsistence activities either at or near a site (Bellier and Cattelain 1990; Pétillon and Letourneux 2008). The identification of 'true' projectile points—those that did, or could have, functioned as tips of hunting weaponry—is therefore crucial. However, given that most of these objects are broken, any attempt to classify them without taking into account both their specific technical and functional history will not yield the required information (Pétillon 2003). Beyond a simple validation (is every point a projectile element?), the real challenge of studies of thrown, launched or thrusted projectiles in osseous materials is to characterize the projectile point type and its launching method (e.g., Carrère and Lepetz 1988; Plisson and Geneste 1989; Geneste and Plisson 1990; Chadelle et al. 1991; Geneste and Plisson

Michelle C. Langley (ed.), *Osseous Projectile Weaponry*, Vertebrate Paleobiology and Paleoanthropology,
DOI 10.1007/978-94-024-0899-7_11

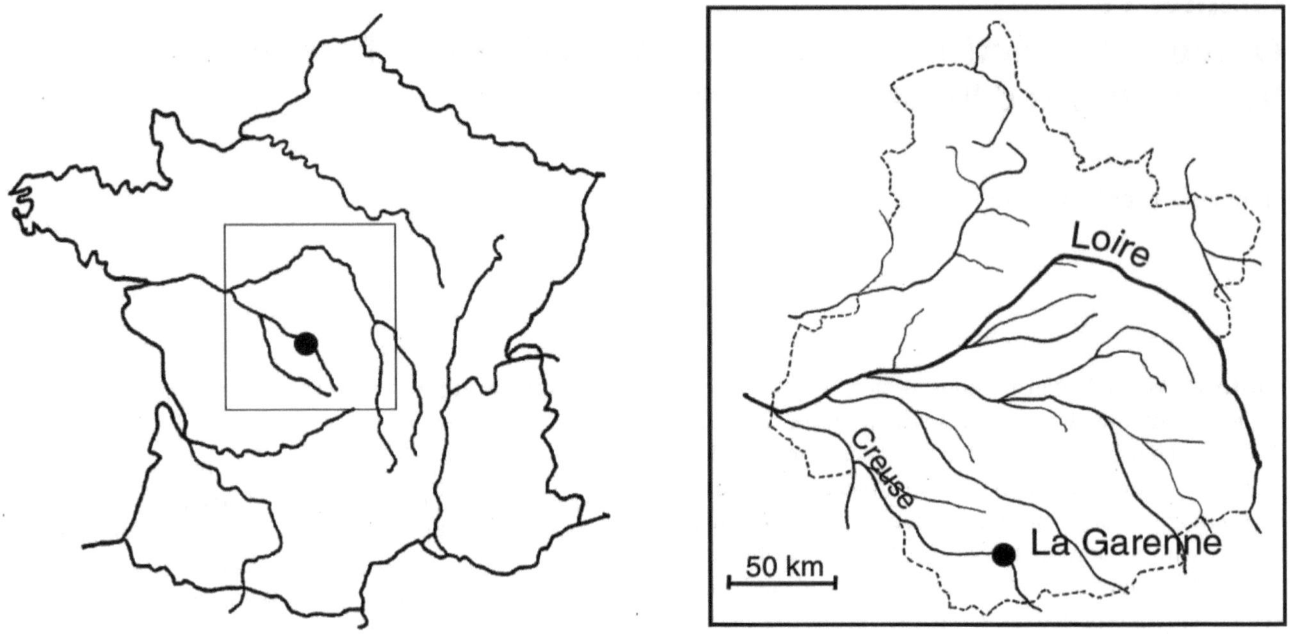

Fig. 11.1 Location of La Garenne (Indre, France)

1993; Knecht 1997, 2000; Stodiek 2000; Pétillon 2005, 2006; Pétillon et al. 2011).

The case of the '*sagaie à base raccourcie*' ('spear point with a shortened base') type serves as a good case study for the hazards of defining projectile weaponry elements based on typological considerations alone. Initially, thought to represent a distinct projectile point type, they were successively interpreted as finished tools, as recycling of pieces broken in use, and as manufacturing waste-products. This 'type' is now challenged by the technical analysis of Magdalenian artefacts from the site '*à navettes*' (meaning 'with spear foreshafts') at La Garenne (Indre, France; Chauvière and Rigaud 2005, 2008; Fig. 11.1), where it was shown that these pieces do not correspond in any way to elements hafted onto projectile shafts, or to points used for other functions. Instead, they are consistent with being waste-products from the manufacture and use of osseous (mostly cervid antler) projectile points. Through a presentation of their characteristics, it will be shown how to distinguish between 'true' projectile points and other pointed artefacts, among them pointed waste-products.

Chronological and Geographic Distribution of Points with a 'Shortened Base'

It has been observed for Magdalenian bone projectile points, that particular care was taken in the realization of the hafting system (single or double bevel, conical base, forked base; Delporte et al. 1988; Pétillon 2006; Pétillon et al. 2011). Thin grooves (striations) that appear to have improved the efficiency of an adhesive are also sometimes visible (Fig. 11.2: 1–4), and indirect evidence for the use of an adhesive in the form of residual ochre has been reported (Allain and Rigaud 1986). These 'true' points, associated with different *chaînes opératoires*, are characterized by a strict size range, a very pointed tip made by extensive shaping, and a straight profile. This morpho-technical variability can aid in the reconstruction of complete points, as well as their launching method: hand thrusted or thrown, spearthrower, bow, etc.

It should be noted that this functional attribution (to a projectile point) is not necessarily as easy as it may seem and the criteria for their identification must be defined. As a first step, the point/shaft/base division is operative for the smallest fragmentary pieces (Delporte et al. 1988). The determination of a projectile point (and its type) is seriously complicated when only shafts (with no point or base), or distal points with no bases (which are most common in assemblages) are recovered (Fig. 11.2: 5–8). In this latter case, there is a high risk of confusing projectile point fragments with 'pointed' manufacturing waste-products.

A chronological status was, sometimes, attributed to 'points with a shortened base', especially for the Upper Magdalenian (Sacchi 1986:186; Fano Martinez et al. 2005:193), but it can only be regional and does not appear to have extended beyond these limits. The fact that we find the same segmenting stigmata on other Magdalenian objects, such as bone needles, pierced batons and *baguettes demi-rondes* in antler makes any attempt at precise geographical and chronological demarcation difficult (Fig. 11.4).

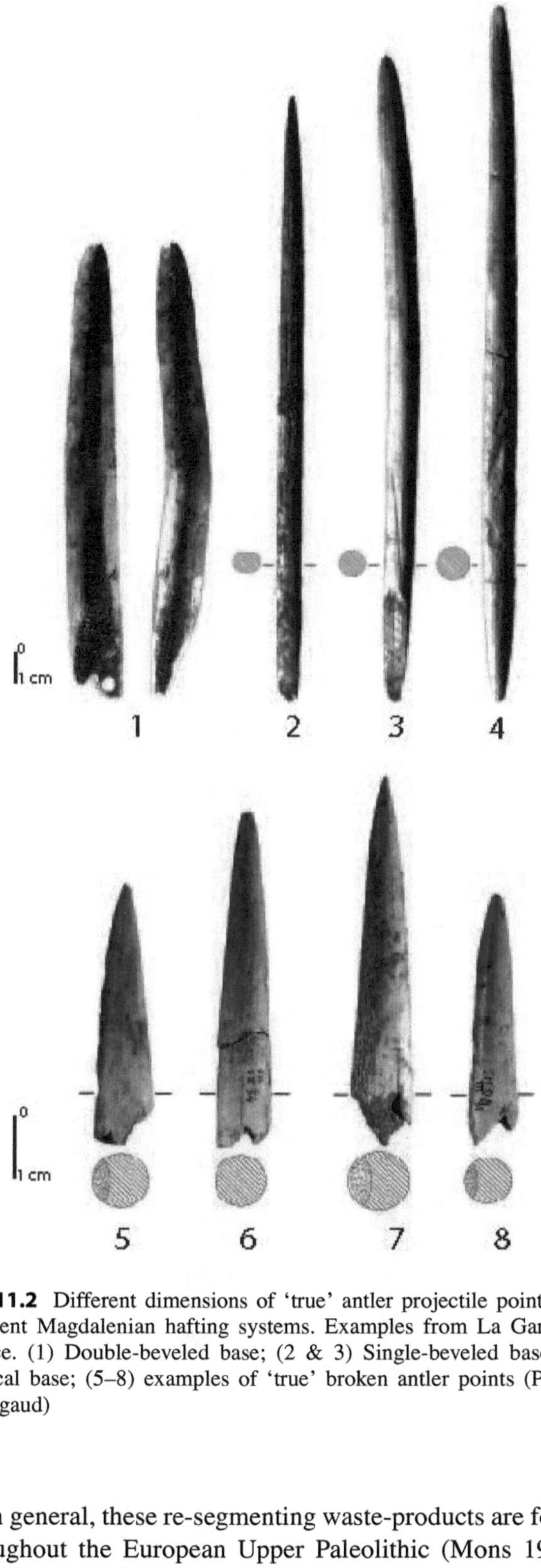

Fig. 11.2 Different dimensions of 'true' antler projectile points and different Magdalenian hafting systems. Examples from La Garenne, France. (1) Double-beveled base; (2 & 3) Single-beveled base; (4) Conical base; (5–8) examples of 'true' broken antler points (Photos A. Rigaud)

In general, these re-segmenting waste-products are found throughout the European Upper Paleolithic (Mons 1988), beginning in the Aurignacian (Otte 1979). They have been identified in assemblages attributed to the Gravettian (Goutas 2004) and in the Lower, Middle and Upper phases of the Magdalenian (Pétillon et al. 2008; Langlais et al. 2010; Ladier 2014; Lefebvre 2014). Segmenting procedures similar to those described here have also been found in Natufian contexts, here in association with the manufacturing of bone awls (Le Dosseur 2003, 2006, 2010).

Morpho-Technical Features

The feature possessed by all 'points with a shortened base' is to have been subject to a specific type of segmentation (Fig. 11.3: 1–3). The definition proposed by Mons (1988): "elongated objects with a piercing distal extremity (pointed, rounded, rarely sharp), with a smooth shaft and proximal extremity roughly shortened by irregular removals" affirms that these 'points with a shortened base' were intended to penetrate a target. According to Mons (1988), André Cheynier was the first to use the designation 'shortened base', with others later employing: with a 'shortened' (Cheynier 1958), 'stepped' (Peyrony and Peyrony 1938:50), 'nibbled' (Allain 1958:544) or 'stemmed' (Leroy-Prost 1975:127) base, or a 'flaked and snapped' (Sacchi 1986:176), or 'unworked' (Bertrand 1995:11) base.

Among the chopping techniques preceding segmentation by flexion, one appears to have been used more than the others: scoring by scraping with the edge of a burin. Mostly applied to rods obtained by double grooving, this method consists of diminishing the diameter by shaving away material, always in the same direction, and restricted to a short area. Experimental data have validated several observations made from the archaeological materials from La Garenne (Rigaud 2004a). For a right-handed person, the rod is held in the left hand, with the thumb above, and the burin in the left hand, pinched between the thumb and bent index finger. The scraping movement involves simultaneous pressure and the right hand both ensures the pressure and guides the tool, while the left thumb moves the burin over a small distance, but with force and precision. It is also possible to proceed as if sharpening a pencil—by scraping toward oneself. After a few passes, the burin begins to 'chatter' and a stepped section is formed. After the piece is broken, one extremity resembles a 'sharpened pencil', while the other appears 'nibbled'. The 'nibbles' are easily identifiable while the traces of scraping can be easily confused with traces made during original shaping of the tool. This method can be used in combination with transverse movements by a flint tool, inappropriately called 'sawing' by others, which aid in the creation of a clear break with no bevel.

For pieces with a medium width section, the simplest method involves only breakage by flexion with no preliminary preparation (Fig. 11.3: 4), though it is not the easiest to identify. This situation is not a problem for rods that have not yet been shaped after extraction, but these cases are rare (Fig. 11.3: 5). Many examples indeed show that rod extraction

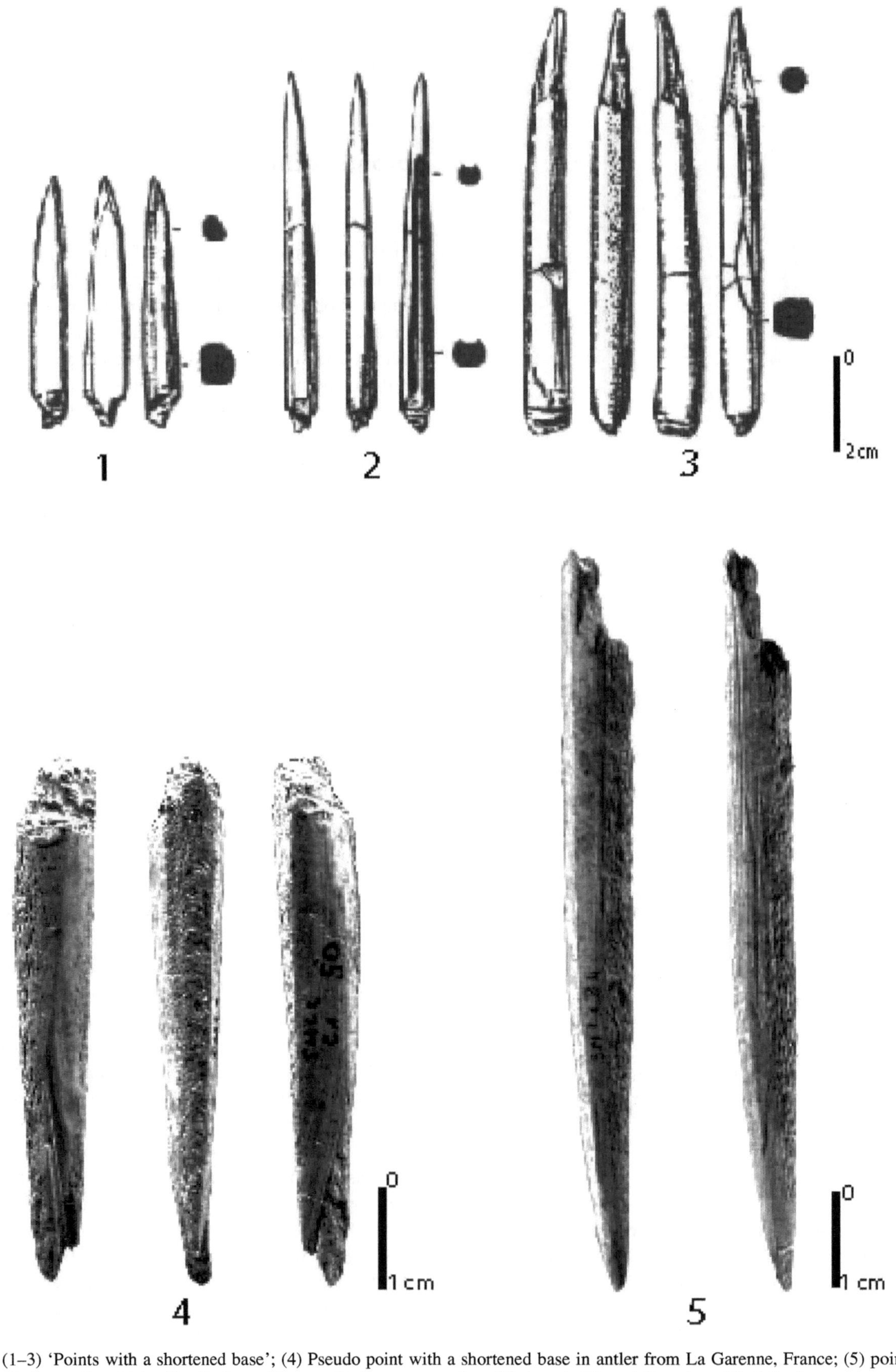

Fig. 11.3 (1–3) 'Points with a shortened base'; (4) Pseudo point with a shortened base in antler from La Garenne, France; (5) pointed waste-product in reindeer antler obtained by flexion with no preliminary preparation. (Photos: A. Rigaud. Drawings: M. Orliac)

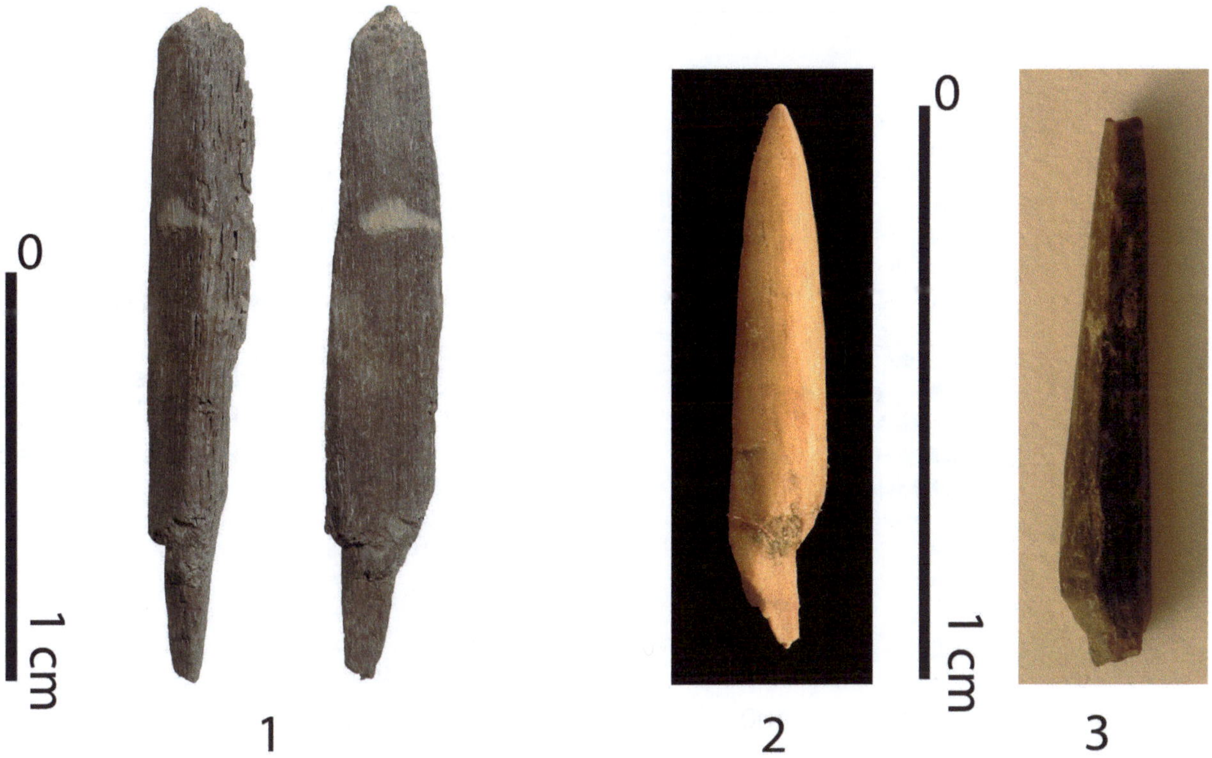

Fig. 11.4 (1–3) 'Points with a shortened base' resulting from the manufacturing of bone needles. Arancou, France. (Photos F. Plassard)

is almost always followed by scraping and sometimes a total or partial grooving of the rod well beyond the limits of the intended object. It is after this shaping phase that the rod is segmented to obtain an object corresponding to the dimensions of the finished tool. The underlying objective of this segmentation is also to obtain straight volumes, without resorting to a hypothetical straightening by soaking, exposure to fire, or with the aid of a pierced baton (Rigaud 2001, 2004b).

The products resulting from this segmentation, with a rectangular or triangular morphology, remain pointed, but can have a curved profile. Their dimensions differ from those of 'true' points (Fig. 11.5). They must not be confused with the triangular pieces described elsewhere, which are also waste-products, but resulting from the multiple grooving of reindeer antler, and which show no evidence of shaping (e.g., Bonnissent 1993; Averbouh 2000).

Waste-Products from the Manufacture of Pointed Objects

The high frequency of segmenting requires us to define its objectives (blank production, shaping, recycling) in each case identified. Segmentation can occur directly following blank extraction or after initial shaping, by scraping or polishing. If the piece is segmented after a more or less complete shaping of the pointed implement, the characteristic traces of blank extraction can be partly, or even completely, erased. The objects are thus sharp, with a more or less settled surface, a polygonal section or an almost circular section on some. They can be confused with finished objects.

To emphasize the technical status of these objects, we now employ the term 're-segmenting waste-products'. This terminology, borrowed from the medical field, corresponds well to the analogy that exists between the process of segmenting bone materials and the surgical act that consists of removing part of an organ (Chauvière and Rigaud 2005). The French term 'résection' was used by Allain et al. (1985) to appoint the voluntary process of segmenting of the bone antler 'navettes' in the "Magdalenian à navettes".

Consequently, these 'points with a shortened base' strongly correspond to waste-products of antler pointed objects resulting from manufacture.

Review of Magdalenian assemblages finds that these artefacts can be quite common. For example, M.C. Langley found a great number of 'Points with a shortened base' during a recent analysis of various Middle to Late Magdalenian assemblages. In these contexts, they were interpreted as produced by the hunter wishing to remove a worn projectile point tip from the larger weapon and thus extend its use life, or to free a weapon tip lodged in a carcass (Langley 2013, 2014).

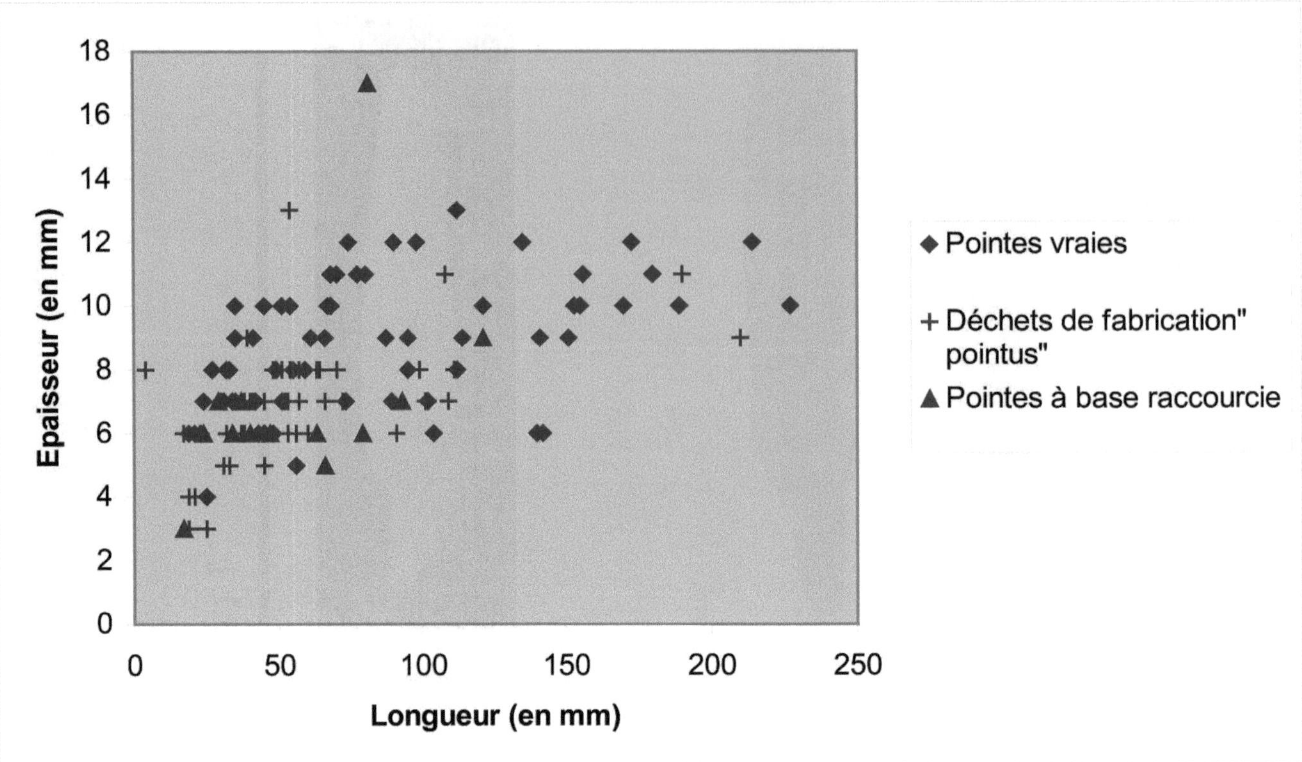

Fig. 11.5 The length/thickness (in mm) of the 'true' points, pointed waste-products and 'points with a shortened base' in antler considered from La Garenne, France (Document F.-X. Chauvière and A. Rigaud)

Discussion

Technical Stigmata: From Observation to Interpretation

Why, when starting from a correct identification of the technical stigmata (segmenting, scraping) and their relative chronology, are these 'points with a shortened base' often falsely interpreted?

Let us first remember that it is essentially the recurring presence of one pointed extremity and a "finished" appearance that lead to the interpretation of these pieces as functional elements capable of inflicting harm to hunted animals. In fact, the adoption of a technical perspective in the analysis of archaeological materials does not necessarily, and immediately, prevent an incorrect interpretation—one which is more related to classifications of archaeologists than to prehistoric reality. In my opinion, this error is linked to an excessively linear and especially 'evolutionary' vision of the *chaîne opératoire* concept, which is all too often described as the sequence from "the raw material to the finished object" (Chauvière 2013). In the end, the idea of producing a waste-product from an object in an advanced stage of shaping is so contrary to the current logic that imagining these pieces as simple waste-products is not evident.

In the same manner, an advanced state of shaping in which the formation of certain parts (point, bevel, ventral groove) and decorations are already 'clear' can precede the true use phase of a projectile point, as is observable on some specimens from La Garenne (Chauvière and Rigaud 2008, 2009). In the literature, the obliteration of decoration by a later creation of a functional part is interpreted as reflecting an abandonment of the "added value" of the object (Bouvier 1987; Chauchat et al. 1999). I, on the other hand, are tempted to see the signs of true anticipation in the realization of an object whose 'organicity' takes precedence over its symbolic dimension, but without this latter becoming obsolete. We must also be wary of our modern understanding of a production sequence, which is not necessarily that of the Magdalenian artisans.

Segmented 'Non-Pointed' Waste-Products

Segmenting procedures similar to those observed for the 'points with a shortened base' have been identified among non-pointed archaeological artifacts such as bases of projectile points (simple or double bevel) or baguettes shaped by freehand percussion, for example.

The hypothesis of the intentional segmenting of a point embedded in a hunted animal to recover the wooden shaft has been proposed by various authors (Leesch 1997; Bertrand 1999). The occasional impossibility of removing the projectile from the animal carcass has been experimentally verified (Leesch 1997:93, 202, after P. Morel). This concords with paleo-environmental data in a periglacial environment, where at an ambient temperature it can be difficult or even impossible to detach the ligature and remove the point to recover the shaft, unless an artificial heat source, such as a fireplace, is used. If points were segmented to recover the shaft, we would expect to find 'shortened' bevels (or points for in bi-conical projectile tips) in archaeological assemblages, which is indeed the case. The points themselves were taken back to the camp while still embedded in the carcass of the hunted animal (Plisson and Geneste 1989; Chadelle et al. 1991; Morel 1993).

The 'shortened' bevels can then be reused as chisels, whether this is verifiable by the presence of a crushed tip or simply their potential, given their dimensions, for this type of reuse. This idea is not new, as in 1910, in his work entitled "*Os, ivoires et bois de renne ouvrés de la Charente. Hypothèses palethnographiques*" (Worked bone, ivory and reindeer antler in the Charente), G. Chauvet illustrated and interpreted three segmented pieces originating from the Magdalenian layers of Le Placard as "*…reindeer antler rod bases with a double bevel deteriorated by extensive use and separated from the shaft to be discarded as waste. These waste-products are perhaps more useful than the beautiful intact pieces, for ethnographic studies…*" He speaks again of one of these bases whose "*…lower part is very worn by extensive use, as if the piece was used as a chisel…*" (Chauvet 1910:82–83).

Nonetheless, considering the data presented above, it is possible that at least some of these 'shortened' bases resulted from an adjustment of their size and axis during an advanced stage of shaping. They would then also be interpreted as waste-products rather than discarded finished objects, or pieces that were recycled or intended to be so. Some Magdalenian specimens from La Garenne (Chauvière and Rigaud 2005), Arancou (Pyrénées-Atlantiques France; Chauchat et al. 1999) and Hauterive-Champréveyres (Switzerland; Leesch 1997:93, 202) can be cited as examples.

Moreover, we find evidence of size and axis adjustments on particular points dating to the Badegoulian at Cuzoul de Vers (Lot, France). Here reindeer antler blanks necessary for tool manufacturing were not extracted by double grooving, but by freehand percussion, with some also being shaped by this technique. At this site, some reindeer antler flakes have been interpreted as resulting from segmenting in order to eliminate an overly curved portion of the not yet pointed blanks (Pétillon and Averbouh 2012; Pétillon and Ducasse 2012). A similar procedure is also suspected for antler blank extraction and shaping in the Lower Magdalenian of Petit Cloup Barrat (Lot, France; Ducasse et al. 2011). Re-segmenting is thus not exclusive to antler beams segmented by double grooving; it can also concern blanks extracted by percussion or splitting.

Conclusion

In archaeological assemblages attributed to the Magdalenian and the Upper Paleolithic in general, we must now question the real quantitative proportions of points made in osseous materials used as weapon tips. The exhaustive analysis of antler pointed elements found in a diverse range of assemblages clearly reduces number of 'true' points, meaning those used as projectile elements, and concomitantly increases the identified number of 'pointed' waste-products.

The large number of these waste-products recovered enables us to precisely identify the various stages in production sequences involved in the transformation of antler from blank to use. The proportion of pointed waste-products and 'true' points can also inform us about the nature of activities carried out at sites and enable us to determine the relative frequency of activities associated with the manufacturing, discard and recycling of used points. We can then consider the role of the archaeological sites at which these artefacts are found.

Acknowledgements I express my gratitude to Michelle C. Langley, Morgane Dachary, Magen O'Farrell, Jean-Christophe Castel, Jean-Pierre Chadelle and Frédéric Plassard. This article is a revised and increased version of a french paper authored by F.-X. Chauvière and A. Rigaud (2005).

References

Allain, J. (1958). Réflexions sur la chronologie du Magdalénien. *Bulletin de la Société Préhistorique Française, 55*(9), 539–545.

Allain, J., Desbrosse, R., Kozlowski, J.-K., & Rigaud, A. (1985). Le Magdalénien à navettes. *Gallia Préhistoire, 28*(1), 37–124.

Allain, J., & Rigaud, A. (1986). Décor et fonction. Quelques exemples tirés du Magdalénien. *L'Anthropologie, 90*(4), 713–738.

Averbouh, A. (2000). *Technologie de la matière osseuse travaillée et implications palethnologiques. L'exemple des chaînes opératoires du bois de Cervidé chez les Magdaléniens des Pyrénées.* Thèse de doctorat, Université de Paris I, Panthéon Sorbonne.

Bellier, C., & Cattelain, P. (1990). *La chasse dans la préhistoire du Paléolithique au Néolithique en Europe, Exposition du Musée du Malgré-Tout, 9 juin – 25 novembre 1990.* Treignes: Editions du Cedarc.

Bertrand, A. (1995). Les armatures de sagaies magdaléniennes en matière dure animale: les collections du Musée des Antiquités Nationales d'Isturitz, Saint-Michel d'Arudy et Espalungue. *Antiquités Nationales, 27*, 11–37.

Bertrand, A. (1999). *Les armatures de sagaies Magdaléniennes en matière dure animal dans les Pyrénées. British Archaeological Reports International Series 773*. Oxford: Archaeopress.

Bonnissent, D. (1993). *Choix et exploitation des bois de renne sur le site de la Madeleine (Dordogne).* Université de Bordeaux I: Mémoire de DEA.

Bouvier, J.-M. (1987). Arts et nature au Paléolithique: propos de naturaliste. In F. Delamanre, T. Hackens, & B. Helly (Eds.), *Datation, caractérisation des peintures pariétales et murales, Cercle Universitaire Européen pour les Biens culturels, Ravello, Valbonne, 30 septembre – 12 octobre 1985, Pact 17* (pp. 11–16). Ravello: Centre Universitaire Européen pour les biens culturels.

Camps-Fabrer, H., Ramseyer, D., Stordeur, D., Buisson, D., & Provenzano, N. (1990). *Poinçons, pointes, poignards, aiguilles, Fiches typologiques de l'industrie osseuse préhistorique, Union internationale des sciences préhistoriques et protohistoriques, Commission de nomenclature sur l'industrie de l'os préhistorique, cahier III.* Aix-en-Provence: Publications de l'Université de Provence.

Carrère, P., & Lepetz, S. (1988). *Etude de la dynamique des pointes de projectiles: élaboration d'une méthode.* Mémoire de maîtrise, Université de Paris I – Panthéon Sorbonne.

Chadelle, J.-P., Geneste, J.-M., & Plisson, H. (1991). Processus de formation des assemblages technologiques dans les sites du Paléolithique supérieur. Les pointes de projectiles lithiques du Solutréen de la grotte de Combe Saunière (Dordogne, France). In Collectif (Ed.), *25 ans d'études technologiques en préhistoire. Bilan et perspectives, XIèmes Rencontres Internationales d'Archéologie et d'Histoire d'Antibes. Actes des Rencontres, 18–20 octobre 1990* (pp. 275–287). Juan-les-Pins: APDCA.

Chauchat, C., Bonnissent, D., Chauvière, F.-X., Dachary, M., Eastahm, A., Fontugne, M., et al. (1999). L'habitat magdalénien du Bourouilla à Arancou (Pyrénées-Atlantiques). *Gallia Préhistoire, 41*, 1–15.

Chauvet, G. (1910). *Os, ivoires et bois de renne ouvrés de la Charente. Hypothèses palethnographiques.* Angoulême: Libr. de la Société archéologique et historique de la Charente.

Chauvière, F.-X. (2013). Le travail des matières dures d'origine animale au Paléolithique supérieur: acquis récents, interprétations neuves? In S. de Beaune (Ed.), *Chasseurs-cueilleurs. Comment vivaient nos ancêtres du Paléolithique supérieur. Méthodes d'analyse et d'interprétation en préhistoire* (pp. 49–57). Paris: CNRS Editions.

Chauvière, F.-X., & Rigaud, A. (2005). Les "sagaies" à base raccourcie ou les avatars de la typologie: du technique au "non-fonctionnel" dans le Magdalénien à navettes de La Garenne. In V. Dujardin (Ed.), *Industrie osseuse et parures du Solutréen au Magdalénien en Europe, Table ronde sur le Paléolithique supérieur récent d'Angoulême (2003)* (pp. 233–242). Paris: Société Préhistorique Française.

Chauvière, F.-X., & Rigaud, A. (2008). Le travail du bois de renne à La Garenne: entre conceptions préhistoriennes et techniques magdaléniennes ou comment séparer ébauches et déchets des pointes *vraies*? *Préhistoire du Sud-Ouest, 16*(2), 163–183.

Chauvière, F.-X., & Rigaud, A. (2009). Oubliées ou stockées? Les ébauches de pointes de projectiles en bois de renne et leur gestion par les Magdaléniens de La Garenne (Indre, France). In S. Bonnardin, C. Hamon, M. Lauwers, & B. Quilliec (Eds.), *Du matériel au spirituel. Réalités archéologiques et historiques des "dépôts" de la Préhistoire à nos jours, XXIXes Rencontres Internationales d'Archéologie et d'Histoire d'Antibes, 16–18 octobre 2008* (pp. 81–88). Antibes: Editions ADPCA.

Cheynier, A. (1958). Impromptu sur les séquences des pointes du Paléolithique supérieur. *Bulletin de la Société Préhistorique Française, 55*(3–4), 190–205.

Christensen, M. (2004). Fiche caractères morphologiques, histologiques et mécaniques des matières dures d'origine animale. In D. Ramseyer (Ed.), *Matières et techniques, Cahier XI* (pp. 79–87). Paris: Société Préhistorique Française.

d'Errico, F., Julien, M., Liolios, D., Baffier, D., & Vanhaeren, M. (2004). Les poinçons en os des couches châtelperroniennes et aurignaciennes de la grotte du Renne (Arcy-sur-Cure, Yonne). Comparaisons technologiques fonctionnelles et décor. In P. Bodu & C. Constantin (Eds.), *Approches fonctionnelles en préhistoire, XXVe Congrès Préhistorique de France, Nanterre 24–26 novembre 2000* (pp. 45–65). Paris: Société Préhistorique Française.

Delporte, H., Hahn, J., Mons, L., Pinçon, G., & de Sonneville-Bordes, D. (1988). *Sagaies, Fiches typologiques de l'industrie osseuse préhistorique, Union internationale des sciences préhistoriques et protohistoriques, Commission de nomenclature sur l'industrie de l'os Préhistorique, Cahier 1.* Aix-en-Provence: Publications de l'Université de Provence.

Ducasse, S., Castel, J.-C., Chauvière, F.-X., Camus, H., Langlais, M., Morala, A., et al. (2011). Le Quercy au coeur du dernier Maximum Glaciaire. La couche 4 du Petit Cloup Barrat et la question de la transition badegoulo-magdalénienne. *Paléo, 22*, 101–154.

Fano Martinez, M. A., d'Errico, F., & Vanhaeren, M. (2005). Magdalenian bone industry from El Horno cave (Ramales, Cantabria, Spain). In V. Dujardin (Ed.), *Industrie osseuse et parures du Solutréen au Magdalénien en Europe. Table ronde sur le Paléolithique supérieur récent d'Angoulême (2003)* (pp. 177–196). Paris: Société Préhistorique Française.

Geneste, J.-M., & Plisson, H. (1990). Technologie fonctionnelle des pointes à cran solutréennes: l'apport des nouvelles données de la grotte de Combe-Saunière (Dordogne). In K. Kozlowski (Ed.), *Les industries à pointes foliacées du Paléolithique supérieur européen, Actes du Colloque de Cracovie 1989* (pp. 293–320). Liège: ERAUL 42.

Geneste, J.-M., & Plisson, H. (1993). Hunting technologies and human behavior: Lithic analysis of Solutrean shouldered points. In H. Knecht, A. Pike-Tay, & R. White (Eds.), *Before Lascaux: The complex record of the early Upper Paleolithic* (pp. 117–135). New York: Telford Press.

Goutas, N. (2004). *Caractérisation et évolution du Gravettien en France par l'approche techno-économique des industries en matières dures animales (étude de six gisements du Sud-Ouest).* Thèse de doctorat, Université Paris I – Panthéon Sorbonne.

Julien, M. (1982). *Les harpons magdaléniens, XVIIe supplément à Gallia Préhistoire.* Paris: Editions du CNRS.

Knecht, H. (1997). Projectiles points of bone, antler and stone: Experimental explorations of manufacture and use. In H. Knecht (Ed.), *Projectiles technology* (pp. 191–212). Interdisciplinary Contributions of Archaeology. New York: Plenum Press.

Knecht, H. (2000). Design strategies of Early Upper Paleolithic bone and antler projectile technologies. In C. Bellier, P. Cattelain, & M. Otte (Eds.), *La chasse dans la préhistoire, Hunting in Prehistory, Actes du Colloque international de Treignes, 3–7 octobre 1990, ERAUL 51* (pp. 28–36). Treignes: Editions du Cedarc.

Ladier, E. (2014). L'industrie en matières dures d'origine animale de Gandil (séries Chaillot et Ladier). In E. Ladier (Ed.), *L'abri Gandil à Bruniquel (Tarn-et-Garonne): un campement magdalénien du temps de Lascaux. Préhistoire du Sud-Ouest, Supplément 13* (pp. 183–228).

Langlais, M., Pétillon, J.-M., De Beaune, S., Chauvière, F.-X., Letourneux, C., Szmidt, C., et al. (2010). Une occupation de la fin du Dernier Maximum glaciaire dans les Pyrénées: le Magdalénien inférieur de la grotte des Scilles (Lespugue, Haute-Garonne). *Bulletin de la Société Préhistorique Française, 107*(1), 5–51.

Langley, M. C. (2013). *Investigating maintenance and discard patterns for Middle to Late Magdalenian antler projectile points: inter-site and inter-regional comparisons.* Unpublished Ph.D. dissertation, University of Oxford.

Langley, M. C. (2014). Magdalenian antler projectile point design: Determining original form for uni- and bilaterally barbed points. *Journal of Archaeological Science, 44*, 104–116.

Le Dosseur, G. (2003). Sens et contre sens. Réflexions concernant l'orientation d'un geste technique observé sur des objets en matières osseuses du Levant. *Préhistoire Anthropologie Méditerranéennes, 12*, 115–127.

Le Dosseur, G. (2006). La fabrication et l'entretien des poinçons en os au Levant sud: investissement technique et règles de production du Natoufien au Néolithique précéramique B récent. In L. Astruc, F. Bon, V. Léa, P.-Y. Milcent, & S. Philibert (Eds.), *Normes techniques et pratiques sociales: de la simplicité des outillages pré- et protohistoriques, Actes des XXVIes Rencontres Internationales d'Archéologie et d'Histoire d'Antibes, octobre 2005* (pp. 345–357). Antibes: ADPCA.

Le Dosseur, G. (2010). Contribution of a craft production, bone industry to the question of the transition from PPNB to PPNC in the southern Levant. In P. Matthiae, F. Pinnock, L. Nigro, & N. Marchetti (Eds.), *Proceedings of the 6th International Congress on the Archaeology of the Ancient Near East. May, 5th–10th 2008, "Sapienza" - Università di Roma, 1* (pp. 703–719). Wiesbaden: Harrassowitz Verlag.

Leesch, D. (1997). *Hauterive-Champréveyres, 10. Un campement magdalénien au bord du lac de Neuchâtel. Cadre chronologique et culturel, mobilier et structures, analyse spatiale (secteur 1)* (Archéologie neuchâteloise, 19). Neuchâtel: Musée cantonal d'archéologie.

Lefebvre, A. (2014). Les industries osseuses des collections anciennes de Brassempouy In A. Simonet (Ed.), *Brassempouy. les collections Dubalen et De La Porterie* (pp. 57–94). Archéologie des Pyrénées occidentales et des Landes, Hors série 6.

Legrand, A. (2002–2003). Le travail expérimental de la peau avec des outils en os. Résultats tracéologiques. *Bulletin de l'Association pour la Sauvegarde du Site Archéologique d'Argentomagus et Amis du Musée, 2*, 22–28.

Leroy-Prost, Ch. (1975). L'industrie osseuse aurignacienne. Essai de classification: Poitou, Charentes, Périgord. *Gallia Préhistoire, 18*(1), 65–156.

Leroy-Prost, Ch. (1979). L'industrie osseuse aurignacienne. Essai régional de classification: Poitou, Charente, Périgord. *Gallia Préhistoire, 22*(1), 205–370.

Liolios, D. (1999). *Variabilité et caractéristiques du travail des matières osseuses au début de l'Aurignacien: approche technologique et économique*. Thèse de doctorat, Université de Paris X-Nanterre.

Mons, L. (1988). Fiche "sagaie à base raccourcie". In H. Delporte (Ed.), *Sagaies, Fiches typologiques de l'industrie osseuse préhistorique, Union internationale des sciences préhistoriques et protohistoriques, Commission de nomenclature sur l'industrie de l'os préhistorique, Cahier 1*. Aix-en-Provence: Publications de l'Université de Provence.

Morel, Ph. (1993). Impacts de projectiles sur le gibier: quelques éléments d'une approche expérimentale. In P. C. Anderson, S. Beyries, M. Otte, & H. Plisson (Eds.), *Traces et fonction: les gestes retrouvés. Actes du Colloque international de Liège, 8-9-10 décembre 1990* (pp. 55–58). Liège: ERAUL 50.

Otte, M. (1979). *Le Paléolithique supérieur ancien en Belgique* (Monographies d'Archéologie Nationale, 5). Bruxelles: Musées royaux d'Art et d'Histoire.

Pétillon, J.-M. (2003). Typologie et utilisation: l'exemple des pointes à base fourchue magdaléniennes. In M. Patou-Mathis, P. Cattelain, & D. Ramseyer (Eds.), *L'industrie osseuse pré et protohistorique en Europe. Approches techniques et fonctionnelles, Actes du colloque 1.6, XIVe Congrès de l'UISPP* (pp. 53–62). Amay: Cercle archéologique Hesbaye-Condroz.

Pétillon, J.-M. (2005). Tir expérimental de pointes à base fourchue en bois de renne. In V. Dujardin (Ed.), *Industrie osseuse et parures du Solutréen au Magdalénien en Europe. Table ronde sur le Paléolithique supérieur récent d'Angoulême (2003)* (pp. 243–256). Paris: Société Préhistorique Française.

Pétillon, J.-M. (2006). *Des Magdaléniens en armes. Technologie des armatures de projectiles en bois de Cervidé du Magdalénien supérieur de la grotte d'Isturitz (Pyrénées-Atlantiques)* (Artefacts 10). Treignes: Cedarc.

Pétillon, J.-M., & Averbouh, A. (2012). Le travail du bois de renne dans les couches badegouliennes. In J. Clottes, J.-P. Giraud, & P. Chalard (Eds.), *Diversités et identités des groupes solutréens et badegouliens en Quercy. L'exemple de l'abri du Cuzoul de Vers (Lot)* (pp. 359–386). Liège: ERAUL 131.

Pétillon, J.-M., Bignon, O., Bodu, P., Cattelain, P., Debout, G., & Langlais, M. (2011). Hard core and cutting edge: Experimental manufacture and use of Magdalenian composite projectile tips. *Journal of Archaeological Science, 38*(6), 1–18.

Pétillon, J.-M., & Ducasse, S. (2012). From flakes to grooves: A technical shift in antler working during the LGM in southwest France. *Journal of Human Evolution, 62*, 1–31.

Pétillon, J.-M., Langlais, M., De Beaune, S., Chauvière, F.-X., Letourneux, C., Szmidt, C., et al. (2008). Le Magdalénien de la grotte des Scilles (Lespugue, Haute-Garonne). Premiers résultats de l'étude pluridisciplinaire de la collection Saint-Perier. *Antiquités Nationales, 39*, 57–71.

Pétillon, J.-M., & Letourneux, C. (2008). Traces d'impacts de projectiles sur le gibier: résultats comparés des tirs expérimentaux à l'arc et au propulseur effectués au Cedarc (Treignes, Belgique) en 2003 et 2004. Projectile impact traces on game: Comparative results of the bow and spearthrower experiments held at the Cedarc (Treignes, Belgium) in 2003 and 2004. *Annales de Paléontologie, 94*, 209–225.

Peyrony, D., & Peyrony, E. (1938). *Laugerie-Haute près des Eyzies (Dordogne)* (Archives de l'Institut de Paléontologie Humaine, no 19). Paris: Masson.

Plisson, H., & Geneste, J.-M. (1989). Analyse technologique des pointes à cran solutréennes du Placard (Charente), du Fourneau-du-Diable, du Pech de la Boissière et de Combe-Saunière (Dordogne). *Paléo, 1*, 65–106.

Rigaud, A. (2001). Les bâtons percés. Décors énigmatiques et fonction possible. *Gallia Préhistoire, 43*, 101–151.

Rigaud, A. (2004a). Fiche débitage du bois de renne au Magdalénien. L'exemple de La Garenne (Indre, France). In D. Ramseyer (Ed.), *Matières et techniques, Cahier XI* (pp. 79–87). Paris: Société préhistorique Française.

Rigaud, A. (2004b). Usures expérimentales sur quatre bâtons percés utilisés comme bloqueurs de câble. *Gallia préhistoire, 46*, 155–169.

Sacchi, D. (1986). *Le Paléolithique supérieur du Languedoc occidental et du Roussillon, XXIe supplément à Gallia Préhistoire*. Paris: CNRS.

Stodiek, U. (2000). Preliminary results of an experimental investigation of Magdalenian antler points. In C. Bellier, P. Cattelain, & M. Otte (Eds.), *La chasse dans la préhistoire, Hunting in Prehistory. Colloque international de Treignes (1990)*, Artefacts 8, ERAUL, 51 (pp. 70–78). Treignes: Cedarc.

Part III
Southeast Asia & Australia

Part II

Chapter 12
Bone Technology from Late Pleistocene Caves and Rockshelters of Sri Lanka

Nimal Perera, Patrick Roberts, and Michael Petraglia

Abstract The site of Batadomba-lena in the Wet Zone of Sri Lanka, yields osseous technologies in association with *Homo sapiens* back to c.36,000 cal years BP. Alongside isolated finds from the nearby site of Fa Hien-lena, these bone tools are the earliest of their kind in South Asia and can contribute to discussions of the adaptive context of osseous technology during Late Pleistocene human dispersals beyond Africa. Here we describe 204 bone points recovered from the Batadomba-lena rockshelter during excavations conducted in the 1980s and 2000s. Contextual analysis, alongside detailed stratigraphic and chronological information, indicates that *Homo sapiens* in Sri Lanka were using osseous technologies as part of a dedicated rainforest subsistence strategy by at least 36,000 cal years BP. Future work on the Sri Lankan material should acknowledge the importance of placing bone toolkits within their wider environmental and social context.

Keywords Batadomba-lena • Bone points • Bipoints • Rainforest • Bone toolkit • South Asia

Introduction

Osseous technologies have featured strongly in archaeological discussions regarding the evolution of *Homo sapiens* and the subsequent dispersal of populations beyond Africa. Bone toolkits have formed a regular part of appraisals of the 'complexity' or 'modernity' of human populations in Pleistocene Africa, Eurasia and Australasia (e.g., McBrearty and Brooks 2000; d'Errico 2003; Mellars 2005; Habgood and Franklin 2008; Langley et al. 2011). This trajectory of research follows the early placement of bone technologies within the 'behaviorally modern' package of Upper Paleolithic Europe that also includes pigment use, complex technological change, subsistence diversification, and symbolic ornamentation (Mellars 1989; Conard 2003). Subsequent systematic analysis of bone technologies from the Still Bay layers of Blombos Cave dated to 78 ka (Henshilwood et al. 2001; Jacobs et al. 2006) and Howiesons Poort layers from Sibudu Cave dated to 61 ka (Backwell et al. 2008) in South Africa served to push these bone indicators of modern human cognition further back in time (see Backwell and d'Errico 2016).

More recently, discussions of early osseous technologies have expanded their geographical scope across Asia and into Sahul. In particular, the finds of an extensive assemblage of osseous tools from the Niah Caves of Borneo have stimulated an increasing number of systematic studies in Southeast Asia (Barton et al. 2009; Piper and Rabett 2009, 2014; O'Connor et al. 2014). The bone tools from contexts dated to c.38 ka in the Niah Caves (38,058 ± 259 cal BP – OxA-13938) make them the oldest in Southeast Asia (Piper and Rabett 2009). This large assemblage is primarily considered to be representative of projectile points, though diverse types, including piercing and digging implements, are also documented (Rabett 2005; Piper and Rabett 2009). Interestingly, in contrast to previous work focused on Africa and Eurasia, recent studies of the Niah Caves and elsewhere in the region, have been primarily focused on the ecological and environmental context of tool-use and subsistence strategies, as opposed to a search for passive indicators of modernity (Rabett 2005, 2012; Rabett and Piper 2012; O'Connor et al. 2014; Piper and Rabett 2014). This situation is perhaps not surprising given the diverse tropical contexts

N. Perera
Postgraduate Institute of Archaeology, University of Kelaniya, Bauddhaloka Mawatha, Colombo 7, Sri Lanka
e-mail: nimalach@hotmail.com

P. Roberts (✉) • M. Petraglia
Research Laboratory for Archaeology and the History of Art, School of Archaeology, University of Oxford, Oxford, UK
e-mail: patrick.roberts@rlaha.ox.ac.uk; michael.petraglia@rlaha.ox.ac.uk

Michelle C. Langley (ed.), *Osseous Projectile Weaponry*, Vertebrate Paleobiology and Paleoanthropology,
DOI 10.1007/978-94-024-0899-7_12

surrounding the early human populations of Southeast Asia (Rabett and Piper 2012).

Paralleling the Niah Caves (Barker et al. 2007), Sri Lanka provides some of the earliest evidence for the human occupation of a rainforest environment, including the earliest reliably dated records of *Homo sapiens* in South Asia (Deraniyagala 1992, 2007). The Wet Zone rainforest rockshelters of Sri Lanka also provides the earliest evidence for bone tool technologies in South Asia (Deraniyagala 1992; Perera 2010; Perera et al. 2011), in association with the emergence of microlithic tools, shell beads, and other articles of personal ornamentation as early as at least c. 36,000 cal BP (Perera et al. 2011). The interaction of mountainous relief and monsoonal climate divide Sri Lanka into climatic Wet, Dry, and Intermediate Zones with corresponding impacts on the vegetation distribution (Perera 1975; Wang et al. 2005; Fleitmann et al. 2007), making the island particularly sensitive to changes in the extent of rainforest coverage and environmental variability. These factors make Sri Lanka an especially interesting setting for analyzing the adaptive context of osseous technologies.

The majority of archaeological research into the early human occupation of Sri Lanka has focused on the Microlithic period best represented by the Wet Zone rockshelters (Deraniyagala 1992; Perera 2010). This period is named after the dramatic appearance of microlithic technologies alongside the earliest evidence for fossil *Homo sapiens*. Bone tools have been uncovered from a number of contexts in the Wet Zone rockshelter site sequences of Fa Hien-lena, Batadomba-lena, and Kitulgala Beli-lena, and have been dated to as early as c. 38,000 cal BP, c. 36,000 cal BP, and c. 13,000 cal BP respectively (Deraniyagala 1992; Wijeyapala 1997; Perera 2010; Perera et al. 2011). They have also been recovered from the Dry Zone Microlithic site of Bellanbandi Palassa (Deraniyagala 1992; Perera 2010) (Fig. 12.1). Remarkably, little systematic work has been undertaken on any of these osseous assemblages to date, and thus, these technologies have not gained the attention of the international community. We therefore provide some of the first descriptions of the osseous technologies of Batadomba-lena. We relate this to chronological, stratigraphic, and subsistence analysis from this site with a view to providing some preliminary insight into the potential of Pleistocene bone tool analysis in Sri Lanka.

Batadomba-lena Rockshelter: Background

Batadomba-lena is a 15 m wide, northeast-facing rockshelter, located above a stream on the foothills of Sri Pada ('Adam's Peak'), Ratnapura District. The rockshelter is approximately 80 km east of Colombo (6°46′N, 80°12′E) and 460 m above sea level at present (Fig. 12.1b). The rockshelter was formed in gneiss bedrock, part of the high-grade metamorphic (quartzites, granulites, schists and gneisses) terrain of the Precambrian Highland Complex that forms the mountainous spine of Sri Lanka (Cooray 1984) (Fig. 12.1b). Batadomba-lena falls within the Wet Zone, which has an annual rainfall of c. 2200–4000 mm, with precipitation peaking during the summer Monsoon but otherwise being evenly distributed throughout the year (Roberts et al. 2015a). The local vegetation is dominated by dense lowland equatorial rainforest comprising evergreen plant communities typical of the Wet Zone of Sri Lanka (Perera 1975; Deraniyagala 1992, 2007).

Batadomba-lena is one of the most intensively researched prehistoric sites in Sri Lanka. The site was first excavated by P.E.P. Deraniyagala in 1938. This initial cursory excavation reached 1.3 m below the surface of a horizon of crystalline nitrate and bat guano dust and uncovered a series of fragmentary human remains and stone artifacts (Deraniyagala 1940, 1943, 1953). On the basis of technological association of the microlithic technology with the 'Balangoda Culture', Deraniyagala reported the site as being of Late Pleistocene-Holocene age. Following the independence of Sri Lanka, the site witnessed a series of more systematic excavations directed by S.U. Deraniyagala of the Department of Archaeology, Government of Sri Lanka between 1979 and 1986 (Deraniyagala 1992). These excavations defined a series of sealed stratigraphic layers numbered 1–10. The sealed habitation layers contained fragmentary and charred skeletal remains of more than 16 'robust' *Homo sapiens* in association with an abundance of microliths, faunal remains, floral remains, and shell beads. A total of 173 bone tools were found throughout the sequence from Layers 2 to 7c during the 1980s excavations (Deraniyagala 1992).

Initially, uncalibrated radiocarbon dating of 10 samples of charcoal confirmed the Late to Terminal Pleistocene age of the site, the samples yielding dates between 28,000 and 10,000 ^{14}C BP (Deraniyagala 1992). The association of a c. 28,000 ^{14}C BP date with fragmentary *Homo sapiens* remains made these some of the oldest known human fossils in South Asia (Kennedy and Deraniyagala 1989). However, stratigraphic detail and chronological resolution of the earliest habitation layers remained limited. In 2005, N. Perera undertook a limited re-excavation of the 1980s excavation profile in order to further resolve the site chronology (Fig. 12.1c). An additional 21 bone tools were uncovered during these excavations. Calibration of 1980s bulk radiocarbon measurements from Layer 7c suggested that the oldest contexts of Batadomba-lena were dated to between 38,873 and 28,075 cal BP. This early date was confirmed by AMS radiocarbon dates from the 2005 season which provided the smaller age range of 35,364–33,895 cal BP (Table 12.1). Micromorphological analysis of

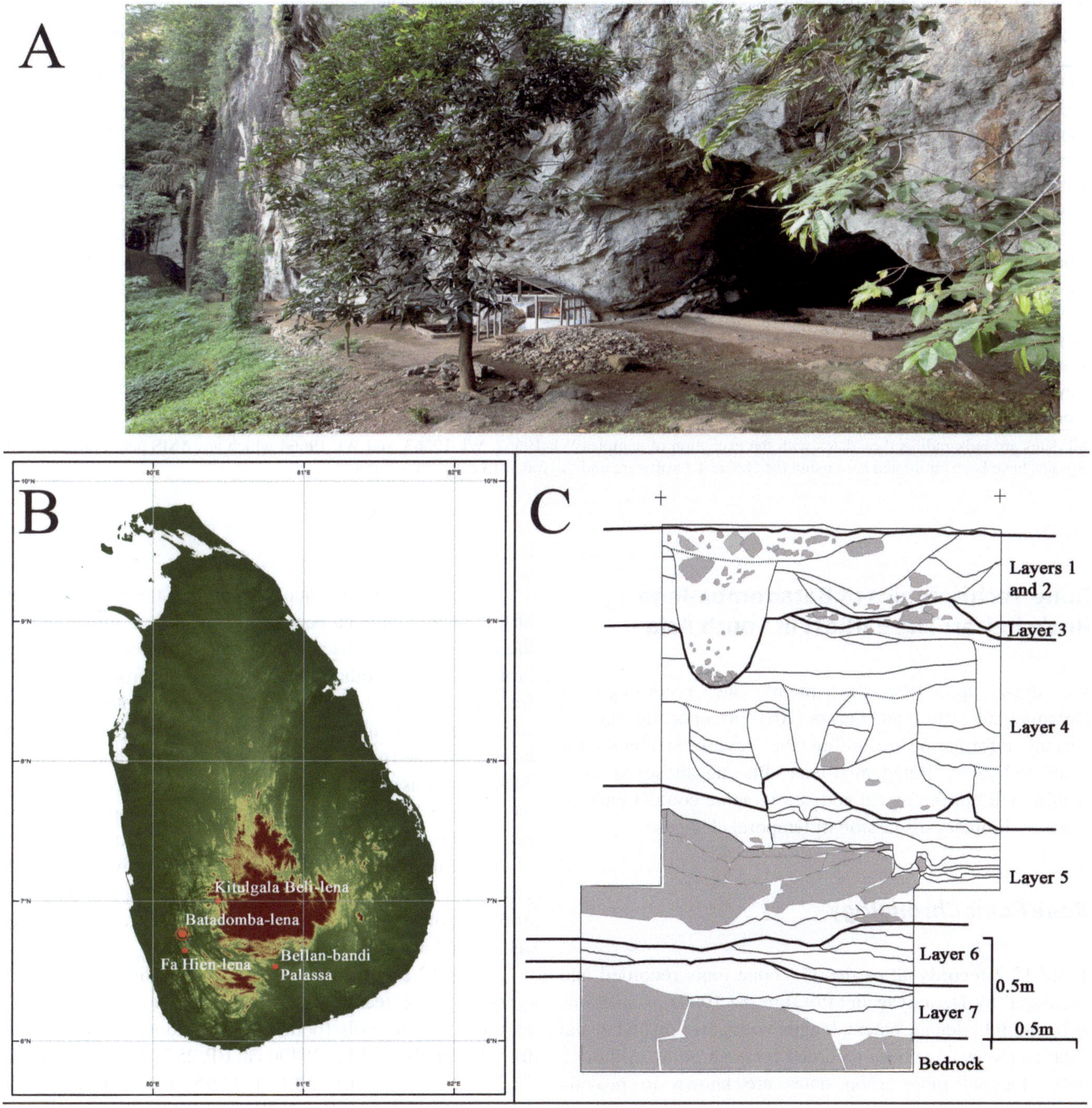

Fig. 12.1 (a) Batadomba-lena rockshelter taken from the northwest of the shelter mouth. (b) Map showing the location of Sri Lankan Microlithic sites with Pleistocene bone tool assemblages discussed in this text. (c) The main stratigraphic features of the 2005 north-facing trench at the site with the main stratigraphic layers described during the 1980s excavation layers superimposed. Redrawn from Perera et al. (2011)

depositional and post-depositional environments confirmed these early dates as reliable (Perera et al. 2011). Together, radiocarbon dating from the 1980s and 2005 excavations provide a robust chronological context for the assemblages of bone tools found throughout the Batadomba-lena sequence. Here, we present all radiocarbon dates from the 1980s and 2005 excavations at Batadomba-lena calibrated using the new IntCal13 calibration curve, which is the first curve to provide a reliable calibration of the earliest dates to 38,873–28,075 and 35,364–33,895 cal BP (Table 12.1).

Table 12.1 Calibrated radiocarbon dates from 1980s and 2005 excavations at Batadomba-lena

Layer/context	Lab code	Material	Conventional (^{14}C BP)	Calibrated (cal BP) (OxCal 4.1, IntCal13)
Context 18 (Layer 4a)	Wk-19965	Charcoal	10,193±57	12,113–11,626
Layer 4a	PRL-855	Charcoal	11,200±330	13,776–12,426
Layer 4b	PRL-856	Charcoal	12,770±470	16,574–13,748
Layer 5	PRL-860	Charcoal	13,130±440	17,046–14,214
Layer 6a	PRL-859	Charcoal	13,880±270	17,580–16,081
Layer 6b	PRL-858	Charcoal	15,390±610	20,239–17,298
Layer 7a	Beta-33281	Charcoal	16,220±300	20,310–18,889
Context 134 (Layer 7b)	Wk-19964	Charcoal	19,350±121	23,634–22,959
Layer 7b	PRL-920	Charcoal	20,150±740	25,960–22,661
Layer 7b	Beta-33282	Charcoal	20,320±500	25,708–23,394
Layer 7b	BS-784	Charcoal	22,360±650	27,851–25,490
Context 104 (Layer 7b)	Wk-19962	Charcoal	22,903±172	27,574–26,788
Layer 7c	PRL-857	Charcoal	27,700±2090	38,873–28,075
Context 71 (Layer 7c)	Wk-19963	Charcoal	30,603±400	35,364–33,894

All dates are bulk radiocarbon dates with the exception of samples Wk-19662, Wk-19663, and Wk 19664 which are AMS determinations. All samples have been calibrated here using the OxCal 4.1 software and the IntCal13 calibration curve

Bone Technologies of Batadomba-lena Rockshelter: The Earliest in South Asia

We draw upon the descriptions and recordings of Deraniyagala (1992) and Perera (2010) to report the state of current information regarding the osseous technologies recovered from Batadomba-lena. Placing these findings within their chronological and stratigraphic context enables some preliminary discussion of temporal changes.

Bone Point Chronology

Table 12.2 records all of the 173 bone tools recorded and excavated by Deraniyagala (1992) in the 1980s excavations at Batadomba-lena. Context, length, width, weight and visual characteristics have been recorded for each point. Although bulk charcoal radiocarbon dates are known to provide approximate ages, the broad picture from Table 12.1, as well as the addition of two AMS-radiocarbon dates from the 2005 excavation, is that the Batadomba-lena osseous technologies span from 38,873–28,075 and 35,364–33,895 cal BP in Layer 7c to 16,574–11,626 cal BP in Layer 4. Analyses of the available radiocarbon calibration curves suggest that the actual date of Layer 7c most likely centres around c. 36,000 cal BP (Perera et al. 2011). Chronological interpretation of bone tools in Layers 3 and 2 is complicated by a lack of radiocarbon dates for these layers and evidence for mixing of prehistoric and historic remains in Layer 2 (Perera 2010).

This chronological sequence confirms the great antiquity of the Batadomba-lena bone toolkits and their importance to the early human archaeological record of South Asia more broadly. The only other bone tools reported from South Asia from this period are a broken harpoon from a Late Palaeolithic site in the Belan Valley of India (Bednarik 2003) and a broken bone point dated to somewhere between 34,000 and 20,000 cal BP and a uniserial harpoon fragment of antler dated to c. 34,000 cal BP from Jwalapuram, Locality 9, India (Clarkson et al. 2009). However, given the rare and fragmentary nature of these finds, the Batadomba-lena record provides the longest and most highly-resolved Pleistocene temporal sequence of bone tool technologies anywhere in South Asia (see James and Petraglia 2005 for a wider discussion).

Figure 12.2a illustrates the frequency of the two main types of bone technology throughout the sequence. The number of bone tools (both single and double points) rises from Layer 7b (27,851–25490 cal BP, 25,708–23,394 cal BP, 25,960–22,661 cal BP, 23,634–22,959 cal BP) to peak in Layer 4 (16,574–11,626 cal BP) during the closing stages of the Pleistocene period. From this point, the number of tools drops to a minimum in Layer 2, which, as stated above, is a mixture of prehistoric, historic, and modern materials (Deraniyagala 1992). As reported from Southeast Asia (Rabett 2012), the use and manufacture of osseous technologies at Batadomba-lena intensifies in the post-Last Glacial Maximum period and into the Terminal Pleistocene, perhaps related to environmental changes seen elsewhere in Sri Lanka, and South and Southeast Asia more widely, at this time (Premathilake and Risberg 2003; Rabett 2012).

Table 12.2 Bone tools from stratigraphic Layer 2 in the 1980s excavations of Batadomba-lena (Deraniyagala 1992)

Layer	Square	Type	Length (mm)	Width (mm)	Weight (g)	Visual characteristics
2	13 G	Single point	22.54	4.55	0.1	Abraded/Flattened
2	14 K	Single point	22.81	3.76	0.2	Abraded
2	12 L	Double point	26.42	5.65	0.4	Abraded
3	12 G	Single point	13.32	2.85	0.3	Split
3	12 G	Single point	16.22	0.60	0.6	Split
3	12 G	Single point	23.98	2.95	0.2	Abraded
3	12 G	Single point	14.41	2.60	0.1	Rounded/Polished
3	12 G	Single point	37.56	4.77	0.7	Abraded/Rounded
3	12 H	Single point	15.62	2.68	0.2	Abraded
3	12 H	Single point	17.74	3.55	0.1	Abraded/Rounded
3	12 J	Single point	12.20	2.45	0.1	Abraded
3	13 H	Single point	11.72	2.35	0.1	Abraded
3	13 H	Single point	15.45	2.25	0.1	Abraded
3	13 H	Single point	25.12	4.45	0.5	Abraded
3	13 J	Single point	19.01	2.76	0.1	Abraded
3	13 J	Single point	19.56	2.76	0.5	Abraded
3	13 K	Single point	15.31	4.43	0.2	Abraded
3	13 K	Single point	15.44	2.45	0.1	Rounded
3	13 K	Single point	16.17	3.55	0.1	Abraded/Polished
3	14 G	Single point	24.11	4.65	0.5	Abraded/Polished
3	14 G	Single point	16.60	2.78	0.2	Abraded/Rounded
3	14 H	Single point	23.62	3.25	0.3	Abraded
3	14 IG	Single point	37.85	5.76	0.5	Split
3	14 J	Single point	12.94	2.75	0.1	Abraded/Polished
3	14 K	Single point	12.99	2.56	0.1	Abraded
3	14 L	Single point	18.79	3.25	0.2	Split/Polished
3	14 K	Single point	26.56	3.55	0.2	Abraded
3	16 K	Single point	17.22	2.65	0.4	Split
3	12 G	Double point	30.91	5.95	0.2	Abraded
3	12 J	Double point	36.04	5.85	0.7	Abraded
3	13 K	Double point	21.88	3.65	0.1	Split
3	15 K	Double point	40.49	5.55	0.9	Rounded
3	16 G	Double point	30.82	3.15	0.4	Split
4	12 G	Single point	16.79	3.24	0.3	Abraded/Polished
4	12 G	Single point	23.81	3.65	0.5	Split/Polished
4	12 G	Single point	12.75	4.90	0.1	Abraded/Rounded
4	12 G	Single point	15.99	2.67	0.2	Abraded/Rounded
4	12 G	Single point	19.26	3.55	0.4	Abraded
4	12 G	Single point	18.92	4.55	0.4	Abraded
4	12 G	Single point	17.01	3.26	0.2	Abraded
4	12 J	Single point	13.71	2.65	0.1	Rounded
4	12 J	Single point	11.60	2.65	0.1	Rounded
4	13 G	Single point	20.12	3.85	0.3	Split
4	13 G	Single point	17.80	4.85	0.3	Abraded/Rounded
4	13 H	Single point	13.11	2.55	0.1	Split
4	13 H	Single point	18.68	3.55	0.1	Polished
4	13 I	Single point	14.37	3.25	0.2	Abraded
4	13 J	Single point	20.75	3.65	0.3	Abraded/Rounded
4	14 G	Single point	19.25	2.55	0.3	Abraded/Rounded
4	14 G	Single point	10.55	2.80	0.2	Abraded/Rounded
4	14 H	Single point	10.35	2.76	0.1	Rounded/Polished
4	14 H	Single point	14.24	2.25	0.1	Abraded
4	14 H	Single point	11.40	3.85	0.1	Polished

(continued)

Table 12.2 (continued)

Layer	Square	Type	Length (mm)	Width (mm)	Weight (g)	Visual characteristics
4	14 H	Single point	50.28	4.76	1.3	Abraded
4	14 K	Single point	23.62	3.54	0.5	Abraded
4	15 G	Single point	24.90	4.90	0.3	Abraded/Rounded
4	15 G	Single point	16.65	5.55	0.2	Abraded/Rounded
4	15 G	Single point	27.73	3.96	0.4	Abraded
4	15 J	Single point	11.72	2.85	0.1	Abraded/Rounded
4	16 G	Single point	27.76	4.28	0.5	Abraded
4	16 G	Single point	20.82	5.65	0.5	Abraded
4	16 J	Single point	24.95	4.92	0.3	Abraded/Rounded
4	16 J	Single point	12.92	2.65	0.1	Abraded
4	16 J	Single point	72.24	5.85	0.3	Abraded
4	16 L	Single point	18.85	2.95	0.3	Abraded
4	17 H	Single point	13.93	4.50	0.1	Abraded/Polished
4	17 H	Single point	21.38	4.65	0.4	Abraded
4	12 G	Double point	21.65	4.55	0.3	Abraded/Rounded
4	12 G	Double point	26.44	4.67	0.5	Abraded
4	13 J	Double point	23.38	3.55	0.3	Abraded/Rounded
4	13 J	Double point	35.54	4.95	0.9	Abraded
4	14 J	Double point	26.68	4.65	0.2	Abraded/Polished
4	15 G	Double point	21.35	4.65	0.4	Abraded/Polished
4	15 G	Double point	54.86	4.85	1.3	Rounded
4	15 G	Double point	21.87	3.45	0.3	Abraded
4	15 G	Double point	27.18	4.35	0.3	Abraded
4	16 I	Double point	37.66	3.98	0.5	Polished
4	16 L	Double point	20.84	4.95	0.2	Abraded/Polished
5	12 G	Single point	23.55	3.55	0.3	Polished
5	12 G	Single point	15.16	3.85	0.2	Abraded/Rounded
5	12 G	Single point	14.93	2.65	0.3	Abraded
5	12 J	Single point	20.55	4.55	0.2	Abraded/Rounded
5	12 J	Single point	12.75	2.34	0.1	Rounded
5	12 K	Single point	11.92	2.55	0.1	Polished
5	13 G	Single point	23.32	3.67	0.3	Abraded/Polished
5	13 G	Single point	11.41	2.65	0.1	Rounded
5	13 H	Single point	21.21	4.95	0.4	Split
5	13 H	Single point	14.44	2.75	0.1	Abraded
5	13 I	Single point	34.95	7.75	0.4	Abraded/Rounded
5	13 J	Single point	26.28	3.95	0.4	Split/Polished
5	13 J	Single point	15.25	2.55	0.2	Polished
5	14 G	Single point	27.73	3.95	0.2	Abraded
5	14 J	Single point	38.16	4.55	0.7	Abraded/Rounded
5	14 J	Single point	14.95	5.50	0.2	Abraded/Rounded
5	14 J	Single point	12.80	3.90	0.1	Abraded/Rounded
5	14 K	Single point	30.71	4.76	0.5	Abraded/Polished
5	14 K	Single point	20.24	3.45	0.2	Abraded
5	15 G	Single point	30.95	6.55	0.3	Abraded/Rounded
5	16 J	Single point	25.60	5.60	0.2	Abraded/Rounded
5	16 J	Single point	14.55	3.50	0.1	Abraded/Rounded
5	16 J	Single point	20.59	4.95	0.2	Abraded/Rounded
5	17 J	Single point	10.52	5.55	0.2	Abraded/Rounded
5	13 G	Double point	27.00	3.95	0.2	Abraded/Polished
5	13 I	Double point	21.84	4.55	0.3	Abraded/Rounded
5	14 H	Double point	29.87	4.25	0.3	Split/Polished
5	14 K	Double point	31.35	5.25	0.4	Abraded

(continued)

Table 12.2 (continued)

Layer	Square	Type	Length (mm)	Width (mm)	Weight (g)	Visual characteristics
6	12 G	Single point	21.12	4.85	0.3	Split/Abraded
6	12 J	Single point	14.35	2.55	0.5	Polished
6	13 G	Single point	12.85	2.56	0.1	Split
6	13 G	Single point	38.28	4.32	0.5	Abraded
6	13 H	Single point	30.07	5.75	0.3	Split/Polished
6	13 H	Single point	34.28	4.20	0.3	Abraded
6	13 H	Single point	18.94	3.75	0.4	Abraded
6	13 J	Single point	31.58	3.65	0.5	Split
6	13 J	Single point	15.98	2.50	0.3	Polished
6	13 J	Single point	21.58	3.67	0.1	Abraded/Rounded
6	14 G	Single point	27.00	4.80	0.4	Abraded/Rounded
6	14 J	Single point	20.94	3.50	0.5	Rounded/Polished
6	14 K	Single point	20.69	4.67	0.2	Abraded/Polished
6	15 G	Single point	32.55	4.87	0.7	Abraded
6	15 G	Single point	30.35	3.75	0.5	Abraded
6	15 G	Single point	23.70	3.65	0.5	Abraded
6	15 H	Single point	12.72	3.54	0.2	Abraded
6	15 J	Single point	36.64	5.50	0.9	Split
6	15 K	Single point	23.07	5.65	0.5	Abraded/Rounded
6	16 H	Single point	20.65	4.95	0.2	Abraded/Rounded
6	16 H	Single point	33.06	5.80	0.6	Rounded
6	16 H	Single point	89.24	2.65	0.3	Abraded
6	12 H	Double point	31.26	5.55	0.5	Abraded
6	13 H	Double point	20.75	3.55	0.2	Abraded/Rounded
6	15 J	Double point	18.22	2.95	0.1	Abraded
7a	12 G	Single point	34.40	3.60	0.5	Split/Polished
7a	12 G	Single point	30.12	3.85	0.1	Abraded/Rounded
7a	12 G	Single point	13.52	2.65	0.1	Abraded
7a	12 J	Single point	18.52	3.45	0.1	Abraded
7a	13 G	Single point	28.35	3.68	0.3	Abraded/Rounded
7a	13 G	Single point	16.23	2.15	0.1	Abraded/Rounded
7a	13 J	Single point	30.01	4.65	0.7	Split
7a	14 G	Single point	12.35	2.65	0.1	Polished
7a	14 G	Single point	26.14	4.25	0.5	Abraded
7a	14 G	Single point	14.27	5.80	0.1	Abraded/Rounded
7a	14 K	Single point	17.10	3.95	0.1	Abraded/Rounded
7a	14 K	Single point	21.55	3.65	0.3	Rounded
7a	15 J	Single point	16.72	6.65	0.1	Abraded/Rounded
7a	16 G	Single point	17.80	3.90	0.2	Abraded/Rounded
7a	12 G	Double point	33.85	4.88	0.5	Rounded/Polished
7a	13 I	Double point	24.84	4.55	0.3	Rounded
7b	13 G	Single point	20.57	4.50	0.3	Abraded/Rounded
7b	13 J	Single point	17.49	3.50	0.5	Polished
7b	14 J	Single point	19.95	3.95	0.2	Abraded/Rounded
7b	15 G	Single point	40.59	4.55	0.4	Abraded/Rounded
7b	16 I	Single point	27.52	4.85	0.5	Abraded
7b	13 H	Double point	17.87	2.65	0.1	Abraded
7c	12 I	Single point	47.25	4.75	0.5	Abraded
7c	12 J	Single point	25.79	2.55	0.5	Split
7c	13 G	Single point	30.79	3.65	0.6	Rounded/Polished
7c	13 G	Single point	9.03	9.50	0.1	Rounded/Polished
7c	13 G	Single point	15.31	3.67	0.1	Abraded/Polished
7c	13 H	Single point	26.98	4.65	0.3	Rounded/Polished

(continued)

Table 12.2 (continued)

Layer	Square	Type	Length (mm)	Width (mm)	Weight (g)	Visual characteristics
7c	13 I	Single point	10.03	2.50	0.1	Abraded/Rounded
7c	13 J	Single point	15.99	2.65	0.1	Abraded/Polished
7c	13 J	Single point	16.65	2.54	0.2	Abraded/Rounded
7c	14 J	Single point	18.99	2.65	0.3	Abraded/Polished
7c	14 J	Single point	23.02	2.65	0.1	Split/polished
7c	14 J	Single point	19.35	2.85	0.1	Rounded
7c	15 J	Single point	15.75	2.55	0.1	Abraded/Polished
7c	15 K	Single point	15.92	3.55	0.2	Abraded
7c	16 G	Single point	8.78	4.55	0.2	Abraded
7c	12 G	Double point	18.96	3.65	0.1	Abraded/Rounded
7c	13 K	Double point	22.11	5.80	0.4	Abraded/Rounded
7c	14 J	Double point	18.89	4.65	0.3	Abraded/Rounded
7c	16 G	Double point	22.75	4.50	0.3	Abraded/Rounded

Bone Point Types and Manufacture

Photographs of the single and double bone points excavated from the Pleistocene Layers 3 to 7c are shown in Figs. 12.3 and 12.4, respectively. Although analysis of this assemblage has so far been limited to basic macroscopic observations, some indications regarding manufacture and functionality have begun to emerge. The entirety of the osseous assemblage is made up of small, light bone pieces, with either a single or double end (Fig. 12.5), sharpened to a point. As discussed by Deraniyagala (1992, 2007), the majority of bone tools appear to have been manufactured through longitudinal splitting of long bones, followed by grinding at one or both ends. This latter process is indicated by longitudinal and transverse striations found on the majority of the tools. Some of the bone artifacts appear to retain striations resulting from manufacture, namely in the primary excision and shaping of long bone fragments. Percussion fractures also appear to have been employed in the initial separation of bone blanks from long bones (Deraniyagala 1992). So far there is no evidence that the groove and splinter technique which characterizes much of the European Upper Paleolithic bone and antler technology exists in Sri Lanka (Deraniyagala 1992). Both single and double points have been recovered from every stratigraphic layer except for Layer 1. Bone tools excavated in the 1980s and 2005 show no divergence in these broad observations of working and manufacture.

The bone points have been further categorized based on one or two of five visual characteristics (Perera 2010). These characteristics include: 'Rounded' in reference to artifacts where grinding has left a rounded surface, 'Polished' which describes smoothly ground bone points, 'Abraded' which denotes bone points with surfaces that show preliminary grinding, 'Split' which signifies those points which retrain traces of their original splitting lines from a long bone and 'Flattened' which refers to the flattening of one part of the tool. Figure 12.6 (a simplification of data displayed in Table 12.2) indicates the relative prevalence of these characteristics in each stratigraphic layer for single and double point points excavated during the 1980s at Batadomba-lena.

The most common classification for both single- and double-ended points from the 1980s excavations is Abraded (n = 58), followed by Abraded/Rounded (n = 46). Considering both single and double points together, 147 of the points (85 %) have traces of abrasion, 75 (43.4 %) have rounding, 49 (28.3 %) demonstrate polishing, and 26 retain evidence for splitting (Perera 2010). Both single and double points demonstrate the highest proportion of polishing in Layer 7c, with a preponderance of splitting or abrasion in the upper Layers, 2–4, perhaps demonstrating some alterations in technological practice following the earliest habitation layers at Batadomba-lena.

The majority of complete osseous tools are light, in general weighing between 0.1 and 0.5 g (Fig. 12.7), and never more than 1.4 g if the heaviest specimen from the 2005 sample (Table 12.3) is included. Although the lengths of complete points vary, ranging from between 9.0 and 89.2 mm, average lengths remain relatively constant (Fig. 12.7). The ranges of the lengths of double-ended points and single-ended points are similar (Fig. 12.7). The only exceptions are one Rounded/Polished double point from Layer 7a, a polished double point from Layer 4 and two Rounded double points from Layers 3 and 4, which are all longer than any single-ended point in the same technological class (Perera 2010). This result might imply that longer bone splinters may have been selected for rounding or polishing at both ends, although it is possible that such use wear could result from the use of the tools rather than their production. However, overall, the form of osseous technologies remains relatively stable from 36,000 cal. BP to c. 12,000 cal BP.

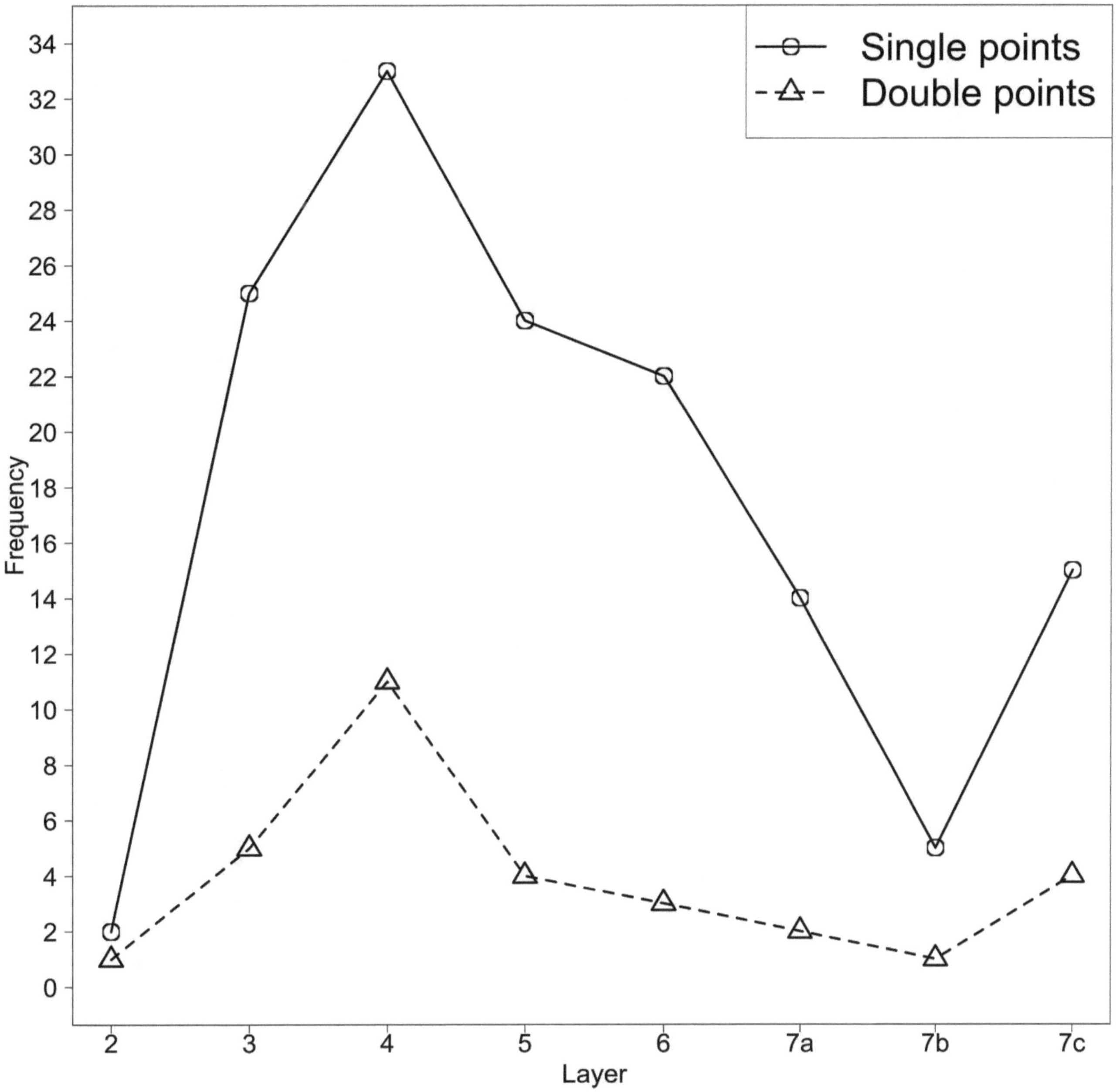

Fig. 12.2 (**a**) Frequency of the number of single and double bone points throughout the Batadomba-lena sequence (from 1980s excavations). (**b**) Changes in weight between the snail assemblages by stratigraphic layer at Batadomba-lena (after Perera et al. 2011) (from 2005 excavation). (**c**) Plot showing changes in the proportion of monkeys, giant squirrels, mongoose, palm civets and snakes, lizards and fish by stratigraphic layer at Batadomba-lena (after Perera 2010) (from 1980s excavations)

Table 12.3 shows slightly more detailed analysis of 21 bone points excavated from Layers 2–6 during the 2005 excavation. The 2005 osseous assemblage has broadly similar characteristics of manufacture, size and weight as discussed for the 1980s assemblage above. However, analysis by faunal specialist J. Perera has enabled the 2005 points to be additionally characterized by the species from which they were made.

Of the 21 bone points excavated in 2005, 19 were manufactured from the long bones of monkeys, while two appear to have come from a jungle fowl bone (Table 12.3). Polishing is noted at a higher rate (13/21) than in the main bone tool assemblage. Four of the bone points showed indications of burning which is close to the ratio for the entire faunal assemblage from the site, implying accidental rather than deliberate burning (Perera 2010).

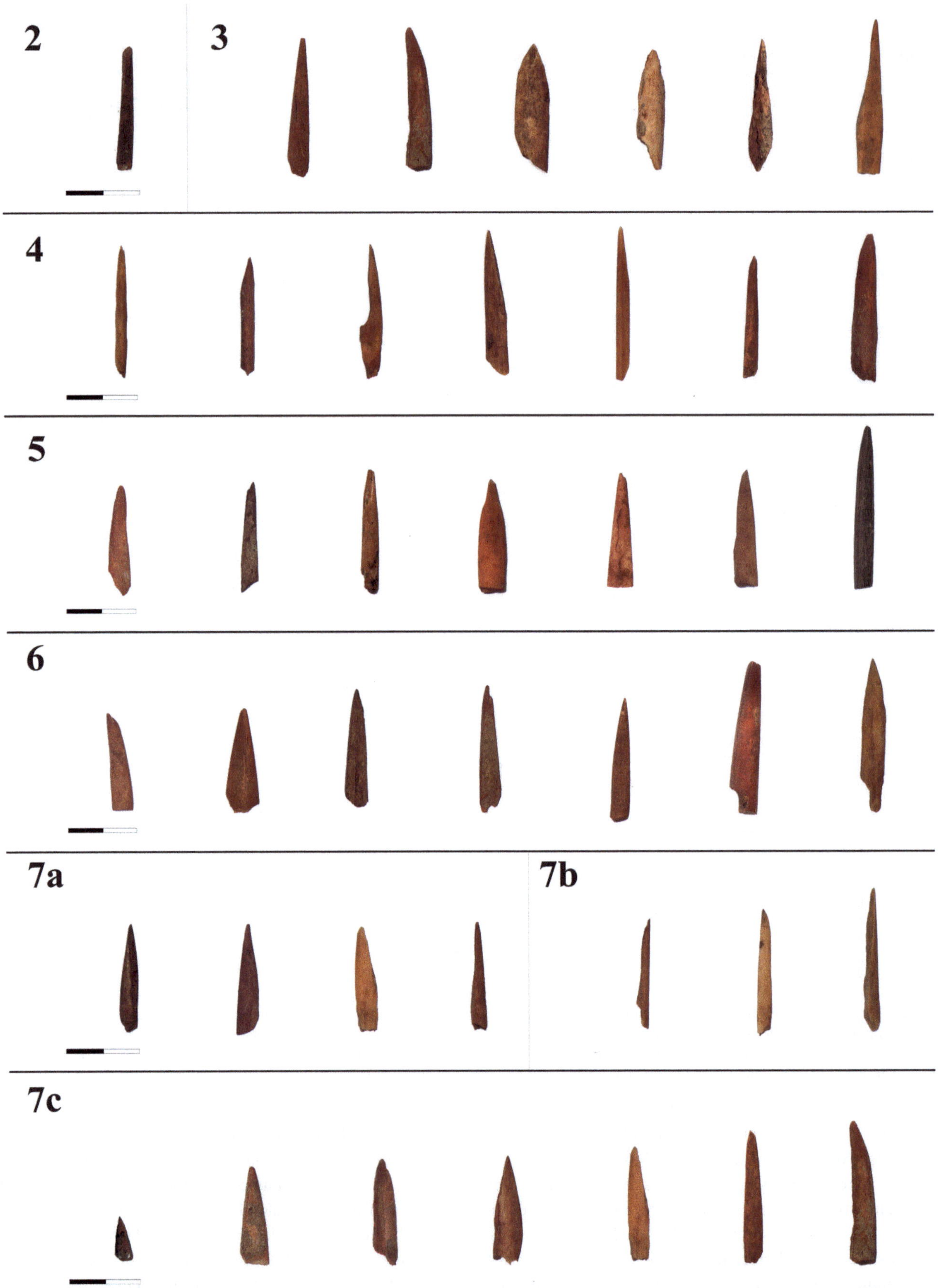

Fig. 12.3 Single bone points excavated from the Pleistocene Layers 3 to 7c from Batadomba-lena (scale=2 cm) (used with permission from R.M. Kushumpriya Rajapaksa)

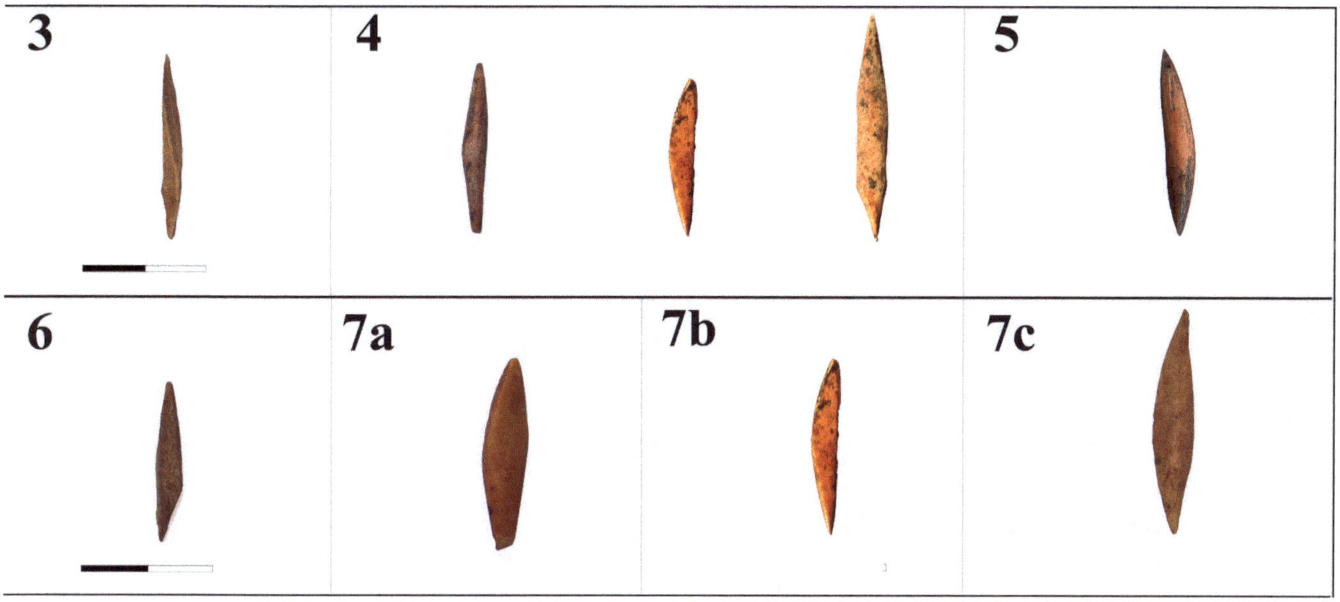

Fig. 12.4 Double bone points excavated from the Pleistocene Layers 3 to 7c from Batadomba-lena (scale = 2 cm) (used with permission from R.M. Kushumpriya Rajapaksa)

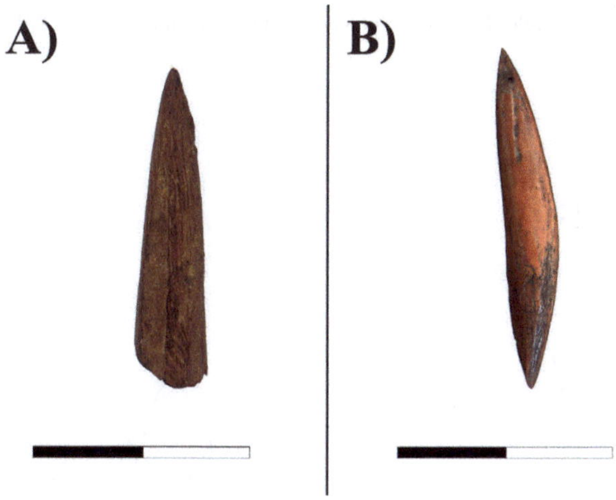

Fig. 12.5 (**a**) Single-ended bone point BD12 from Layer 6 of Batadomba-lena (scale = 2 cm) (used with permission from R.M. Kushumpriya Rajapaksa). (**b**) Double-ended bone point BD41A from Layer 5 of Batadomba-lena (scale = 2 cm) (used with permission from R.M. Kushumpriya Rajapaksa)

Ecological and Environmental Context of the Batadomba-lena Assemblage

Following excavations in the 1980s, Deraniyagala (1992) suggested the bone tools from Batadomba-lena could have been used for a wide range of different functions including: points (arrowheads or spearheads), fishing spearheads, blowpipe darts, winkle to extract the abundant molluscan remains also found at the site, and large picks. The wide metrical range of the bone points seen in Tables 12.2 and 12.3 would arguably fit the idea of multiple uses (Perera 2010).

Although specific uses of the bone tools may remain elusive for now, a discussion of faunal, floral, and environmental analysis at Batadomba-lena provides an additional context to their manufacture and use. Perera and colleagues (2011) report that molluscan remains from the site include large quantities of *Acavidae* landsnails, *Pleuroceridae*, and other freshwater snails from the earliest cultural layers (Layer 7c). This assemblage is indicative of a forested landscape with freely flowing freshwater conditions throughout site occupation. These snails occur in middens associated with hearths and habitation debris. Between 13 and 85 % of snail shells show evidence for burning throughout the sequence, perhaps implying dietary use. The body whorl of many *Acavus* shells have been bored which may also reflect the extraction of edible parts of the snail or perhaps symbolic uses. Although sample size is limited, there is a sharp increase in snail consumption/use (namely freshwater species) in Layers 4 and 3 (after 16,000 cal years BP) (Perera et al. 2011) (Fig. 12.2b). Given the increase in bone tool presence at this time (Fig. 12.2a) it is possible that bone tools may have played an important role in the exploitation of local rainforest molluscs.

An alternative association can arguably also be made with the mammalian faunal assemblage. The identified faunal assemblage is dominated by small and medium-bodied taxa (c. <20 kg) of arboreal and mixed terrestrial arboreal habitat. The bones of monkeys make up the majority of the faunal assemblage (often >70 %), followed by squirrels, civets and

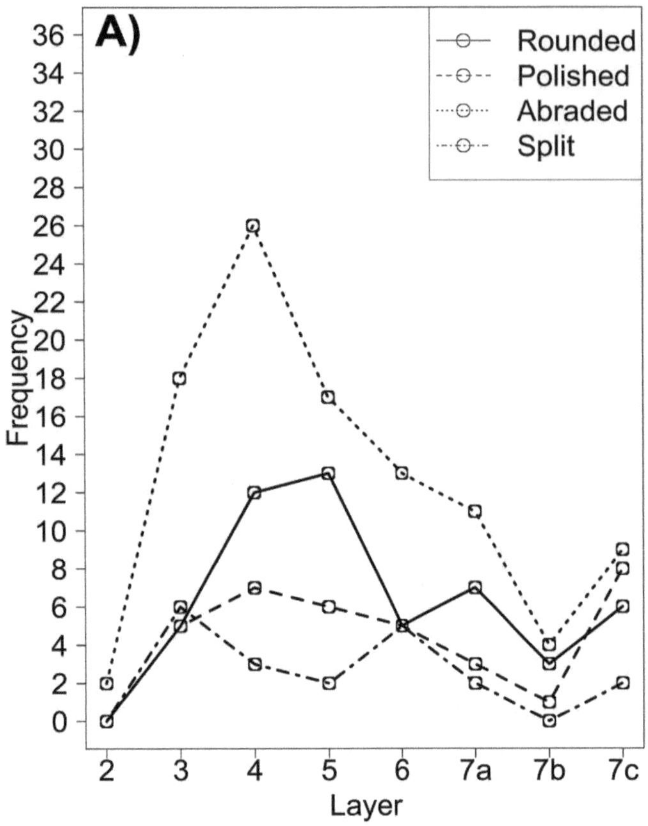

Fig. 12.6 Plots showing the relative frequency of visual categories of double (**a**) and single (**b**) bone point treatment through the Batadomba-lena sequence (from 1980s excavations)

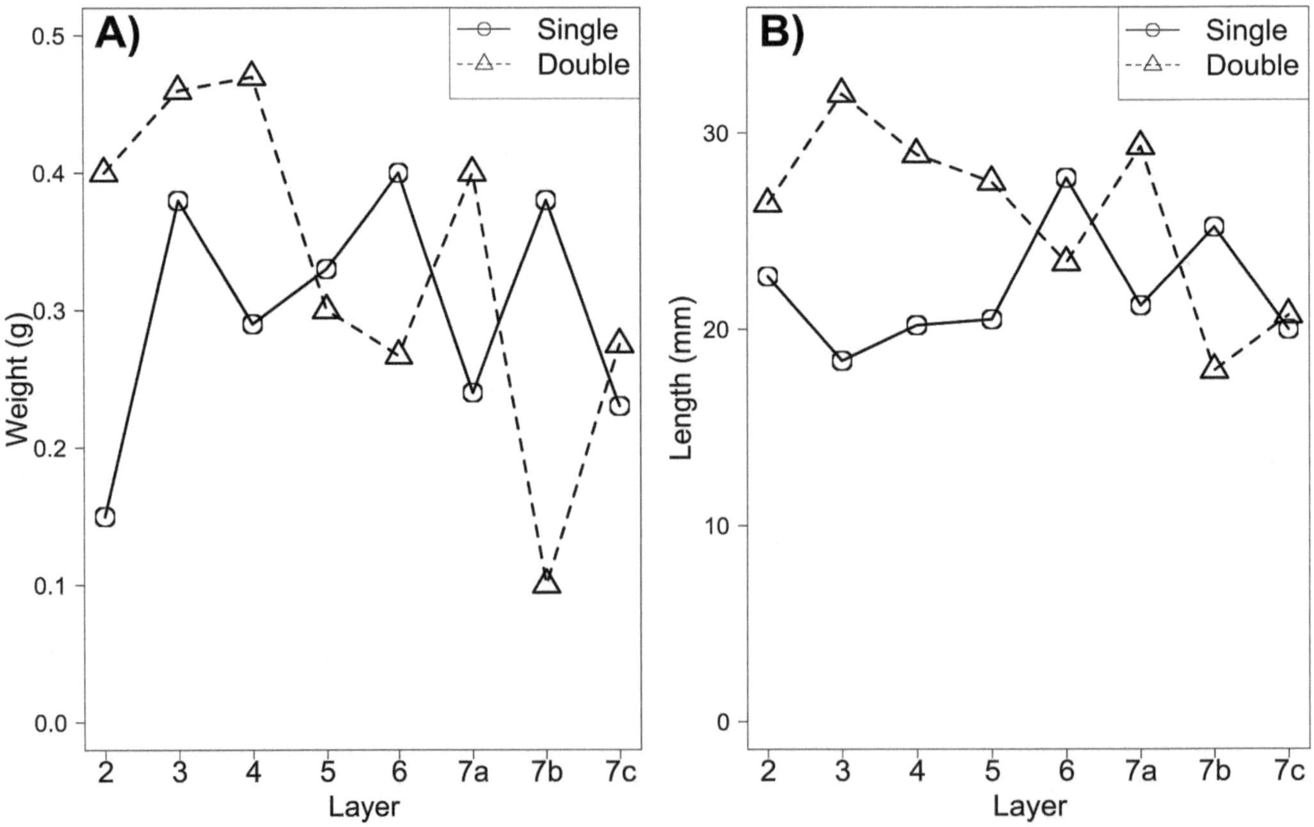

Fig. 12.7 Plots showing the average weight (**a**) and length (**b**) of the bone assemblages (single and double points) from each stratigraphic layer in the Batadomba-lena sequence (from 1980s excavations)

Table 12.3 Bone points excavated during the 2005 excavations of Batadomba-lena (Perera 2010)

Layer	Context	Type	Bone origin	Description	Length (mm)	Width (mm)	Thickness (mm)	Weight (g)
2	5	Single point	Monkey	Distal end broken	11	5	4	
2	16	Single point	Monkey	Surface polished	35	7	3	
2	16	Single point	Monkey	Small fragment with point	10	4	3	
2	5	Double point	Monkey	Surface polished	30	5	4	
3	8	Double point	Monkey	Surface polished	27	–	3	0.4
4	9	Single point	Jungle fowl	Surface polished, distal end broken	23	4	2	0.1
4	11	Single point	Monkey	Surface polished, distal end broken	37	5	4	0.7
4	11	Single point	Monkey	Surface polished, distal end broken	31	7	5	0.8
4	23	Single point	Monkey	Surface polished, distal end broken	61	6	5	1.4
4	23	Single point	Monkey	Surface polished; distal end broken	52	5	5	0.9
4	25	Single point	Monkey	Burning on surface	23	4	3	0.2
4	38	Single point	Monkey	Surface polished; distal end broken	23	5	4	0.1
4	100	Double point	Monkey	Surface polished	32	4	3	0.2
4	102	Double point	Monkey	–	40	6	5	0.6
4	102	Double point	Monkey	Surface polished	25	7	4	0.3
5	49	Single point	Jungle fowl	Burning on surface; distal end broken	26	5	3	0.1
5	56	Single point	Monkey	Surface polished	30	7	3	0.1
5	80	Double point	Monkey	Surface polished	32	1	4	0.1
6	72	Single point	Monkey	Distal end broken	52	7	5	1.1
6	78	Single point	Monkey	Burning on surface	45	5	4	0.8
6	78	Single point	Monkey	Burning on surface	32	4	3	0.4

mongoose. Ground-dwelling mammals are much rarer; largely represented by mouse deer and wild boar. Non-mammalian vertebrate remains mainly comprise jungle fowl, monitor lizards and snakes. Abundant butchery marks, association with hearths, and evidence for exposure to fire is plentiful throughout the assemblage (Perera et al. 2011). Interestingly, although the taxonomic composition of the mammalian assemblage varies only subtly through the Batadomba-lena sequence, the terminal Pleistocene Layers (3 and 4) demonstrate a decrease of monkeys in favor of squirrels, civets and mongoose (Fig. 12.2c). The corresponding increase in the frequency of bone tools in Layer 4 may indicate an association between osseous technologies and small vertebrate exploitation at Batadomba-lena.

The use of bone tools in the exploitation of molluscan assemblages, fishing, and projectile hunting has been well documented elsewhere, both archaeologically and ethnographically (Henshilwood et al. 2001; Rabett 2005; Backwell et al. 2008; Piper and Rabett 2009). The hafting of larger, yet similarly formed, rhomboidal double points was utilized at Oenpelli, Northern Territory, Australia for fish spearing (Mulvaney 1975; Deraniyagala 1992). Similarly, osseous technologies found in association with *Homo sapiens* in the Late Pleistocene layers of the Niah Caves have been interpreted as projectile points (Piper and Rabett 2009). However, the association of bone tools at Batadomba-lena with changes in the percentage of small, ground-dwelling mammals in the faunal assemblage that would be difficult to catch using projectile technologies, may indicate that these bone points could have formed part of snare or trapping technologies rather than projectile tips. Although evidence for trapping and snaring in the Late Pleistocene archaeological record is often limited to indirect interpretations based on the composition of faunal assemblages (Wadley 2010), it has been argued to have occurred in Europe with the arrival of *Homo sapiens* (Stiner et al. 2000) and in Africa during the Middle Stone Age (Wadley 2010). The shape and weight of the Batadomba-lena bone tools would make them appropriate for use within gorge-traps or as part of a trigger mechanism in a potential trap system (Klint Janulis, personal communication). If this is indeed the case, evidence for delayed-return hunting strategies within Sri Lanka from c. 36,000 years cal BP would have significant implications for our understanding of early human subsistence strategies in a rainforest context.

Detailed experimental, microscopic, and residue analyses of the Batadomba-lena osseous assemblage in the near future will facilitate a more specific insight into the particular functions of these tools. However, it is already clear that the Batadomba-lena bone technology was part of a dedicated rainforest subsistence strategy (Roberts et al. 2015b). In addition to the faunal remains already discussed, floral remains indicate the presence of *Canarium* sp. nutshells from Layers 3–5, that frequently form a significant part of the diets of ethnographically known Vedda populations in Sri Lanka (Deraniyagala 1992). The phytolith assemblage also indicates the presence of *Canarium* sp., *Musa* sp., and *Artocarpus* sp. Indicating that tree fruits and nuts constituted a substantial component of Late Pleistocene Sri Lankan diets (Perera et al. 2011). Furthermore, Roberts et al. (2015b) applied stable carbon and oxygen isotope analysis to human and faunal tooth enamel samples from Batadomba-lena Layer 6. The results demonstrate that Sri Lankan human foragers relied primarily on rainforest resources from at least c. 20,000–17,000 years ago, while ongoing work is looking to extend this methodology further back in time. The association of the Batadomba-lena bone tools with this concentrated and stable exploitation of rainforest floral and faunal resources, from c. 36,000–12,000 cal years BP, suggests they played an important role in the subsistence strategies of some of the earliest human rainforest hunter-gatherers.

Implications and Potential of Sri Lankan Osseous Technologies

In addition to the Batadomba-lena sequence, bone technologies have also been mentioned from three other Microlithic sites in Sri Lanka: Fa Hien-lena dated to c. 38,000 cal BP, Kitulgala Beli-lena (Deraniyagala 1992) dated to c. 13,000 BP and Bellan-bandi Palassa, dated to c. 12,000 BP (Deraniyagala 1992; Perera 2010).

Wijeyapala (1997) reported bone tools at Fa Hien-lena, alongside a geometric microlith and shell beads, from layers c. 60 cm below a stratigraphic context dated to c. 38,000 cal BP. Bone tools are not found throughout the remainder of the Fa-Hien sequence, however, and until more detailed stratigraphic information and description is presented for these finds they should be approached with caution. Excavation and post-excavation work at the site has been renewed and should provide additional chronological and stratigraphic resolution in the near future.

The bone tools from Kitulgala Beli-lena, as at Batadomba-lena, comprise single points (c. 2–12 cm long, 0.4–0.7 cm wide) and double points (c. 1.5–4.0 cm, 0.4 cm wide). Kitulgala Beli-lena is a rockshelter site also found in the Wet Zone of Sri Lanka and has yielded a similar faunal assemblage of molluscs and arboreal/semi-arboreal taxa to that found at Batadomba-lena (Deraniyagala 1992). Unique to Kitulgala Beli-lena are a form-ground spatula, with nicks cut into the lateral edges to produce a serrated appearance (c. 2 cm long by 1.5 cm wide), and barrel-beads on segmented long bones (c. 2 cm long, 1.5 cm diameter) (Deraniyagala 1992). Bellan-bandi Palassa, is a Microlithic site found in the Dry Zone of Sri Lanka, and is now dated to c.12,000 cal BP (Perera 2010). The bone artifacts at this site include small single and double points, bone picks, a bone spatula (a large c. 20 cm specimen made from elephant bone), a potential awl made from the canine of a civet mandible, as well as a notched *Muntiacus* antler (Deraniyagala 1992; Perera 2010). The potential to compare the archaeological contexts and functions of bone tools at these sites, from two different contemporaneous environmental zones, presents another area for future research.

More generally, the Sri Lankan osseous assemblages add to the evidence from the Niah Caves, Borneo (Barker et al. 2007; Rabett 2012) and, more recently, Matja Kuru 2, Timor-Leste (O'Connor et al. 2014) in demonstrating that complex bone tool technologies accompanied early humans into the novel rainforest environments of Asia. Even from preliminary analyses, it is clear that these technologies are different in manufacture methods and morphology when compared to the Upper Paleolithic bone toolkits of Europe (Deraniyagala 1992). Bone technologies were clearly able to facilitate a number of different practices, be they symbolic or part of subsistence strategies, in a range of novel climates and environments encountered by *Homo sapiens* during expansion beyond Africa.

Conclusion

We have presented evidence from the rockshelter of Batadomba-lena, Sri Lanka that indicates the early appearance of osseous technologies with *Homo sapiens* in Sri Lanka by at least c. 36,000 cal BP. The Sri Lankan evidence offers both the oldest and longest sequence of bone toolkits anywhere in South Asia and holds much potential for future systematic and experimental research. Macroscopic description and photographic presentation of the Batadomba-lena bone points indicates relative stability of bone tool technologies at the site through time. Contextual analysis of associated environmental and subsistence changes indicate that they may have been important in the exploitation of freshwater molluscan resources, and potentially the diversification of the diet towards small prey species, during the Late Pleistocene period. Overall, it is clear that the Batadomba-lena bone points were firmly bound up within a specialised rainforest exploitation strategy that was adopted by the earliest Microlithic human populations in Sri Lanka (Roberts et al. 2015a, b). As with the evidence from Southeast Asia (Piper and Rabett 2014), this indicates that bone tool analysis is much more fruitful when placed within the wider environmental and social context of early human expansion into new and diverse environments, rather than as passive indications of supposed 'modernity'.

Acknowledgments We would like to thank Dr. S.U. Deraniyagala, Dr Senarath Dissanayaka, L.V. De Mel, Jude Perera, and all of the members of the Excavations Branch at the Department of Archaeology, Sri Lanka, and Prof. Jagath Weerasinghe, Postgraduate Institute of

Archaeology, Sri Lanka, for their technical and financial support in the excavation and interpretation of the Batadomba-lena sequence and its material culture. We would also like to extend our thanks to R.M. Kushumpriya Rajapaksa (Polonnaruwa Project of the Cultural Triangle, Central Cultural Fund of Sri Lanka) for his help with photography. Finally, we would like to thank Klint Janulis, Michelle Langley, and two anonymous reviewers for their constructive comments on an earlier draft of this chapter. N.P. was supported by the Australian National University and Dr David Bulbeck during his 2005 excavation and interpretation of Batadomba-lena. P.R. is currently supported by a doctoral research award from the Natural Environmental Research Council (NERC) (grant no. 1322282), UK. M.P. acknowledges the support of the European Research Council (grant no. 295719).

References

Backwell, L. R., & d'Errico, F. (2016). Osseous projectile weaponry from early to late Middle Stone Age Africa. In M. C. Langley (Ed.), *Osseous projectile weaponry: Towards an understanding of Pleistocene cultural variability* (pp. 15–29). Dordrecht: Springer.

Backwell, L., d'Errico, F., & Wadley, L. (2008). Middle Stone Age bone tools from the Howiesons Poort layers, Sibudu Cave, South Africa. *Journal of Archaeological Science, 35*, 1566–1580.

Barker, G., Barton, H., Bird, M., Daly, P., Datan, I., Dykes, A., et al. (2007). The 'human evolution' in lowland tropical Southeast Asia: The antiquity and behaviour of anatomically modern humans at Niah Cave (Sarawak, Borneo). *Journal of Human Evolution, 52*, 243–261.

Barton, H., Piper, P. J., Rabett, R., & Reed, I. (2009). Composite hunting technologies from the Terminal Pleistocene and Early Holocene, Niah Cave, Borneo. *Journal of Archaeological Science, 36*, 1708–1714.

Bednarik, R. G. (2003). The earliest evidence of palaeoart. *Rock Art Research, 20*, 89–135.

Clarkson, C., Petraglia, M., Korisettar, R., Haslam, M., Boivin, N., Crowther, A., et al. (2009). The oldest and longest enduring microlithic sequence in India: 35 000 years of modern human occupation and change at the Jwalapuram locality 9 rockshelter. *Antiquity, 83*, 326–348.

Conard, N. J. (2003). Palaeolithic ivory sculptures from southwestern Germany and the origins of figurative art. *Nature, 426*, 830–32.

Cooray, P. G. (1984). *The geology of Sri Lanka*. Colombo, Sri Lanka: National Museum Publications.

Deraniyagala, P. E. P. (1940). The Stone Age and cave men of Ceylon. Part I. *Journal of the Royal Asiatic Society, Ceylon Branch, 34*, 351–373.

Deraniyagala, P. E. P. (1943). The Stone Age and cave men of Ceylon. Part II. *Journal of the Royal Asiatic Society of Ceylon, 35*, 159–162.

Deraniyagala, P. E. P. (1953). Some aspects of the prehistory of Ceylon. Part 3: The Balangoda Culture. *Spolia Zeylanica, 27*, 125–131.

Deraniyagala, S. U. (1992). *The prehistory of Sri Lanka: An ecological perspective* (2nd ed.). Colombo, Sri Lanka: Department of Archaeological Survey.

Deraniyagala, S. U. (2007). The prehistory and protohistory of Sri Lanka. In P. L. Prematilleke, S. Bandaranayake, S. U. Deraniyagala, & R. Silva (Eds.), *The art and archaeology of Sri Lanka* (Vol. 1, pp. 1–96). Colombo, Sri Lanka: Central Cultural Fund.

d'Errico, F. (2003). The invisible frontier. A multiple species model for the origin of behavioural modernity. *Evolutionary Anthropology, 12*(188), 202.

Fleitmann, D., Burn, S. J., Mangini, A., Mudelsee, M., Kramers, J., Villa, I., et al. (2007). Holocene ITCZ and Indian monsoon dynamics recorded in stalagmites from Oman and Yemen (Socotra). *Quaternary Science Reviews, 26*, 170–188.

Habgood, P. J., & Franklin, N. R. (2008). The revolution that didn't arrive: A review of Pleistocene Sahul. *Journal of Human Evolution, 55*, 187–222.

Henshilwood, C. S., Sealy, J. C., Yates, R., Cruz-Uribe, K., Goldberg, P., Grine, F. E., et al. (2001). Blombos Cave, Southern Cape, South Africa: Preliminary report on the 1992–1999 excavations of the Middle Stone Age levels. *Journal of Archaeological Science, 28*, 421–448.

Jacobs, Z., Duller, G. A. T., Wintle, A. G., & Henshilwood, C. S. (2006). Extending the chronology of deposits at Blombos Cave, South Africa, back to 140 ka using optical dating of single and multiple grains of quartz. *Journal of Human Evolution, 51*, 255–273.

James, H. A., & Petraglia, M. (2005). Modern human origins and the evolution of behavior in the later Pleistocene record of South Asia. *Current Anthropology, 46*(S5), S3–S27.

Kennedy, K.A.R., Deraniyagala, S.U. 1989. Fossil remains of 28,000 year old hominids from Sri Lanka. *Current Anthropology 30*, 394–399.

Langley, M., Clarkson, C., & Ulm, S. (2011). From small holes to grand narratives: the impact of taphonomy and sample size on the modernity debate in Australia and New Guinea. *Journal of Human Evolution, 61*(2), 197–208.

McBrearty, S., & Brooks, A. (2000). The revolution that wasn't: A new interpretation of the origins of modern human behaviour. *Journal of Human Evolution, 39*, 453–63.

Mellars, P. (1989). Technological changes and the Middle-Upper Palaeolithic transition: Economics, social and cognitive perspectives. In P. Mellars & C. Stringer (Eds.), *The human revolution* (pp. 338–365). Edinburgh, England: Edinburgh University Press.

Mellars, P. (2005). The impossible coincidence. A single-species model of the origins of modern human behaviour in Europe. *Evolutionary Anthropology, 14*, 12–27.

Mulvaney, D. J. (1975). *The prehistory of Australia* (2nd ed.). Melbourne, Australia: Pelican Books.

O'Connor, S., Robertson, G., & Aplin, K. P. (2014). Are osseous artefacts a window to perishable material culture? Implications of an unusually complex bone tool from the Late Pleistocene of East Timor. *Journal of Human Evolution, 67*, 108–119.

Perera, N. H. (2010). *Prehistoric Sri Lanka: Late Pleistocene Rockshelters and an Open Air Site. British Archaeological Reports International Series 2142*. Oxford: Archaeopress.

Perera, N., Kourampas, N., Simpson, I. A., Deraniyagala, S. U., Bulbeck, D., Kamminga, J., et al. (2011). People of the ancient rainforest: Late Pleistocene foragers at the Batadomba-lena rockshelter, Sri Lanka. *Journal of Human Evolution, 61*, 252–269.

Perera, N. P. (1975). A physiognomic vegetation map of Sri Lanka (Ceylon). *Journal of Biogeography, 2*, 185–203.

Piper, P. J., & Rabett, R. (2009). Hunting in a tropical rainforest: Evidence from the Late Pleistocene and Early Holocene at Lobang Hangus, Niah Cave, Borneo. *International Journal of Osteoarchaeology, 19*, 551–565.

Piper, P. J., & Rabett, R. (2014). Late Pleistocene subsistence strategies in Island Southeast Asia and its implications for understanding the development of modern human behaviour. In R. Dennell & M. Porr (Eds.), *Southern Asia, Australia, and the search for human origins* (pp. 118–135). Cambridge, England: Cambridge University Press.

Premathilake, R., & Risberg, J. (2003). Late quaternary history of the Horton plains, central Sri Lanka. *Quaternary Science Reviews, 22*, 1525–1541.

Rabett, R. (2005). The early exploitation of Southeast Asian mangroves: Bone technology from caves and open sites. *Asian Perspectives, 44*, 154–179.

Rabett, R. (2012). *Human adaptation in the Asian Palaeolithic: Hominin dispersal and behaviour during the Late Quaternary*. Cambridge, England: Cambridge University Press.

Rabett, R., & Piper, P. J. (2012). The emergence of bone technologies at the end of the Pleistocene in Southeast Asia: Regional and evolutionary implications. *Cambridge Archaeological Journal, 22*(1), 37–56.

Roberts, P., Boivin, N., & Petraglia, M. (2015a). The Sri Lankan 'Microlithic' Tradition c. 38,000 to 3,000 years ago: Tropical technologies and adaptations of *Homo sapiens* at the Southern Edge of Asia. *Journal of World Prehistory*. doi:10.1007/s10963-015-9085-5.

Roberts, P., Perera, N., Wedage, O., Deraniyagala, S. U., Perera, J., Eregama, S., et al. (2015b). Direct evidence for human reliance on rainforest resources in late Pleistocene Sri Lanka. *Science, 347*, 1246–1249.

Stiner, M. C., Munro, N. D., & Surovell, T. A. (2000). The Tortoise and the Hare. Small-game use, the broad spectrum revolution and paleolithic demography. *Current Anthropology, 41*(1), 39–79.

Wadley, L. (2010). Were snares and traps used in the Middle Stone Age and does it matter? A review and a case study from Sibudu, South Africa. *Journal of Human Evolution, 58*(2), 179–192.

Wang, P., Clemens, S., Beaufort, L., Braconnot, P., Ganssen, G., Jian, Z., et al. (2005). Evolution and variability of the Asian monsoon system: State of the art and outstanding issues. *Quaternary Science Reviews, 24*, 595–629.

Wijeyapala, W. H. (1997). *New light on the prehistory of Sri Lanka in the context of recent investigations at cave sites*. Unpublished Ph.D. thesis, University of Peradeniya.

Chapter 13
The Walandawe Tradition from Southeast Sulawesi and Osseous Artifact Traditions in Island Southeast Asia

Ken Aplin, Sue O'Connor, David Bulbeck, Philip J. Piper, Ben Marwick, Emma St Pierre, and Fadhila Aziz

Abstract This chapter describes a sample of points and other osseous artifacts recovered from Holocene contexts at three sites in Walandawe, Southeast Sulawesi, Indonesia. Microscopic observations of use traces and manufacturing techniques are presented as well as metrical observations and morphological classifications. The points show a suite of temporal trends apparently related to a shift from a predominant use as hafted projectile points to their growing use as penetrative tools. Trends include a higher incidence of wear and decline in tip damage, a decrease in bipoint production, an increased focus on unipoints, and a manufacturing shift from predominantly scraping cortical bone to frequently grinding suid incisors and long-bone shafts. Notwithstanding these changes, the Walandawe osseous artifacts constitute an identifiable tradition with systematic differences from other Island Southeast Asian assemblages located in southwest Sulawesi and especially Borneo, the Aru Islands, the northern Moluccas and the New Guinea Bird's Head.

K. Aplin • S. O'Connor (✉) • D. Bulbeck
Department of Archaeology and Natural History, College of Asia and the Pacific, The Australian National University, Canberra, ACT, Australia
e-mail: aplin.ken@gmail.com; sue.oconnor@anu.edu.au; david.bulbeck@anu.edu.au

P.J. Piper
School of Archaeology and Anthropology, Australian National University, Canberra, Australia
e-mail: philip.piper@anu.edu.au

B. Marwick
Department of Anthropology, University of Washington, Seattle, WA, USA
e-mail: bmarwick@uw.edu

E.S. Pierre
School of Social Science, The University of Queensland, Brisbane, QLD, Australia
e-mail: e.stpierre@uq.edu.au

F. Aziz
Department of Prehistory, National Center for Archaeological Research and Development, Jakarta, Indonesia
e-mail: wadhila@yahoo.co.id

Keywords Sunda • Holocene prehistory • Bipoints • Unipoints • Suidae materials

Introduction

Artifacts manufactured from the bone, antler, and teeth of vertebrates—collectively termed 'osseous artifacts'—have attracted widespread archaeological attention for what they can reveal about the technological capacity, subsistence behavior, and symbolic faculties of prehistoric peoples. One impetus has been the significance attached to sophisticated osseous artifacts in the recent debate over the definition of 'full modernity' in human behavior (McBrearty and Brooks 2000; Ambrose 2001; Henshilwood and Marean 2003; Klein 2009), in particular those implements incorporated into composite tools through hafting technologies. This purposeful behavior is contrasted with a much older pattern of the 'casual' use of split bones documented for even the earliest hominins (Backwell and d'Errico 2001) and the occasional working of bone and ivory using percussion methods by Middle Pleistocene European *Homo* populations (Cassoli and Tagliacozzo 1994; Rosell et al. 2011). Claims for the coeval systematic fashioning of bone and ivory by cutting, shaving and polishing remain unsubstantiated (Villa and d'Errico 2001), even though these techniques were already used for fashioning wooden spears (Thieme 1997).

This paper focuses on osseous artifacts from Island Southeast Asia (ISEA), an ecologically and culturally diverse region comprising thousands of islands (Fig. 13.1). To the west, the large islands of Borneo, Sumatra and Java, as well as many smaller ones, are sited on the continental Sunda Shelf; during glacial episodes these coalesced to form an expanded Sundaic landmass continuous with adjacent Mainland Southeast Asia. By contrast, to the east of the Sunda Shelf, the Philippine Archipelago, the large island of Sulawesi, and the Lesser Sunda and North and South Moluccan island groups, are all isolated by deep water and remained so even during glacial maxima.

Michelle C. Langley (ed.), *Osseous Projectile Weaponry*, Vertebrate Paleobiology and Paleoanthropology,
DOI 10.1007/978-94-024-0899-7_13

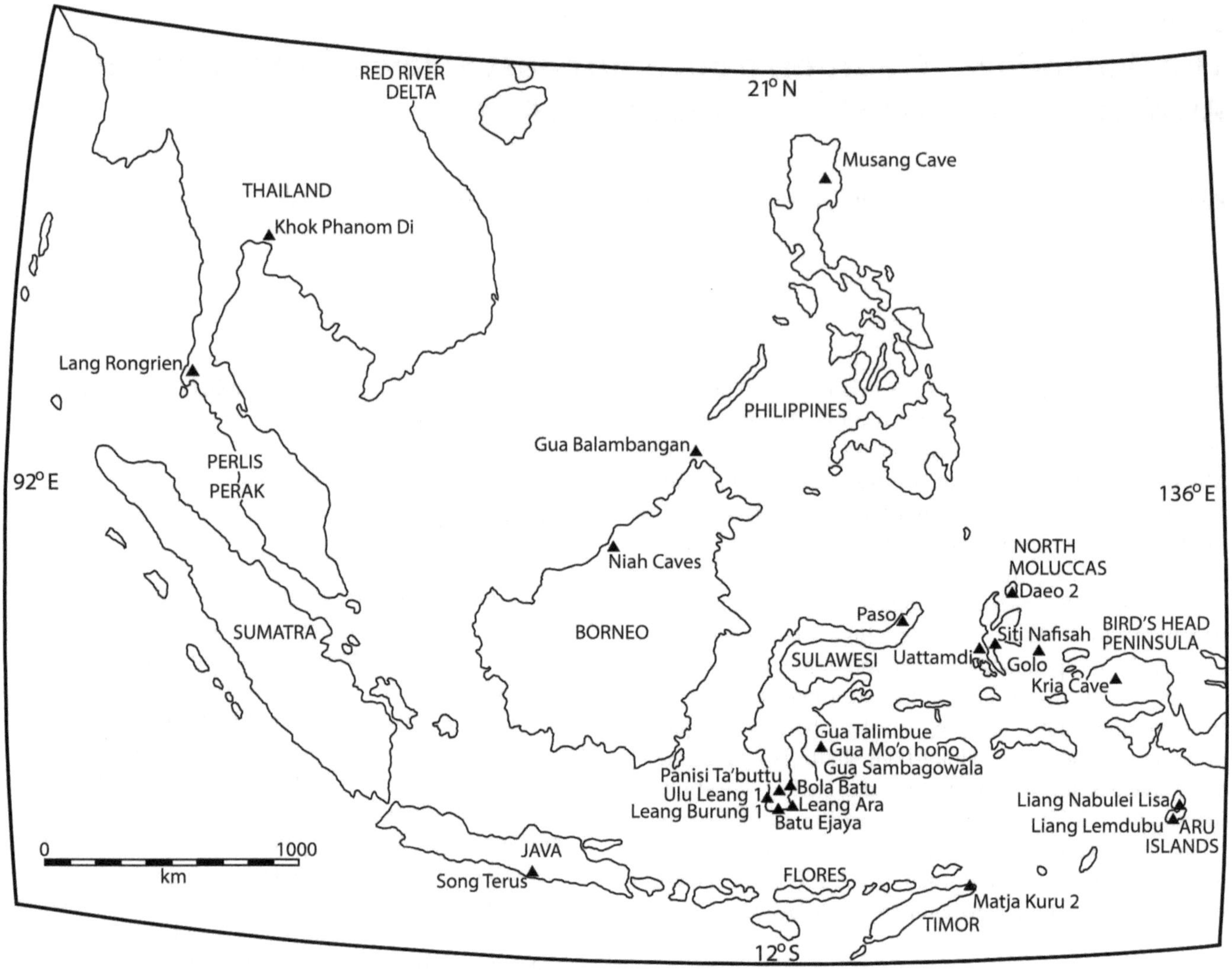

Fig. 13.1 Mainland and Island Southeast Asia, showing localities mentioned in the text

The Sundaic landmass was first colonized by early hominins more than 1.8 Ma (Dennell and Roebroeks 2005), and one species managed to cross water barriers to colonize Flores in the Lesser Sundas by 1.0 Ma (Morwood et al. 2004; Brumm et al. 2010), conceivably via Sulawesi (Morwood and Jungers 2009). However, the majority of the eastern oceanic islands appear to have been colonized by *Homo sapiens* and only within the last 40–50 ka (O'Connor 2007; O'Connor and Aplin 2007), as part of a region-wide dispersal that included for the first time, the combined landmass of New Guinea and Australia (O'Connor and Chappell 2003).

Osseous artifacts occur in small numbers in the earliest known modern human occupation levels in Southeast Asia (Rabett 2005; Rabett and Piper 2012). The oldest confirmed examples, including modified and utilized teeth and bones, date to ca. 45 cal kBP at the Niah Caves on Borneo (Rabett et al. 2006; Higham et al. 2009). A slightly younger example, associated with a date of 42.4 ± 0.9 cal kBP, is a bone fragment from Lang Rongrien in Thailand worked with the groove-and-snap technique (Anderson 1990, 1997). Two potentially early bone artifacts from the 'Tabuhan' layers (ca. 30–80 ka) of Song Terus in East Java are illustrated by Kusno (2006). Further to the south, Australian archaeological sites dated to 20–30 cal kBP have yielded a variety of osseous artifacts, including 'spatulas' manufactured in a standardized fashion from wallaby fibulae (Webb and Allen 1990; Allen et al. 2016) and decorative bone beads (Dortch 1979).

A remarkably complex bone artifact, discovered recently in a ca. 34 cal kBP level at Matja Kuru 2 in East Timor, is interpreted as the base of a projectile point that was hafted through combined use of mastic and cordage (O'Connor et al. 2014). Currently unique within the wider regional context, this object shows that sophisticated hafting technologies were in use much earlier in the region than previously documented. To be sure, the existence of these technologies has been posited as a necessary prerequisite for the systematic colonization of ISEA, Melanesia and Australia by modern humans (Balme et al. 2009; Balme 2013; Balme and O'Connor 2014), and is consistent with multiple lines of evidence for advanced symbolic behavior in the Australian archaeological record (e.g., Langley et al. 2011).

The quest remains for ISEA archaeologists to determine where, when, and under what circumstances osseous artifacts became sufficiently standardized to constitute recognizable traditions, in the context of underlying factors—demographic, social and/or economic (Bentley and Shennan 2003; Brumm and Moore 2005; Collard et al. 2006; Powell et al. 2009; Rabett and Piper 2012)—that influence the variable expression of technological complexity. We aim to contribute to this emerging narrative of osseous artifact use in Southeast Asia in two ways. First, we describe a pan-Holocene assemblage of osseous artifacts from recent excavations at Walandawe in Southeast Sulawesi. Second, we compare this assemblage with other, well-documented osseous artifact assemblages from ISEA.

A New Regional Osseous Artifact Sequence from Southeast Sulawesi

Bone points, together with backed microliths and stone arrowheads with a hollowed base ('Maros' points), are regarded as a hallmark of the Toalean culture of Holocene South Sulawesi (Van Heekeren 1972:106–125; Bellwood 1997:193–196). Although numerous osseous artifacts have been recovered from cave excavations in this area, those derived from colonial period excavations remain unreported. Olsen and Glover (2004) described two assemblages totaling 160 artefacts from the excavations of Ulu Leang 1 (Glover 1976, 1979) and Leang Burung 1 (Mulvaney and Soejono 1971). The prehistoric archaeology of other parts of Sulawesi has not been well studied, though the Paso shell mound at Minahasa, North Sulawesi (Bellwood 1976) has a small bone artifact assemblage (17 'bone awls'). The samples of osseous artifacts reported here come from three newly excavated caves in the Walandawe area of Kecamatan Routa, Kabupaten Konawe, Southeast Sulawesi. These sites contain rich cultural assemblages comprised of stone and osseous artefacts with earthenware pottery recovered in the upper levels. Here we focus exclusively on the osseous artifact assemblages.

Gua Talimbue

Gua Talimbue is a limestone cliff foot cave located at 320 m asl (Fig. 13.2). Excavation in two adjoining 1 m^2 test pits, B and E, reached a depth of approximately 4.1 m below the surface. Excavation ceased at this depth owing to time constraints. Preliminary dating of the sequence suggests an age of ca. 19 cal kBP for the lower excavated levels, ca. 9.5 cal kBP for the imbricated shell layer, and an age range of ca. 9.5–3.5 cal kBP for the main cultural unit, though with a likely hiatus or slowdown in sedimentation between ca. 6.5 and 4.5 cal kBP.

The preservation state of the vertebrate fauna varies systematically through the deposit. In the upper half of the deposit the bone is well preserved and the majority is unburnt. Progressively more of the preserved bone is burnt below this point and by 2.5 m depth, virtually all of the bone is burnt. This progressive loss and deterioration of unburnt bone is usually a good indicator of post-depositional degradation of the biologically more active unburnt remains. The quantity of recovered bone declines sharply at around 3.2 m depth. Only occasional teeth and fragments of highly resistant calcined bone were found in the basal clay horizon. Detailed fauna identifications are yet to be made.

The Talimbue sample consists of 95 osseous artifacts identified during preliminary field sorting of the vertebrate fauna from Pit B. Many more osseous artifacts will likely be discovered when detailed analysis of the fauna from Pit B is undertaken, as smaller broken point fragments are likely to have been overlooked during the initial sort. Pit E can also be expected to produce a similar number of artifacts made on teeth and bone. The majority of the osseous artifacts from Square B come from levels dating to between ca. 7 and 9.5 cal kBP (N = 65), with smaller numbers from three other time slices: ca. 6–7 cal kBP (N = 17); ca. 3.5–4.5 cal kBP (N = 12); and <3.5 cal kBP (N = 1).

Gua Mo'o hono

Gua Mo'o hono is located at 344 m asl (Fig. 13.2). This limestone rockshelter runs approximately 20 m from the northwest to the southeast, with a width of up to 5 m at the drip line. The 2.6 m deep test pit excavated in Mo'o hono contains three chronostratigraphic units; as yet unexcavated deposits continued below this level. Dating of the sequence indicates that the basal silts predominantly accumulated in two pulses from ca. 6.0–7.0 cal kBP and 3.4–4.5 cal kBP. The upper hearth complex probably accumulated within the last 1500 years.

Vertebrate faunal remains amount to ca. 16 kg of bone with an estimated 80,000 fragments. Bones from the upper hearth complex are predominantly unburnt and well preserved, while those from the silty unit are predominantly burnt to calcined and many have an evenly worn and polished appearance owing to water transport and chemical corrosion. Many fragments in the lower deposit are eroded, resulting in loss of surface morphology. Identifications indicate a predominance of larger game including the Babirusa (*Babyrousa babyrussa*), Celebes Warty Pig (*Sus celebensis*), and Anoa (*Bubalus*

Fig. 13.2 Mo'Ohono (*Left*) and Talimbue (*Right*)

depressicornis). Smaller animals are also represented including Sulawesi bear cuscus (Ailurops ursinus), Sulawesi dwarf cuscus (Strigocuscus celebensis), endemic Sulwesi macaques, brown palm civet (Macrogalidia musschenbroekii), rodents, birds, reptiles, frogs, and fish (restricted to the upper units).

After allowing for fragments from the same item, a total of 49 osseous artifacts were identified, distributed across depositional phases as follows: ca. 6.0–7.0 cal kBP (N=24), 4.5–6.0 cal kBP (N=3), 3.4–4.5 cal kBP (N=8), and <3.4 cal kBP (N=14).

Gua Sambagowala

Gua Sambagowala is a dolomite rockshelter located at 344 m asl. The floor area measures ca.19 m by 3–7 m wide to the drip line. The deposit was excavated to a depth of around 2.4 m, with as yet unexcavated deposits continuing below this level. A consistent series of C14 dates suggests an accumulation at fairly constant rate between ca. 6.0 and 3.5 cal kBP. A small quantity of sediment represents the last few thousand years.

Moderate quantities of well-preserved faunal remains were recovered but these are yet to be identified. Five bone points were found during excavation and preliminary sorting of the faunal remains; these come from levels dating between ca. 4.5 and 6.0 cal kBP (N=4) and between 3.5 and 4.5 cal kBP (N=1).

A Five Phase Chronology

For analysis of temporal trends the various samples were allocated to five time intervals or 'phases': Phase A: <3.4 ka; Phase B: 3.4–4.5 ka; Phase C: 4.5–6.0 ka; Phase D: 6.0–7.0 ka; and Phase E: 7–9.5 ka (represented only at Talimbue). Delineation of the time slices is arbitrary insofar as it is based on the major phases of deposition of sediments and cultural materials in the excavated sites. The extent to which these correspond to transformations in the wider environment and/or cultural landscape remains to be determined through analysis of other components of the excavated assemblages. One break that does correspond with clear changes in other materials is that between Phase B to Phase A—this corresponds with the appearance in the stratigraphic sequences of pottery and polished stone artifacts. Phase A thus corresponds to the 'Neolithic' and 'Metal Phase' as recognized for ISEA (Bellwood 1997), although we currently have no basis to date the onset of the Neolithic/Metal Phase at Walandawe any earlier than ca. 1.5 ka.

The Walandawe Osseous Artifacts

The combined sample of 149 osseous artifacts from Walandawe is diverse, being variably made on bones and teeth from a variety of taxa, and displaying a wide range of morphologies. It also exhibits distinctive kinds of use-related wear and damage. Before covering the points, which dominate the assemblage (Table 13.1), we describe the rarer spatula, tusk tool and ground-surfaced morphotypes.

Spatulas

Three fragmentary examples were identified, all made from cortical bone, perhaps the split long-bone shafts of large mammals. All come from Gua Talimbue spits 54–56 (9.0–9.5 cal kBP), and so three of the 31 osseous artifacts from spits 54 and lower in Gua Talimbue are spatulas, whereas none of the 118 osseous artifacts from higher Phase E and Phase A–D contexts are of this type.

The most complete example measures 32.2 mm in length and is flattened, with a maximum thickness of 3.8 mm and a basal width of 14.5 mm. It tapers towards the heavily worn tip where it is 10.1 mm wide. The tip was shaped by coarse oblique grinding and the working edge is rounded in both plan and cross-section. A zone of intense polish extends back from the working edge on both sides of the artifact. Coarse scratches running transversely and obliquely to the working edge overlie the polish, while some earlier score marks are overlain by polish, indicative of either prolonged use or multiple episodes of use.

A second specimen is a spall from a working edge very similar in form to that described above. The spall has a maximum width of 12.7 mm at the heavily worn end and a maximum thickness of 3.4 mm. The working edge is rounded, and it features oblique coarse scratches both prior and subsequent to the major episode of polish.

A third specimen is very likely a mid-shaft fragment of a spatula. This 4.3 mm thick piece has a zone of polish that extends along margins that have been rounded by scraping, and a flattened facet on its outer surface shaped by shaving with a stone tool followed by grinding over its full length on a fine abrasive surface. The considerable investment of time expended in working the outer layers of bone suggests that the thickness of the artifact was an important attribute.

The three pieces were clearly all used to perform similar tasks that involved highly repetitive penetration of a fibrous medium.

Expedient Tools, Including a Tusk Tool

Four of the 36 osseous artifacts from Phases A and B, but none of the 113 osseous artifacts from Phases C to E, are classified as expedient tools. Of particular interest is a tusk tool from Gua Mo'o hono A/11, probably dating to ca. 3.5 cal kBP.

Table 13.1 Counts of Walandawe osseous artifacts classified to morphotype/material

Morphotype/material	Phase A	Phase B	Phase C	Phase D	Phase E	Total
Cortex bipoint	0	4	1	5	14	24
Dentin bipoint	0	1	0	1	1	3
Cortex unipoint	3	8	1	10	17	39
Dentin unipoint	0	2	1	5	2	10
Long-bone shaft unipoint	1	0	0	3	0	4
Cortex point tip	3	2	2	6	16	29
Dentin point tip	1	0	2	3	3	9
Cortex point shaft fragment	3	1	0	5	5	14
Dentin point shaft fragment	1	0	0	2	2	5
Cortex pre-form	0	1	0	0	1	2
Cortex ground fragment	1	0	0	1	0	2
Cortex expedient tool	1	0	0	0	0	1
Shaft expedient tool	1	0	0	0	0	1
Tusk expedient tool	0	1	0	0	0	1
Other tooth expedient tool	0	1	0	0	0	1
Tooth tool with use wear	0	0	0	0	1	1
Cortex spatula	0	0	0	0	3	3
Total	15	21	7	41	65	149

Note: The uni- or bi-points shown in Figs. 8–11 are all probably unipoints based on their elongation index, and so are included here with the unipoints

The artifact is formed from the terminal section of a babirusa lower canine. The basal end is roughly fractured and shows no other modification. The tip is essentially unmodified except through use and some probable resharpening over several phases (Fig. 13.3a, b). The Gua Mo'o hono tusk tool was clearly used for cutting. The outer enamel surface has a high gloss and fine oblique striations which are sub-parallel along most of the preserved length. The cutting edge is rounded from use, and relatively blunt except where subject to resharpening (see below). The inner surface lacks gloss or striations over most of the surface. It is clear that usage of the tool was highly systematic and repetitive, with the tool held in one position relative to the worked medium (Fig. 13.3c, d).

The tip area of the artifact has bilateral spalling. External spalling is concentrated on the chisel-like end of the tool and parallels the direction of the use-wear striations (Fig. 13.3e). It was probably produced in an attempt to resharpen the cutting edge, which may have broken the artifact. Internal spalling on the tip of the artifact is more extensive and extends not only around the tip but back along the blade for ca. 10 mm (Fig. 13.3f). It appears to predate the external spalling and shows a light abrasion gloss over spall boundaries indicative of use after the spall removal. However, the pattern of use differed from that which produced the high gloss on the external surface.

We suspect the artifact was used in two different ways. The original use was systematic and repetitive, and may have involved the cutting of a plant material that contained abundant siliceous phytoliths. Perhaps owing to the blunting of the natural cutting edge of the tooth, spalls were struck from one side and then the other to allow continued use of the artifact, albeit most likely in a different fashion with more focus on the tip than on the cutting edge.

Ground Pieces: Fragments of an 'Edge Tool'?

Four fragments of burnt cortex bone from Gua Mo'o hono A/4 and A/5 (Phase A) probably represent pieces of a single original artifact manufactured from a long-bone shaft fragment of a large mammal. On each fragment the outer surface of the bone has at least one flat, glossy facet. The striations on these facets are very fine showing they were ground on a very fine-grained stone such as a mudstone, rather than on the sand-sized texture of the surfaces used to grind the spatulas and most of the points.

The original form of the artifact is uncertain. However, one fragment 25 mm in length has a narrow grinding facet on the external surface as well as a second facet that descends to what might have been a beveled tip. If this interpretation is correct, this singular artifact may have originally resembled some of the potential wood-working tools from Pulau Balambangan and other sites categorized by Rabett (2005) as 'edge tools'.

Points

A total of 137 artifacts are classified as points or point fragments (Table 13.1). Included within this category are artifacts that are relatively slender and elongate, and terminate in one or two sharp ends (where preserved). Points in this broad sense account for the clear majority (75–100 %) of osseous artifacts in each of the five phases.

As detailed below, the Walandawe bipoints and unipoints show little differentiation by raw material, blurring the boundary between the bipoint and unipoint categories. More or less complete bipoints with two preserved tips are obvious enough, while other specimens show clear, manufactured taper toward both ends. However, any single pointed artifact could be a finished unipoint, an unfinished bipoint, or a fragment of a bipoint, broken off prior to the point of taper.

Intuitively, the longer a single-tapered unipoint is relative to its width, the less likely it is to be a fragment of a bipoint. Accordingly, we calculated an 'elongation index' (EI) for each point, under the formula L/(WxT)—the ratio of extant length to the product of maximum width and thickness. In the case of bipoints—identified by tapering at both ends—there was minimal difference in EI values between complete bipoints (N = 11, mean 2.0, range 1.2–3.2) and bipoints lacking one of their tips (N = 12, mean 2.2, range 1.1–4.9). Thus, it is probable that single-pointed artifacts tapering at just one end with EI values of 1.1 or more are sufficiently complete not to be confused with bipoint fragments. Another useful criterion derives from the fact that unipoints with use-related wear or damage to the tip cannot be regarded as unfinished bipoints.

Using these criteria, samples of 53 attenuate unipoints (Fig. 13.4b–d) and 27 bipoints (Fig. 13.3g–i) are identified. All other 57 tip and shaft fragments from points remain unassigned.

There appears to be a chronological transition from a fairly even representation of bipoints and unipoints in Phase E (44 % and 56 % respectively), to a reduced representation of bipoints (29 % vs. 71 %) in Phases B–D, and finally their absence from Phase A contexts. However, no statistically significant differences can be demonstrated from the available sample sizes.

Raw Material Usage

Three raw materials were used for the manufacture of points—cortex bone slivers, narrow unsplit shaft bones, and dentin of incisor roots. Bone from small mammals was rarely used as a raw material, although one of the shaft-bone points appears to be made on the radius of a small mammal, most likely a rodent. The other three were made on long bones from pig- or Anoa-sized mammals. Similarly, the majority of the points made from slivers of cortex bone were presumably

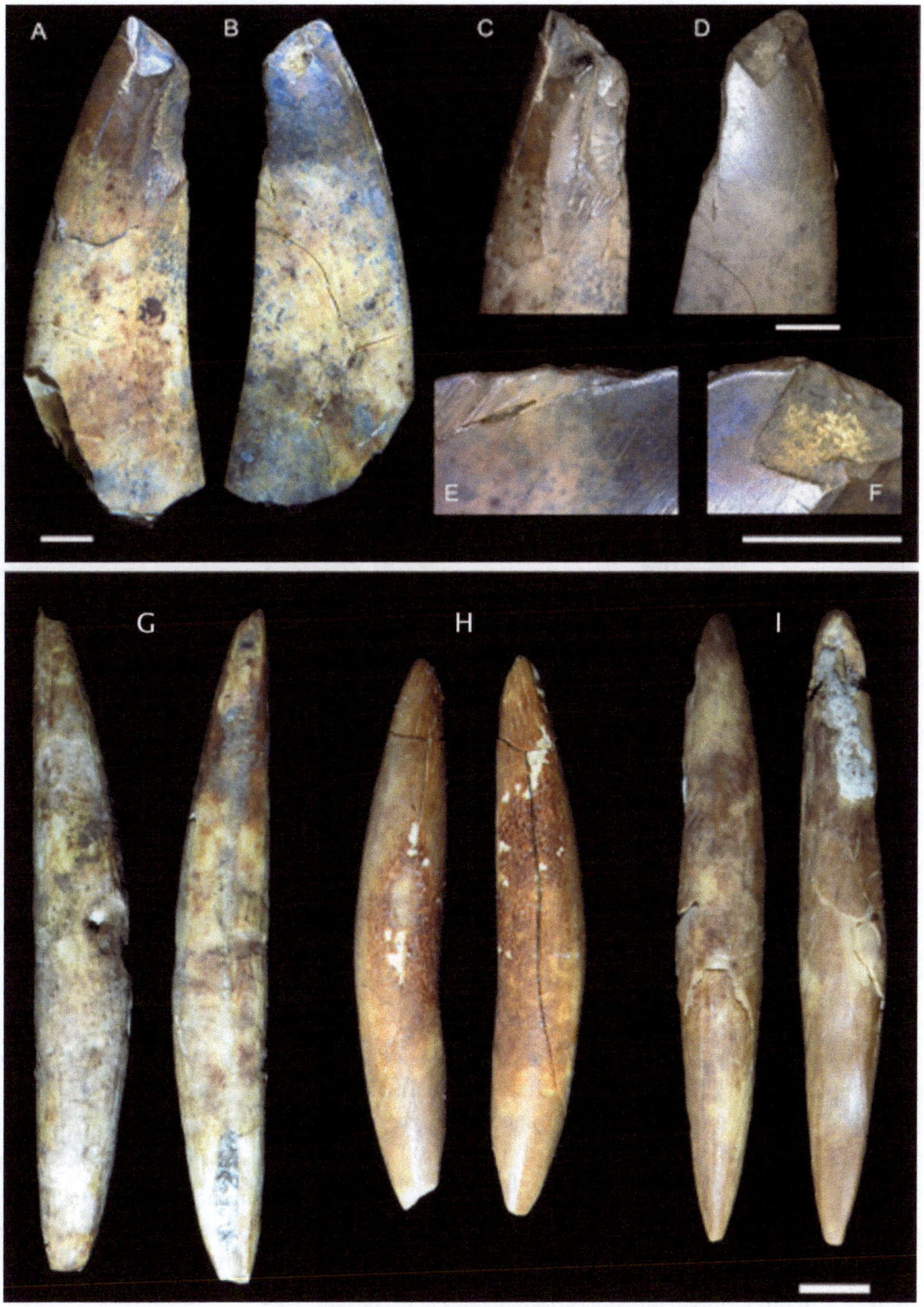

Fig. 13.3 (**a–f**) Tusk tool from Gua Mo'o hono A/11 (**g–i**) Examples of Walandawe bipoints; all three are manufactured from dentin, most likely incisor roots of Suidae

created by breaking up long bones or metapodials from large ungulates. None of the artifacts made from dentin retain an enamelled crown that would assist with taxonomic identification. However, from the size and cross-sectional proportions of these artifacts, the most likely parent teeth are the lower incisors of suids (Sulawesi warty pig and/or Babirusa). One metrically distinctive example may be manufactured from the canine of a dog (see below).

The majority of points (106/137, 77 %)—both unipoints (39/53, 74 %) and bipoints (24/27, 89 %)—were made from cortex bone slivers. Smaller proportions were made from dentin (19 % of unipoints, 11 % of bipoints, 20 % of all points) or shaft bone (8 % of the unipoints, 3 % of all points). Cortex was particularly the predominant material for Phase E points (52/60, 87 %) compared with Phases A–D (54/77, 70 %), and this difference is statistically significant (Chi-square = 4.37, 1 *d.f.*, $p < 0.05$).

Over half of the points (80/137, 58 %) are burnt to some degree, though only a small proportion (8/137, 6 %) are burnt to the level of calcination. The proportions in each burning

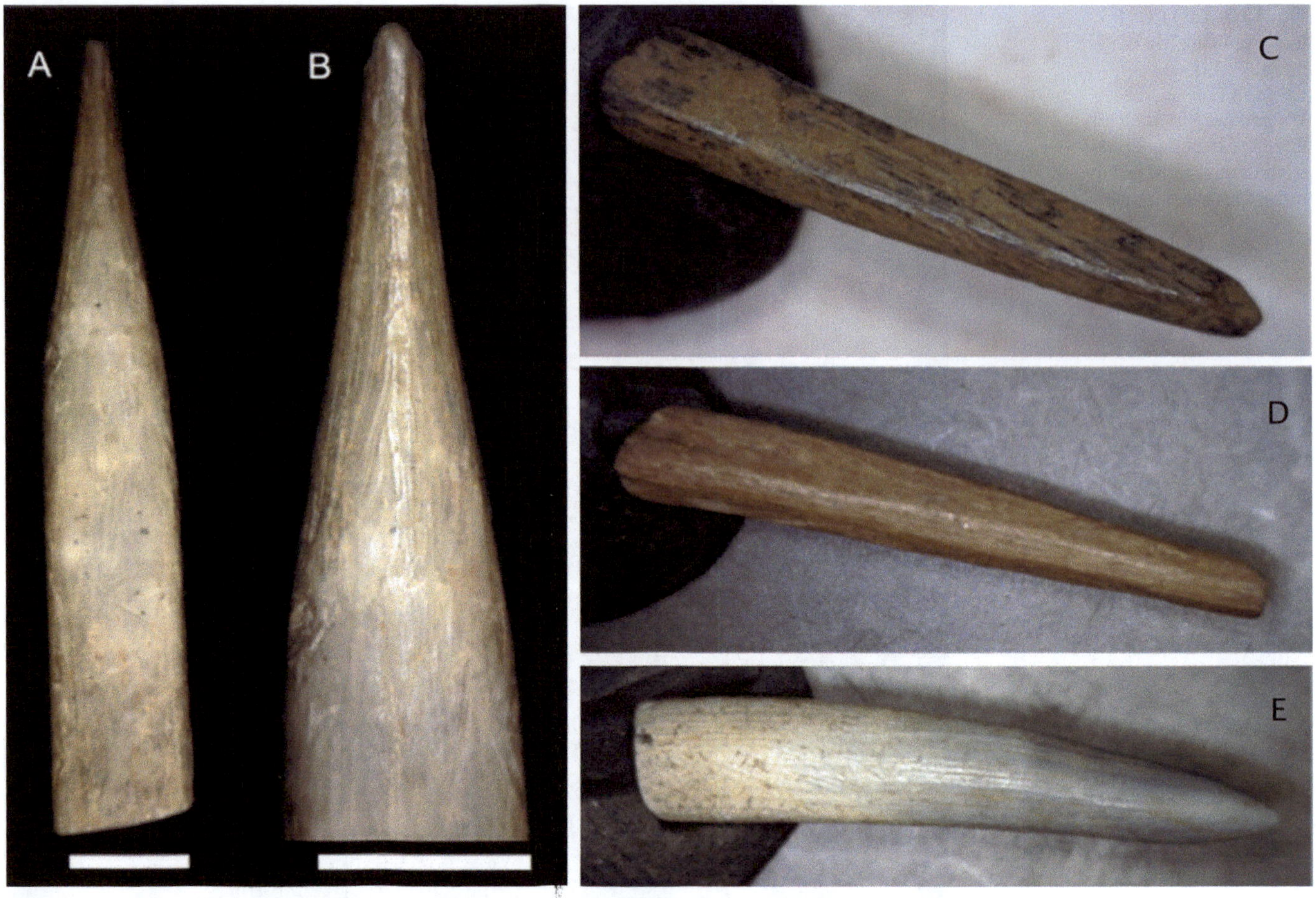

Fig. 13.4 (**a**, **b**) A fine point from Talimbue Square B Spit 42 created solely by scraping. The tip is rounded by wear and light polish extends back from the tip. (**c**–**e**) Examples of Walandawe attenuate unipoints manufactured from cortex bone (*upper* two) and dentin (*bottom*), most likely Suidae incisor root

class are similar for each raw material type, and while bipoints differ from unipoints in that over half of them (16/27, 59 %) are unburnt, the difference is not significant. The Phase E sample shows a higher proportion of burnt points (42/60, 70 %) than the samples from the later phases (28–67 %). This can be attributed to a tendency for the degradation of unburnt examples in the older layers of the sites, as observed for the faunal remains generally.

It is difficult to know whether burnt examples were originally made on burnt raw materials or whether the burning occurred subsequent to discard; and further, whether or not any use of burnt raw materials was a deliberate act of selection. In the case of calcined examples, the raw material is extremely brittle and it is unlikely that it would be suitable for point manufacture. Accordingly, these examples are likely to have been subject to secondary burning after discard. However, there is no clear consensus over the relative merits of less heavily burnt bone or dentin compared with fresh unburnt material for point manufacture and use (Pasveer 2004:182).

Points' Metrical Variability

The points' metrical variability is examined here in relation to raw material type and broad morphotype. Because point length is often not available due to breakage, we focus on the cross-sectional dimensions and tip length. Where the tips of bipoints are discussed, we follow Pasveer (2004) in distinguishing between the 'primary end' which shows the more pronounced evidence of use or damage, and the opposing 'secondary end'.

Dentin Points

Width and thickness are strongly correlated for most dentin points. Their minimal variation in cross-sectional proportions would reflect shape uniformity in the predominant raw material—Suidae lower incisors. The single specimen that falls off the linear regression line is consistent in size and

Table 13.2 Metric attributes of bipoints (mm) through the Walandawe chronological sequence

Phases for which data are available	N	Mean	Standard deviation	Coefficient of variation	Range
Point length					
Phase B	1	48.7	–	–	–
Phase D	2	44.0	–	–	39.0–49.7
Phase E	8	33.0	10.99	0.33	24.1–53.7
Shaft width					
Phase B	5	4.4	1.59	0.36	2.8–6.9
Phase C	1	5.1	–	–	–
Phase D	6	6.0	0.43	0.07	5.4–6.6
Phase E	15	5.5	1.78	0.32	3.3–9.1
Shaft thickness					
Phase B	5	3.6	1.41	0.39	2.2–5.9
Phase C	1	3.6	–	–	–
Phase D	6	3.9	0.77	0.20	3.1–5.2
Phase E	15	3.6	1.02	0.29	2.5–5.9
Tip length					
Phase B	5	10.8	3.70	0.34	5.6–15.5
Phase D	5	14.9	3.41	0.23	10.8–18.8
Phase E	13	13.5	4.68	0.35	8.4–24.1

shape with being made from the root of a dog canine. However, the apparent age of this specimen (6–7 ka), which comes from a Phase D context in Gua Mo'o hono, and the recent determination that the canines of the Sulawesi endemic brown palm civet are of a similar size to those of prehistoric Southeast Asian dogs urges caution in accepting this identification without further corroboration.

Cortex Points

Points made from cortex slivers have much more variable cross-sectional proportions than those made from dentin. The linear relationship between width and thickness is moderate ($R^2=0.34$) with an overall tendency for points, especially bipoints, to become more flattened in shape with increasing width. The width–thickness relationship is stronger for bipoints ($R^2=0.57$) and attenuate unipoints ($R^2=0.54$) considered separately, suggesting that bipoints and unipoints may differ in design and function.

Tip length of cortex points overall shows a moderately strong relationship ($R^2=0.31$) with shaft width. For bipoints the relationship is very strong ($R^2=0.74$) and this remains true even when two outlier points with point lengths >21 mm are excluded from the calculation ($R^2=0.60$). This result reinforces the notion that bipoints comprise a discrete category of tools. By contrast, the sample of attenuate unipoints (N=26) shows only a weak relationship ($R^2=0.22$) between point and length shaft width, which raises questions over the unity of this group of artifacts. This issue is further addressed below.

Cross-Sectional Comparison of Points Made on Different Raw Materials

Bipoints made on dentin have a thicker shaft than recorded for any of the bipoints made on cortical bone. Unipoints made on dentin also have a thicker shaft than cortical-bone unipoints of a similar width, with the exception of one unusually thick cortical-bone unipoint. The four unipoints made from shaft bone include two that cluster with the dentin unipoints and two that cluster with the cortical-bone unipoints.

Temporal Variation in Metric Attributes

Five of the 15 Phase E bipoints have shaft widths greater than those recorded for any Phase B–D point, and three of eight have a shorter point length than recorded for any later bipoint. However, both of these variability aspects may reflect the relatively large available sample size available for Phase E, combined with the uniformity of the Phase D bipoints on their shaft widths (Table 13.2). As for the unipoints, their average size reduced over time, with even the small sample from Phase C (N=2) conforming to the trend. The Phase A unipoints have the smallest average on all four measured attributes, the Phase B averages are always less than the Phase D averages, while the Phase E averages are always the largest or, in the case of shaft thickness, virtually the same as the Phase D average (Table 13.3).

Finally, when the tip category is considered, the metric contrast appears to be between Phase E and Phases A–D,

Table 13.3 Metric attributes of unipoints (mm) through the Walandawe chronological sequence

Phases	N	Mean	Standard deviation	Coefficient of variation	Range
Point length					
Phase A	4	16.6	3.04	0.18	13.8–20.9
Phase B	10	17.9	4.32	0.24	10.9–23.9
Phase C	2	17.2	–	–	16.0–18.3
Phase D	18	22.6	7.49	0.33	12.7–36.5
Phase E	19	24.3	8.87	0.36	12.3–40.8
Shaft width					
Phase A	4	3.4	0.86	0.25	2.4–4.5
Phase B	10	3.7	1.40	0.38	1.7–6.4
Phase C	2	5.0	–	–	4.7–5.3
Phase D	18	4.9	0.98	0.27	3.1–6.3
Phase E	18	5.6	1.38	0.25	2.9–8.1
Shaft thickness					
Phase A	4	2.7	0.53	0.20	2.1–3.3
Phase B	9	3.0	1.28	0.42	1.5–5.6
Phase C	2	3.3	–	–	2.5–4.1
Phase D	18	3.6	0.98	0.27	1.8–5.1
Phase E	17	3.5	0.95	0.27	2.0–5.4
Tip length					
Phase A	4	7.6	3.56	0.47	5.1–12.4
Phase B	9	9.1	3.81	0.42	5.4–16.0
Phase C	2	9.6	–	–	6.1–13.1
Phase D	18	9.4	2.87	0.30	4.2–16.3
Phase E	17	11.3	2.67	0.24	7.5–18.3

with the Phase E tips on average showing the widest but the thinnest dimensions and the longest tip length. This may be a reflection of the higher proportion of Phase E points that are bipoints compared with Phases A–D.

Point Manufacturing Techniques

There is evidence for use of three different techniques in the manufacture of points—scraping, grinding, and shaving—and many examples show the application of two or even three of these techniques.

Scraping marks ("(called 'cutting' by Pasveer 2004; Pasveer and Bellwood 2004; Pasveer 2005) typically follow the long axis of the points. They are usually multi-layered with somewhat wavy, cross-cutting courses. Comparable scraping marks have been illustrated for points from South Sulawesi (Olsen and Glover 2004:Fig. 16), Borneo (Barton et al. 2009:Fig. 2), the North Moluccas (Pasveer and Bellwood 2004:Fig. 2b) and West Papua (Pasveer 2004:Fig. 6.2). A general conclusion, supported by some experimental work (Olsen and Glover 2004:Fig. 17), is that these marks result by scraping against the edge of a relatively steep angled, flaked stone artifact. That scraping alone is a sufficient technique to produce quite delicate points is illustrated by several fine examples with intact tips (Fig. 13.4a, b).

Shaving marks (called 'cutting' by Pasveer 2004, 2005; Pasveer and Bellwood 2004) typically follow the long axis of the point. These are usually wide concave excisions, often with an undulating surface and/or chattering. Essentially similar shaving marks have been illustrated for points from West Papua by Pasveer (2004:Fig. 6.2). Shaving would have presumably been carried out with a more acutely angled stone artifact and would require the application of greater force than scraping. Intuitively, shaving might provide less precise control and incur a greater risk of breakage than any other manufacturing technique.

Grinding marks are distinguished by the presence of parallel, relatively even-sized striations. In the case of the Walandawe collection, the great majority of grinding facets have coarse striations, with an average striation width of ca. 0.1 mm, corresponding to the particle size of 'fine sand'. Grinding against a finer medium is observed in one point made on a narrow shaft bone; in this specimen an oblique grinding facet created a working edge that resembles a hypodermic needle.

The grinding surfaces vary from flat planes to more rounded surfaces. Surfaces that are flattened in two dimensions are created by restricting the movement between the

Table 13.4 Comparison of ground mid-shaft and tip fragments from Walandawe points for their direction of grinding

Direction of grinding	Mid-shaft fragments (N=8)	Tip fragments (N=17)
Exclusively oblique	3 (37.5 %)	8 (47.1 %)
Exclusively transverse	1 (12.5 %)	7 (41.2 %)
Exclusively longitudinal	0 (0 %)	1 (5.9 %)
Oblique and transverse	2 (25.0 %)	1 (5.9 %)
Longitudinal and transverse	2 (25.0 %)	0 (0 %)

Statistical significance testing: 4 of 8 mid-shaft fragments and 1 of 17 tip fragments show a change in grinding direction, Fisher exact test=0.022

abraded object and the abrasive surface to a single plane. A rounded surface is created by rotating the abraded object as it moves across a fixed abrasive surface, or vice versa. Among the Walandawe points, flat ground surfaces are more common in smaller diameter examples, whereas rounded ground surfaces are more common in larger diameter examples.

Where grinding is the only observed technique used in point manufacture, its direction relative to the long axis is variably longitudinal, oblique or transverse (Table 13.4). In a sample of eight mid-shaft fragments, half of them show an altered direction of grinding during manufacture; with transverse and oblique grinding on five examples apiece and longitudinal grinding on two. In a sample of 17 detached tips, grinding was exclusively oblique in eight specimens, transverse in seven and longitudinal in one, plus one combination of transverse and oblique. Thus, ground tips appear to differ from ground mid-shaft fragments by the more common use of exclusively transverse grinding, and the less frequent changing of direction.

Where grinding appears in combination with one of the other methods, it was invariably used as a 'finishing' method. There is no example of a ground point subsequently shaped by scraping or cutting, while the reverse is commonplace. The honing of points to a fine tip through grinding of multiple facets is a hallmark of the Walandawe collection.

No evidence was observed of cutting (nearly vertical incision into a flat surface, including the 'groove-and-snap' technique) as a method of creating pre-forms or 'blanks' from within larger pieces of bones.

Approximately 50 % of the Walandawe sample of points and point fragments with determinate manufacturing show evidence of a single method. This is most often scraping (35 %), followed by grinding (14 %), with exclusive use of shaving being a rare occurrence (1 %). Use of two or more techniques on a single point is commonly observed, with scraping followed by grinding (43 %) being the most commonly observed sequence. Scraping followed by shaving alone or followed by shaving and grinding each account for 3 % of points, and shaving followed by grinding for 1 % of points.

There was a replacement of scraping with grinding, over time, as applied as a single method for point manufacture. In the oldest sample (Phase E), almost half of the points were made by scraping only and none by grinding only. With the sample of points from Phases C–D, around one-third were made by scraping only and about one-tenth by grinding only, and in the youngest samples (Phases A–B), the proportion made solely by scraping declined to about one-tenth while the proportion made solely by grinding rose to one-half. Grinding was employed in point manufacture throughout the Holocene but predominantly as a finishing technique in the early days, as shown by the 39 % of Phase E points, and over 50 % of Phases C–D points, produced by scraping followed by grinding.

The question arises as to whether chronological changes in technique could be explained by changes in morphotype and/or utilized material over time. These latter changes do not appear to provide an explanation, based on calculations of the 'expected' number of bipoints, unipoints and tip/shaft fragments made by the various permutations of technique and material for the Walandawe collection as a whole. This is because there are no statistically significant differences between the observed and expected numbers for any of these permutations. The only statistically significant differences (all involving unipoints made from cortex slivers) reflect the higher incidence of scraping and lower incidence of grinding in Phase E compared with later phases. Accordingly, the manufacture of the Walandawe points appears to have involved a chronological shift from scraping to grinding that was independent of the contemporary changes in morphotype and utilized material.

The manufacturing process shows systematic variation in relation to the artifacts' cross-sectional area (CSA: maximum width x thickness). The most delicate points (CSA < 10 mm^2) show similar numbers of exclusive scraping, exclusive grinding, and scraping combined with grinding, with rare recourse to shaving. In all larger size classes, grinding is rarely ever the only method used, while exclusive scraping becomes more common up to the CSA 20–30 mm class. The small sample of points with CSA > 30 mm^2, compared with the less robust point classes, show a predominance of scraping combined with grinding, and a higher proportion of points with supplementary use of shaving.

The relatively high frequency of grinding of the most delicate points may reflect the risk of breakage using other methods. With the more robust classes of points, scraping on its own or in combination with other methods would have been less labor-intensive and adequately safe.

Use Traces

Use traces are visible on the tips of many points and, less frequently, on the shafts. The wear and damage categories recognized here follow Pasveer (2004; Pasveer and Bellwood 2004).

Polish or gloss is the most common kind of use-related wear (Fig. 13.5). Of 115 points that retain at least one intact tip, 62 were scored as having polish. Most specimens were scored as having 'light polish' in which abrasion is restricted to projections and ridges where contact would be expected during penetration of the point into a rigid worked material. On 14 points the tip is enveloped in a more general surface polish (scored as 'high polish') indicative of prolonged use. Polish of this kind is almost certainly caused by repeated penetration of a finely abrasive material such as a phytolith-rich plant material.

Many points with polished tips also show use-related striations around the tip. These are produced either by contact with harder particles within the worked medium or with sand or silt particles introduced during penetration. Visible striations were broadly categorized as fine (silt-sized; Fig. 13.5a, c) or coarse (sand-sized; Fig. 13.5a), and the direction scored as predominantly transverse or oblique to the long axis of the point. Oblique striations are indicative of a twisting action during active penetration, while transverse

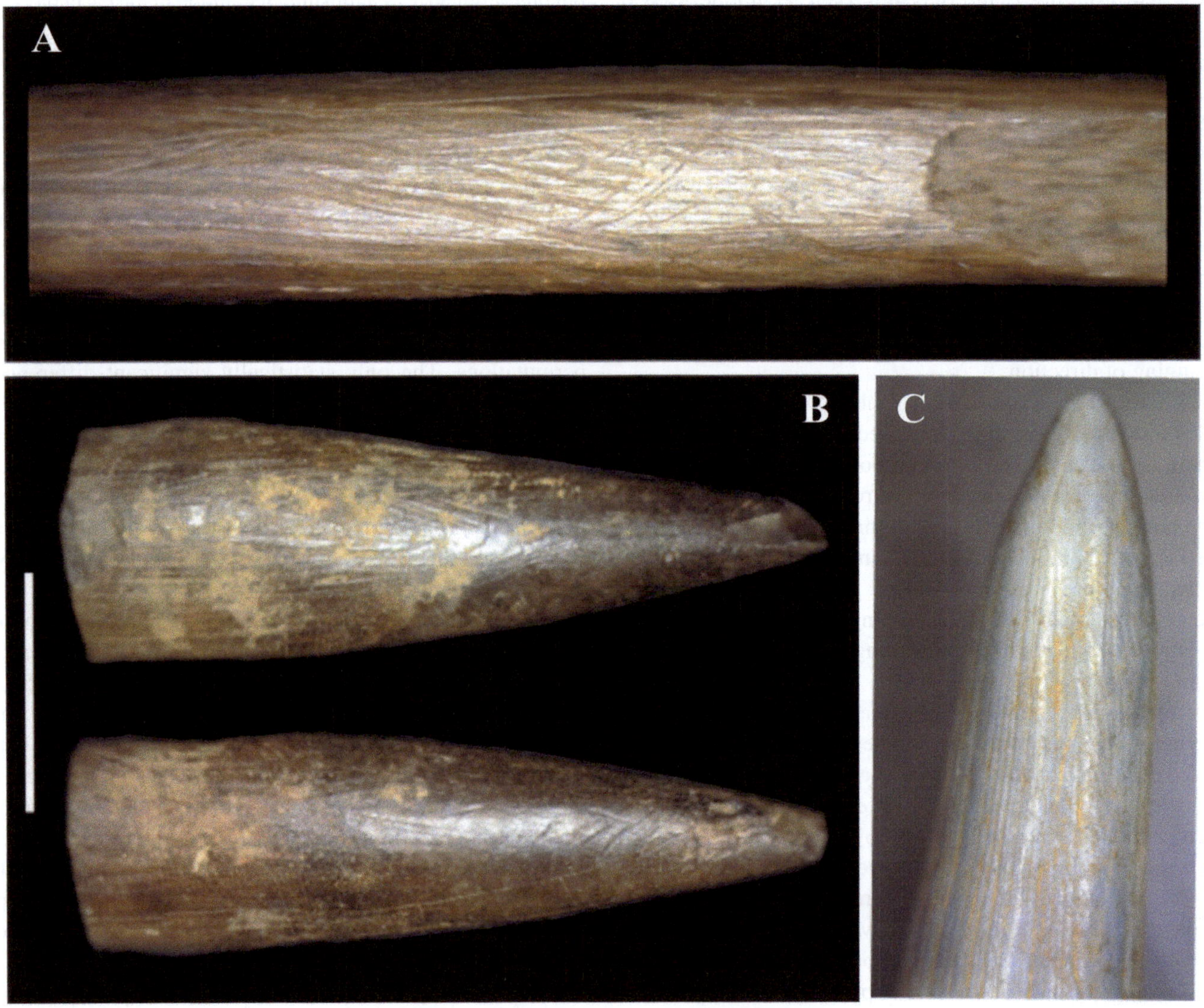

Fig. 13.5 (**a**) Hafting cordage marks on the central region of a Walandawe bipoint. (**b**) Walandawe dentin point showing use-related polish and both fine and coarse oblique striations, as well as step fracture spalling on both sides of the tip. (**c**) Walandawe cortex bone point showing combination of polish and fine oblique striations

striations are indicative of rotation without much penetration. Longitudinal striations resulting from a pushing action without rotation were rarely observed and never dominant.

In a smaller number of points the polish extends back from the tip onto the mid-shaft. Light polish is present on the mid-shaft in 32 out of 87 points that preserve the shaft segment, and it forms a heavy gloss on five points. Polish on the mid-shaft of points most likely indicates a function in which points were either pulled through, or inserted deep into, a fibrous and abrasive material.

Use-related damage is more difficult to distinguish from post-depositional damage. Step-fracturing is probably the best indicator of use, as it represents failure under compressive forces. Crushing and transverse step-fracturing are less diagnostic; either might occur during use but they might also occur after discard. Snap fracturing of the tip may also occur during manufacturing.

Use Wear

Polish and associated striations are present on all major categories of points, and on all raw material types, though at varying frequencies.

The unipoint ends show the highest proportion of observations involving use-related wear (76 %, compared with 52 % for bipoint primary ends, 25 % for bipoint secondary ends, 63 % for tips and 58 % for shaft sections). Chi-square tests show the difference is significant comparing unipoint ends with bipoint primary ends (4.24, 1 *d.f.*, $p<0.05$), bipoint secondary ends (5.10, 1 *d.f.*, $p<0.025$) and shaft sections (10.46, 1 *d.f.*, $p<0.005$). While the difference between unipoint ends and point tips is not significant (Chi-square=2.29, 1 *d.f.*, $p>0.1$), broken-off unipoint ends probably dominate the tip class. Accordingly, this last result is consistent with the inference of a systematically higher rate of use wear on unipoint ends than bipoint ends.

The data also suggest that bipoint secondary ends have the lowest rate of use-related wear, but the difference is not statistically significant with respect to bipoint primary ends, tips or shaft sections.

Use-related wear most often takes the form of light polish for all end/tip classes and for shaft sections, followed by fine striations on bipoint primary ends, and by coarse striations and high polish on unipoint ends and shaft sections. No significant differences could be found between the end/tip/shaft classes in terms of their form of use-related wear.

With respect to the utilized material, the small sample of shaft-bone ends (all from unipoints) show the highest rate of use-related wear (100 %) and the highest rate of coarse striations as a form of use wear (33 %). Dentin tips and ends are similar to shaft-bone ends in these regards, with 82 % of observations involving use-related wear and 16 % of observations involving coarse striations. Tips and ends made from cortex slivers show the lowest rate of observations of use-related wear (57 %) and the difference is significant when compared with shaft-bone ends (Fisher exact test=0.038) and with dentin ends/tips (Chi-square=5.87, 1 *d.f.*, $p<0.025$). The available data also suggest that use wear on cortex-bone tips and ends more often takes the form of light polish than with dentin and shaft-bone tips and ends, although the differences are not significant. Dentin has a higher compressive strength (ca. 275–300 MPA; Craig and Peyton 1958) than bone (ca. 170 MPA; Schmidt-Nielsen 1984:6), and the greater durability of points made from dentin rather than cortex slivers may explain the former's higher use-wear rates.

Use Damage

Approximately three-quarters of the points show some form of damage, observed at a lower rate on tips (58 % of 38 cases) than the extant point ends (79 % of 89 cases), a difference that is statistically significant (Chi-square=4.75, 1 *d.f.*, $p<0.05$). The explanation here may be that the damaged tips are more difficult to recognize as (fragmentary) points than points with intact damaged ends.

The most commonly observed form of damage was crushing, observed on 39 % of cases (including ends/tips with step damage as well as crushing), and between 29 and 50 % of every morphotype class and material class. Snaps were observed on 21 % of cases (16–29 % of every morphotype class and material class) and step damage on 17 % of cases (including ends/tips with crushing as well as steps). Considering just the ends/tips with damage, we could find no statistically significant differences between ends and tips, between morphotypes or between materials in the form that the damage took.

Hafting or Cordage Marks

Mid-shaft regions generally do not show striations of the kind seen around the tips, even where polish is present. However, in ten examples, the mid-shafts bear small clusters of oblique to transverse striations such as might be left by hafting cordage (Fig. 13.5c). Cordage striations are most likely to form when grains of sand or silt get caught between the cordage material and the bone surface. The frequency of such marks in an assemblage represents a minimum estimate of the extent of hafting within an osseous industry.

Eight of the 11 points with possible cordage marks are bipoints, including one made on dentin and the rest on cortex slivers. These are spread through the chronological sequence, with one example in each of Phases B, C and D, and five in Phase E contexts. Three specimens classified as unipoints, one from Phase D and two from Phase E contexts, all made on cortex slivers, also show possible hafting cordage marks.

Regional Occurrence of Major Osseous Artifact Types

Each of the major osseous artifact morphotypes found in the Walandawe sites—spatulas, suid tusk tools, edge tools, and points—is present regionally in a variety of similarly aged contexts.

Spatulas

Spatulas have been reported from a number of Toalean sites in South Sulawesi, including Leang Ara, Panisi Ta'buttu, Batu Ejaya, Bola Batu (Van Heekeren 1972:110–114) and Ulu Leang 1 (Glover 1976:141). The examples from Ulu Leang I were subsequently discounted as products of water abrasion by Olsen and Glover (2004:287–288) who also urged caution in accepting other examples from this region. The recovery of several initial Holocene spatulas from Gua Talimbue confirms their presence in the archaeological record of Sulawesi but it does not diminish the need for re-examination of other claimed examples of this distinctive tool type.

The Gua Talimbue spatulas appear qualitatively different from Rabett's (2005) category of 'edge tools' which have sharpened, chisel-like working edges. The Talimbue spatulas instead have working edges that appear to be rounded through repeated use, and show no evidence of resharpening as might be expected if a sharp, chisel-like edge was desired. These working edges closely resemble those found on cassowary bone 'knives' still in common use across Melanesia to separate the drupes from the cone of Marita Pandanus (*Pandanus conoides*). These knives are manufactured from the lower leg bones of adult cassowaries and their use involves insertion of the pointed end of the tool between the edible endocarp and the pyrene or 'stone' of the Pandanus cone, followed by wedging away of the endocarp (Sillitoe 2008:61–64). The working edge of these tools is rounded rather than sharp or beveled and commonly shows a high polish from prolonged use. The levering action involved in their use occasionally results in transverse fracturing or spalling away of the tip; if the damage is not too extreme, the point can be shaped to allow continued use of the tool (Aplin, pers. observ.).

According to Sillitoe's (2008) observations on the Wola people of Papua New Guinea, most adult men possess a cassowary bone knife, but they are highly durable items, manufactured only infrequently. Sillitoe (2008:61) documented 12 examples ranging in age from 2.5 to 30 years, with an average age of 14.5 years. Other than when they break, these highly valued and curated artifacts are unlikely to enter the archaeological record. Accordingly, they will be grossly underrepresented in an osseous assemblage in comparison to their importance. The intensity of wear observed on the Talimbue spatulas, coupled with the fact that all recovered examples appear to be broken fragments, is highly suggestive of a similar explanation for their rarity in the Talimbue assemblage.

Spatulas are also reported from the North Moluccas (Pasveer and Bellwood 2004) and Aru Islands (Pasveer 2005) in eastern Indonesia, and various sites in Borneo, Java and Peninsular Malaysia to the west (Olsen and Glover 2004).

Tusk Tools

Rabett (2004, 2005) has described ethnographic examples of pig canines from New Guinea hafted to a handle and used as cutting or digging tools. Aplin has examined similar tools still in use as digging implements in Papua New Guinea. However, these show a general rounding and polish of all sharp edges and do not reproduce the combination of strongly directional striations and high gloss that occurs on the babirusa tusk tool from Gua Mo'o hono.

Pig tusks with variable degrees of modification were reported from Niah Cave by Harrisson and Medway (1962; see also Rabett 2004) who distinguished three main kinds of tusk tools. The 'knives' were made from canines split longitudinally and with the exposed dentin ground away at an angle to the enamel to create a sharp edge. The 'chisels' and 'points' both show modification to the tip of unsplit canines—transverse grinding of the tip to create a strong gouging edge for the chisels, and grinding of the tip to a sharp apex for the points. Notwithstanding certain potential similarities of the Niah knives and chisels to the Gua Mo'o hono tusk tool, in terms of use wear, the systematic use of this tool for cutting silica-rich plant matter (noted above) may reflect a function unmatched in the Niah Cave assemblage.

Edge Tools

According to Rabett's (2005) review, 'edge tools' are recorded from four main areas: Java, Borneo, the Malay Peninsula, and Indochina (Thailand and Vietnam). In some cases (e.g., Java—Morwood et al. 2008:Fig. 7; Kusno 2006) it is unclear whether these are chisel-like tools or spatulas of the kind described above. However, from published descriptions and illustrations it is clear that true edge tools occur at Niah Cave on Borneo (Harrison and Medway 1962), at Perlis and Perak sites in Peninsular Malaysia (Collings

1937:Fig. 4; Van Stein Callenfels and Noone 1940), and at Khok Phanom Di in south-central Thailand (Rabett 2005:Fig. 4). Few of these tools are well dated but edge tools from Khok Phanom Di would date to 3.5–4.0 cal kBP (Higham and Thosarat 1998). Whether or not all of these tools were functionally equivalent is unknown.

Points

Points are the most commonly reported osseous artifacts throughout the wider Asia-Pacific region (Rabett 2005). Here, most archaeological sites of terminal Pleistocene to Holocene age that have been studied in detail contain one or more bone fragments with signs of use of a naturally sharp point or modification to produce a sharp point. These are usually treated as expedient tools of limited significance and they are either given brief mention in site reports (e.g., Glover 1986:190; David et al. 2010:46) or else go unremarked. In a smaller number of sites, these artifacts occur in sufficient numbers to attract some degree of characterization, though rarely accompanied by quantitative description.

Rabett (2005) identifies a number of geographic areas and time periods where bone points were manufactured in moderate to large numbers. Included are the early to mid-Holocene Toalean sites of South Sulawesi, the early Holocene sites of Paso in northeast Sulawesi (Bellwood 1976) and Musang Cave in the Philippines (Thiel 1990), the terminal Pleistocene to mid-Holocene in northwest Borneo (Harrisson and Medway 1962; Rabett 2005; Barton et al. 2009), the early to late Holocene in the North Moluccan sites of Golo, Daeo 2, Siti Nafisah and Uattamdi (Pasveer and Bellwood 2004), the terminal Pleistocene to late Holocene of the Aru Islands (Pasveer 2005), the mid-Holocene in the Bird's Head Peninsula of West Papua (Pasveer 2004), the terminal Pleistocene to late Holocene of East Java (Van Heekeren 1972; Prasetyo 2002; Morwood et al. 2008), the early to mid-Holocene on the Thai-Malay Peninsula (Collings 1937; Van Stein Callenfels and Noone 1940; Tweedie 1953; Anderson 1988, Higham and Thosarat 1998), and the terminal Pleistocene to late Holocene of the Red River Delta of Vietnam (Viet 2005). Only a few of these assemblages have been described in sufficient detail to support detailed comparisons with the Walandawe points.

Comparison with Other Sulawesi Point Assemblages

Olsen and Glover (2004:287–288) classified the osseous points from Ulu Leang 1, Leang Burung 1 and other South Sulawesi sites into two morphotypes—'bipoints' and 'awls'. Apart from providing some measurement ranges for bipoints (length: 25–67 mm; width: 2–7 mm) and some graphs of tip dimensions, no metrical data are provided. The size range for bipoints are similar to Walandawe (length: 24.1–53.7 mm; width: 2.8–9.1 mm.

Olsen and Glover (2004) describe both classes of points as manufactured on fortuitous splinters of bone. However, several of the illustrated bipoints (Olsen and Glover 2004:Figs. 18b and 8e) and one awl (Fig. 20e) appear to be hollow and these were more likely made on tooth roots or segments of unsplit long bone. Manufacturing of both bipoints and awls is described as primarily by longitudinal scraping. Light diagonal or transverse grinding is also mentioned but it is evidently uncommon, drawing the remark that "only in one case was it particularly extensive" (Olsen and Glover 2004:290). No evidence for hafting was observed; however, the fact that some bipoints display crushing and longitudinal flaking to the tip was seen to support a function as hafted projectiles (Olsen and Glover 2004:295).

Further grounds for identifying the bipoints as projectile tips, either in arrows or darts used in blowpipes, involve the presence of 'incised lines' that may have held poison. These were observed in "at least six of the bipoints from Ulu Leang 1" (Olsen and Glover 2004:294) and one extraordinary specimen has a regular mesh of incised lines cut into the point from the tip and covering about 70 % of the shaft (Olsen and Glover 2004:Fig. 19). There is no evidence of complex incision like this in the Walandawe collection. However, some of the coarse scratches observed around the tip of Walandawe points, interpreted here as a component of use wear, may be comparable to the simpler examples of 'incised lines' reported by Olsen and Glover.

Points classified as awls made up a small proportion of the assemblage at Ulu Leang 1 and Leang Burung 1. This class is said to be highly variable in form and made on different elements. The only modification these points show is longitudinal scraping of the edges and tip, with polish described as poorly developed. Tips are fine and rounded in cross-section and some have a shoulder which is tentatively interpreted as resulting from use with a piercing action (Olsen and Glover 2004:291–292, Fig. 20).

Bone points are regarded as one of the characteristic artifacts of the Toalean culture of South Sulawesi, including 'classic' Toalean sites with their backed microliths and hollow-based stone arrowheads in the southwest of the peninsula (Bulbeck 2006). Bulbeck (2004) hypothesized that there was a possible relationship between the classic Toaleans' typologically diverse artifacts, intensive Celebes boar culling and strongly monsoonal climate. If this is correct, the Walandawe sites would bear a closer relationship to South Sulawesi's non-classic Toalean sites on the basis of a less seasonal climate, a lack of typologically specialized stone projectile components, and (to judge by Gua Mo'o hono) a faunal assemblage that is not dominated by the Celebes boar.

Comparison with Niah Cave Points

The earlier excavations at Niah Cave produced most of its point assemblage, which thus lacks precise temporal control. Harrisson and Medway (1962) distinguished 'awls' from 'simple points'. Awls were defined as "Medium-sized bone shafts worked to a sharp point. The angle between the worked face and the long axis of the bones is small" (Harrisson and Medway 1962:352). Five measured examples had point lengths of 45–69 mm, long by the standards of any of the Walandawe points, and shaft widths of 12–13 mm, which fall beyond the Walandawe range. Simple points were defined as "All either small shaft slivers split from large bone shafts or small bone shafts sharpened without splitting" (Harrisson and Medway 1962:352).

Within the simple point category, 'flat points' were further described (Harrisson and Medway 1962:352–353) as:

> Slivers from the long-bones of medium-sized animals ... worked to a point at one end. The edges are smoothed, either parallel or convex, converging slightly toward the base as well as the tip. The base is in most cases an unfinished fracture, but in two cases has also been smoothed. Working of the bone was achieved by grinding; the finish is rough and marked by conspicuous parallel scratches from the fabricating surface...

These points are comparable in general form to the Walandawe attenuate unipoints of cortex bone. However, a metrical comparison reveals minimal overlap in shaft dimensions, as the Niah flat points are wider than the Walandawe attenuate unipoints of similar shaft thickness. Barton et al. (2009) make a case for at least some of the Niah Cave points being hafted projectile armature, most likely used for hunting monkeys. Flattening of these Niah Cave points may be a purposeful feature aimed at more efficient hafting.

A second category of simple points, 'rounded points', were formed from "Whole small shaft bones, round in cross-section, either worked to a taper or exploiting the natural pointed shape (e.g., tortoise rib, fish spine)" (Harrisson and Medway 1962:353). The lengths for a sample of 26 flat points ranged from 36 to 56 mm, with shaft widths of 5.0–8.5 mm and shaft thicknesses of 1.2–3.2 mm. These points tend to be longer than the Walandawe unipoints although the shaft dimensions are similar.

Harrisson and Medway (1962) did not report any bipoints from Niah Cave. However, Barton et al. (2009) describe a single bipoint of early Holocene age measuring 100.3 mm in length—much longer than any recorded from Sulawesi—and 4.1 mm in shaft width. The Niah Cave bipoint has similar dimensions to stingray spines recovered from the same stratigraphic levels, and may represent an attempt to replicate a stingray spine.

Comparison with Bird's Head Points

The Bird's Head assemblage of 92 points, from Kria Cave, dates to ca. 7.0–4.3 cal kBP (Pasveer 2004). The assemblage comprises bipoints and unipoints, the former made almost entirely on cortex bone slivers, and the latter almost entirely from unsplit sections of long-bone shaft. The method of manufacturing of both categories resembles that of the Walandawe points insofar as scraping was almost always employed to form the shaft, sometimes accompanied by shaving, with grinding often used to finish the tips. However, whereas the Walandawe points typically have transverse to oblique grinding of the shaft section, this was not observed in the Kria Cave points.

The kinds of use-related wear and damage in the Bird's Head and Walandawe point assemblages are similar. However, the Bird's Head bipoints show a higher incidence of primary end polish than the unipoints, the opposite of the Walandawe points. The Bird's Head bipoints were very likely hafted onto the end of handles and used as engravers or perforators of wood or tough fibrous materials, while the unipoints were probably hand held and used in perforating, penetrating or separating fibrous material (Pasveer 2004:175–176).

The Walandawe bipoints are on average larger for every dimension, and also more variable (as measured by the coefficient of variation) except for tip length, where any difference is minimal. Both samples are manufactured from cortex bone slivers and the source bones are from similarly sized mammals—mainly suids in the Walandawe case and a forest wallaby (*Dorcopsis muelleri*) in the Kria Cave case. The size contrasts probably reflect the postulated use of the Walandawe bipoints predominantly as projectile points, compared with the engraving and perforating use of the Kria Cave bipoints, rather than any structural differences in raw material.

The Walandawe unipoints are on average shorter for their point length and tip length but have slightly wider and thicker shafts. Again the Walandawe unipoints are more variable except for tip length. These differences most likely reflect the greater difficulty of making unipoints from cortex bone slivers, as characterizes the majority of the Walandawe sample, compared with unipoint manufacture from unsplit, naturally elongated long-bone shafts, which is a feature of the Kria Cave sample.

Comparison with North Moluccan Points

The largest sample of points from the North Moluccan islands, from Golo Cave (N = 108), dates to two broad time periods—around 7.4 cal kBP and post-ceramic, after ca.

3.2 cal kBP. The points include bipoints and unipoints with sharply pointed tips, and spatula-like tools with broad, bluntly rounded tips. The bipoints are made predominantly on cortex bone slivers (93 %), with a few examples made from narrow, unsplit long-bone shafts. The unipoints are made on a more even mix of narrow, unsplit long-bone shafts (59 %) and cortex bone slivers (41 %). The spatula-like artifacts are made from specific skeletal elements—wallaby fibulae and fruit bat humeri (Pasveer and Bellwood 2004).

All three categories of Golo points resemble the Kria Cave points in their method of manufacture. Scraping was almost always employed to form the shafts, sometimes after initial shaving, with localized grinding often used to finish the tips. Grinding of the shafts was not observed.

The pattern of use-related wear and damage in the North Moluccan point assemblages is very similar to that recorded for Kria Cave and led to the same inferences: the bipoints were mainly hafted and used as borers or engravers, and the attenuate unipoints were probably hand held and used as perforators (Pasveer and Bellwood 2004). This is confirmed by the metrical similarity of the Golo and Kria Cave bipoints and unipoints, both in terms of average dimensions, and a similarly low coefficient of variation on shaft dimensions and bipoint length.

Comparison with Aru Islands Points

Two cave sites in the Aru Islands—Liang Lemdubu and Liang Nabulei Lisa—produced a small assemblage of 47 points ranging in age from ca. 25 ka to the late Holocene (Pasveer 2005). The largest sample (N = 34) comes from the upper layers of Liang Lemdubu and probably dates to within the last 2 ka.

The point collection includes one bipoint of 43 mm length, four unipoints of 28–38 mm length, and tip and shaft fragments (Pasveer 2005:Table 11.9). The most commonly observed manufacturing method is grinding, especially in the Liang Lemdubu assemblage (72 %), but many show evidence of prior scraping and a few of shaving. Unsplit long-bone shafts account for 70 % of the sample of points from Liang Lemdubu and 80 % of those from Liang Nabulei Lisa, with the remainder produced from cortex bone slivers.

Use-related wear and damage suggest mainly hand-held use with some artifacts used for drilling or engraving work that produced localized tip polish and fractures, and others in more perforative tasks that produce more general polish along the tip and shaft.

The Aru Island points tend to have small shaft dimensions compared with the Walandawe points, and show minimal overlap with the Niah Cave flat points.

Discussion and Conclusions

The Walandawe sites broaden our knowledge of the osseous artifacts of prehistoric Sulawesi in several ways. First, they document the manufacture of bipoints and other osseous artifacts in Sulawesi by 9.5 cal kBP. Second, they demonstrate that carefully manufactured bipoints, long regarded as one of the hallmark artifacts of the Toalean culture of South Sulawesi (Olsen and Glover 2004), were in use in the Walandawe area without any of the classic Toalean cultural markers such as backed microliths and hollow-based stone points. Third, they broaden the range of osseous artifact morphotypes known with certainty from Sulawesi, including unipoints and bipoints made from incisor roots of Suidae, spatulas, tusk tools, and possible edge-tools.

As reviewed above, each of the major osseous artifact morphotypes found in Sulawesi is recorded elsewhere in the terminal Pleistocene to Holocene record of ISEA. Unipoints and to a lesser degree bipoints can occur in large numbers, whereas the other morphotypes are represented by only a modest number of specimens. However, as argued earlier on the basis of the ethnographic example of cassowary bone knives from New Guinea, the more specialized kinds of artifacts were highly curated and more likely to be repaired than discarded upon breakage. Items of this kind are usually poorly represented in the archaeological record, even if they were in everyday use. But curation and sampling issues are unlikely to explain certain shared absences across most or all of ISEA. These include the lack of evidence for use of the groove-and-snap technique, in contrast with Thailand (Anderson 1990) and elsewhere in Eurasia (Pétillon and Ducasse 2012), and the absence of decorative items, which can feature prominently in Eurasian 'Upper Paleolithic' assemblages (e.g., Derevianko et al. 2005; Lbova 2010). Similarly, harpoon-like morphotypes are reported from Indochina, Thailand and Java (Olsen and Glover 2004) but no further east. Thus, ISEA appears to have harbored a distinctive, regional complex of osseous artifact production following the Late Glacial Maximum.

Rabett and Piper (2012) identify a shift in the Southeast Asian archaeological around 15 ka in the pattern of use of osseous materials: bone came into more frequent use as a raw material; and the forms tended to become more standardized and, for the first time, were clearly identifiable as hafted components of composite tools. They further observe that these developments appear to have occurred earlier in ISEA than on the mainland, and possibly in the eastern, oceanic islands before the continental Sundaic islands. They posit that the increased prominence of osseous technologies was one expression of a more general surge of inventive behavior "demanded of foraging communities as they adapted to the far-reaching environmental and demographic changes that were reshaping this region at that time" (Rabett and Piper 2012:37), including the emergence of more specialized hunting activities. Such a scenario would appear highly relevant for the Walandawe assemblages, at least for the 9500 years for which we have a robust faunal record.

The recent description of a complex hafted artefact from a 34 ka context on East Timor (O'Connor et al. 2014) does not refute the general notion of a terminal Pleistocene flourish in osseous artifact production and usage. However, it does suggest that the earliest modern humans to enter ISEA were fully equipped, both conceptually and technically, to manufacture complex osseous artifacts and to combine them into composite tools using a variety of hafting technologies. No doubt, the same basic production methods were applied to other media, whether in the production of wooden, stone or shell artifacts (Szabó et al. 2007), or the construction of housing and watercraft (Balme et al. 2009; Balme and O'Connor 2014), thereby ensuring continuity of cognitive and physical skill irrespective of whether any particular method was applied specifically to osseous raw material.

Within ISEA, osseous points are common and widespread enough to investigate the possibility of a multiplicity of traditions. The Walandawe and Maros assemblages represent similar traditions on the basis of sharing a relatively high frequency of bipoints, the preferential manufacture of both unipoints and bipoints on cortical bone slivers, and the application of scraping as the primary shaping technique. However, the available documentation also indicates certain specializations at Walandawe, including the more frequent and intensive use of grinding as a method of shaping shaft and tip, the relatively high incidence of cordage marks on bipoints, and the long-term use of suid tooth roots for point manufacture. A second identifiable tradition includes the bone points from the Bird's Head of New Guinea and from the North Moluccan islands. As reviewed above, these are similar to each other—and distinct from Walandawe—in the selection of raw material, manufacturing methods employed, patterns of use-related wear and damage, and metric attributes. The point assemblages from the Niah Caves on Borneo and, less emphatically, the Aru Islands are also distinctive. While they are similar in their emphasis on grinding rather than scraping as a shaping mechanism and their scarcity of bipoints, the Niah 'flat points' stand out morphologically and metrically from other ISEA unipoint assemblages.

The characteristics that distinguish the traditions described above in each case persisted over millennia. They reflect practices transmitted through time by active teaching and/or passive observational learning between generations, and through space by the sharing of information between neighboring populations (Shennan 2001, 2002). The existence of these separate traditions documents cultural continuity in each of the areas for which we have detailed characterizations. However, as emphasized by Gosden's (1994) concept of 'dynamic traditionalism', cultural continuity does not abnegate change but sets the course for its development. This is exemplified by the Walandawe points with their suite of temporal changes that apparently reflect a shift from a predominant use as projectile points to an increasing role in penetrating fibrous materials. The apparent phasing out of spatulas after the early Holocene, and the appearance of expedient tools during the late Holocene, also reflect change within a single tradition.

Within the ISEA context, Walandawe is currently unique in revealing statistically significant transformation through time in an osseous assemblage. This is most evident for the points with their declining presence of tip damage and increasing signs of use wear to the tips and shafts, decreased bipoint production with a growing focus on unipoints, a transition in the manufacturing focus from scraping cortex slivers to grinding dentin and long-bone shafts as well as cortex slivers, and a reduction in tip length as well as unipoint dimensions generally. This is an exciting development in osseous artifact studies in ISEA, which we shall further explore and contextualize with integrated analysis of the other remains excavated from the Walandawe sites.

Acknowledgments The Walandawe excavations were undertaken as part of "The archaeology of Sulawesi: a strategic island for understanding modern human colonization and interactions across our region" project funded by the Australian Research Council (DP110101357). The Indonesia's government permission for the research was granted by the Bureau of Research and Technology, and facilitated by Bambang Sulistyanto (Director) and Bagyo Prasetyo (Head of the Department of Prehistory) at the National Center for Archaeological Research and Development, Jakarta, and by Made Sudarmika, Director of the Office for Archaeology at Makassar. The authors also extend their thanks to the Indonesian and Australian archaeologists who assisted the excavations.

References

Allen, H., Langley, M. C., & Taçon, P. S. C. (2016). Bone projectile points in prehistoric Australia: Evidence from implements, ethnography, and rock art. In M. C. Langley (Ed.), *Osseous projectile weaponry: Towards an understanding of Pleistocene cultural variability* (pp. 209–218). Dordrecht: Springer.

Ambrose, S. H. (2001). Paleolithic technology and human evolution. *Science, 291*, 1748–1753.

Anderson, D. D. (1988). Excavations of a Pleistocene rockshelter in Krabi and the prehistory of southern Thailand. In P. Charoenwongsa & B. Bronsen (Eds.), *Prehistoric studies: The Stone and Metal Ages in Thailand* (pp. 43–59). Bangkok: Thai Antiquity Working Group.

Anderson, D. D. (1990). *Lang Rongrien Rockshelter: A Pleistocene–early Holocene archaeological site from Krabi, southwestern Thailand* (University Museum Monograph 71). Philadelphia: University of Pennsylvania.

Anderson, D. D. (1997). Cave archaeology in Southeast Asia. *Geoarchaeology: An International Journal, 12*, 607–638.

Backwell, L. R., & d'Errico, F. (2001). Termite gathering by Swartkrans early hominids. *Proceedings of the National Academy of Science, 98*, 1358–1363.

Balme, J. (2013). Of boats and string: The maritime colonisation of Australia. *Quaternary International, 285*, 68–75.

Balme, J., & O'Connor, S. (2014). Early modern humans in Island Southeast Asia and Sahul: Adaptive and creative societies with simple lithic industries. In R. Dennell & M. Porr (Eds.), *Southern Asia, Australia and the search for human origins* (pp. 164–174). Cambridge: Cambridge University Press.

Balme, J., Davidson, I., McDonald, J., Stern, N., & Veth, P. (2009). Symbolic behaviour and the peopling of the southern arc route to Australia. *Quaternary International, 202*, 59–68.

Barton, H., Piper, P. J., Rabett, R., & Reeds, I. (2009). Composite hunting technologies from the terminal Pleistocene and early Holocene, Niah Cave, Borneo. *Journal of Archaeological Science, 36*, 1708–1714.

Bellwood, P. (1976). Archaeological research in Minahasa and the Talaud Islands, northeastern Indonesia. *Asian Perspectives, 19*, 240–288.

Bellwood, P. (1997). *Prehistory of the Indo-Malaysian Archipelago* (revised edition). Honolulu: University of Hawai'i Press.

Bentley, R. A., & Shennan, S. J. (2003). Cultural transmission and stochastic network growth. *American Antiquity, 68*, 459–485.

Brumm, A., Jensen, G. M., van den Bergh, G. D., Morwood, M. J., Kurniawan, I., Aziz, F., et al. (2010). Hominins on Flores, Indonesia, by one million years ago. *Nature, 464*, 748–752.

Brumm, A., & Moore, M. W. (2005). Symbolic revolutions and the Australian archaeological record. *Cambridge Archaeological Journal, 15*, 157–175.

Bulbeck, D. (2004). Divided in space, united in time: The Holocene prehistory of South Sulawesi. In S. G. Keates & J. M. Pasveer (Eds.), *Quaternary research in Indonesia (Modern Quaternary Research in Southeast Asia 18)* (pp. 129–166). Leiden: A.A. Balkema.

Bulbeck, D. (2006). Economic and technological change during the middle and late Holocene in the Lamoncong highlands, South Sulawesi, Indonesia. In A. Bacus, I. C. Glover, & V.C. Pigott (Eds.), *Uncovering Southeast Asia's past. Selected papers from the 10th International Conference of the European Association of Southeast Asian Archaeologists* (pp. 393–410). Singapore: National University of Singapore Press.

Cassoli, P. F., & Tagliacozzo, A. (1994). Considerazioni paleontologiche, paleoecologiche e archeozoologiche sui macromammiferi e gli uccelli dei livelli del Pleistoceno Superiore del Riparo di Fumane (VR) Scavi 1988–1991. *Bolletino del Museo Civico di Storia Naturale di Verona, 18*, 349–445.

Collard, M., Shennan, S. J., & Tehrani, J. J. (2006). Branching, blending, and the evolution of cultural similarities and differences among human populations. *Evolution and Human Behavior, 27*, 169–184.

Collings, H. D. (1937). An excavation at Bukit Chuping, Perlis. *Bulletin Raffles Museum (Series B), 1*, 94–119.

Craig, R. G., & Peyton, F. A. (1958). Elastic and mechanical properties of human dentin. *Journal of Dental Research, 37*(4), 710–718.

David, B., Geneste, J.-M., Aplin, K., Delannoy, J.-J., Araho, N., Clarkson, C., et al. (2010). The Emo site (OAC), Gulf Province, Papua New Guinea: Resolving long-standing questions of antiquity and implications for the history of the ancestral Hiri maritime trade. *Australian Archaeology, 70*, 39–54.

Dennell, R., & Roebroeks, W. (2005). An Asian perspective on early human dispersal from Africa. *Nature, 438*, 1099–1104.

Derevianko, A. P., Postnov, A. V., Rybin, E. P., Kuzmin, Y. V., & Keates, S. G. (2005). The Pleistocene peopling of Siberia: A review of environmental and behavioural aspects. *Indo-Pacific Prehistoric Association Bulletin, 25*, 57–68.

Dortch, C. E. (1979). 33,000 year old stone and bone artifacts from Devil's Lair, Western Australia. *Records of the Western Australian Museum, 7*, 329–367.

Glover, I. C. (1976). Ulu Leang Cave, Maros: A preliminary sequence of post-Pleistocene cultural development in South Sulawesi. *Archipel, 11*, 113–154.

Glover, I. C. (1979). The effects of sink action on archaeological deposits in caves: An Indonesian example. *World Archaeology, 10*, 302–317.

Glover, I. (1986). *Archaeology in Eastern Timor, 1966–67. Terra Australis 11*. Canberra: Australian National University.

Gosden, C. (1994). *Social being and time*. Oxford: Blackwell.

Harrisson, T., & Medway, L. (1962). A first classification of prehistoric bone and tooth artifacts (based on material from Niah Great Cave). *Sarawak Museum Journal, 10*(n.s. 19–20), 335–362.

Henshilwood, C. S., & Marean, C. W. (2003). The origin of modern human behaviour: Critique of the models and their test implications. *Current Anthropology, 44*, 627–651.

Higham, C., & Thosarat, R. (1998). *Prehistoric Thailand*. London: Thames & Hudson.

Higham, T. F., Barton, H., Turney, C. S., Barker, G., Ramsey, C. B., & Brock, F. (2009). Radiocarbon dating of charcoal from tropical sequences: Results from the Niah Great Cave, Sarawak, and their broader implications. *Journal of Quaternary Science, 24*, 189–197.

Klein, R. (2009). *The human career* (3rd ed.). Chicago: Chicago University Press.

Kusno, S. F. (2006). *L'industrie osseuse de l'horizon Keplek, Holocène de la grotte de Song Terus, Punung, Java Est (Indonésie)*. Unpublished M.A. Dissertation, Muséum National d'Histoire Naturelle.

Langley, M. C., Clarkson, C., & Ulm, S. (2011). From small holes to grand narratives: The impact of taphonomy and sample size on the modernity debate in Australia and New Guinea. *Journal of Human Evolution, 61*, 197–208.

Lbova, L. (2010). Evidence of modern human behavior in the Baikal Zone during the Early Upper Paleolithic Period. *Journal of Indo-Pacific Archaeology, 30*, 9–13.

McBrearty, S., & Brooks, A. S. (2000). The revolution that wasn't: A new interpretation of the origin of modern human behaviour. *Journal of Human Evolution, 39*, 453–563.

Morwood, M. J., & Jungers, W. L. (2009). Conclusions: Implications of the Liang Bua excavations for hominin evolution and biogeography. *Journal of Human Evolution, 57*, 640–648.

Morwood, M. J., Seojono, R. P., Roberts, R. G., Sutikna, T., Turney, C. S. M., Westaway, K. E., et al. (2004). Archaeology and age of a new hominin from Flores in eastern Indonesia. *Nature, 431*, 1087–1091.

Morwood, M. J., Sutikna, T., Saptomo, E. W., Westaway, K. E., Awe Due, R., Yuniawati, D. W., et al. (2008). Climate, people and faunal succession on Java, Indonesia: Evidence from Song Gupuh. *Journal of Archaeological Science, 35*(7), 1776–1789.

Mulvaney, D. J., & Soejono, R. P. (1971). Archaeology in Sulawesi, Indonesia. *Antiquity, 45*, 26–33.

Nguyen Viet. (2005). The Da But culture: Evidence for cultural development in Vietnam during the middle Holocene. *Bulletin of the Indo-Pacific Prehistory Association, 25*(3), 89–93.

O'Connor, S. (2007). New evidence from East Timor contributes to our understanding of earliest modern human colonisation east of the Sunda Shelf. *Antiquity, 81*, 523–535.

O'Connor, S., & Aplin, K. P. (2007). A matter of balance: An overview of Pleistocene occupation history and the impact of the Last Glacial Phase in East Timor and the Aru Islands, eastern Indonesia. *Archaeology in Oceania, 42*, 82–90.

O'Connor, S., & Chappell, J. (2003). Colonisation and coastal subsistence in Australia and Papua New Guinea: Different timing, different modes. In C. Sand (Ed.), *Pacific archaeology: Assessments and prospects* (pp. 17–32). Noumea: Service des Musées du Patrimoine.

O'Connor, S., Robertson, G., & Aplin, K. P. (2014). Are osseous artefacts a window to perishable material culture? Implications of an unusually complex bone tool from the Late Pleistocene of East Timor. *Journal of Human Evolution, 67*, 108–119.

Olsen, S. L., & Glover, I. C. (2004). The bone industry of Ulu Leang 1 and Leang Burung 1 rockshelters, Sulawesi, Indonesia, in its regional context. In S. G. Keates & J. M. Pasveer (Eds.), *Quaternary research in Indonesia (Modern Quaternary Research in Southeast Asia 18)* (pp. 273–299). Leiden: A.A. Balkema.

Pasveer, J. M. (2004). *The Djief Hunters: 26,000 years of rainforest exploitation on the Bird's Head of Papua, Indonesia (Modern Quaternary Research in Southeast Asia 17)*. Lisse: A.A. Balkema.

Pasveer, J. (2005). Bone artefacts from Liang Lemdubu and Liang Nabulei Lisa, Aru Islands. In S. O'Connor, M. Spriggs, & P. M. Veth (Eds.), *The archaeology of the Aru Islands, Eastern Indonesia. Terra Australis 22* (pp. 235–254). Canberra: The Australian National University.

Pasveer, J. M., & Bellwood, P. (2004). Prehistoric bone artefacts from the northern Moluccas, Indonesia. In S. G. Keates & J. M. Pasveer (Eds.), *Quaternary research in Indonesia (Modern Quaternary Research in Southeast Asia 18)* (pp. 301–359). Leiden: A.A. Balkema.

Pétillon, J. M., & Ducasse, S. (2012). From flakes to grooves: A technical shift in antlerworking during the Last Glacial Maximum in southwest France. *Journal of Human Evolution, 62*, 435–465.

Powell, A., Shennan, S., & Thomas, M. G. (2009). Late Pleistocene demography and the appearance of modern human behavior. *Science, 324*, 1298–1301.

Prasetyo, B. (2002). The bone industry. In H. T. Simanjuntak (Ed.), *Gunung Sewu in prehistoric times* (pp. 181–194). Yogyakarta: Gadjah Mada University Press.

Rabett, R. J. (2004). The ones that come ready made: The identification and use of *Sus* tusks as tools at prehistoric cave sites in Malaysia. *Archaeofauna, 13*, 131–142.

Rabett, R. J. (2005). The early exploitation of Southeast Asian mangroves: bone technology from caves and open sites. *Asian Perspectives, 44*, 154–179.

Rabett, R. J., & Piper, P. J. (2012). The emergence of bone technologies at the end of the Pleistocene in Southeast Asia: Regional and evolutionary implications. *Cambridge Archaeological Journal, 22*, 37–56.

Rabett, R. J., Piper, P. J., & Barker, G. (2006). Bones from Hell: Preliminary results of new work on the Harrisson faunal assemblage from the deepest part of Niah Cave, Sarawak. In A. Bacus, I. C. Glover, & V. C. Pigott (Eds.), *Uncovering Southeast Asia's past: Selected papers from the 10th International Conference of the European Association of Southeast Asian Archaeologists* (pp. 46–59). Singapore: National University of Singapore Press.

Rosell, J., Blasco, R., Campeny, G., Díez, J. C., Alcalde, R. A., Menéndez, L., et al. (2011). Bone as a technological raw material at the Gran Dolina site (Sierra de Atapuerca, Burgos, Spain). *Journal of Human Evolution, 61*, 125–131.

Schmidt-Nielsen, K. (1984). *Scaling: Why is animal size so important?* Cambridge: Cambridge University Press.

Shennan, S. (2001). Demography and cultural innovation: A model and its implications for the emergence of modern human culture. *Cambridge Archaeological Journal, 11*(1), 5–16.

Shennan, S. J. (2002). *Genes, memes and human history*. London: Thames and Hudson.

Sillitoe, P. (2008). *Made in Niugini: Technology in the Highlands of Papua New Guinea*. London: British Museum.

Szabó, K., Brumm, A., & Bellwood, P. (2007). Shell artefact production at 32,000–28,000 bp in Island Southeast Asia: Thinking across media? *Current Anthropology, 48*(5), 701–723.

Thiel, B. (1990). Excavations at Musang Cave, northeast Luzon, Philippines. *Asian Perspectives, 28*(1), 61–81.

Thieme, H. (1997). Lower Paleolithic hunting spears from Germany. *Nature, 385*, 769–771.

Tweedie, M. W. F. (1953). The Stone Age in Malaya. *Journal of the Malay Branch of the Royal Asiatic Society, 26*(2), 1–90.

Van Heekeren, H. R. (1972). *The Stone Age of Indonesia* (revised edition). The Hague: Martinus Nijhoff.

Van Stein Callenfels, P. V., & Noone, H. D. (1940). Report on an excavation in the rockshelter of Gol Ba'it, near Sungai Siput (Perak). In F. N. Chasen & M. W. F. Tweedie (Eds.), *Proceedings of the Third Congress of Prehistorians of the Far East* (pp. 119–130). Singapore: Government Printer.

Villa, P., & d'Errico, F. (2001). Bone and ivory points in the Lower and Middle Palaeolithic of Europe. *Journal of Human Evolution, 41*, 69–112.

Webb, C., & Allen, J. (1990). A functional analysis of Pleistocene bone tools from two sites in southwest Tasmania. *Archaeology in Oceania, 25*, 75–78.

Chapter 14
Bone Projectile Points in Prehistoric Australia: Evidence from Archaeologically Recovered Implements, Ethnography, and Rock Art

Harry Allen, Michelle C. Langley, and Paul S.C. Taçon

Abstract While osseous projectile points are frequently recovered and well understood in African and European contexts, those from Pleistocene Australia remain vaguely reported. This chapter outlines the current evidence for prehistoric osseous projectile technology on the Australian continent through the integration of data from archaeologically recovered implements, rock art, and ethnography. Organic implements are recovered only rarely from Pleistocene archaeological contexts in Australia, however, in ethnographic times, a wide range of both bone and wooden projectile technologies were used for hunting and defense. Spears played a significant part in Aboriginal economies, mythological traditions, and in the reproduction of gender roles. This chapter will show that while the evidence for osseous prehistoric projectile technology in Australia is less rich than in other regions of the world, owing to a variety of reasons including taphonomic processes and the ready availability of alternative materials, the Australian data nevertheless contributes to a greater understanding of Pleistocene technological choices as well as cultural variability during this period.

Keywords Kangaroo bone • Wooden projectile weaponry • Dynamic figures • Weapon stencils • Ethno-archaeology

H. Allen
Department of Anthropology, University of Auckland, Auckland, New Zealand
e-mail: h.allen@auckland.ac.nz

M.C. Langley (✉)
Institute of Archaeology, University of Oxford, Oxford, UK

Archaeology and Natural History, College of Asia and the Pacific, Australian National University, Canberra, ACT, Australia
e-mail: michelle.langley@anu.edu.au

P.S.C. Taçon
PERAHU, School of Humanities, Languages and Social Science, Griffith University, Gold Coast, QLD, Australia
e-mail: p.tacon@griffith.edu.au

Introduction

In terms of an understanding of world Pleistocene archaeology, knowledge of Australian archaeology and ethnography offers a number of opportunities. The search for regularities in the hunter-gatherer existence during the 1960s–1980s saw ethno-archaeologists venturing out to study contemporary hunter-gatherers in Australia (as well as South America, Africa, and India) (e.g., Hayden 1979; Lee 1979; Gould 1980). While these studies demonstrated the usefulness of higher order generalizations concerning foraging, time and energy use, and diet breadth in understanding such societies, we have come to understand that comparing populations with different historical trajectories operating under very different environmental circumstances must always be undertaken with the greatest of caution (e.g., Hiscock 2008).

Despite these caveats, examination of ethnographic bone and other organic projectile weaponry in various cultural contexts has provided useful insights into the selection of raw materials for manufacturing these technologies as well as their use in food gathering and social activities. For example, the ethno-archaeology of Australian Aboriginal bone point use has found that these objects were utilized in several, sometimes very different, contexts: spearthrower pegs; projectile points and barbs for land and marine hunting; fishing gorges; awls, needles and fasteners associated with clothing; needles and spikes used with fiber for making nets, bags, baskets and traps; fabricators for fine secondary and pressure flaking of stone projectile points; and, finally, as implements capable of directing sorcery towards a victim. This range of use contexts for morphologically similar bone artefacts on the same continent (albeit over a very large area) indicates that archaeologists in Australia, as elsewhere, must utilize all available data (depositional context, microscopy, residues, etc.) when investigating their use histories.

Also as elsewhere, the Pleistocene Australian toolkit has been suggested to have been formed largely from organic

Michelle C. Langley (ed.), *Osseous Projectile Weaponry*, Vertebrate Paleobiology and Paleoanthropology,
DOI 10.1007/978-94-024-0899-7_14

materials (bone, shell, wood, grass, bark, etc.) (Balme et al. 2009), though at the extreme end of organic material use in comparison to other similarly aged material culture repertoires. However, while organic technologies, including osseous projectile elements are dominant in the Australian ethnographic record, they are far less common in the Pleistocene archaeological record, and consequently, the small assemblages that have been recovered have received significantly less attention from researchers than their lithic counterparts. As is seen in Fig. 14.1 below, almost all Pleistocene sites which have preserved bone points are located in the south east and west of Australia. However, there are a multitude of rock art panels located throughout the north in which a diverse range of organic material culture is depicted, particularly wooden artefacts such as boomerangs and other weaponry and fiber objects such as bags and forms of body adornment. Not only are these aspects of material culture painted in association with human figures but there are also hundreds of stencils of perishable objects that convey their real-life dimensions (e.g., see Chaloupka 1993; Walsh 2000). Thus, we know that such organic material culture was present in the north and has simply not survived owing to differential preservation (Langley et al. 2011).

The significant impact taphonomic processes have had on this record, along with the potential for a hunting toolkit almost entirely organic in composition, has resulted in an archaeological record which is unique in character and challenging for researchers in many respects. Despite these setbacks, the Australian archaeological record provides multiple examples of bone points utilized as part of weapon systems. These artefacts will be discussed in the context of available ethnographic data and parietal imagery.

Bone Projectile Weaponry in Australian Ethnography and Archaeological Sites

A review of the available literature on Australian prehistoric osseous projectile technology revealed that these implements are particularly understudied. Often they are simply noted as 'bone points' with no further description and which may include any number of tools with vastly different functions. A notable exception to this trend is the work of Webb (1987; Webb and Allen 1990; but also see Bird and Frankel 2001 and Brockwell and Akerman 2007), described below.

The use of bone points as the tips or barbs of projectile weapons must be interpreted within the wider context of Australian spear technology. In this context spear shafts were manufactured from wood either in the form of single piece spears or as composite, multi-component spears, with the latter consisting of combinations of shaft, or shaft and foreshaft, and heads firmly attached with sinew/cord, often with the use of a resin or wax adhesive. Single piece spears might simply be straight pointed sticks or else have barbs carved in the solid. In central Australia, where suitable shaft materials were scarce, spear shafts might consist of two or more pieces of wood spliced together, so a composite spear could consist of a single or multi-piece shaft with a plain hardwood blade head often with an attached hardwood barb. Up to three or four types of wood from different species of trees may be used in the construction of such spears, i.e., tail-shaft, main shaft, head, and barb (Gould 1970). Composite spears might have heads made from wood, stone, or bone. Davidson (1934), reports that spears with detachable bone barbs were used through much of northern Queensland and parts of southern New South Wales and northern Victoria (see Fig. 14.1).

Where Australian spears possessed intricate barbing, hardwood was the material used. Australia has many hardwood species (ironwoods, eucalypts, wattles), which can be worked into intricate barb forms when green, either carved in the solid on single piece spears or as heads which can be attached to heavy or light shafts (see Figs. 14.2 and 14.3b). Spears with detachable wooden spearheads worked well with spearthrowers, and were relatively easy to make and to repair if broken (Allen 2011a).

Spears with intricately carved wooden heads, whether in the solid or detachable, were common across most of northern Australia (Davidson 1934:63–70). Thomson (nd. fieldnotes 1280–2, 1290–2 and 1306–8; [1936 and 1942]) and Warner (1937:487) provide Aboriginal terms for a wide variety of hardwood heads, whether barbed on one or both margins, unbarbed and blade-like, bifurcated or double pronged, or even, whether the barbs are fully or partially cut, as in the 'eyelet' or 'lace' types. The complex variations in the design of carved barbs suggests that these have to do with socio-ideological rather than purely technical factors, probably as a demonstration of carving skill or as an indication of an individual craftsperson (Cundy 1989:109).

Examples of these carved wooden projectile weapons have been recovered for contexts dating as far back as the terminal Pleistocene. At Wyrie Swamp (South Australia), an assemblage of 25 artefacts made from Sheoak, including digging sticks, boomerangs, one-piece spears with points or barbs, and pointed stakes were recovered from water-logged deposits (Luebbers 1975, 1978). The inclusion of boomerangs in this extraordinary assemblage indicates that this projectile technology dates back at least to the terminal Pleistocene in Australia. Some of the oldest rock art of northern Australia, including the Gwion Gwion (Bradshaw) paintings of the Kimberley and the Dynamic Figures of Arnhem Land believed to be terminal Pleistocene or older, both feature boomerangs and there are associated boomerang stencils (e.g., see Chaloupka 1993; Walsh 2000) (see Fig. 14.3e). Unfortunately, no bone tipped spears were included in the Wyrie Swamp terminal Pleistocene

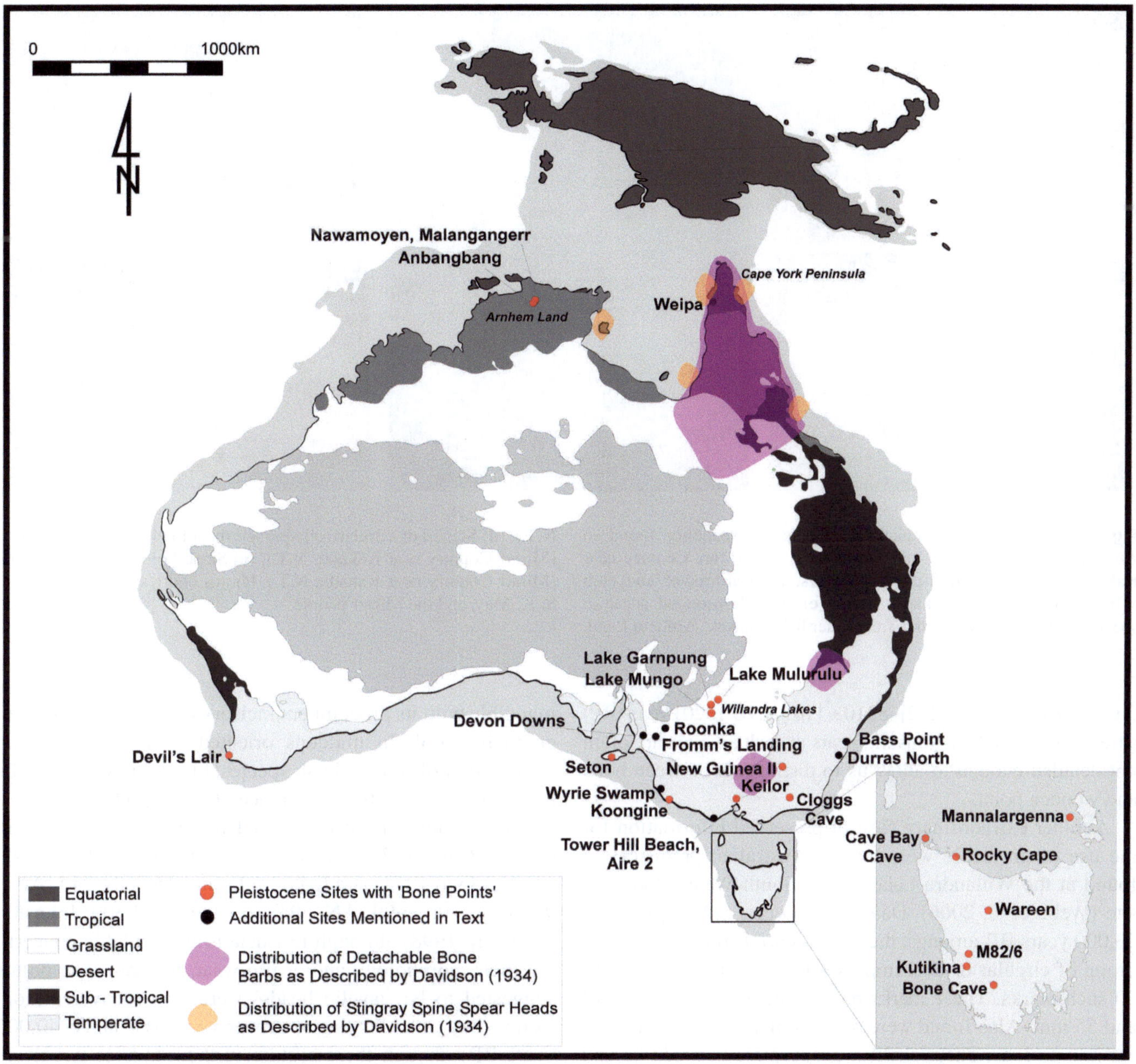

Fig. 14.1 The location of Pleistocene sites with 'bone points' cited in their inventory and the ethnographic distribution of detachable bone barb points and stingray spine spear heads documented by Davidson (1934)

Fig. 14.2 Multi-barbed spearhead from the Pilbara with dovetail hafting junction resin adhering at *left* (Photo courtesy of K. Akerman)

Fig. 14.3 Examples of depictions of organic weaponry found in Australian rock art: (**a**) A multi-pronged spear (Mirarr Country near Kakadu, N.T.); (**b**) Multi-pronged spears, a spearthrower and club (Mirarr Country near Kakadu, N.T.); (**c**) A multi-pronged spear in spearthrower associated with a freshwater fish (Injakak, Arnhem Land, N.T.); (**d**) Stencil of a unilaterally barbed spear likely carved from wood (Mirarr Country near Kakadu, N.T.); (**e**) Stencil of a large boomerang (Mirarr Country near Kakadu, N.T.). Photos courtesy of P.S.C. Taçon, S. K. May, and the Mirarr people

(10,200±150 [ANU 1292]–8210±110 [ANU 1320] years BP) assemblage and depictions of spears at rock art sites are often not detailed enough to allow us to distinguish between bone and wooden points.

Another extraordinary case of preserved information for the use of projectile weaponry in Australian prehistory is found at the Willandra Lakes (New South Wales) footprint site (Webb et al. 2006). Dated to between c. 19,000 and c. 23,000 years BP, amongst the 123 footprints recorded were a group of circular or oval impressions including a cluster of 20 such marks. These marks have diameters of between 42 and 55 mm. Also found were a series of 1.2–4.5 m-long linear impressions. These shallow grooves are between 43 and 55 mm wide and are suggestive of poles or similar objects being dragged across the surface. These latter marks were interpreted as spears ricocheting across the surface of the mud flat (Webb et al. 2006). It is possible that the spear throwing event recorded at this site was undertaken with points made from either wood and/or bone.

As stated above, all of the 'bone points' of Pleistocene age have been mostly recovered from sites in the southeast and southwest of the continent where preservation conditions are best. Where raw material is identified for archaeologically recovered points, it is usually macropodid (kangaroo family) long bones (ulna, tibia, and fibula). These skeletal elements experience high loading and stress during the life of the animal and are understandably, therefore, the most suitable bones for the production of weaponry, though the lower mandible with its procumbent incisors was also utilized as an engraving tool. In middens oriented towards the coastal exploitation of bird and fish species, bird bone will often be the main material utilized in place of macropod.

'Bone points' have been cited for Pleistocene levels in Cloggs Cave (17,720±840 [ANU 1044]–13,690 [ANU 1182]; Flood 1974) and Lake Mulurulu (15,560±240 [ANU 948b]–12,800±990 [ANU 948a]; Allen 1972:309; Johnston and Clark 1998) although few details are available concerning their morphology and/or use wear. Several bone points suggested to be 'javelin heads' were also recovered from Keilor (Dry Creek) dating to between c. 30,000 and c. 40,000 years BP—though these require further study to positively identify their function (Gallus 1970a, b, 1972).

Pleistocene bone points were also recovered from Devil's Lair (south western Australia) (Fig. 14.4a). A double bevelled bone, probably made from kangaroo fibula (B3693) (Dortch 1979; Turney et al. 2001). Other fragments of bone points recovered from this site display fractures consistent with use as projectile points (bevel and splinter fractures) (see Figures in Dortch 1984:59), though these fractures could be produced in other uses, such as awls.

While we have seen that 'bone points' have been reported from deposits dating as far back as c. 40,000 BP, the first set of Pleistocene bone points which have been suggested to have been used as projectile tips through examination of use wear (rather than solely artefact morphology) were recovered from M82/6 and Bone Cave in Tasmania. Manufactured

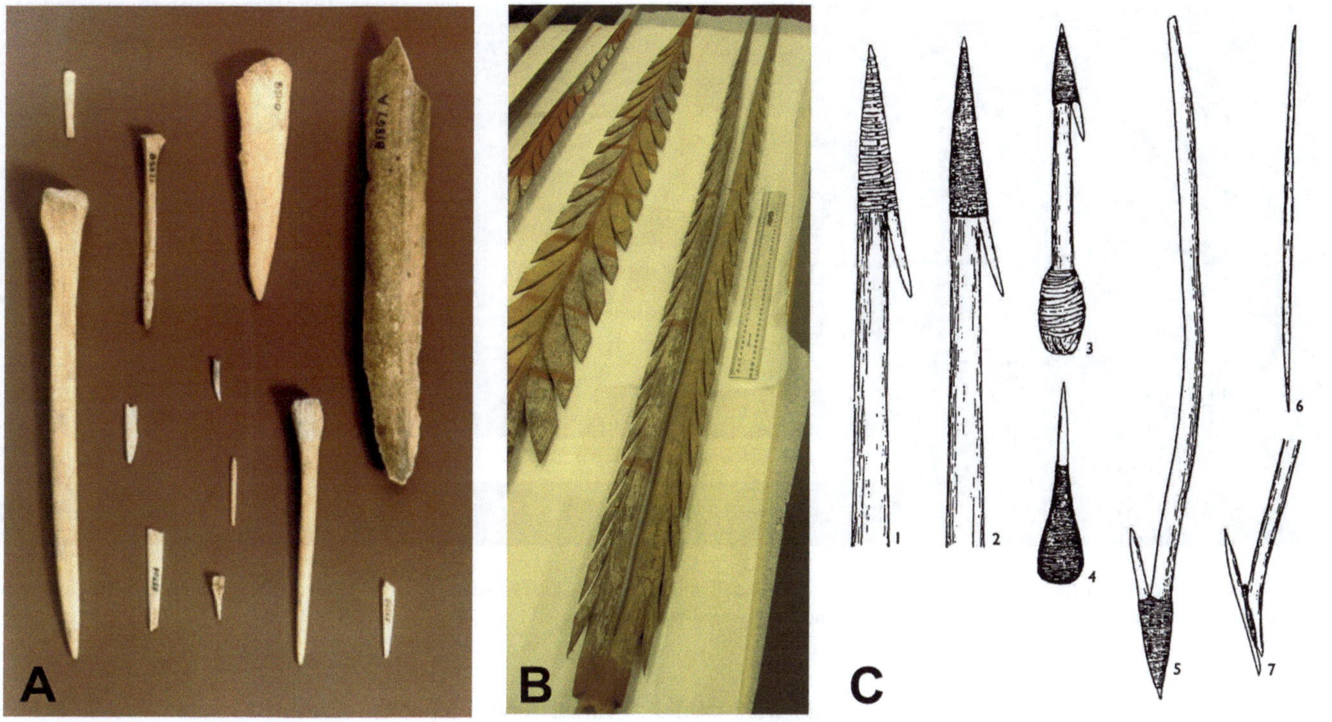

Fig. 14.4 Examples of archaeologically recovered 'bone points' and ethnographically documented organic projectile weaponry: (**a**) 'bone points' recovered from Devil's Lair (WA) (Photo courtesy of the Western Australian Museum); (**b**) Examples of wooden 'Tiwi spears' (Photo courtesy of H. Allen); (**c**) Thomson's (1939:210) drawings of bone points used to tip spears, as awls and fish hooks

from *Macropus rufogriseus*, Webb (1987; Webb and Allen 1990) identified a number of what she refers to as 'fine points' recovered from these two sites, with the Bone cave examples dating to between 29,000±520 (Beta 29987) and 13,700±860 BP (Beta 26509) (Webb and Allen 1990). Use wear identified on these implements suggested a number of uses, including the spearing of furry mammals—presumably red-necked wallaby whose skeletal remains were the dominant fauna in the studied sites. The tools exhibit impact fractures and two examples display possible evidence for hafting. Tasmanian bone points have also been recovered from Wareen, dating to between 33,610±370 (Beta 46873) and 22,370±470 years BP (Beta 26962), at Cave Bay Cave (c. 20,800–c. 18,500 years BP), Kutikina (19,770±850 [ANU 2785]–14,840±930 years BP [ANU 2781]), and Mannalargenna (c. 18,000–c. 15,000 years BP) (Allen et al. 1988; Brown 1993; Sim 1998).

Use wear analysis has also been undertaken on a number of implements recovered from Koongine Cave (Victoria). Bird and Frankel (2001) analyzed two bone points (dated to between 9240±100 (Beta 15996) and 9710±180 (Beta 14861) years BP, and identified use wear consistent with piercing fresh skin, dry skin, fish bark, kelp, shellfish, and birds, as well as hooking fish and scraping or incising bark or fish skin. It is acknowledged in the literature that points could be multifunctional tools and the use wear on the Koongine Cave points supports with this idea while also showing that these implements were employed in hunting.

While most of these previously mentioned points are assumed to constitute (at least partially) components of weaponry for use against terrestrial fauna, there is a great deal of associative evidence linking bone points with fishing in Australia. This evidence comes from mention of their use in fishing spears in ethnographic studies, the examination of fishing spears in museum collections and, finally, their presence in archaeological shell midden sites, where fishing was the major activity represented (Hale and Tindale 1930; Thomson 1939; McCarthy 1940; Lampert 1966; Sutton 1994; Mulvaney and Kamminga 1999:288).

Ethnographic examples of fishing spears may consist of a single-piece, unbarbed wooden spear (as in the Kimberley and along most of the Western Australian coast and adjacent Northern Territory) or may be composite with either a single bone or wood point or a barb and point combination. On multi-pronged spears (leisters), the bone points are gummed and/or bound onto prongs made of split wood and attached to a light wood shaft which floats after thrown. These light weight weapons were often associated with the use of spearthrowers.

Fishing gear in northern Australia also included spears with two wooden prongs, the barbs carved either on the outside or inside of each prong (Hugo 1983:104–113), and in Central

Fig. 14.5 Adelaide River (N.T.) bone points studied by Brockwell and Akerman (2007) (Adelaide River Collection, Museum and Art Gallery of the Northern Territory; Photography by Darren Boyd and reproduced with permission)

Australia hand thrown unbarbed one piece spears were used (Allen 2011b:259–61). Multi-pronged fishing spears have been recorded at rock art sites in Kakadu National Park and Arnhem Land in styles which date to less than 5000 years of age but, once again, it is near impossible to distinguish depictions of bone points from wooden ones (see Fig. 14.3c; Lewis 1988:399; Chaloupka 1993:146–151; Brockwell and Akerman 2007:89–90). Brockwell and Akerman (2007) provide a comparative analysis of ethnographically recorded fishing spears from north Australia (Fig. 14.5).

In 1934, Thomson provided an extensive record of technology associated with dugong (*Dugong dugon*) hunting on eastern Cape York Peninsula. Dugongs are large marine mammals and, along with marine turtles, their capture requires a complex technology: a stable platform from which to launch a harpoon, in this case a dugout canoe with outrigger; a harpoon with a detachable head; all connected by a stout rope to a float (Thomson 1934). Thomson (1934:258) provides information on ten harpoon heads from Cape York. Of these artefacts, six were of iron or wire, three were of hardwood with barbs carved in the solid or with a detachable barb of wood and there was only a single example which had a wooden point with a bone barb. Of the 55 harpoon heads in the Thomson collection at the Melbourne Museum, 40 were identified as dugong harpoons, 11 as turtle harpoons and two as either/both dugong and turtle harpoons. The majority of these (47 or 85 %) have hardwood heads, mostly spike-like, with low barbs carved in the solid, while the remaining heads were made of iron or wire. No harpoon heads with detachable bone barbs are identified (Hugo 1983:119–148). On the other hand, Thomson (1939:210, Fig. 1) provides a figure of an acacia harpoon head with a detachable bone barb from western Cape York.

Small bone bipoints recovered from Devon Downs Rockshelter, on the Murray River not far from Adelaide, were interpreted by Tindale (Hale and Tindale 1930:205–6) as fish gorges, based on a report that similar tools were used as fish gorges in Geelong, Victoria (Brough Smyth 1878:lvii, 202, Fig. 227:391). However, the use of small bone points as fish gorges is not well supported in the Australian literature, and Pretty (1977) argues that they are better suited for use as projectile tips. Tindale further argued that the presence of these small points indicated a change in bone tool technology at the site, a change ascribed to the arrival of a 'Mudukian' culture (after *muduk*, 'bone' in the Murray River Murundian language). Both Mulvaney (1961:83–4) and McCarthy (1940) criticized Tindale's use of small bipoints to define his Mudukian culture on the grounds that these artefacts were not restricted in either time or space in Australia.

Both uni- and bipointed bone points are commonly found in shell middens in several regions of Australia. One such example is the Weipa (Cape York, Queensland) shell mounds which date to between c. 3500 and 750 years BP. At this site, amongst the millions of shell fragments, ground and polished bone points of the type bound to wooden handles for use as spear barbs, as well as stingray barbs, presumably for the same purpose, were found. Several wallaby incisor teeth which exhibited artificially split form cutting edges (forming a toothed scraper), probably used for sharpening spear tips, were also recovered (Bailey 1977; Bailey et al. 1994).

Another example is Durras North (New South Wales), which provided an assemblage almost entirely made up of non-lithic technologies. Dating to 500 years BP, both fish-hooks and bone points (22 small bipoints and 196 unipoints of which 98 % were made from bird bone) were recovered, including large, slightly curved examples (Lampert 1966). Lampert (1966:112–3) used this collection in addition to similar examples found in museums (and which ranged in length between 70 and 165 mm) to argue that these artefacts were likely to have been used as point and barb combinations on single spears (that is, as leisters or similar such composite points).

Also recovered from this site and of interest to this discussion, were broken macropodid incisors which are argued by a number of researchers to have been engraving and/or resharpening tools (e.g., Balme 1979). These artefacts were noted by Thomson to have used ethnographically to make both bone points and fishhooks (Pickering 1980), and thus their appearance in sites such as Durras North, while unsurprising, is of interest to those wishing to understand the *chaîne opératoire* of this particular bone technology. Furthermore, the appearance of these particular artefacts in an archaeological site may indicate the manufacturing and/or maintenance of osseous or wooden weapons at that location. Indeed, broken macropod incisors, and use worn incisors still in their mandibular sockets, have been recorded at Rocky cape, Devil's Lair, Anbangbang, Malangangerr and Nawmoyn (Lampert 1966:97; Jones 1971:497; Dortch and Merrilees 1973; Schrire 1982).

These particular tools are almost certainly under-recognized in faunal assemblages recovered from archaeological sites, with an example of this situation neatly presented by Balme (1979). Balme makes the case that a higher than expected number of macropod broken incisors in the collection from Devil's Lair (south western Australia) may represent the discarding of engraving tools which had lost their usefulness.

Moving back to the association of bone uni- and bipoints with fishing middens, these artefacts have also been recorded at archaeological sites such as Fromm's Landing on the Murray River; Tower Hill Beach and the Aire 2 site on the Otway peninsula in Victoria; and in middens in Western Arnhem Land and the Adelaide River Plains in the Northern Territory (Hale and Tindale 1930; Mitchell 1958; Mulvaney 1960, 1962; Schrire 1982; Brockwell and Akerman 2007). In midden sites in coastal Queensland and New South Wales bone points are also found to be associated with shell fishhooks. This association, however, is patchy in both time and space, with each type of fishhook having different distributions. For instance, 'J' shaped fishhooks of wood, or wood with bone barbs, were used in Arnhem Land and Cape York Peninsula (Thomson 1936, 1939:Fig. 1; Schrire 1972), while circular shell fishhooks, similar to those which were widespread across Polynesia, occur in central Queensland and on the central and southern coasts of New South Wales (Walters 1988:Fig. 2). On the basis of ethnographic evidence, Gerritsen (2001) extends this distribution to the Murray River, southern Victoria and south coastal South Australia. Lampert and Hughes (1974:229) argue that both the shell fishhooks and the bone-tipped, multi-pronged fishing spear came into use at the same time in southern New South Wales, shortly after the sea reached its current position during the mid-Holocene. However, the widespread occurrence of shell fishhooks in coastal New Guinea, together with indications of an earlier date for their presence in central Queensland, suggests that they diffused into eastern Australia and spread along the east coast during the recent past (c. 1000–500 years BP) (Walters 1988:Table 1).

Shell fishhooks have featured in debates on wider issues in Australian prehistory. For example, in examining the Bass Point (New South Wales) midden site, Bowdler (1976) argued that the introduction of fishhooks around 600 years ago suggested that there had been a shift in fishing activities at the site, with women's shell hooks and lines augmenting the men's use of barbed fishing spears, the latter being indicated by small bone points. The argument that the introduction of a new fishing technology (i.e., hook and line) between 700 and 500 years ago heralded a change in women's coastal gathering strategies has been extended by Walters (1988).

Somewhat related, is the debate over another bone tool which may have been involved in fishing activities. At Rocky Cape (Tasmania), Jones (1971:522–3, 1978, 1995:427–8) interpreted pointed and spatulate bone artefacts as skin-working tools on the basis of their similarity to tools used for this purpose on the Australian mainland. Sharpened bone fibulae occur in levels dated between 8000 and 3500 years BP at this site, after which they drop out of the archaeological record, shortly after the evidence for fishing in the form of fish bones in middens (1978:32–3). At the time of first European observations, the Tasmanians did not eat fish, nor did they use bone tools for any purpose—including to sew skin garments—although they did wear animal skins draped loosely about their shoulders (Jones 1971:522). Jones interpreted both losses as a simplification of the Tasmanian toolkit and a diminution in the range of foods eaten (1977:202–3).

The interpretation that these fibulae tools were used for skin working is one which has been countered by Bowdler and Lourandos (1982:123–126) and Bowdler (1984:125–7). These authors note evidence that fibulae were used as netting needles in the manufacture of hunting, fishing and fowling nets, baskets and a wide variety of woven bags (Lamond 1950:169). Observing that the bone points at the Rocky Cape site are present at the same time as large numbers of Labrid (Wrasse) fish bones, and that both drop out of the record at about the same time, Bowdler suggests that the bone spatula and points represent part of a coastal fishery based on basket fish traps or nets. She also argues that the dropping of fish from the diet is a related event indicating a re-organisation of Tasmanian hunting strategies along gendered lines. However, Colley (1987) counters that it is very difficult to reconstruct

Aboriginal fishing methods on the basis of archaeological fish remains. The debate regarding the interpretation of the Tasmanian bone tools and their loss at c. 3500 years BP continues and cannot be resolved on the evidence currently available. The reliable interpretation of bone points as potential armatures remains difficult and more experimental work needs to be done.

One of the most interesting uses of bone projectile technology for which there is archaeological evidence in Australia is their use in interpersonal violence (ritualized or otherwise). Roonka Flat (on the Murray River, South Australia) is an open site dating to between 3930 ± 120 (ANU 407) and 18,150 ± 340 (ANU 406) years BP (Pretty 1977). This site contains numerous burials, perhaps the most well-known of which is a double burial of an adult male and small child (Grave 108) dating to the last phase of the site (c. 3000 years BP). The adult wore a chaplet of matched wallaby teeth in two parallel strands, a band of wallaby incisors around his forearm and probably a skin cloak (or similar) owing to the placement of bone pins and mammal tarsals over the body, while bird bones over and around the torso of the child suggest a garment fringed with bird part-limbs. Fifty-three bone tools were recovered from burials at this site, among them bone projectile points. Most interestingly, Grave 45 (also dating to the last phase of the site) contained a contracted burial of a man whose rib cage had been pierced by a bone point, causing the shattering and displacement of ribs. This weapon is made from mammal long bone, is concave in section and tapers to a conical tip.

Interestingly, 'bone points' have also been recorded ethnographically as being used in projective magic. Walshe (2008) reviewed the classification of 706 bone implements in the collections of the South Australian Museum, and noted that more than 20 sub-categories have been used by researchers, indicating both confusion and inconsistency in their recording and analysis. Excluding bone used for ornaments, there are a total of 555 bone points or utilized bone pieces in the collection. 450 (81 %) of these have entered the catalogue as pointing bones used in sorcery, probably indicating collector bias and possibly Aboriginal manufacturing to satisfy collector demand (Walshe 2008:172). However, these records also note the presence of resin and human hair string. Such indications, together with ochre and deep incisions, are present on 83 macropod fibula uni-points in the collection, providing strong evidence that these artefacts were used for magical purposes (Walshe 2008:181–4). Walshe also notes evidence that under certain circumstances, bone points could move from their mundane function, as awls or gouges, to the magical realm—increasing the ambiguity of museum and archaeological classifications.

Finally, both small hardwood pegs and bone points were used as hooks in spearthrowers (commonly termed *woomera* in Australia) (see Cottrell and Kamminga 1992) and there are many rock painting depictions of these tools in northern Australia, especially in the Kakadu-Arnhem Land region (Fig. 14.3b, c). The small bone pegs were light and strong and relatively easy to replace if they failed, an example of a maintainable technology (Bleed 1986). When the spear-thrower first appeared in Australia is currently unknown, though as we have seen above there are numerous paintings of these implements in use in rock art dating to between 4000 and 6000 years ago. Additionally, skeletal evidence from an individual buried at Lake Mungo (WLH 3) displays severe osteoarthritis in his right elbow, which is suggestive of 'spear-thrower elbow', although Webb (1989) concludes that its cause is uncertain and may have been an infection exacerbated by spear-throwing, with or without a spearthrower, though stone knapping might also be a possible cause of repetitive strain damage. The tiny nature of the bone spear-thrower peg makes it uncertain that these artefacts would survive for very long in archaeological record.

Summary and Conclusions

Despite prehistoric bone technologies being rare outside the southern regions of Australia, they have nonetheless provided insights into technological and social changes which have occurred throughout Australian prehistory. Rock art, ethnography, and archaeologically recovered implements have shown that bone, stingray spines, and shell have been used in a number of areas to manufacture projectile weaponry, particularly over the past 10,000 years, though we cannot rule out that osseous projectile weaponry has a much greater antiquity on this continent owing to the advancing coastline during the LGM. Given the strong association in both ethnography and archaeological sites between osseous weaponry and the exploitation of coastal resources, we must keep this latter event in mind when investigating the use of organic weaponry (and other tools) in Australia.

Additionally, the choice made by many groups of Indigenous Australians to use various woods to manufacture both projectile points and spearthrowers despite suitable bone being available, is an interesting cultural decision. Waguespack et al. (2009) have speculated that the choice to manufacture projectile points out of stone, antler or bone rather than wood may not only be owing to a slight advantage in lethal efficiency, but also that "they require greater effort and skill to produce", thereby providing "a medium for expressing self and/or group identity, essentially a form of costly signaling" (Waguespack et al. 2009:797; also see O'Connor et al. 2014 for similar comments).

Both ethnographic examples and depictions (including stencils) of barbed spears found in northern Australia indicate that these weapons were carefully and skillfully crafted, with a wide variety of barb styles observed over the continent. It therefore appears that, while crafted primarily from wood, hunting weaponry in Australia (as elsewhere) had a dual function: as a food gathering implement and as a social signaling technology. Furthermore, their depiction in rock art argued to date to the terminal Pleistocene suggests that this dual function may have a great antiquity.

To conclude, while bone projectile technology is rarely recovered from Australian early prehistoric sites—perhaps owing to both the advancing coastline during the LGM and the preference to use wood by ancient and ethnographic Australians—those that have been identified have nevertheless provided insights into changing technologies and social structures during prehistory. Australian implements have been particularly understudied and, therefore, our knowledge of their importance in prehistoric lifeways remains extremely limited. Future work should aim to address this situation, and should include a focused study on why wood is preferred over bone, stingray spines (etc.) for the manufacture of projectile weaponry in this region.

Acknowledgements The University of Auckland, the University of Oxford and Griffith University are thanked for supporting this research. The Mirrar community (Northern Territory, Australia) are thanked for permission to use photographs of their rock art, as is the Western Australian Museum. We would also like to thank Kim Akerman for his many useful comments and observations. All dates provided in this chapter are as originally reported in the literature.

References

Allen, H. (1972). *Where the crow flies backwards: Man and land in the Darling Basin*. Unpublished Ph.D. Dissertation, The Australian National University.

Allen, H. (2011a). Thomson's spears: Innovation and change in eastern Arnhem Land projectile technology. In Y. Musharbash (Ed.), *Ethnography and the production of anthropological knowledge: Essays in honour of Nicolas Peterson* (pp. 69–88). Canberra: ANU EPress.

Allen, H. (2011b). The space between: Aboriginal people, the Victorian Exploring Expedition and the relief parties. In E. B. J. D. McCann (Ed.), *Burke and Wills: The scientific legacy of the Victorian exploring expedition* (pp. 245–274). Collingwood: CSIRO Publishing.

Allen, J., Cosgrove, R., & Brown, S. (1988). New archaeological data from the Southern Forests region, Tasmania. *Australian Archaeology, 27*, 75–88.

Bailey, G., Chappell, J., & Crib, R. (1994). The origin of *Anadara* shell mounds at Weipa, North Queensland, Australia. *Archaeology in Oceania, 29*, 69–80.

Bailey, G. N. (1977). Shell mounds, shell middens and raised beaches in the Cape York Peninsula. *Mankind, 11*, 132–143.

Balme, J. (1979). Artificial bias in a sample of kangaroo incisors from Devil's Lair, Western Australia. *Records of the Western Australian Museum, 7*, 229–244.

Balme, J., Davidson, I., McDonald, J., Stern, N., & Veth, P. (2009). Symbolic behaviour and the peopling of the southern arc route to Australia. *Quaternary International, 202*, 59–68.

Bird, C., & Frankel, D. (2001). Excavations at Koongine Cave: Lithics and land use in the terminal Pleistocene and Holocene of South Australia. *Proceedings of the Prehistoric Society, 67*, 49–83.

Bleed, P. (1986). The optimal design of hunting weapons: Maintainability or reliability? *American Antiquity, 51*, 737–747.

Bowdler, S. (1976). Hook, line and dilly bag: An interpretation of an Australian coastal Shell midden. *Mankind, 10*, 248–258.

Bowdler, S. (1984). *Hunter Hill, Hunter Island: Archaeological investigations of a prehistoric Tasmanian Site. Terra Australis* 8. Canberra: Department of Prehistory, Research School of Pacific Studies, The Australian National University.

Bowdler, S., & Lourandos, H. (1982). Both sides of Bass Strait. In S. Bowdler (Ed.), *Coastal archaeology in eastern Australia: Proceedings of the 1980 Valla Conference on Australian Prehistory* (pp. 121–132). Canberra: Department of Prehistory, Research School of Pacific Studies, The Australian National University.

Brockwell, S., & Akerman, K. (2007). Bone points from the Adelaide River, Northern Territory. *Australian Aboriginal Studies, 2007*(1), 83–97.

Brough Smyth, R. (1878). *The Aborigines of Victoria with notes relating to the habits of the natives of other parts of Australia and Tasmania*. Melbourne: John Ferres, Government Printer.

Brown, P. (1993). Mannalargenna Cave: A Pleistocene site in Bass Strait. In M. Spriggs & B. Fankhauser (Eds.), *Sahul in review: Pleistocene archaeology in Australia, New Guinea and Island Melanesia* (pp. 258–274). Canberra: Department of Prehistory, Research School of Pacific Studies, Australian National University.

Chaloupka, G. (1993). *Journey in time*. Sydney: Reed Books.

Colley, S. M. (1987). Fishing for facts: Can we reconstruct fishing methods from archaeological evidence? *Australian Archaeology, 24*, 16–26.

Cottrell, B., & Kamminga, J. (1992). *Mechanics of pre-industrial society*. Cambridge: Cambridge University Press.

Cundy, B. J. (1989). *Formal variation in Australian Spear and Spearthrower Technology. British Archaeological Reports International Series 546*. Oxford: Oxbow.

Davidson, D. S. (1934). Australian spear-traits and their derivations. *Journal of the Polynesian Society, 43*(41–72), 142–162.

Dortch, C. E. (1979). 33,000 year old stone and bone artifacts from Devil's Lair, Western Australia. *Records of the Western Australian Museum, 7*, 329–367.

Dortch, C. E. (1984). *Devil's Lair: A study in prehistory*. Perth: Western Australian Museum.

Dortch, C. E., & Merrilees, D. (1973). Human occupation at Devil's Lair, Western Australia during the Pleistocene. *Archaeology and Physical Anthropology in Oceania, 8*, 89–115.

Flood, J. (1974). Pleistocene man at Cloggs Cave: His tool kit and environment. *Mankind, 9*, 175–188.

Gallus, A. (1970a). The Keilor Project. *The Artefact, 17*, 1–8.

Gallus, A. (1970b). The Keilor Project. *The Artefact, 17*, 1–6.

Gallus, A. (1972). Excavations at Keilor: Report No. 2. *The Artefact, 27*, 9–18.

Gerritsen, R. (2001). Aboriginal fish hooks in southern Australia: Evidence, arguments and implications. *Australian Archaeology, 52*, 18–28.

Gould, R. A. (1970). Spears and spear-throwers of the Western desert Aborigines of Australia. *American Museum Novitiates, 2403*, 1–41.

Gould, R. A. (1980). *Living archaeology*. Cambridge: Cambridge University Press.

Hale, H. M., & Tindale, N. B. (1930). Notes on some human remains in the Lower Murray Valley, South Australia. *Records of the South Australian Museum, 4*, 145–218.

Hayden, B. (1979). *Paleolithic reflections: Lithic technology and ethnographic excavation among the Australia Aborigines*. New Jersey: Humanities Press.

Hiscock, P. (2008). *Archaeology of ancient Australia.* London: Routledge.
Hugo, D. F. (1983). *Fishing technology from the Donald Thomson collection, National Museum of Victoria: A contemporary re-analysis.* Unpublished MA Thesis, University of Queensland.
Johnston, H., & Clark, P. (1998). Willandra Lakes archaeological investigations 1968–98. *Archaeology in Oceania, 33*(3), 105–119.
Jones, R. (1971). *Rocky cape and the problem of the Tasmanians.* Unpublished Ph.D. Dissertation, University of Sydney.
Jones, R. (1977). The Tasmanian paradox. In R. V. S. Wright (Ed.), *Stone tools as cultural markers: Change, evolution and complexity* (pp. 189–204). Canberra: Australian Institute of Aboriginal Studies.
Jones, R. (1978). Why did the Tasmanians stop eating fish? In R. A. Gould (Ed.), *Explorations in ethnoarchaeology* (pp. 11–48). Albuquerque: University of New Mexico Press.
Jones, R. (1995). Tasmanian archaeology: Establishing the sequences. *Annual Review of Anthropology, 24*, 423–446.
Lamond, H. G. (1950). Aboriginal net making. *Mankind, 4*, 168–169.
Lampert, R. J. (1966). An excavation at Durras North, New South Wales. *Archaeology and Physical Anthropology in Oceania, 1*, 83–118.
Lampert, R. J., & Hughes, P. (1974). Sea level change and Aboriginal coastal adaptations in southern New South Wales. *Archaeology and Physical Anthropology in Oceania, 9*, 226–235.
Langley, M. C., Clarkson, C., & Ulm, S. (2011). From small holes to grand narratives: The impact of taphonomy and sample size on the modernity debate in Australia and New Guinea. *Journal of Human Evolution, 61*, 197–208.
Lee, R. B. (1979). *The !Kung San: Men, women and work in a foraging society*. Cambridge: Cambridge University Press.
Lewis, D. (1988). *The rock paintings of Arnhem Land, Australia: Social, ecological and material culture changes in the post-glacial period. British Archaeological Reports International Series 415.* Oxford: Oxbow.
Luebbers, R. A. (1975). Ancient boomerangs discovered in South Australia. *Nature, 253*, 39.
Luebbers, R. A. (1978). *Meals and menus: A study of change in prehistoric coastal settlement in South Australia.* Unpublished Ph.D. Dissertation, Australian National University.
McCarthy, F. D. (1940). The bone point, known as *muduk*, in eastern Australia. *Records of the Australian Museum, 20*, 313–319.
Mitchell, S. R. (1958). An Aboriginal bone industry. *Mankind, 5*, 194–199.
Mulvaney, D. J. (1960). Archaeological excavations at Fromm's Landing on the lower Murray River, South Australia. *Proceedings of the Royal Society of Victoria, 72*, 53–85.
Mulvaney, D. J. (1961). The stone age of Australia. *Proceedings of the Prehistoric Society, 27*, 56–107.
Mulvaney, D. J. (1962). Archaeological excavations on the Aire River, Otway Peninsula, Victoria. *Proceedings of the Royal Society of Victoria, 75*, 1–15.
Mulvaney, D. J., & Kamminga, J. (1999). *Prehistory of Australia.* St Leonards: Allen and Unwin.
O'Connor, S., Robertson, G., & Aplin, K. P. (2014). Are osseous artefacts a window to perishable material culture? Implications of an unusually complex bone tool from the Late Pleistocene of East Timor. *Journal of Human Evolution, 67*, 108–119.
Pickering, M. (1980). A technique of bone tool manufacture: From photographs in the Donald F. Thomson Collection, National Museum of Victoria. *The Artefact, 5*, 93–97.
Pretty, G. (1977). The cultural chronology of Roonka Flat: A preliminary consideration. In R. V. S. Wright (Ed.), *Stone tools as cultural markers* (pp. 288–331). Canberra: Australian Institute of Aboriginal Studies.
Schrire, C. (1972). Ethno-archaeological models and subsistence behaviour in Arnhem Land. In D. L. Clarke (Ed.), *Models in archaeology* (pp. 653–670). London: Methuen.
Schrire, C. (1982). *The Alligator Rivers: Prehistory and ecology in Western Arnhem Land. Terra Australis* 7. Canberra: Department of Prehistory, research School of Pacific Studies, The Australian National University.
Sim, R. (1998). *The archaeology of isolation? Prehistoric occupation in the Furneaux Group of Islands, Bass Strait, Tasmania.* Unpublished Ph.D. dissertation, Australian National University.
Sutton, P. (1994). Material culture traditions of the Wik People, Cape York Peninsula. *Records of the South Australian Museum, 27*, 31–52.
Thomson, D. F. (nd). *Fieldnotes.* Typescripts, Thomson Collection, Melbourne Museum.
Thomson, D. F. (1934). The dugong hunters of Cape York. *Journal of the Royal Anthropological Institute, 64*, 237–262.
Thomson, D. F. (1936). Notes on some bone and stone implements from north Queensland. *Journal of the Royal Anthropological Institute, 66*, 71–74.
Thomson, D. F. (1939). The seasonal factor in human culture. *Proceedings of the Prehistoric Society, 5*, 209–221.
Turney, C. S. M., Bird, M. I., Fifield, L. K., Roberts, R. G., Smith, M., Dortch, C. E., et al. (2001). Early human occupation at Devil's Lair, southwestern Australia 50,000 years ago. *Quaternary Research, 55*(1), 3–13.
Waguespack, N. M., Surovell, T. A., Denoyer, A., Dallow, A., Savage, A., Hyneman, J., et al. (2009). Making a point: Wood- versus stone-tipped projectiles. *Antiquity, 83*, 786–800.
Walsh, G. (2000). *Bradshaw art of the Kimberley*. Toowong: Takarakka Nowan Kas Publications.
Walshe, K. (2008). Pointing bones and bone points in the Australian Aboriginal Collection of the South Australian Museum. *Records of the South Australian Museum, 33*, 167–203.
Walters, I. (1988). Fish hooks: Evidence for dual social systems in southeastern Australia? *Australian Archaeology, 27*, 98–114.
Warner, W. L. (1937). *A black civilization: A social study of an Australian tribe*. Revised Edition. Chicago: Harper and Brothers Publishers.
Webb, C. (1987). *Use-wear on bone tools: An experimental programme and three case-studies from South-East Australia.* Unpublished BA Hons thesis, La Trobe University.
Webb, C., & Allen, J. (1990). A functional analysis of Pleistocene bone tools from two sites in Southwest Tasmania. *Archaeology in Oceania, 25*, 75–78.
Webb, S. (1989). *The Willandra Lakes hominids.* Canberra: Department of Prehistory, Research School of Pacific Studies, Australian National University.
Webb, S., Cupper, M. L., & Robins, R. (2006). Pleistocene human footprints from the Willandra Lakes, southeastern Australia. *Journal of Human Evolution, 50*, 405–413.

Part IV
The Americas

Chapter 15
A Review of Late Pleistocene North American Bone and Ivory Tools

Michael J. O'Brien, R. Lee Lyman, Briggs Buchanan, and Mark Collard

Abstract Osseous (bone and ivory) rods dating to the Early Paleoindian period (ca. 13,300–11,900 calendar years before present) have been found over much of North America. Previous researchers have attributed several possible functions to these artifacts, including use as projectile points, as foreshafts, as pressure-flaker handles, as sled shoes, and as levered hafting wedges. Considering the important link that osseous rods provide between the late Pleistocene cultures of North America and the Upper Paleolithic cultures of Europe and Asia, it is crucial that archaeologists define the range of variation and possible functions represented in the North American osseous rods. In this chapter we provide an up-to-date review of the distribution of late Pleistocene osseous rods across North America; describe the range of variation in morphology and attributes associated with this sample; and discuss the possible range of functions represented.

Keywords Bone rods • Ivory • Early Paleoindian period • Function

Introduction

Bone, ivory, and antler tools dating to the Early Paleoindian period (ca. 13,300–11,900 calendar years before present [cal BP]) have been found over much of North America, especially the Pacific Northwest and Columbia Plateau, the northern Plains, and Florida. Researchers have attributed numerous functions to these tools, including use as projectile points, foreshafts for spears or darts, shoes for sled runners, handles for pressure flakers, levered hafting wedges, fish hooks, atlatl nocks, billets, awls and punches, and shaft wrenches. Considering the significant functional link that the tools might provide with the Upper Paleolithic cultures of Europe and eastern Asia, it is important that archaeologists define the range of variation and possible uses represented in the North American sample. Here our interest is primarily on bone and ivory tools, which represent by far the largest percentage of Early Paleoindian non-stone tools recovered to date. Our discussion builds on earlier work (e.g., Lyman et al. 1998; Pearson 1999; Webb and Hemmings 2001; Redmond and Tankersley 2005), but it is not an exhaustive survey of all known tools, many of which still have not been measured and described in the public record.

Background

Archaeological interest in early prehistoric bone and ivory tools from North America can be traced back to 1937, when Cotter (1937:14) reported the discovery of a "cylindrical shaft of bone" in association with mammoth remains in the excavations at what became known as Blackwater Draw Locality No. 1, just outside Clovis, New Mexico (Hester 1972; Saunders and Daeschler 1994; Boldurian and Cotter 1999). There actually were two specimens, found within 2 days of each other (Boldurian and Cotter 1999). Within a few decades, similar items were found in Saskatchewan, Canada (Wilmeth 1968), and the states of Alaska (Rainey 1939, 1940), Washington (Daugherty 1956; Irwin and Moody 1978), Oregon (Cressman 1942, 1956), California (Riddell 1973), Montana (Lahren and Bonnichsen 1974), and Florida (Jenks and Simpson 1941).

M.J. O'Brien (✉) • R.L. Lyman
Department of Anthropology, University of Missouri, Columbia, MO, USA
e-mail: ObrienM@missouri.edu; LymanR@missouri.edu

B. Buchanan
Department of Anthropology, University of Tulsa, Tulsa, OK, USA
e-mail: bbuchana@sfu.ca

M. Collard
Human Evolutionary Studies Program and Department of Archaeology, Simon Fraser University, Burnaby, BC, Canada
e-mail: mcollard@sfu.ca

Michelle C. Langley (ed.), *Osseous Projectile Weaponry*, Vertebrate Paleobiology and Paleoanthropology,
DOI 10.1007/978-94-024-0899-7_15

Many of the early discoveries prompted suggestions of typological similarity among the specimens, leading to speculation that the ages of the newly discovered pieces were similar to that of the Blackwater Draw specimens. For example, specimens from Alaska were said to be "similar [to] long bone points in direct association with mammoth bones" at Blackwater Draw (Rainey 1939:394), and the specimens from Blackwater Draw were said to be "very much like the [Saskatchewan] specimen" in terms of "width and thickness" (Wilmeth 1968:101). When Cressman consulted Cotter on the typological identity of some specimens recovered from southern Oregon, Cotter thought one of them was identical to the specimens from Blackwater Draw (Cressman 1942). Similarly, Jenks and Simpson (1941:318) stated that their specimens from Florida were "typologically the same" as those from Blackwater Draw.

Thus by the early 1940s, numerous bone and ivory tools from varied contexts across the United States were being assessed as "belong[ing] to a long extinct culture, probably of closely approximating age, namely, of late glacial or early post-glacial time" (Jenks and Simpson 1941:318). These tools became a hallmark artifact of the Clovis culture (e.g., Sellards 1952) and remain so today (e.g., Bonnichsen et al. 1987; Pearson 1999; Redmond and Tankersley 2005; Stanford and Bradley 2012), albeit not as well documented and studied as the thin, fluted projectile points that gave the culture its name (Fig. 15.1) and which are sometimes found in caches alongside bone shafts, or rods (Lahren and Bonnichsen 1974; Gramly 1993).

Despite this status, there are relatively few radiocarbon dates that directly tie the tools to the Clovis period (13,300–12,800 cal BP [Haynes 2002; Collard et al. 2010]). Two bone rods from Sheriden Cave in north-central Ohio (Redmond and Tankersley 2005) were recovered from a cultural horizon with radiocarbon dates in the 12,900–12,500 cal BP range; bone collagen from one of the specimens was subsequently dated to 13,025–12,925 cal BP (Waters et al. 2009). Similarly, two bone-collagen samples from bone rods found at the Anzick site in Montana yielded dates in the 13,000–12,800 cal BP range (Morrow and Fiedel 2006), which mirrors the range of a bone-collagen sample from an ivory rod from Sloth Hole in Florida (Hemmings 2004).

However, it is clear that not all bone and ivory tools date to the Clovis period. For example, a bone rod from the Sheaman site in Wyoming (Frison 1982), which was long assumed to be Clovis in age based on context and radiocarbon dates (Haynes et al. 2004), may actually postdate Clovis based on three collagen dates on the rod that average 12,175 ± 155 cal BP (Waters and Stafford 2007). These dates place it in the Folsom period (ca. 12,800–11,900 cal BP). Three rods from the Agate Basin site in Wyoming (Frison and Zeimens 1980; Frison and Craig 1982) also date to the Folsom period. Three specimens from the Lind Coulee site in Washington may date several thousand years later (Daugherty 1956; Irwin and Moody 1978), but recently obtained AMS dates on collagen from associated bone suggest they may be as old as 12,000–11,200 cal BP (Craven 2004). Bone and ivory tools are also known from sites in Alaska (Rainey 1939; Ackerman 1996; Holmes 1996) that predate and postdate Clovis.

Today, Paleoindian bone and ivory tools are known from at least 11 continental U.S. states and the province of Saskatchewan (Fig. 15.2). Their distribution is highly uneven, with over half the known specimens—many of them in private collections (Dunbar and Waller 1983; Wagers 1986)—coming from Florida. There is no reason to suspect that the distribution is attributable to anything other than preservation. The large number of specimens from Florida is tied to the postglacial rise in sea level that submerged low-lying archaeological and paleontological sites located in and around Florida's extensive karst system—Page-Ladson (Webb 2006) and Sloth Hole (Hemmings 1999, 2004), for example—thus preserving Paleoindian organic remains (Willis 1988; Dunbar et al. 1989; Hemmings 1999; Webb and Hemmings 2001; Bradley et al. 2010). Compared to the sample of known Clovis points from North America, which ranges into the thousands (Anderson et al. 2010), the extant sample of osseous tools is "severely impoverished" (Redmond and Tankersley 2005:504).

Variation in Form and Function

One noticeable characteristic of North American rods is the presence of beveling on one or both ends, the obvious conclusion being that bevels had something to do with how the tools were used. Cotter (1954), for example, referred to the rods as "beveled bone foreshaft portions or spear tips," which he assumed were derived from the familiar *sagaie*, or javelin, points of bone or reindeer horn from the European Upper Paleolithic. Similarly, Jenks and Simpson (1941) referred to the bone rods as "beveled artifacts" and thought that at least one of the three Florida specimens they described represented a "hunting point." Cressman (1942) referred to his specimens as bone "points" or "foreshafts," later describing one specimen as a "long beveled end projectile point" because it was "found in the lower left abdominal part of [a human] skeleton"; the beveled end was said to be "for hafting to a shaft" Cressman (1956:431).

The emerging designation of these artifacts as foreshafts received some formality in a report by Lahren and Bonnichsen (1974) on the Anzick materials from Montana (Fig. 15.2)—a deposit of flakes, bifaces, bone rods, and eight Clovis points buried in a collapsed rock shelter. The authors provided a description of the bone rods—two complete and nine fragmented, together representing an assemblage of perhaps as few as four to six rods (Jones 1996) or as many as eight

Fig. 15.1 Clovis points from various North American sites. Photo by Charlotte D. Pevny; courtesy Michael D. Waters

(Lassen 2005)—and a model of how they thought the specimens served as foreshafts to which Clovis points were hafted. Frison (1982:156) later stated that the "true function of [these] objects…remains an open question; they are postulated as having been both foreshafts and actual projectile points." Still later, Wilke et al. (1991) argued on the basis of experimental work that the Anzick specimens were handles to which an antler bit was hafted, thereby producing a composite tool for pressure flaking.

Another site—East Wenatchee (also known as the Richey–Roberts Clovis cache) in eastern Washington (Fig. 15.2)—received considerable archaeological attention in the late 1980s when a cache of 14 large Clovis blades and 14 bi-beveled bone rods was unearthed in an apple orchard.

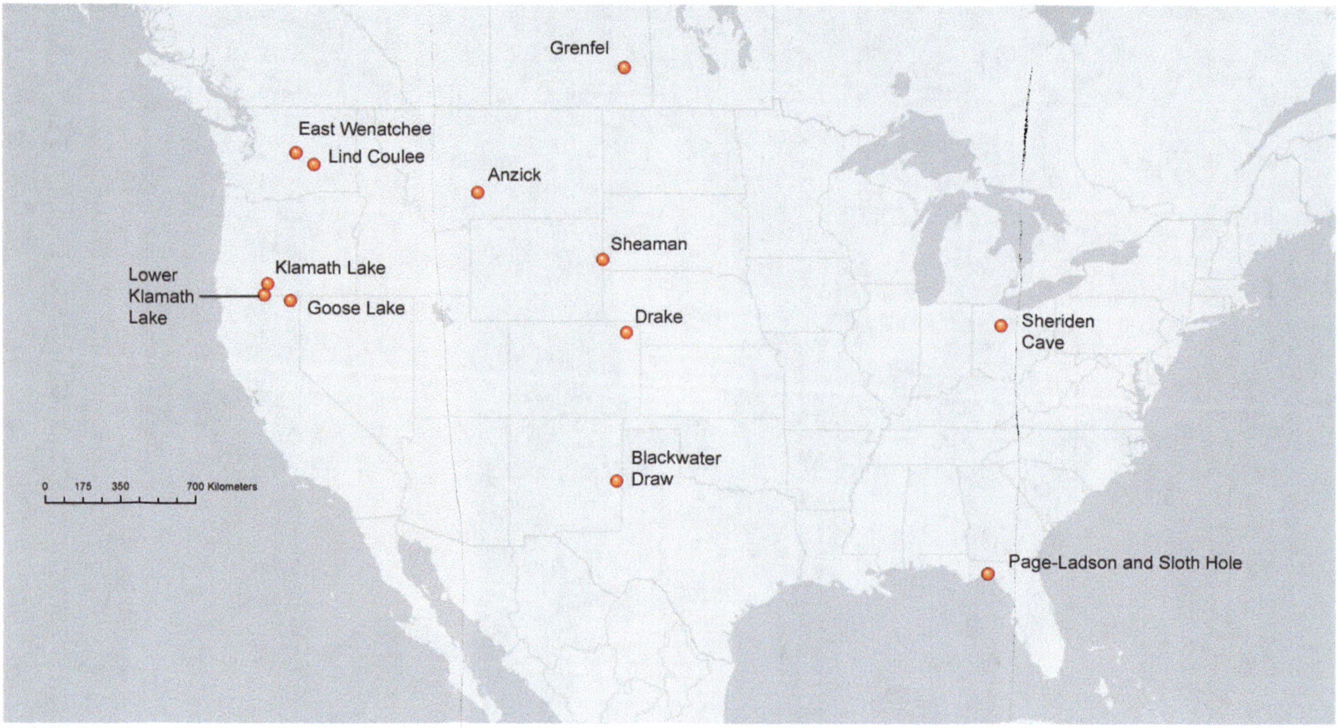

Fig. 15.2 Locations of some North American sites that have produced bone or ivory rods

Mehringer (1988a, b, 1989) indicated that it was speculative whether the rods were foreshafts, pressure flakers, or "wedges for splitting wood." Gramly (1993:8) later noted that "the rods are paired by size" and expressed a preference for the hypothesis that the size-paired sets of specimens once served as shoes for sled runners.

Part of the continuing puzzlement over what functions the beveled rods might have served originates in the fact that the two original Blackwater Draw specimens appeared similar to each other and thus were thought to comprise a single type of artifact (Cotter 1937). Some analysts later argued that all beveled rods, irrespective of geographic origin, were of the same "type" (Cressman 1942; Cotter 1954; Cressman 1956; Lahren and Bonnichsen 1974), whereas others (e.g., Riddell 1973) identified two types on the basis of whether beveling occurs on only one end or on both ends. Table 15.1 lists a sample of bone and ivory rods from several locales. Despite the fact that many data are missing—a result of the fragmentary nature of some specimens or a lack of recording—note the wide variation in measurements. Length, for example, ranges from 112 to 281 mm, maximum width from 8 to 30 mm, and maximum thickness from 10 to 22 mm. Figure 15.3 plots length versus maximum width for specimens in Table 15.1. Note that 11 of the 12 measurable rods from East Wenatchee (another rod was too fragmentary to measure and still another was left in the ground [Gramly 1993]) are in a grouping well outside that of other specimens in Table 15.1 as a result of their larger maximum widths.

Possible Functions

Given the variation in size and number of beveled ends, it perhaps is predictable that different specimens or sets thereof would have been interpreted differently in terms of their suspected function. The question begged by these observations concerns the relevance of the attributes considered for determining artifact function. We examine below several possible functions, focusing on the mechanical efficiency of particular attribute combinations displayed by the tools when serving a particular use.

Foreshafts

One function commonly found in the archaeological literature with respect to bone and ivory rods is that of *foreshaft* (e.g., Bonnichsen et al. 1987; Stanford 1991; Wilke et al. 1991)—the middle piece of a compound weapon between the projectile point and main shaft. Foreshafts should be rod-like, given the intended purpose of making retooling and game killing more efficient (e.g., Frison 1974, 1978), but why are the ends beveled? If the bevel were a mechanically critical attribute, then we might wonder why some rods are beveled on both ends and others on only one end. We doubt that this variation is the result of some rods not yet being completely manufactured or finished products. This assessment is based

Table 15.1 Descriptive data for a sample of osseous rods from North America

Specimen	Material	Length	Width	Thickness	Bevel	Bevel incised?	Bevel length†
Anzick-37	Bone		17	12	1		49
Anzick-38	Bone		19	13	1		
Anzick-39	Bone				1		48
Anzick-67	Bone	228	15	12	2?		58
Anzick-94	Bone		18	13	1		44
Anzick-95	Bone		18	13	1		44
Anzick-117	Bone		15	10			
Anzick-118/119	Bone	281	18	14	2	Yes	46/51
Anzick-120	Bone		19	11			
Anzick-122	Bone		20	13			
Anzick-123	Bone		20	14			
Florida-A	Bone	182+	12.3	12	1	Yes	58
Florida-B	Ivory	91+	8.5		1	Yes	25
Florida-C	Ivory	150.5+	10.1			Yes	
Blackwater Draw-9	Bone	252	15		2	On 2	
Blackwater Draw-10	Bone	234	17		2	On 1	
Lind Coulee-178	Bone	134	13.4		1		61.6
Lind Coulee-140	Bone	251+	16.4	10.4			
East Wenatchee-A	Bone	263	24	18	2	On 2	59/35
East Wenatchee-B	Bone	209	24	17	2	On 2	
East Wenatchee-C	Bone	252	24	18	2	On 2	70/50
East Wenatchee-D	Bone	242	29	19			
East Wenatchee-E	Bone	231	28	20			
East Wenatchee-F	Bone	190	26	18	2?	On 1	50/83(?)
East Wenatchee-G	Bone	232	30	22	1	Yes	
East Wenatchee-H	Bone	177	26	18	1	Yes	46
East Wenatchee-I	Bone	215	30	21			
East Wenatchee-J	Bone	171	27	19	1	Yes	42
East Wenatchee-K	Bone	193	28	20	1	Yes	50
East Wenatchee-L	Bone	115	13	12			
Sheaman	Ivory	203	12.1	10	1	Yes	74.7
Lower Klamath Lake	Bone	250±	13±		1		
Klamath Lake	Bone	190	15	12	1	No(?)	70
Saskatchewan-1	Bone	207	15	12.5		Yes‡	
Goose Lake-1d	Bone	133	10		1		
Goose Lake-1e	Bone	168	11		1		
Goose Lake-1f	Bone	197	13		1		
Goose Lake-2a	Bone	112	8		1		
Goose Lake-2b	Bone	198	12		2		
Goose Lake-2c	Bone	180	9		2		

All measurements are in mm. See text for most references; Florida is Jenks and Simpson (1941).
†If two bevels are present, two measurements are listed, separated by "/".
‡Cut groove encircles an end, but there is no bevel

on the fact that there are other attributes of the beveled ends that have not been discussed in the literature but which could be critical to correctly determining the function of the rods. Unfortunately, bevel length is often not reported, and with few exceptions (e.g., Jones 1996; Redmond and Tankersley 2005), neither is the angle of beveling.

Lahren and Bonnichsen (1974), following earlier workers (e.g., Cotter 1937; Hester 1972), presented a model of how the bevels might have served the hafting function (Fig. 15.4). Part of this model probably grew out of Cotter's (1954) earlier-noted remark that the North American rods resembled "sagaie or javelin points" from the European Upper Paleolithic. It is true that there are some resemblances—the European specimens are beveled, and some but not all bevels of the sagaie points have a pattern of grooves (e.g., Bordes 1968, Fig. 58 #1)—but that pattern is unlike the one on most

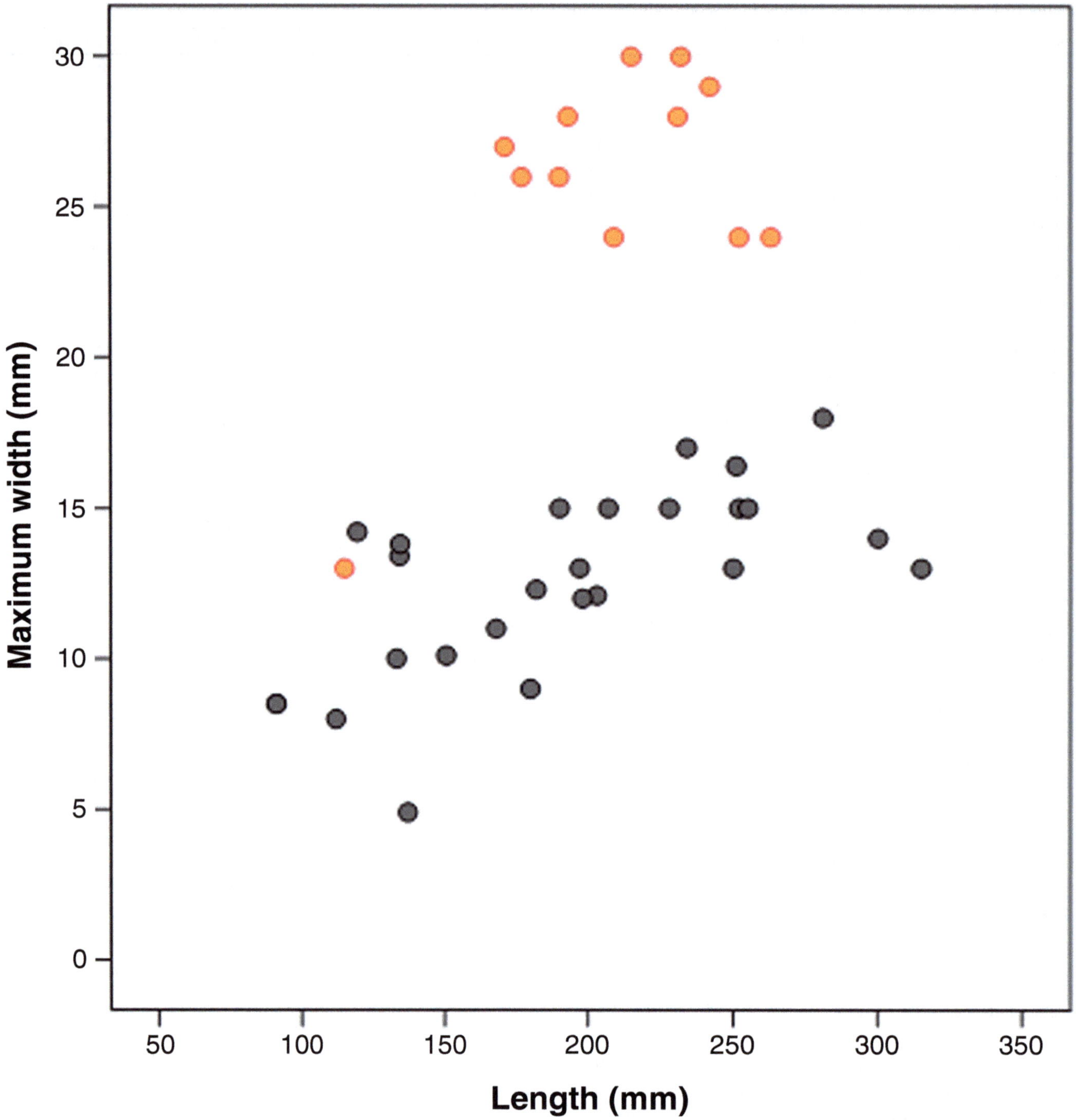

Fig. 15.3 Bivariate scatterplot of length versus maximum width for rods listed in Table 15.1. Specimens from East Wenatchee, Washington, are shown in *orange*

North American specimens. Further, unlike various North American specimens, examples of sagaie points such as those to which Cotter referred (1) all display single bevels, (2) taper from the distal end of the bevel more or less consistently to a point, and (3) have a straight rather than a convex face opposite the bevel (e.g., de Sonneville-Bordes 1963, Figs. 3 #8 and 7 #2; Bordes 1968, Figs. 55 #4, 56 #11, and 58 #2).

Experiments by Callahan (1994) indicate that a bevel-to-bevel haft works well and avoids the problem of limited penetration found with a socket haft. We note that a bevel-to-bevel haft avoids problems of penetration only if the face opposite the bevel is straight and if there is a smooth transition in diameter from the foreshaft to the shaft. Specimens from Anzick (Lahren and Bonnichsen 1974; Wilke et al. 1991)

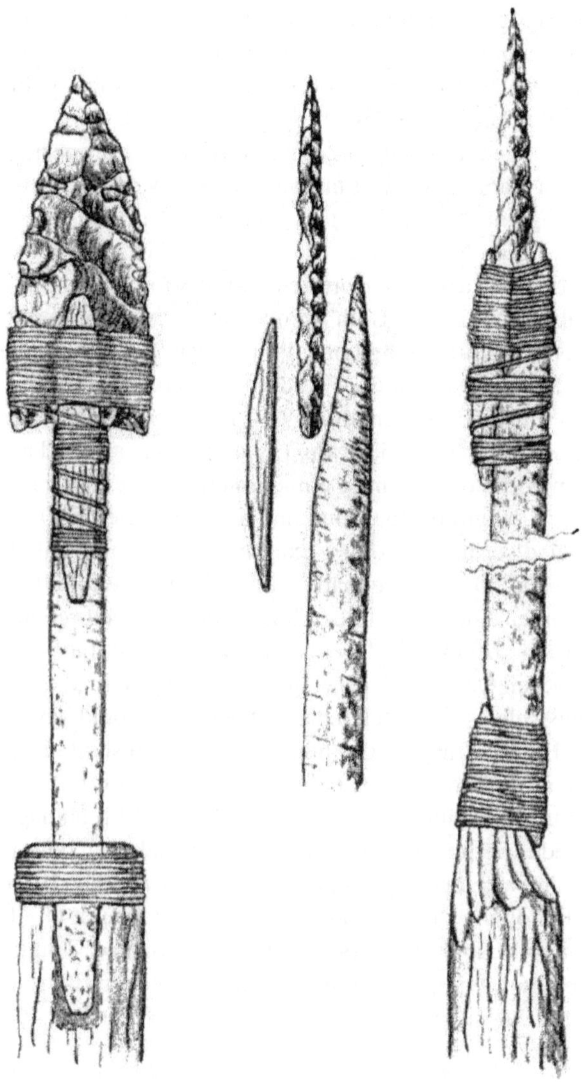

Fig. 15.4 Lahren and Bonnischsen's (1974) model of bi-beveled bone rods functioning as foreshafts

have this attribute, as does one rod from Blackwater Draw (Boldurian and Cotter 1999) and the ivory specimen from the Sheaman site in eastern Wyoming (Frison and Zeimens 1980; Frison 1982).

Lahren and Bonnichsen's model (Fig. 15.4) might be reasonable if the manner of hafting foreshafts to shafts is modified to a bevel-to-bevel haft, but there are three other attributes of their model that warrant comment. First, no bone or wood "splints" like that shown in Fig. 15.4 have ever been found. If they were made of wood, perhaps they did not preserve, but if they *were* made of wood, why were the main foreshafts made of bone? Second, the bases of Clovis points almost always were ground (e.g., Woods and Titmus 1985), which is unnecessary given the hafting model in Fig. 15.4 because the base of the point is not resting on anything. Such basal grinding would perhaps be necessary, however, if a point were hafted in and seated on the base of a wooden nock. If the base were sharp, or if there were not a strip of, say, hide between the point base and the nock base, the point would serve as an efficient wedge and split the shaft when the point met resistance during penetration. Third, experiments by Lyman et al. (1998) indicate a point hafted between a splint and a foreshaft as in Fig. 15.4 would result in poor alignment of the point, foreshaft, and shaft. Callahan (1994) used this hafting system successfully, but his version was rather different than that shown in Fig. 15.4.

Some experimenters (e.g., Frison 1974; Huckell 1982; Frison 1986, 1989; Smallwood 2015) have used as analogs wooden foreshafts (with points attached) recovered from late prehistoric contexts (Frison 1962, 1965). Frison (1989), Huckell (1982), and Callahan (1994) replicated such wooden foreshafts, hafted lithic points in a nock in the distal ends of the foreshafts, and seated the proximal ends of the foreshafts in sockets (of various shapes) in the ends of the main shafts. The replicate wooden foreshafts described by Frison (1989) and Huckell (1982) averaged 19.7 mm in diameter and ranged from 13.9 to 24 mm in diameter. The smallest one appeared to be too small (Frison 1989:769) to function properly, as it "broke in two places when used with a thrusting spear." Thus, Frison concluded that the optimum diameter of wooden foreshafts was 17–18 mm.

Stanford (1996) proposed a different foreshaft model that consists of two unique aspects. First, the bi-beveled rods are viewed as composite pieces that fit together to create a lengthened foreshaft "capable of lethal penetration into a mammoth" (Stanford 1996:45). If correct, one might wonder why some of the rod faces opposite the bevel are straight and others convex relative to the long axis of the rod (Lyman et al. 1998). Also, basic principles of geometry dictate that the angles of the bevels facing one another be identical to ensure a straight shaft. As well, we note that bone rods, even when the bevels are scored to increase friction, would be difficult to lash together and not have them come apart as a result of flexion. The second aspect of Stanford's (1996) model consists of an antler "foreshaft socket." The blunt end of an osseous rod serving as a foreshaft would sit in one end of the antler socket, which has a nock-like slot at both ends, and a Clovis point would be seated in the other end. Boldurian and Cotter (1999) proposed a similar model for the two bone rods from Blackwater Draw.

Pearson (1999) also thought of bi-beveled rods as foreshafts and proposed that two rods of equal length were glued and lashed together on their ventral surfaces around a projectile point and a main shaft. A bonding mastic was then applied to their midsections and to the cross-hatched bevels. A projectile point and a wedge-shaped shaft were then inserted into the mastic-covered "V" openings created by the facing bevels. The projectile point would have been tied to the narrower slot, whereas the shaft would have been secured to the opening with the wider angle. Lashings were then used to bind both pieces tightly around the projectile point and shaft. Parallel hatching on the surface opposite the bevels prevented the lashings from slipping during use. Depending on the manner in which the foreshaft was fastened to the main shaft, it may or may not have been detachable. Pearson (1999) proposed that the advantages of double bi-beveled-rod foreshafts over single-piece "clothes-pin" foreshafts—similar to Stanford's (1996) antler foreshaft socket—include faster repair time, high curation rate, increased resiliency under impact, and greater versatility in accommodating points of different dimensions (Pearson 1999). Unlike with regular clothes-pin foreshafts, it would be unnecessary to manufacture a whole new armature when one of the tangs broke.

Projectile Points

Some analysts (e.g., Cressman 1956; Frison and Zeimens 1980; Guthrie 1983; Stanford 1991; Redmond and Tankersley 2005; Waters et al. 2009, 2011) think some of the beveled rods from North America served as projectile points. In terms of one performance characteristic—penetration—experiments using antler, bone, and wooden projectiles indicate that antler—caribou (*Rangifer tarandus*), in particular—penetrates better than bone and that both of them perform better than wood (Butler 1980; Guthrie 1983). Penetration, however, is only one part of performance, others being the limits of material morphology, durability, and difficulty to repair. Experimental replication and use as reflected in the ethnographic record have documented other performance characteristics as they relate to flaked-stone points versus organic points (Ellis 1997; Knecht 1997; Elston and Brantingham 2002), but we are unaware of detailed studies that have focused on those same characteristics within the organic-point group (bone, ivory, and antler).

As noted in the discussion on foreshafts, European sagaie have a single bevel and taper more or less continuously to a sharp point. By far the most detailed descriptions of similar points from North America come from two bone rods from Sheriden Cave in north-central Ohio (Redmond and Tankersley (2005). Both points were made from split sections of mammal long bone. Point 1 was 134.2 mm long and had a maximum width of 13.8 mm, a maximum thickness of 10.6 mm, a bevel length of 46.0 mm, and a bevel angle of 7.0°. Point 2 (Fig. 15.5) was 119.4 mm long and had a maximum width of 14.2 mm, a maximum thickness of 11.6 mm, a bevel length of 46.9 mm, and a bevel angle of 10.5°. The beveled surface of point 1 exhibited deep, fine parallel incisions that likely were made with a chert-flake tool. On each specimen, additional fine oblique incisions occurred just distal to the beveled surface as well as on the surface opposite the bevel. On point 2, carving around the proximal end of the nonbeveled surface left a distinct node that could have facilitated the hafting of the bone rod to a foreshaft. Scanning electron microscopy of point 1 revealed minor impact damage to the pointed tip that closely matched tip damage produced experimentally on replicated Magdalenian bone points that impacted bone targets (Arndt and Newcomer 1986).

Following the methods outlined by Lyman et al. (1998), Redmond and Tankersley (2005) compared length and maximum-width measurements of complete rods from several locales, including 12 from East Wenatchee, 6 from

Fig. 15.5 Three views of bone point no. 2 from Sheriden Cave, Ohio. Note the scoring on the beveled end (*top view*) and the small purposely made protuberance on the opposite side from the bevel, which could have facilitated the hafting of the point to a wood or bone foreshaft. Photo courtesy Peter A. Bostrum, Lithic Casting Lab, Troy, Illinois

Florida, 2 from Anzick, the 2 Blackwater Draw specimens, and the 2 Sheriden Cave specimens. To this collection, they added nine complete single-bevel specimens (points) from the Upper Paleolithic Aurignacian V and Protomagdalenian levels in Laugerie-Haute rock shelter in southwestern France (Knecht 1991), which date to 26,400–24,900 cal BP. Redmond and Tankersley's bivariate plot of length versus maximum width, shown in Fig. 15.6 (top) with slight modification (a few of the specimens they included are not shown), shows that the Sheriden Cave specimens cluster tightly with the French points, as do two of the Florida specimens and the smallest specimen from East Wenatchee. The other North American specimens are longer and wider.

Redmond and Tankersley also compared maximum thickness and maximum width of the specimens for which thickness measurements were available (Fig. 15.6 [bottom]). Sheriden Cave points are more similar to the French points in terms of shaft morphometrics. The French points maintain a tight cluster of their own, with one slight outlier. Eight of the nine French specimens measure less than 11.0 mm in maximum width and less than 9.0 mm in maximum thickness. These relatively slender points are clearly set off from the larger rods, especially those from East Wenatchee, while the two specimens from Sheriden Cave are just to the upper right of the French specimens. The Sheriden Cave specimens are also similar to the Upper Paleolithic specimens

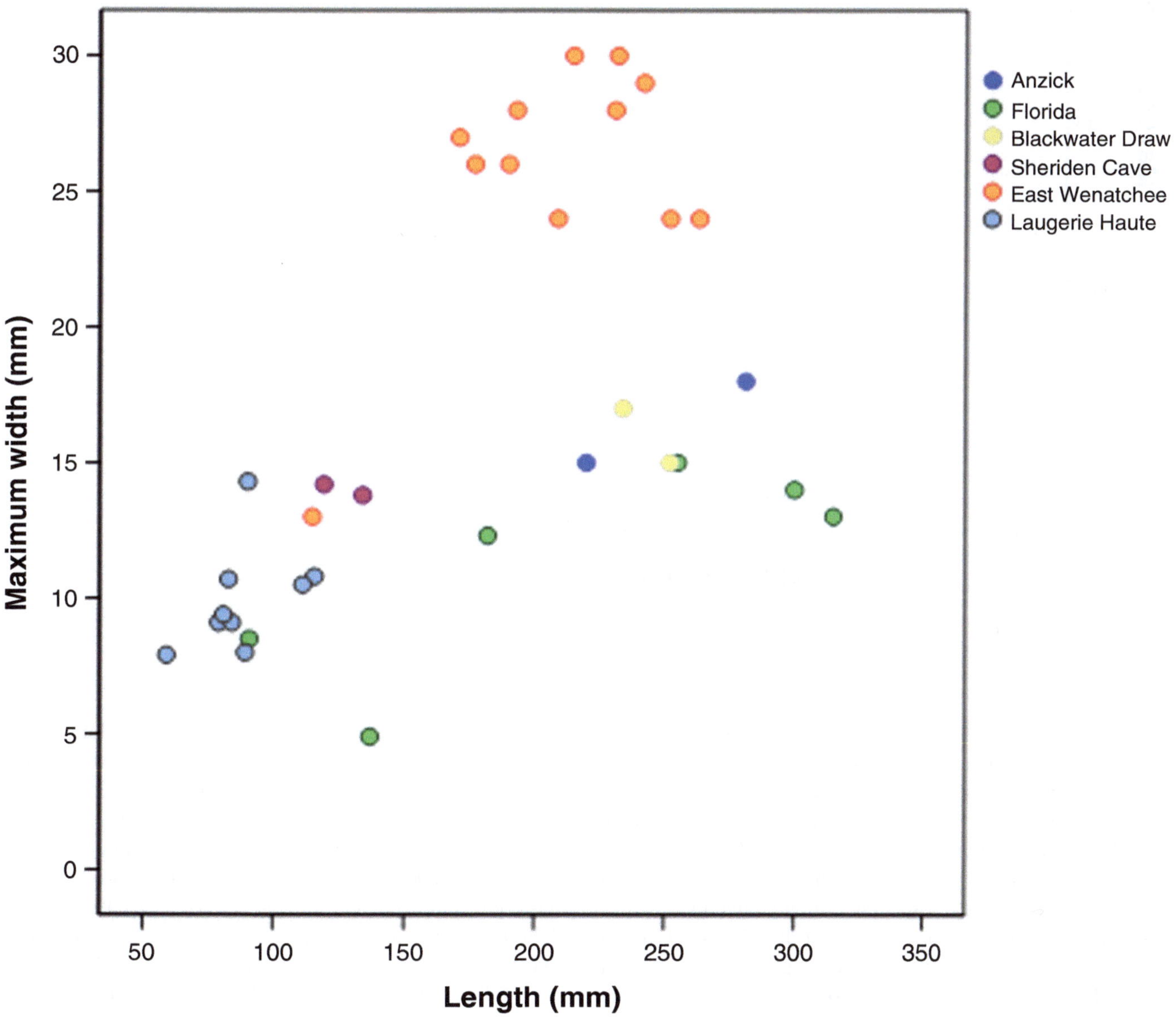

Fig. 15.6 Bi-variate scatterplots of length versus maximum width (*top*) and maximum thickness versus maximum width (*bottom*) of select osseous rods from various North American localities and from Upper Paleolithic levels in Laugerie-Haute rockshelter in southwestern France. After Redmond and Tankersley (2005)

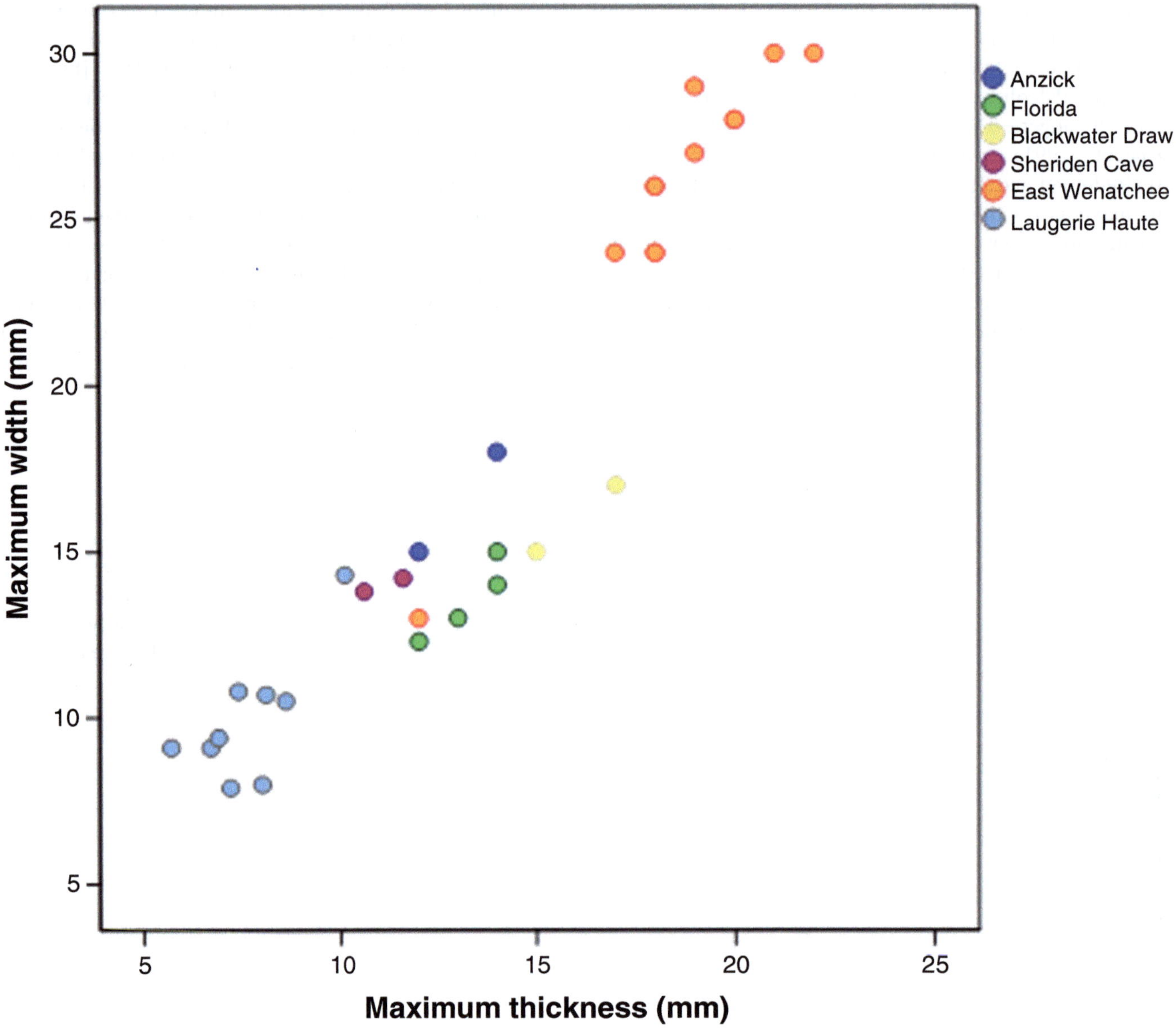

Fig. 15.6 (continued)

(e.g., Knecht 1991, 1993) in terms of manufacture: the use of bone over antler, longitudinal scraping of the shaft to form a tapered tip, scoring of the bevel with incisions, and the presence of striations on lateral margins and opposite the beveled surface (Redmond and Tankersley 2005).

One of the more interesting examples of a Paleoindian bone projectile point is a specimen found embedded in a rib of a single disarticulated mastodon at the Manis site on the Olympic Peninsula of western Washington. Excavations in 1977–1979 at the base of a kettle pond unearthed mastodon bones that showed evidence of spiral fracturing and cut marks (Gustafson et al. 1979; Gilbow 1981; Petersen et al. 1983; Gustafson 1985). Two little-known bone specimens displayed evidence of what was interpreted as use-related "polish" (Runnings et al. 1989). The single best-known artifact found associated with the mastodon was a foreign bone fragment, interpreted as the tip of a bone or antler projectile point, embedded in a mastodon rib (Gustafson et al. 1979; Waters et al. 2011). Radiocarbon ages of around 14,000 cal BP were obtained from organic matter associated with the mastodon, putting it earlier than Clovis (Gustafson 1985). Confidence in both the age of and evidence for human involvement with the Manis mastodon waned rapidly in the years following discovery (e.g., Grayson and Meltzer 2002; Haynes 2002); Dincauze (1984), for example, does not mention Manis in her overview of evidence for pre-Clovis archaeological materials in the western hemisphere. Subsequent AMS radiocarbon dating by Waters et al. (2011)

showed that the age of the mastodon was between 13,860 and 13,765 cal BP, still earlier than Clovis.

In addition, high-resolution X-ray computed tomography (CT) scanning revealed that the osseous object embedded in the rib was dense bone shaped to a point. Waters et al. (2011:352) note that "the point would have penetrated the hair and skin and about 25 to 30 cm of superficial epaxial muscles. … Thus it was at least 27 to 32 cm long, comparable with the known length of later, Clovis-age thrown and thrust bone points." Perhaps, but if the dimensions are correct, the Manis point was considerably longer than the points from Sheriden Cave (Redmond and Tankersley 2005). Waters and colleagues found no evidence of bone growth around the point, indicating that the mastodon died soon after the point entered it. Questions remain, however, as to whether the foreign object embedded in the rib is in fact a projectile point (Largent 2012).

Pressure-Flaker Handles

Wilke et al. (1991:258) suggest on the basis of ethnographic documentation that the rods from Anzick "represent a type of hand-held tool that once had an additional part attached to the beveled end with pitch and sinew. Such an implement would be a pressure flaker, with an antler bit bound to the beveled end or ends of a bone or ivory handle." Experiments they performed indicated that "pitch was necessary to keep the bit from slipping on the bevel, and incisions on the lateral and dorsal surfaces of the beveled ends of the handles were necessary to keep the sinew from slipping toward the [distal] end" (p. 259). Lahren and Bonnichsen (1974:149) state that "black material, probably resin, is still apparent on the beveled ends of six of the seven specimens." Damage on one end of one of the Anzick rods is thought to have been produced when the bit wore down and was not rehafted to extend beyond the end of the handle. Wilke et al. (1991:266) state that all rod specimens from Anzick are broken and note that the reason(s) for this "cannot be determined."

Sled-Runner Shoes

Gramly (1993) believes that the beveled bone rods from East Wenatchee were used as "sled shoes"—coverings for sled runners. Those specimens, however, are nothing like archaeological specimens of what have been called bone and ivory sled shoes associated with the Western Thule culture (ca. 1000 BP) of Alaska (Giddings and Anderson 1986) or with the Dorset culture (ca. 2500–1000 BP) of the eastern Arctic (Maxwell 1985). This is not to say that Paleoindian sled shoes had to resemble those made nine or ten millennia later; rather, the point is that those later archaeological specimens are not morphologically similar to the East Wenatchee bone rods. The Arctic specimens have wide, thin cross sections, relatively flat surfaces, and perforations that apparently were used in lashing the shoes to sled runners. If the East Wenatchee specimens had been so used, one would expect use-wear in the form of striae parallel to the long axis of the rods and distributed on only one face of each rod, but Gramly (1993) does not report evidence of this sort of wear. No evidence of such wear shows up on the precise replicas made by Peter Bostrum of the Lithic Casting Lab.

Wedges/Prybars

That rods of one sort or another were used prehistorically to butcher carcasses is indicated by some of the butchering marks on remains of late Pleistocene proboscidians in North American sites (e.g., Fisher, 1984a, b; Shipman et al. 1984; Fisher et al. 1994). Based on an analysis of mammoth remains from Blackwater Draw that were associated with two bi-beveled rods, Saunders and Daeschler (1994) proposed that the rods acted as wedges to dismember at least the feet of the animals. They made this proposal based on indentations observed on the foot bones of two mammoths that matched the dimensions of the rods. Experimental evidence supports the usefulness of bone rods during dismemberment (Park 1978). Reports on ethnoarchaeological research among modern African foragers who still hunt proboscidians do not mention the use of wedges or prybars to assist with butchering the carcass (Crader 1983; Fisher 1992), but the hunters employ metal tools, including hatchets and machetes, which may negate the necessity of a prybar to help hold joints apart for dissection with a stone knife.

Staffs

Bradley (1995) proposed that bi-beveled rods were placed bevel to bevel to form meter-long staffs that held some symbolic function—a proposal based on the fact that rods have been found in association with ochre-covered artifacts and skeletal remains, such as those from Anzick (Lahren and Bonnichsen 1974). We do not see much utility in this speculative scenario. As we noted with respect to the hypothesis of a lengthy foreshaft made of multiple bi-beveled rods, the bevel angles would have to be identical to create a straight staff. If the staff were, say, about 1 m long, then using the East Wenatchee specimens, which average about 20 cm in length, four bevel-to-bevel joints between five specimens would be required. A longer staff would require more joints. The sturdiness of such a composite tool is unknown.

Spears

Painter (1986) proposed that bi-beveled rods were midsections of broken tools that were then beveled and lashed together and finally outfitted with a pointed bone at the distal end and a blunted bone at the proximal end to create a composite spear. One problem with this proposition is that if bevels are the locations of repair, we would not expect a pair of bevels on one rod to always be located on the same side of the rod (Pearson 1999).

Hafting Wedges

Based on several lines of evidence—technological, contextual, and experimental—Lyman et al. (1998) proposed that some bi-beveled rods served a primary function as levered hafting wedges used to tighten sinew binding on saw-like Clovis implements. Their analysis centered on the 14 bone bi-beveled rods and 14 large Clovis points from East Wenatchee (Fig. 15.2). In their model, a wooden handle was nocked, a groove was cut in the handle to accommodate a bi-beveled rod, and a large Clovis-style biface was placed in the nock and wrapped with sinew. The groove for the rod extended onto one tang of the nock, and the rod was slid into the groove and levered down to tighten the overwrapped sinew. This leverage helped retighten the sinew as it became moist and stretched during butchering and precluded the necessity of unwrapping, remounting, and rewrapping to tighten the haft. Grooves cut on the bevels of the rod helped hold it in place when it is levered down. Lyman and colleagues proposed that the hafting-wedge function of the rods readily accounts for why they were beveled on both ends. Should the beveled end being used as a binding wedge fracture, one has but to merely turn the rod 180°, insert the intact edge under the haft binding, and lever the rod down to maintain a tight binding. The fact that 15 of the 18 preserved beveled ends of the East Wenatchee specimens display fractures across the bevel precisely like an experimentally broken hafting wedge lends strength to this interpretation of the East Wenatchee rods. Beveling of the proximal end—toward the handle—allows the binding holding it down to be more easily slipped on and off the levered-down end of the rod. The thick cross section of the East Wenatchee rods would have made for a larger cross section under the bevel, where the most force was concentrated when the rod was being used to tighten the haft binding.

One stimulus to the experimental work by Lyman and colleagues (Lyman et al. 1998; Lyman and O'Brien 1999) was the recovery context of many of the bi-beveled bone rods—caches containing large bifaces, fluted bifaces, and occasionally other items (Kilby 2008). These include Anzick (Jones and Bonnichsen 1994), East Wenatchee (Mehringer 1988a, b, 1989; Gramly 1993), and probably Drake, in northeastern Colorado (Stanford and Jodry 1988). It was the fact that 14 rods and 14 Clovis points—one rod per point—were found at East Wenatchee that prompted Lyman and colleagues to wonder if that correspondence might be significant. Incidentally, if Lassen (2005) is correct in his assessment of the number of rods represented among the pieces from Anzick—eight—then there are now two caches that contain equal numbers of rods and Clovis points.

Discussion

As Lyman et al. (1998:904) point out, "the archaeological record of Clovis-era rods is not what one might hope for." Of the specimens listed in Table 15.1, which represent only a small fraction of the number of specimens that have been found in North America, fewer than half were recovered from well-reported primary contexts. For example, it is unclear as to the precise nature of the recovery contexts of many of the specimens from Florida, but based on our review of the literature (e.g., Dunbar and Waller 1983; Dunbar et al. 1989; Dunbar and Webb 1996; Hemmings 1999; Webb and Hemmings 2001), few were in primary contexts. Instead, they came from sinkholes, rivers, and beaches. Aquatic environments provide protection for ivory and bone artifacts much more than do other depositional regimes except perhaps for peat bogs, limestone-enriched sediments, rockshelters, and xeric settings (Pearson 1999).

Even in the rare instances where rods have been recovered from primary contexts, there often is a lack of consensus as to the nature of the context. For example, Gramly (1993) stated that the specimens from East Wenatchee were located in a shallow 1.1-by-1.5-m pit—a conclusion based on observations of slightly darker, looser sediment above the artifacts. Mierendorf (1997) disputes the claim. At Anzick, initial thinking was that the rods and lithic tools were burial offerings interred with the remains of two juveniles (Lahren and Bonnichsen 1974; Wilke et al. 1991). Lahren (1999) revised this account to include only one of the two individuals with the cache (see Lassen 2005). Chronological reassessment by Morrow and Fiedel (2006) and Stafford (Waters and Stafford 2007) suggests that the human remains postdate deposition of the Clovis cache (see Lahren 2006 for historical details on the site).

Another issue that has stymied the study of Paleoindian bone and ivory rods is inconsistency in how data have been reported (Lyman et al. 1998; Moore and Schmidt 2009). For example, there is minimal consistency in the specific attributes chosen to describe particular specimens, with the exception that it is typically, but not always, noted that a particular specimen is beveled on one or both ends, made of bone or ivory, and is long relative to width and thickness. As a result of the quality of the published record, it is unclear if, for example, variation in length and maximum width displayed by a sample of these specimens (Fig. 15.3) represents morphological variation that is somehow functionally significant. Further, inferring typological identity of specimens cannot be accomplished with any reliability because there is no agreed-on set of necessary and sufficient conditions for type membership. An additional stumbling block here is that minimal discussion has been offered as to the analytical purpose of the types. Are they for descriptive purposes, are they index fossils indicative of age or cultural affiliation, or are they meant to facilitate interpretation of the function(s) of the specimens? We suggest that what might be referred to as *principles of systematics* is where future studies of these fascinating items must begin (e.g., Lyman and O'Brien 2002; O'Brien and Lyman 2002).

Chronology is also problematic for bone and ivory technology in North America because with few exceptions, such as Sheriden Cave (Redmond and Tankersley 2005; Waters et al. 2009, 2011) and Anzick (Morrow and Fiedel 2006), temporal affiliation is based solely on (1) association with Clovis points, such as at East Wenatchee, or (2) morphological similarity with specimens from those associations, such as the Florida finds. We currently do not know if or how single- or bi-beveled rods changed over time in terms of function or when they dropped out of use.

Based on what we *do* know about beveled rods, it appears that, as Taylor (2006) listed them, the following characteristics generally apply. The rods are:

- made from mammoth bone or ivory
- 150–250 mm long
- 10–30 mm wide
- 10–22 mm thick
- beveled on one or both ends
- scored on the beveled surface with cross hatching
- found with Clovis points and bifaces in cache and kill sites

With very few exceptions, these characteristics line up well with those specimens listed in Table 15.1.

If form and function are related, it appears that bone and ivory rods served a variety of purposes. Our best guess, backed up by experimental data, is that single-bevel pieces served as projectile points—certainly the evidence from one of the specimens from Sheriden Cave (Redmond and Tankersley 2005) indicates as much—whereas bi-beveled rods could have served as foreshafts and perhaps as hafting wedges. Certainly any piece at any time could have been multipurpose, including serving as a prybar, as Saunders and Daeschler (1994) propose for the specimens from Blackwater Draw. As more specimens are described, with an eye to the kind of detail that Redmond and Tankersley (2005; Waters et al. 2009) noted for the specimens from Sheriden Cave, our knowledge of how Paleoindian osseous rods were made and used should increase considerably. We echo a point made by Moore and Schmidt (2009:57): Given appropriate attention to such things as microtraces of manufacturing and use-wear, "organic implements can provide a more than adequate means of developing and testing hypotheses concerning prehistoric technological organization, social interaction, and settlement distributions."

Note Radiocarbon dates discussed here appear in the literature in various forms, but irrespective of whether a date was reported in raw radiocarbon years, as a calibrated date, or both, we (re)calibrated all dates using CalPal ver.1.5 (http://www.calpal-online.de) to create uniformity.

Acknowledgments We thank Michelle Langley for inviting us to participate in the volume. We also thank Michael R. Waters and the Center for the Study of the First Americans, Texas A&M University; Peter A. Bostrom and the Lithic Casting Laboratory; and two anonymous reviewers, both of whom helped clarify various points of discussion.

References

Ackerman, R. E. (1996). Lime Hills, Cave 1. In F. H. West (Ed.), *American beginnings: The prehistory and palaeoecology of Beringia* (pp. 470–477). Chicago: University of Chicago Press.

Anderson, D. G., Miller, D. S., Yerka, S. J., Gillam, J. C., Johanson, D. T., Goodyear, A. C., et al. (2010). Paleoindian database of the Americas 2010: Current status and findings. *Archaeology of Eastern North America, 38*, 63–90.

Arndt, S., & Newcomer, M. (1986). Breakage patterns on prehistoric bone points. In D. A. Roe (Ed.), *Studies in the Upper Paleolithic of Britain and Northwest Europe. British Archaeological Reports International Series S296* (pp. 165–173). Oxford: Archaeopress.

Boldurian, A. T., & Cotter, J. L. (1999). *Clovis revisited: New perspectives on Paleoindian adaptations from Blackwater Draw, New Mexico*. Philadelphia: University Museum/University of Pennsylvania.

Bonnichsen, R., Stanford, D., & Fastook, J. L. (1987). Environmental change and developmental history of human adaptive patterns: The Paleoindian case. In W. F. Ruddiman & H. E. Wright Jr. (Eds.), *North America and adjacent oceans during the last deglaciation: The geology of North America* (Vol. K-3, pp. 403–424). Boulder: Geological Society of America.

Bordes, F. (1968). *The old stone age*. New York: McGraw-Hill.

Bradley, B. A. (1995). *Clovis ivory and bone rods*. www.primtech.net/ivory/ivory.html.

Bradley, B. A., Collins, M. B., & Hemmings, A. (2010). *Clovis technology*. Ann Arbor: International Monographs in Prehistory.

Butler, W. B. (1980). Penetrating elephant hide with wood atlatl darts. *Plains Anthropologist, 25*, 353–356.

Callahan, E. (1994). A Mammoth undertaking. *Bulletin of Primitive Technology, 1*, 23–29.

Collard, M., Buchanan, B., Hamilton, M. J., & O'Brien, M. J. (2010). Spatiotemporal dynamics of the Clovis–Folsom transition. *Journal of Archaeological Science, 37*, 2513–2519.

Cotter, J. L. (1937). The occurrence of flints and extinct animals in pluvial deposits near Clovis, New Mexico, part IV, report on excavation at the Gravel Pit, 1936. *Proceedings of the Academy of Natural Sciences of Philadelphia, 89*, 1–16.

Cotter, J. L. (1954). Indications of a Paleo-Indian co-tradition for North America. *American Antiquity, 20*, 64–67.

Crader, D. C. (1983). Recent single-carcass bone scatters and the problem of "butchery" sites in the archaeological record. In J. Clutton-Brock & C. Grigson (Eds.), *Animals and archaeology: 1. Hunters and their prey. British Archaeological Reports International Series 163* (pp. 107–141). Oxford: Archaeopress.

Craven, S. L. (2004). New dates from the Lind Coulee site (45GR97), Washington. *Current Research in the Pleistocene, 21*, 28–30.

Cressman, L. S. (1942). Archaeological researches in the northern Great Basin. *Carnegie Institution of Washington, Publication, 538.*

Cressman, L. S. (1956). Klamath prehistory. *Transactions of the American Philosophical Society, 46*, 374–513.

Daugherty, R. D. (1956). Archaeology of the Lind Coulee site, Washington. *Proceedings of the American Philosophical Society, 100*, 223–278.

de Sonneville-Bordes, D. (1963). Upper Palaeolithic cultures in Western Europe. *Science, 142*, 347–355.

Dincauze, D. F. (1984). An archaeological evaluation of the case for pre-Clovis occupations. In F. Wendorf & A. E. Close (Eds.), *Advances in world archaeology* (Vol. 3, pp. 275–323). New York: Academic Press.

Dunbar, J. S., & Waller, B. (1983). Interview with Ben Waller. *Florida Anthropologist, 36*, 31–39.

Dunbar, J. S., & Webb, S. D. (1996). Bone and ivory tools from Paleoindian sites in Florida. In D. G. Anderson & K. E. Sassaman (Eds.), *The Paleoindian and Early Archaic Southeast* (pp. 331–353). Tuscaloosa: University of Alabama Press.

Dunbar, J. S., Webb, S. D., & Cring, D. (1989). Culturally and naturally modified bones from a Paleoindian site in the Aucilla River. In R. Bonnichsen & M. Sorg (Eds.), *Bone modification* (pp. 473–497). Orono: Center for the Study of the First Americans.

Ellis, C. J. (1997). Factors influencing the use of stone projectile tips. In H. Knecht (Ed.), *Projectile technology* (pp. 37–74). New York: Plenum.

Elston, R. G., & Brantingham, P. J. (2002). Microlithic technology in northern Asia: A risk-minimizing strategy of the Late Paleolithic and Early Holocene. In R. G. Elston & S. L. Kuhn (Eds.), *Thinking small: Global perspectives on microlithization* (pp. 103–116). Washington, DC: Archeological Papers of the American Anthropological Association, No. 12.

Fisher, D. C. (1984a). Mastodon butchery by North American Paleo-Indians. *Nature, 308*, 271–272.

Fisher, D. C. (1984b). Taphonomic analysis of late Pleistocene mastodon occurrences: Evidence of butchery by North American Paleo-Indians. *Paleobiology, 10*, 338–357.

Fisher, D. C., Lepper, B. T., & Hooge, P. E. (1994). Evidence for butchery of the Burning Tree mastodon. In W. S. Dancey (Ed.), *The first discovery of America: Archaeological evidence of the early inhabitants of the Ohio area* (pp. 43–57). Columbus: Ohio Archaeological Council.

Fisher, J. W., Jr. (1992). Observations on the Late Pleistocene bone assemblage from the Lamb Spring site, Colorado. In D. J. Stanford & J. S. Day (Eds.), *Ice Age hunters of the Rockies* (pp. 51–81). Boulder: University Press of Colorado.

Frison, G. C. (1962). Wedding of the Waters Cave: A stratified site in the Bighorn Basin of northern Wyoming. *Plains Anthropologist, 7*, 246–265.

Frison, G. C. (1965). Spring Creek Cave, Wyoming. *American Antiquity, 31*, 81–94.

Frison, G. C. (1974). *The Casper site: A Hell Gap bison kill on the High Plains*. New York: Academic Press.

Frison, G. C. (1978). *Prehistoric hunters of the High Plains*. New York: Academic Press.

Frison, G. C. (1982). The Sheaman site: A Clovis component. In G. C. Frison & D. Stanford (Eds.), *The Agate Basin site: A record of the Paleoindian occupation of the northwestern High Plains* (pp. 143–157). New York: Academic Press.

Frison, G. C. (1986). Mammoth hunting and butchering from a perspective of African elephant culling. In G. C. Frison & L. C. Todd (Eds.), *The Colby Mammoth site* (pp. 115–134). Albuquerque: University of New Mexico Press.

Frison, G. C. (1989). Experimental use of Clovis weaponry and tools on African elephants. *American Antiquity, 54*, 766–784.

Frison, G. C., & Craig, C. (1982). Bone, antler, and ivory artifacts and manufacture technology. In G. C. Frison & D. Stanford (Eds.), *The Agate Basin site: A record of the Paleoindian occupation of the northwestern High Plains* (pp. 157–173). New York: Academic Press.

Frison, G. C., & Zeimens, G. M. (1980). Bone projectile points: An addition to the Folsom cultural complex. *American Antiquity, 45*, 231–237.

Giddings, J. L., & Anderson, D. D. (1986). *Beach ridge archaeology of Cape Krusenstern: Eskimo and pre-Eskimo settlements around Kotzebue Sound, Alaska*. Washington, DC: National Park Service Publications in Archaeology, No. 20.

Gilbow, D. W. (1981). *Inference of human activity from faunal remains.* MA thesis, Washington State University.

Gramly, R. M. (1993). *The Richey Clovis cache*. Buffalo: Persimmon Press.

Grayson, D. K., & Meltzer, D. J. (2002). Clovis hunting and large mammal extinction: A critical review of the evidence. *Journal of World Prehistory, 16*, 313–359.

Gustafson, C. E. (1985). The Manis Mastodon site. *National Geographic Research Reports, 20*, 283–295.

Gustafson, C. E., Gilbow, D., & Daugherty, R. (1979). The Manis mastodon site: Early man on the Olympic Peninsula. *Canadian Journal of Archaeology, 3*, 157–164.

Guthrie, R. D. (1983). Osseous projectile points: Biological considerations affecting raw material selection and design among Palaeolithic and Paleoindian peoples. In J. Clutton-Brook & C. Grigson (Eds.), *Animals and archaeology, British Archaeological Reports International Series 163* (pp. 273–294). Oxford: Archaeopress.

Haynes, C. V., Kornfeld, M., & Frison, G. C. (2004). Short contribution: New geochronological and archaeological data for the Sheaman Clovis site, eastern Wyoming, U.S.A. *Geoarchaeology, 19*, 369–379.

Haynes, G. (2002). *The early settlement of North America: The Clovis era*. Cambridge: Cambridge University Press.

Hemmings, C. A. (1999). *The Paleoindian and Early Archaic tools of Sloth Hole (8JE121): An inundated site in the lower Aucilla River, Jefferson County, Florida*. MA thesis, University of Florida.

Hemmings, C. A. (2004). *The organic Clovis: A single continent-wide cultural adaptation.* Ph.D. Dissertation, University of Florida.

Hester, J. J. (compiler). (1972). *Blackwater Locality No. 1: A stratified early man site in eastern New Mexico*. Fort Burgwin Research Center Publication, No. 8. Dallas.

Holmes, C. E. (1996). Broken Mammoth. In F. H. West (Ed.), *American beginnings: The prehistory and palaeoecology of Beringia* (pp. 312–318). Chicago: University of Chicago Press.

Huckell, B. B. (1982). The Denver elephant project: A report on experimentation with thrusting spears. *Plains Anthropologist, 27*, 217–224.

Irwin, A. M., & Moody, U. (1978). *The Lind Coulee site (45GR97)*. Washington Archaeological Research Center, Project Report, No. 56. Pullman.

Jenks, A. E., & Simpson, H. H. (1941). Beveled artifacts in Florida of the same type as artifacts found near Clovis, New Mexico. *American Antiquity, 4*, 314–319.

Jones, J. S. (1996). *The Anzick site: Analysis of a Clovis burial assemblage*. MA thesis, Oregon State University.

Jones, J. S., & Bonnichsen, R. (1994). The Anzick Clovis burial. *Current Research in the Pleistocene, 11*, 42–44.

Kilby, J. D. (2008). *An investigation of Clovis caches: Content, function, and technological organization.* Ph.D. Dissertation, University of New Mexico.

Knecht, H. D. (1991). *Technological innovation and design during the early Upper Paleolithic: A study of organic projectile technologies*. Ph.D. dissertation, New York University.

Knecht, H. D. (1993). Early Upper Paleolithic approaches to bone and antler projectile technology. In G. L. Peterkin, H. M. Bricker, & P. Mellars (Eds.), *Hunting and animal exploitation in the later Palaeolithic and Mesolithic of Eurasia* (pp. 33–47). Washington, DC: Archeological Papers of the American Anthropological Association, No. 4.

Knecht, H. D. (1997). Projectile points of bone, antler, and stone: Experimental explorations of manufacture and use. In H. Knecht (Ed.), *Projectile technology* (pp. 191–212). New York: Plenum.

Lahren, L. A. (1999). Human skeletal remains from the Anzick site (24PA506). Manuscript on file, Center for the Study of the First Americans, Texas A&M University, College Station.

Lahren, L. A. (2006). *Homeland: An archaeologist's view of Yellowstone Country's past*. Livingston: Cayuse Press.

Lahren, L. A., & Bonnichsen, R. (1974). Bone foreshafts from a Clovis burial in southwestern Montana. *Science, 186*, 147–150.

Largent, F. (2012). Reconsidering the Manis mastodon. *Mammoth Trumpet, 27*(4), 17–20.

Lassen, R. D. (2005). *A comparison of Clovis caches*. MA thesis, Texas A&M University.

Lyman, R. L., & O'Brien, M. J. (1999). Prehistoric osseous rods from North America: Arguments on function. *North American Archaeologist, 20*, 347–364.

Lyman, R. L., & O'Brien, M. J. (2002). Classification. In J. P. Hart & J. E. Terrell (Eds.), *Darwin and archaeology: A handbook of key concepts* (pp. 69–88). Westport: Bergin & Garvey.

Lyman, R. L., O'Brien, M. J., & Hayes, V. (1998). A mechanical and functional study of bone rods from the Richey–Roberts Clovis cache, Washington, U.S.A. *Journal of Archaeological Science, 25*, 887–906.

Maxwell, M. S. (1985). *Prehistory of the eastern Arctic*. Orlando: Academic Press.

Mehringer, P. J., Jr. (1988a). Clovis cache found: Weapons of ancient Americans. *National Geographic, 174*, 500–503.

Mehringer, P. J., Jr. (1988b). The Richey–Roberts Clovis cache, East Wenatchee, Washington. *Northwest Science, 62*, 271–272.

Mehringer, P. J., Jr. (1989). Of apples and archaeology. *Universe, 1*, 2–8. Pullman: Washington State University.

Mierendorf, R. R. (1997). Comments on the East Wenatchee Clovis site (45DO482), Washington state, as reported on by Richard M. Gramly. *Current Research in the Pleistocene, 14*, 57–60.

Moore, C. R., & Schmidt, C. W. (2009). Paleoindian and Early Archaic organic technologies: A review and analysis. *North American Archaeologist, 30*, 57–86.

Morrow, J. E., & Fiedel, S. J. (2006). New radiocarbon dates for the Clovis component of the Anzick site, Park County, Montana. In J. E. Morrow & C. Gnecco (Eds.), *Paleoindian archaeology: A hemispheric perspective* (pp. 123–138). Gainesville: University of Florida Press.

O'Brien, M. J., & Lyman, R. L. (2002). The epistemological nature of archaeological units. *Anthropological Theory, 2*, 37–56.

Painter, F. (1986). Pointed weapons of wood, bone and ivory: Survival tools of early man in North America. *Central States Archaeological Journal, 33*, 62–76.

Park, E. (1978). The Ginsberg caper: Hacking it as in Stone Age. *Smithsonian, 9*, 85–96.

Pearson, G. A. (1999). North American Paleoindian bi-beveled bone and ivory rods: A new interpretation. *North American Archaeologist, 20*, 81–103.

Petersen, K. L., Mehringer, P. J., Jr., & Gustafson, C. E. (1983). Late-glacial vegetation and climate at the Manis Mastodon site, Olympic Peninsula, Washington. *Quaternary Research, 20*, 215–231.

Rainey, F. (1939). Archaeology in central Alaska. *American Museum of Natural History, Anthropological Papers, 36*, 351–405.

Rainey, F. (1940). Archaeological investigation in central Alaska. *American Antiquity, 5*, 299–308.

Redmond, B. G., & Tankersley, K. B. (2005). Evidence of early Paleoindian bone modification and use at the Sheriden Cave site (33WY252), Wyandot County, Ohio. *American Antiquity, 70*, 503–526.

Riddell, F. A. (1973). Fossilized California bone artifacts. *Masterkey, 47*(1), 28–32.

Runnings, A. L., Gustafson, C. E., & Bentley, D. (1989). Use-wear on bone tools: A technique for study under the scanning electron microscope. In R. Bonnichsen & M. H. Sorg (Eds.), *Bone modification* (pp. 259–266). Orono: Center for the Study of the First Americans.

Saunders, J. J., & Daeschler, E. B. (1994). Descriptive analyses and taphonomical observations of culturally-modified mammoths excavated at "The Gravel Pit," near Clovis, New Mexico in 1936. *Proceedings of the Academy of Natural Sciences of Philadelphia, 145*, 1–28.

Sellards, E. H. (1952). *Early man in America*. Austin: University of Texas Press.

Shipman, P., Fisher, D. C., & Rose, J. J. (1984). Mastodon butchery: Microscopic evidence of carcass processing and bone tool use. *Paleobiology, 10*, 358–365.

Smallwood, A. M. (2015). Building experimental use-wear analogues for Clovis biface functions. *Archaeological and Anthropological Sciences, 7*, 13–26.

Stanford, D. J. (1991). Clovis origins and adaptations: An introductory perspective. In R. Bonnichsen & K. L. Turnmire (Eds.), *Clovis: Origins and adaptations* (pp. 1–13). Corvallis: Center for the Study of the First Americans.

Stanford, D. J. (1996). Foreshaft sockets as possible Clovis hafting devices. *Current Research in the Pleistocene, 13*, 44–46.

Stanford, D., & Bradley, B. (2012). *Across Atlantic ice: The origin of America's Clovis culture*. Berkeley: University of California Press.

Stanford, D. J., & Jodry, M. (1988). The Drake Clovis cache. *Current Research in the Pleistocene, 5*, 21–22.

Taylor, J. (2006). *Projectile points of the High Plains*. Buffalo: Jeb Taylor Artifacts.

Wagers, C. S. (1986). Florida river bone points. *Prehistoric Artifacts of North America, 20*, 20–21.

Waters, M. R., & Stafford, T. W., Jr. (2007). Redefining the age of Clovis: Implications for the peopling of the Americas. *Science, 315*, 1122–1126.

Waters, M. R., Stafford, T. W., Jr., McDonald, H. G., Gustafson, C., Rasmussen, M., Cappellini, E., et al. (2011). Pre-Clovis mastodon hunting 13,800 years ago at the Manis site, Washington. *Science, 334*, 351–353.

Waters, M. R., Stafford, T. W., Jr., Redmond, B. G., & Tankersley, K. B. (2009). The age of the Paleoindian assemblage at Sheriden Cave, Ohio. *American Antiquity, 74*, 107–111.

Webb, S. D. (2006). *First Floridians and last mastodons: The Page–Ladson site in the Aucilla River*. Gainesville: University Press of Florida.

Webb, S. D., & Hemmings, C. A. (2001). Ivory and bone tools from late Pleistocene deposits in the Aucilla and Wacissa River, north-central Florida. In B. A. Purdy (Ed.), *Enduring records: The environmental and cultural heritage of wetlands* (pp. 1–8). Oxford: Oxbow Books.

Wilke, P. J., Flenniken, J. J., & Ozbun, T. L. (1991). Clovis technology at the Anzick site, Montana. *Journal of California and Great Basin Anthropology, 13*, 242–272.

Willis, C. (1988). Controlled surface collection of the Little River Rapids site (8Je603): A stratigraphically deflated site in the Aucilla River, north Florida. *Florida Anthropologist, 41*, 453–470.

Wilmeth, R. (1968). A fossilized bone artifact from southern Saskatchewan. *American Antiquity, 33*, 100–101.

Woods, J. C., & Titmus, G. L. (1985). A review of the Simon Clovis collection. *Idaho Archaeologist, 8*, 3–8.

Chapter 16
Hunter-Gatherers of the Old and New Worlds: Morphological and Functional Comparisons of Osseous Projectile Points

Marianne Christensen, Dominique Legoupil, and Jean-Marc Pétillon

Abstract The osseous projectile points and tools of hunter-gatherers from the European Pleistocene compare surprisingly well with the equipment of hunters from other continents, including the New World. This is especially true for harpoon heads and barbed spear points. These fundamental hunting and fishing weapons are common to the prehistoric populations of the Old World and to hunter-gatherers of the northern and southern regions of the New World. In southernmost South America, osseous projectiles have survived the millennia since the first human occupations, about 6200 years ago, until modern times. Beyond certain typological, and likely functional, constants that are commonly found among cold region hunters, they also display specific features (size, morphology of the proximal ends, raw materials) that seem to reflect techniques and hunting strategies associated with particular species. In this chapter, our intention is to examine potentially meaningful similarities between this equipment from Patagonia and Tierra del Fuego, known through both archaeological and ethnological documents, and that of European Pleistocene hunters. We will emphasize certain morphological and technical features, such as proximal shapes for hafting mechanisms or line attachment systems, and the number and type of barbs, along with their functional causes and consequences. Our results indicate that in the current state of the debate, fishing, fowling and small mammal hunting is the most plausible hypothesis for the use of barbed elements in the terrestrial context of most Upper Magdalenian sites. Though we cannot exclude the possibility that Upper Magdalenian groups were among the very few hunter-gatherers to use detachable harpoons to hunt larger terrestrial species, such as ungulates when crossing rivers, specific evidence is currently missing.

Keywords New World • Old World • Morphology • Functionality • Ethnoarchaeology

M. Christensen (✉)
Institut d'Art et d'Archeologie, Université de Paris I Pantheon-Sorbonne, UMR 7041 – ArScAn, Équipe Ethnologie Préhistorique, Paris, France
e-mail: marianne.christensen@univ-paris1.fr

D. Legoupil
CNRS - UMR 7041 – ArScAn, Équipe Ethnologie Préhistorique, Maison de l'Archéologie et de l'Ethnologie, Nanterre, France
e-mail: dominique.legoupil@mae.cnrs.fr

J.-M. Pétillon
CNRS - UMR 5608, Laboratoire TRACES, Université Toulouse Jean-Jaurès, Toulouse, France
e-mail: petillon@univ-tlse2.fr

Introduction

Regardless of the material they are made from (metal, stone, osseous materials, wood), projectile points correspond to a concept as universal as the knife and are essential to fulfilling the needs of hunter-gatherer societies. The feature they share is a sharp point (rarely a sharp edge) that enables them to penetrate the prey or enemy. The penetration of the weapon depends on the shape of the point and the delivery mode. It can be hand-held or launched, either directly by the sheer strength of the hunter, or indirectly with a launching device (spearthrower, bow, crossbow, etc.).

Projectile points have been given many names that vary more or less depending on the language: pike points (*pointe d'hast*), casting points (*pointe de jet*), arrows (*pointe de trait*), projectile points (*pointe d'armature*); as have the weapons to which they refer: spear, pike, javelin, assegai, harpoon, bolt, arrow, etc. (*lance, pique, javelot, dard, sagaie, harpon, carreau, flèche*). In the case of osseous points made from bone, ivory and antler, the range also includes the harpoon, a point with barbs that is nearly impossible to make from a lithic material and that is designed to keep the prey from escaping.

Michelle C. Langley (ed.), *Osseous Projectile Weaponry*, Vertebrate Paleobiology and Paleoanthropology,
DOI 10.1007/978-94-024-0899-7_16

Present among prehistoric hunters of the Old World, as well as those of the northern (Arctic and sub-Arctic regions) and southern regions of the New World, these weapons are the subject of our discussion. We will focus specifically on the equipment of sea-nomads from the southernmost region of America. These hunter-gatherers, less known than others, perhaps due to their smaller number and limited territory, continuously used harpoons from the mid-Holocene to the early twentieth century. Based on this case study, for which ethnographic records exist, we will seek elements that can contribute to our understanding of the function of these osseous points, whose use is frequently debated in archaeological contexts, including the Magdalenian period.

We will examine possible correlations between hunting tools in different parts of the world and attempt to identify functional trends. We will emphasize certain morphological characteristics and try to distinguish specific features (size, barb and hafting mechanism morphologies, and raw materials) and their role in the hunting strategies associated with different animal species. Finally, insofar as possible, we will attempt to clarify the definition of harpoons.

The Bone Points of the Patagonian and Tierra del Fuego Indians: Ethnographic Data

Detachable Harpoons and Spears: A Clearly Distinct Function

The iconic weapons of hunter-gatherers from southern South America, known as "harpoon spears", or rather their tips, are among the objects most often collected by the navigators that travelled through the Magellan Strait to pass from the Atlantic to the Pacific, or who sought to avoid Cape Horn, sometimes making stops on the southern islands or in the Beagle Channel (Fig. 16.1).

These weapons were often exchanged for axes, nails or clothes and travellers reported them as curiosities. We thus find them in many European and North American museums (Borrero and Borella 2010; Estévez and Vila 2013). They were usually made of marine mammal bone (cetaceans and pinnipeds), or exceptionally of wood. They were sometimes made with metal tools. These harpoon heads are characteristic of the "*Indios canoeros*" (Indians in a canoe), Yamana (or Yaghan), Alakaluf (or Kaweskar) and Chono, who occupied the labyrinth of islands and channels along the Pacific facade ending in Cape Horn and Tierra del Fuego. They are very rare among terrestrial hunters, the Patagonian giants of Magellan (or Tehuelches) on the continent and Selk'nam (or Ona) on the big island of Tierra del Fuego, who instead used the "*boleadora*" (three linked stone balls) or bows to hunt guanaco (*lama guanicoe*) on the steppes of the Atlantic highlands.

In fact, within the category "harpoon" two kinds of weapons coexist:

(1) The harpoon—*stricto sensu*—with a detachable head, mobile and connected to a line; and
(2) a kind of weapon that some call "harpoon", with a fixed head, designated by others as lance or spear, depending on the language (Latin or Anglo-Saxon) of the numerous navigators and ethnologists who observed the equipment of Patagonian Indians in the nineteenth (Hyades and Deniker 1891; Spears 1895; Bridges 1998; etc.) and early twentieth centuries (Gallardo 1910; Cooper 1917; Gusinde 1986 [1937], Lothrop 2002 [1928], etc.).

The confusion is increased by the multifaceted nature of the "fixed harpoon", which is occasionally single-barbed, but usually multi-barbed, sometimes unilateral, sometimes bilateral. This harpoon type can easily be transformed into a composite by-product when armed with two or three points (or prongs), known as a leister spear.

The functioning of these two weapons (Fig. 16.2 below) is fundamentally different, however, as clearly described by Hyades, doctor and ethnologist for the Scientific Mission of Cape Horn, who spent a year with the Yamana on Hoste Island in 1882–1883:

> "*Les harpons constituent l'arme la plus usitée pour la chasse. Il y en a de plusieurs formes; les plus communes sont en os et de deux espèces: l'une à pointe fixe avec une rangée de dents..., l'autre à pointe mobile avec une seule entaille... Dans les deux cas, la pointe est adaptée à un manche de 3m à 4m de long, en bois de Fagus betuloides ou de Drimys winteri. Les harpons mobiles servent à la chasse des otaries, que les Fuégiens guettent dans leur pirogue et sur lesquelles ils lancent leur harpon, qui, pénétrant dans le corps de l'animal, se détache du manche: celui-ci sert ainsi de bouée, et guide le chasseur qui peut poursuivre sa proie et lui infliger de nouveaux coups jusqu'à ce qu'elle soit tuée. Les harpons en dents de scie, fixes sur le manche, servent à la chasse aux oiseaux, aux loutres, et représentent le plus commun des harpons fuégiens. Lovisato*[1] *leur donne le nom de lance, parce que, dit-il, autant qu'il a pu le voir cette arme pénètre par sa pointe, tandis que l'extrémité du manche est tenue en main par le Fuégien*[2]*; il est vrai, ajoute-t-il, que l'amincissement du manche, qui finit en pointe du côté opposé au harpon, donne tout à croire que ce soit une arme de jet : il ne l'a jamais vu employer ainsi, mais il admet cependant comme très probable que les Fuégiens puissent s'en servir quelquefois de cette manière... D'après ce que nous avons vu, et d'après les réponses invariables des Fuégiens à nos questions sur ce sujet, posées de mille façons différentes, nous*

[1] Lovisato (1883).

[2] The term "Fuegian" is not used to describe the terrestrial hunters of Tierra del Fuego; it is restricted to sea-nomads living on the southern coast and on the numerous islands in the southernmost part of the Strait of Magellan.

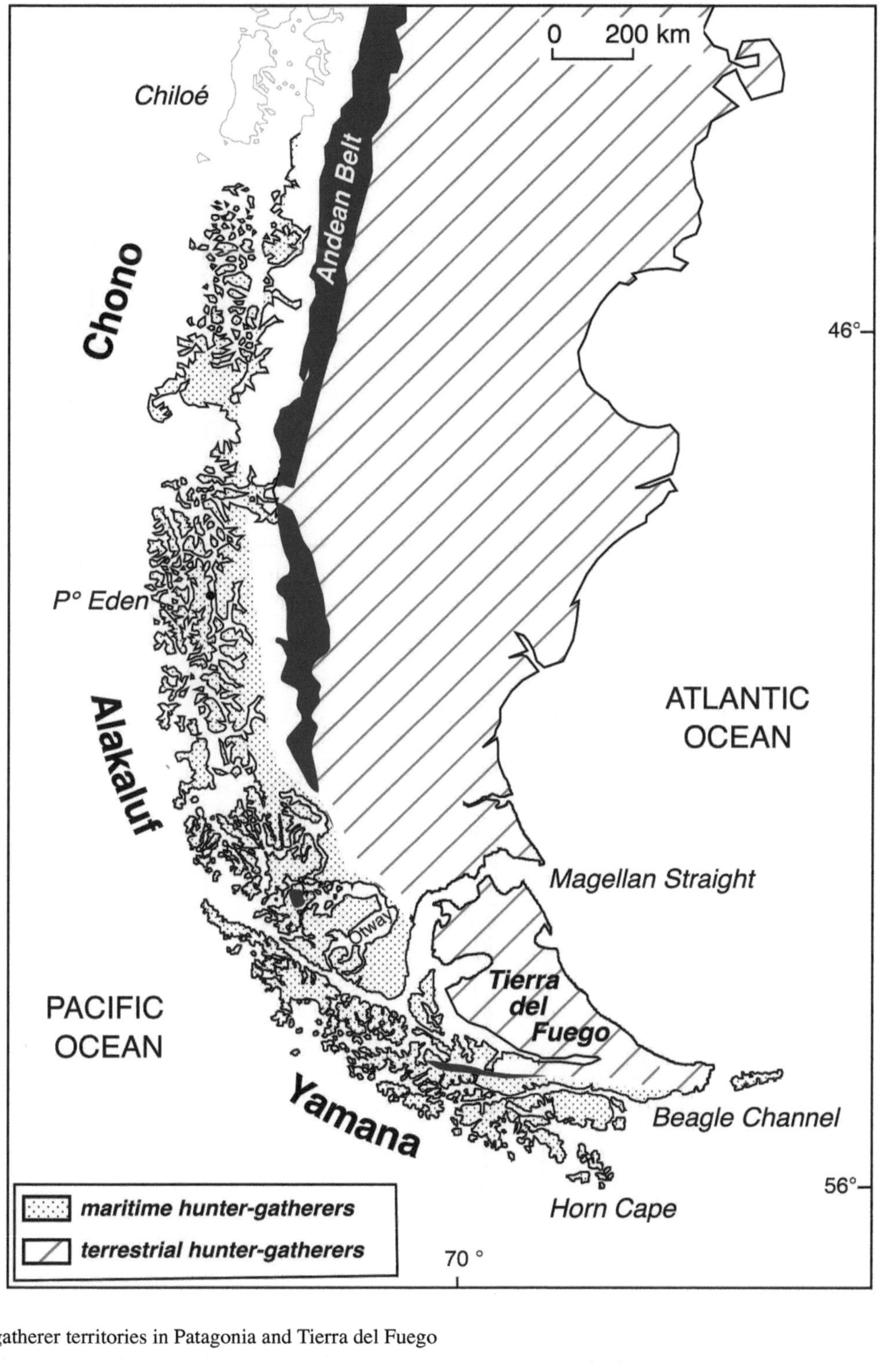

Fig. 16.1 Hunter-gatherer territories in Patagonia and Tierra del Fuego

admettons que le harpon fixe est une arme de jet… tout comme le harpon mobile, et que ce n'est qu'exceptionnellement que les Fuégiens l'emploient à la manière d'une lance ou d'une pique.[3] *Si l'on voulait adopter un nom spécial pour ce harpon fixe, il faudrait plutôt accepter celui de javelot proposé par Bove. Mais il nous paraît que le nom de harpon fixe exprime bien mieux la réalité, sans recourir à des appellations plus prétentieuses. Nous n'avons pas vu non plus la particularité indiquée par Lovisato au sujet de l'amarrage du harpon mobile sur le manche; d'après cet auteur, la pointe en os de baleine est assujettie au manche, au moyen d'une lanière en peau de phoque d'une*

[3]Which is a hand-held N.D.A.

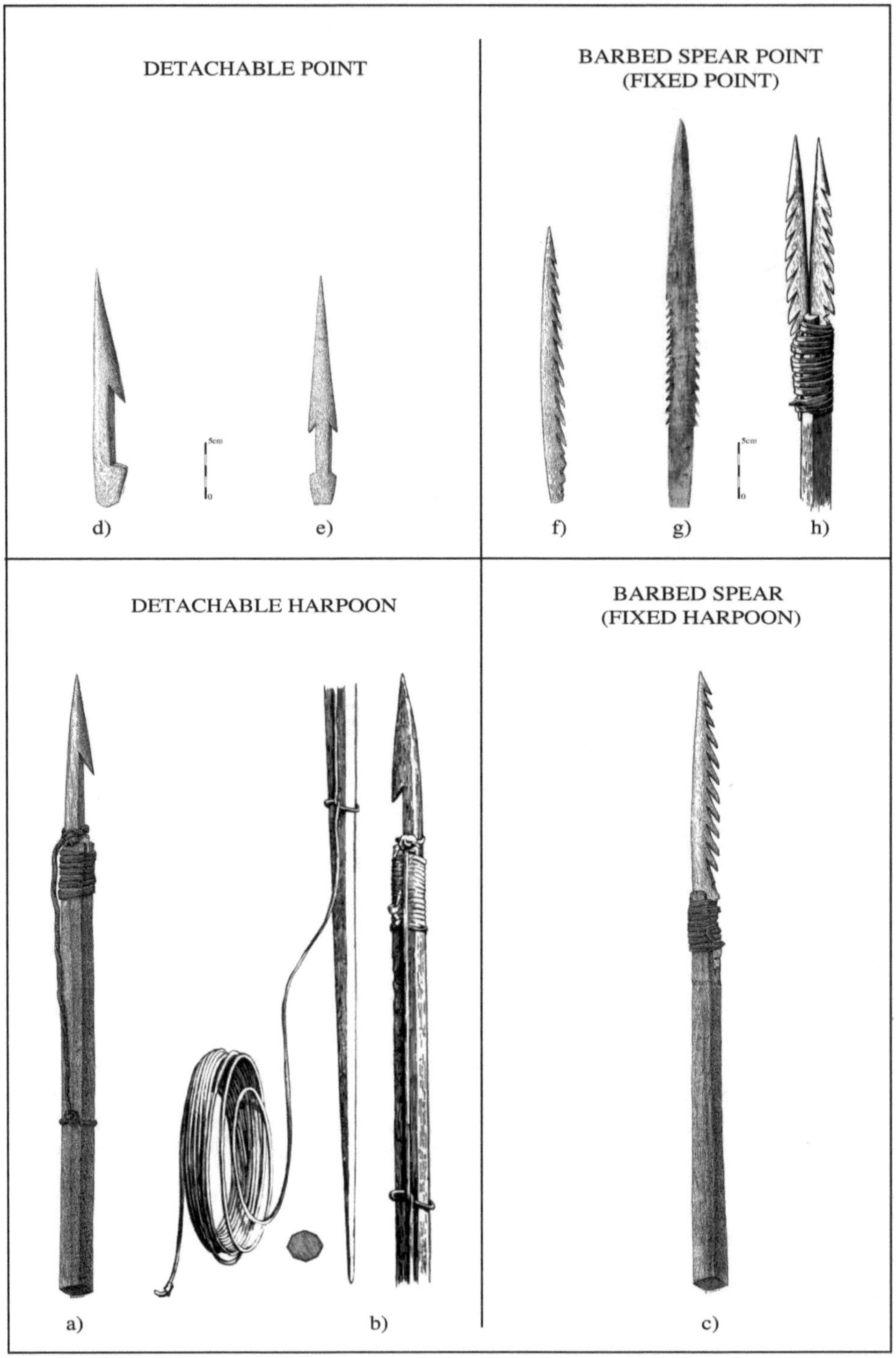

Fig. 16.2 Characteristic harpoons and harpoon heads of the Canoeros. *Below:* Yamana mobile harpoon heads and shafts: (**a**) "small harpoon" after Hyades and Deniker (1891), pl. XXXI-6; (**b**) "large harpoon" after Gusinde 1986:Fig. 34; (**c**) fixed harpoon after Hyades and Deniker (1891), pl. XXXI-5. *Above*: Ethnographical harpoon heads: (**d**) single-barbed with simple-shouldered base after Hyades and Deniker (1891), pl. XXXI-7 (**e**) double-barbed with double-shouldered base after Hyades and Deniker (1891), pl. XXXII-2, (**f**) unilateral multibarbed head after Hyades and Deniker (1891), pl. XXXII-1; (**g**) bilateral Huemul harpoon head (after Laming-Emperaire 1972:Fig. 6); (**h**) leister (after Gusinde 1986:Fig. 36)

longueur de plus de 20m. Nous avons toujours constaté que la partie mobile de la lanière n'avait pas plus de 1m de longueur : ce qui est suffisant d'ailleurs pour que la pointe étant enfoncée dans le corps de l'animal poursuivi et détachée du manche, celui-ci reste flottant et serve de balise ou de bouée aux Fuégiens pour continuer leur chasse" [The harpoons are the most used hunting weapon. There are several forms. The most common are made of bone and of two kinds: a fixed point with one row of teeth… the other with a mobile point with a single notch… In both cases, the point is adapted to a 3 to 4 m long shaft made of wood (*Fagus betuloides* or *Drimys winteri).* Detachable harpoons are used to hunt pinnipeds. The Fuegians watch for them in the canoe and launch their harpoon into them. When it penetrates into the body of the animal it comes off the handle, which then serves as a floater and guides the hunter who pursues the prey and imposes new blows until it is killed. Sawtooth harpoons, fixed to the handle, are used to hunt birds and otters, and this is the most common Fuegian harpoon. Lovisato[1] calls them lances, because he says that as far as he has seen this weapon pierces the prey with its point, while the extremity of the shaft is held in the hand by the Fuegian hunter.[2] It is true, he adds, that the thinning of the shaft, which ends in a point on the side opposite the harpoon, indicates that it could be used as a throwing weapon. Though he never saw it used in this way, he admits that it is very likely that the Fuegians sometimes used it in this manner… From what we saw, and based on the invariable responses of the Fuegians to our questions on this subject, asked a thousand times in different ways, we assume that the fixed harpoon is a hurled weapon… like the detachable harpoon, and that the Fuegians rarely used it in the manner of a spear or pike.[3] If we were to give a specific name to this fixed spear, it should be called a javelin, as proposed by Bove. But it seems to us that the name fixed harpoon is closer to the reality without resorting to more pretentious names. We also did not see the particularity indicated by Lovisato concerning the detachable attachment of the harpoon on the shaft. According to this author, the harpoon head in whale bone is fastened to the shaft by means of a strap of sealskin more than 20 m long. We have always observed that the detachable part of the strap is no longer than 1 m, which is sufficient for the head to stay inserted into the body of the animal and detached from the handle; the latter floats and serves as beacon or buoy so that the Fuegians can continue their hunt] (Hyades and Deniker 1891:353).

Gusinde, an Austrian ethnologist, author of the main synthesis of ethnographic data on the Indians of the region, retains the distinction between detachable and fixed harpoons (Gusinde 1986 [1937]). However, he distinguishes two sub-types of the detachable harpoon: the small harpoon whose head is linked to the shaft by a short line (Fig. 16.2a), and the large harpoon (Fig. 16.2b), whose head is loosely linked to the shaft with a long line, about twenty meters long, according to the description of Lovisato. The shaft of the large harpoon could easily be released from the head by a few jolts, the latter then remaining directly connected to the hunter who can easily slack or pull back the line, as in fishing. The hunter would recover the shaft later, as it was undoubtedly a valuable object given the investment needed for its manufacture. In both cases, the head is detachable, but in the first, it remains linked to the shaft, and in the second to the hunter. What changes is the length of the line. Both harpoons, the small and the large, have harpoon heads that are comparable in size.

According to Gusinde, the Yamana used both of these detachable harpoon types: the small harpoon, (with a short line) was especially suitable to hunting at sea, and the large harpoon (with a long line) was better for use on land. They also used a "harpoon" *sensu lato* with a fixed head and no line, which Gusinde called a spear, or more precisely, a "barbed spear" (spear with a barbed point) (Fig. 16.2c).

The Detachable Harpoon: An Essential Weapon for Hunting Marine Mammals

According to most observers, detachable harpoons were mainly used to hunt pinnipeds. Two species inhabited the region: fur seals (*Arctocephalus australis*) with adult females weighing around 50 kg and adult males that could attain 150 kg, and sea lions (*Otaria byronia*), with a strong sexual dimorphism, females weighing roughly 150 kg and some males exceeding 300 kg. *Delphinidae* were sometimes hunted, and exceptionally, a very large phocidae, the elephant seal (*Mirounga leonina*) which could weigh as much as 600 kg, along with whales under certain conditions.

The "small harpoon", with a short line, was generally single-barbed with a simple-shouldered base, both located on the same side (Fig. 16.2d). This asymmetry, characteristic of the late period, favours an at least partial toggling of the head relative to the shaft when hurled into the animal. One can presume that it was launched with the barb turned upward (Legoupil 1981), as with the spur on the toggle harpoon used by modern sperm whale hunters, so that the shaft and barb favours the failover and hocking in the animal's muscles. This "small harpoon" was mostly used at sea. The hunter stood in the front of the fragile bark canoe with his weapons placed in the stem, ready to harpoon prey encountered by chance, while the woman paddled in the stern. Once an animal was harpooned, the harpoon head was detached from the shaft but still linked to it by the short line. According to Gusinde, the line attachment was located in the first third of the shaft, so that it would maintain an oblique position in the water: this created a very high tensile strength for the animal. The animal was then killed with blows from cudgels, paddles, or thrusted with spears. According to Gusinde, this hunt in open water was particularly effective near dense seaweed beds (*Macrocystis pyrifera*) or along rocky shorelines where sea lions frequently fish. The reflex of the harpooned animal is to take refuge in seaweed, where the shaft became tangled in the very tough seaweed that was sometimes several meters long, quickly stopping the animal.

In locations with no seaweed, the "large harpoon" was preferred in order to continue pursuing the wounded animal still connected to the canoe. This more dangerous process was probably reserved for smaller animals, though, the "large harpoon" was mainly used for hunting on land, *"Ils tuent les phoques*[4] *dans des cavernes*[5] *ou sur les plages à coups de masse, ou bien ils les harponnent du bord de la mer avec de fortes lances ou des harpons attachés par un lien léger, mais solide, autour du corps du chasseur; ils tuent de la même manière les marsouins et de grands poissons"* [They kill seals[4] in caves[5] or on beaches with a sledgehammer, or they harpoon them on the shore with strong spears or harpoons attached by a light but solid link around the body of the hunter; they kill porpoises and large fish in the same way] (Hyades and Deniker 1891:9). Curiously, Hyades, seems to mention here the use of a harpoon with a long line whose existence he had previously denied. The line allowed the hunter by sheer strength to keep hold of his prey. This means that the hunter had to be stronger than the prey, or at least that he managed to retain it sufficiently to wound it and finally kill it.

Gusinde states that the long lined harpoon could be armed with a bilaterally doubled barbed head and double-shouldered base (Fig. 16.2e), which allowed it to pass through the thick layer of fat on these marine mammals. According to Lothrop (2002 [1928]), these bilateral points constitute a regional feature, characteristic of the Yamana. Actually, these are very occasionally found in Alakaluf territory as evidenced by two of these points preserved in the collections of the "Instituto de la Patagonia" in Punta Arenas. One is recent owing to its stratigraphic position in a shell midden on the lower terrace of Bahia Colorada in the Otway Sound; the other comes from the southern coast of Isla Dawson (Caleta Cono).

Ashore harpooning was used to prevent pinnipeds from taking flight in the sea. But Gusinde notes that the harpoon could easily be replaced by the cudgel, used in the manner of North America sealers. However, it is likely that this method is better adapted to small pinnipeds as it could be dangerous with larger animals, such as male sea lions.

The Barbed Spear (or Fixed Harpoon): A Versatile Weapon

The barbed spear could be armed as a detachable harpoon with a single or double barbed head, but most frequently a multi-barbed head was used, usually unilateral (Fig. 16.2c, f), though occasionally bilateral (Fig. 16.2g). As for the detachable harpoon, the length of the shaft could reach 3–4 m. Sometimes 2 or 3 multi-barbed heads was hafted together thus constituting a "leister" (Fig. 16.2h). The lower edge of the barbs was either straight or V-shaped, depending on the manufacturing procedure. Sawing was the technique used in both cases, though it was implemented in a different manner. In the first case it was done bifacially, so that the barb emanated from two opposite grooves located on each the side of the blank, while in the second, a single groove was produced by oblique sawing from the edge of the blank, and was undoubtedly facilitated by the use of metal tools obtained from the European sailors. This operation was also done with a shell knife, which they used to "*cut bones of extraordinary hardness, to make spears for killing fish*" (Drake 1578, quoted by Hyades and Deniker 1891:3). On ethnographic specimens, the proximal end of the multi-barbed head was often notched unilaterally (see Fig. 16.2f), favoring hafting by tying. This notched base seems predominant among the Yamana samples (Estévez and Vila 2013). The rigidity of the hafting system was obtained by inserting it into a slot that was three-quarters closed on the shaft, leaving only a lateral split open so that the notched portion of the base would protrude out. This weapon was used as a spear (hand-held and thrusted) or as javelin or spear (hand-held and launched).

The barbed spears were used to kill marine mammals and to hunt birds, otters, and even artiodactyls or large cetaceans. They were also used for fishing. Finally, they were also weapons feared by sailors and were probably used in internal conflicts and brawls.

There is little evidence for bird hunting with barbed spears. Other hunting tactics were used ashore for common bird species, including slings, goose snares, and catching cormorants by hand at night on the cliffs. However, according to Gusinde, barbed spears were very effective for hunting penguins, whose skin is difficult to penetrate, however, they are surprisingly under-represented in archaeological sites. Spearing birds appears to have been a relatively opportunistic technique used to catch specific species with individual behaviors, such as steamer ducks (*Tachyeres pteneres*) or large procellariiforms, sometimes discovered in archaeological sites. Barbed spears would probably have been less effective (and less efficient) for small gregarious birds that are easily frightened in their colonies.

Barbed spears were also used to catch fish, a practice that is well documented in southernmost America. In this context, they could be armed with one or two prongs: "*Lorsque les Fuégiens veulent harponner des poissons de très grande taille, ils attachent ordinairement deux pointes de harpon en dents de scie sur le même manche, l'extrémité libre de ces pointes divergeant légèrement*" [When the Fuegians want to spear very large fish, they usually attach two sawtoothed spear points to the same shaft, whose tips slightly diverge] (Hyades and Deniker 1891:356). According to Spears (1895), the assembly could even be triple. This artifact is clearly a leister.

Though otters are poorly represented in the archaeological record, they were found in some later sites including Herschel Island in the archipelago of Cape Horn (Legoupil 1993–1994), and at Punta Baja, dated to the sixteenth century, in the Otway Sea (Legoupil 1989). Their fur, used by

[4] sic ! In the XIX° and even in the XX° century, phocidae and otariidae were commonly confused. Here Hyades is talking about otaridae the only pinnipeds present in Patagonia.

[5] Marine caves accessible with canoe.

the *Canoeros* to cover their young children, was a popular barter product used by fur traders in the nineteenth century. The otter was driven out with the help of dogs, which were introduced by the first navigators "*le chien... déloge la loutre et le Fuégien, aux aguets, la harponne au débûcher. Souvent notre chasseur casse son harpon*" [the dog... drives out the otter and the Fuegian keeping watch, harpoons the animal when it is startled. Often our hunter breaks his spear] (Hyades and Deniker 1891:364). They were also be caught with a leister by the last Alakaluf of Puerto Eden using "*une sorte de foëne à deux pointes pour la loutre*" [A kind of spear with two points for the otter] (Emperaire 1955:196). Despite appearances, otter hunting did not occur only during the late period. We found traces of this activity at the site of Dawson which dates to more than 3000 years old (Legoupil et al. 2011).

Artiodactyls were rarely hunted with barbed spears. Still, according to Emperaire, a multi-barbed "fixed harpoon" was used to hunt huemul (*Hippocamelus bisulcus*), a type of small deer present in the foothills of the Andes and on some large mountainous islands. He describes the "*harpon à huémuls, le plus long de tous, à deux rangs de barbelures*" [huemul harpoon, the longest of all with two rows of barbs] (Emperaire *ibid.*). One of the few known examples measures 37 cm long. It is bilaterally barbed and the proximal end has no specific layout (see Fig. 16.2g).

The Yamana also occasionally used a "barbed spear" for hunting guanaco along the southern coast of Tierra del Fuego: "*Les chasseurs se placent derrière les arbres avec des harpons, prêts à les lancer avec force sur le guanaque au moment où il passera devant eux*" [The hunters are placed behind the trees ready to hurl the harpoon into the guanaco when it passes by] (Hyades and Deniker 1891:356). This kind of game is nonetheless typical for terrestrial hunters and exceptional for maritime hunters. The harpoon was then probably a makeshift, ineffective version of the Selk'nam bow uncommon among Canoeros. Hyades and Deniker do not specify the kind of harpoon used, but Lothrop indicates that the Yamana in the eastern part of the Beagle Channel used a spear without a line attachment projection at the base: "*It resembles the seal-spear* ["detachable harpoon"] *in having a single barb but lacks the projection seen on the seal-spear's tang*" (Lothrop 2002 [1928]:153).

Inland, osseous harpoon heads are very rare: only a few examples have been reported in northern Patagonia (Molina 1966–1971) and at Tierra del Fuego (Scheinsohn 1997). They were nonetheless used occasionally by the Selk'nam to hunt pinnipeds on the shore when they were not practicing navigation, or to capture fish caught in natural rocky traps "*...lo más usado es el pequeño arpón con punta dentada de hueso, con el que clavan á los peces que quedan en sitios con poca agua o a los que están escondidos bajo las piedras cuando la marea se retira*" [the most used is the small harpoon with a barbed bone point to catch fish in places with little water or among stone when the tide ebbs] (Gallardo 1910:203) "*...en la hoja se ven muescas destinadas a desgarar las carnes ... Estos arpones no sólo sirven para matar lobos, sino tambien para pescar, y se hacen de diferente tamaño según el uso á que preferencia se les destinará*" [... on the point one can see slots intended to flesh the meat... These harpoons are not only used to kill pinnipeds, but also serve for fishing, and they are made in different sizes according to their use] (*Ibid*:283). Very few single-barbed harpoons on metapodial bones from the site of Punta Santa Maria 2 lack line attachment projections: they were probably also used for fishing (Torres 2009).

Finally, the hunting of large cetaceans is widely attested in ethnohistoric data. This activity consisted mostly of scavenging a naturally killed or sick whale easily killed and towed to the beach by one or more canoes with one or several harpoons. Evidence for true whale hunting is rare, "*nadie dared acercarse has an animal adulto sano*" [nobody dared to brave a sane, adult animal] (Gusinde 1986:503). But indications do exist, especially in the Beagle Channel by the end of the nineteenth century: "*this evening some ten or more canoes came each having a share of the poor whale, which literally was killed by inches, having received into its body somewhere about hundred spears...*" (Bridges 1874 quoted by Gusinde 1986). The same behaviour is observed in the Magellan Strait during the late seventeenth century: "*Ils font la chasse à la baleine de cette manière : ils vont cinq ou six canots ensemble et, lorsqu'ils en ont trouvé une, ils la poursuivent, la harponnent avec de grandes flèches*[6] *qui ont le bout qui entre, d'os ou de pierre à fusil taillées fort industrieusement, ensuite la laisse perdre son sang et quand elle est morte, la marée l'échoue sur la côte où ils la vont chercher quelques jours après*" [They hunt whales in this way: five or six canoes went together and when they found one, they pursued it and harpooned it with large arrows[6] whose tips, made of bone or flint, penetrated it and then left it shedding its blood, and when it died, the tide brought it up the shore where they came back to get it a few days later] (Duplessis 2003 [1698–1701]).

In both cases, spears or javelins were used to kill the animal, but there was no harpooning in the strict sense, with a line. In fact there is no evidence of "true" harpooning to hunt whales; this would have been nearly impossible given the size of these giant sea mammals and the speed and the depth at which they can dive, reaching several dozens or even hundreds of meters. Nothing in the technical equipment of the *Canoeros* would enable them to resist the traction of a large whale; neither the line—which would have to be several dozens or even hundreds of meters long—nor their fragile bark canoes, nor their human force. On the other hand, towing a dead whale would certainly require the use of the

[6]One should understand projectile point.

detachable harpoon, the only weapon able to penetrate the thick layer of fat of these animals and to set deeply into the muscles so that it could withstand the pulling of the canoe.

The Bone Points of the Patagonian and Tierra del Fuego Indians: Archaeological Data

In post-Magellan times, a strong dichotomy exists among the bone points of the Canoeros groups: single-barbed points with a shouldered base for fastening a line are characteristic of detachable harpoons, and multi-barbed points (rarely single-barbed) with a smooth or notched base, are characteristic of barbed spears. Both types were present during the first maritime adaptation, more than 6000 years ago. Though they are the cultural markers of these sea nomads throughout the chronological sequence, their shape, size and decoration changes depending on the period and the region.

During the Early Period (±6300/5500 BP), detachable harpoons heads, or at least those considered as such by analogy with ethnographic materials, were mostly armed with a single barb (Fig. 16.3a, b), and in exceptional cases, 2 or 3 barbs that were either placed in a line (Fig. 16.3c, d) or next to each other—the so-called "*vulpicéphales*" harpoons (Piana 1984) from the Beagle Channel (Fig. 16.3e), also called "V-shaped" by V. Scheinsohn (2010). Their shaft usually had a biconvex to ovoid cross-section and the characteristic attribute of this period is the cross-shaped base formed by two protruding bilateral spurs. The protrusion, and their well-marked anterior angle, undoubtedly helped to fasten a line.

A little later, during the Intermediate period (± 4500/2000 BP), bone points became less abundant and also less standardized, but the archaeological data is scarce. The morphology of the proximal ends varies; it can take the form of buttons, sometimes odd and asymmetric (Fig. 16.3f, g), or both buttons and bead type (Fig. 16.3g). A new type appears in several sites; its short, more or less conical base is equipped with two small perpendicular spurs that are not concentric with the barb (Fig. 16.3h).

The harpoon heads from the Early Period are much smaller than the ethnographic ones, as shown by an analysis performed for 40 entire harpoon heads from both ends of the chronological sequence (Fig. 16.4): 19 cross-shaped harpoon heads from major ancient sites[7] measured an average of 15.1 cm long, while 27 ethnographic harpoon heads[8] measured 36.4 cm long for single-barbed with single-shouldered based examples (see Fig. 16.2d), and 21.6 cm for double-barbed and double-shouldered bases respectively (see Fig. 16.2e).

A similar observation was reported by V. Scheinsohn for another assemblage of 39 pieces: archaeological harpoon heads from the Beagle Channel (all periods) measured between 14 and 18 cm long, while the ethnographic ones (Museum of La Plata) measured around 31–32 cm in length (after Scheinsohn 2010:Fig. 5). L. A. Borero and F. Borella also observed this striking size difference between single-shouldered harpoon heads from ethnographic (collections from Museo Etnográfico « Juan Bautista Ambrosetti » of the University of Buenos Aires and from The British Museum in London) and various archaeological specimens (2010:Fig. 4 & Table 3). They also observed that the proximal ends of the harpoon heads were often unfinished. Concerning ethnographic specimens, it is comforting to note that J. Estévez and A. Vila obtained from their European collections (164 harpoons of which 8 were mounted on shafts) the same dimensional characteristics as we did on harpoons mostly from South American collections: that is, that single-shouldered based ones are in average longer that the double-shouldered specimens and the latter do not exceed 35 cm in length (2013).

The small size of archaeological harpoon heads in general, and especially within the Early Period, could be related to the size of the hunted prey: the small *Arctocephalus australis*, largely dominant in the assemblages of ancient sites (Legoupil 1997; San Román 2004, 2010; Orquera et al. 2011), and the rare exploitation of male sea lions and whales during this period.

According to Scheinsohn (2010) the larger size of ethnographic points is linked to their use as "exchangeable goods for trading with European travellers" (*Ibid*: 299). However, there are also large archaeological harpoon heads, particularly among those from the Intermediate period, which in some cases reach 30 cm (see Fig. 16.3f–h). What could explain the large size of the ethnographic points? First, the manufacturing of large harpoons could have been facilitated in the modern period by the adoption of metal tools. Or the size of the harpoon heads could be related to the available raw materials, since large heads were indeed always made on whale jawbones or ribs, which are less common than pinniped bones, from which it would be impossible to extract a straight, non-porous blank 30 cm long. Another possibility is that the size of the harpoon could be adapted to that of the game: large harpoons were then used mostly to hunt male sea lions, elephant seals and whale-calves, or to tow whales. Estevez and Vila (2013:Fig. 16) highlight the relationship between point size and the hunted prey. They argue, based on ethnographic data that harpoons longer that 40 cm were used for whale hunting. Finally, the development of large harpoon heads in modern times could also reflect the influence of European and North American whalers who began to frequent the region from the late eighteenth century. This

[7] Englefield, Bahia Colorada, Punta Santa Ana in the strait of Magellan/Otway sea; and Túnel I et Immiwaïa the Beagle Channel.

[8] Assemblages: Musée du Quai Branley et Musée d'Archéologie National (France), National History Museum de Montevideo (Uruguay) et de Santiago du Chili, Musée de la Merced (Santiago du Chili).

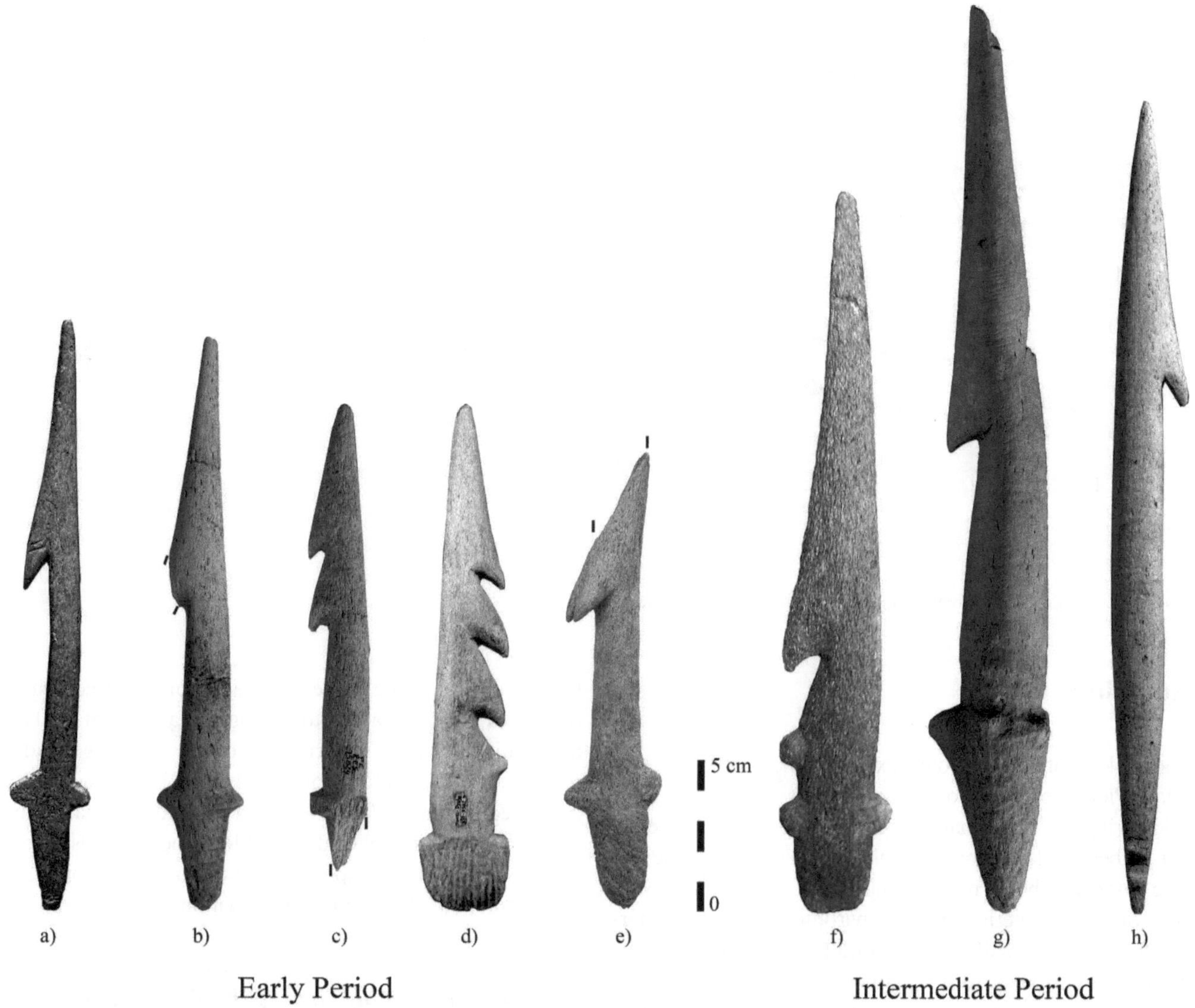

Fig. 16.3 Archaeological harpoon heads: (**a**) Punta Santa Ana (col. Inst. Patagonia, cl. Christensen); (**b**) Bahía Colorada (col. Inst. Patagonia, cl. Christensen); (c) Punta Carrera (col. Inst. Patagonia, cl. Christensen); (**d** et **e**) Túnel I (col. CADIC Ushuaia, cl. Legoupil and Christensen); (**f**) Bahia Valentin (cl. Legoupil, cortesy Zangrando); (**g**) and (**h**) Offing 2 (cl. Mae, Oboukhoff and Barroche)

situation may particularly be the case for harpoons with bilateral double barbed heads and double-shouldered bases which are morphologically similar to European whaling harpoons made of metal. They are known only within post-European assemblages and could be "imitation of the European harpoons that were used since the fifteenth century … rather that an indigenous development" (*Ibid*:291).

The multi-barbed points are more complex to understand. The archaeological samples are mostly broken, but generally quite small. However, like detachable harpoon heads, their size is larger during Post-Magellan time. Their barbs are less pronounced than those of detachable harpoons, and their number is highly variable (up to several dozens). The samples from the Early Period are almost always unilaterally barbed (Fig. 16.5a–g) and have no specific features at their base, unlike most ethnographic samples. For the Intermediate period, the data is still scarce and disparate, but the known samples are quite different from those from the Early Period with relatively few and small barbs (Fig. 16.5h, i). The base of one bulky multi-barbed head discovered on an island in the Strait of Magellan is equipped with two small perpendicular spurs (Fig. 16.5i), identical to the single-barbed harpoon head from the site of Offing (*supra*, Fig. 16.3h). Only the number and morphology of barbs differentiate the two weapons. This line attachment system is also documented on some barbless spears with a nearly circular cross-section found in an undated site in the Beagle Channel region, Rio Chico 3 (coll. Territorial Museum of Ushuaia), as well as on several broken pieces from the site of Ponsonby near the Otway Sea and dated to 4000 BP (Legoupil 2003).

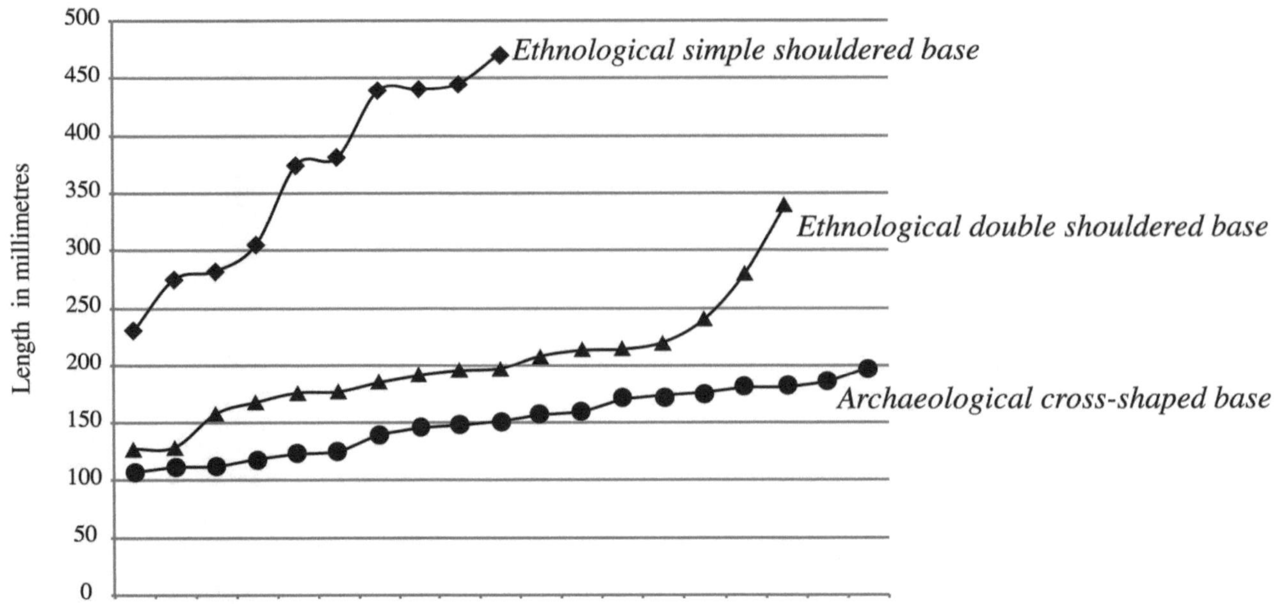

Fig. 16.4 Length of archaeological and ethnographical harpoon heads

Discussion

The same kind of osseous points existed in the Old World since the late Pleistocene, and on the outermost ends of the New World during the second half of the Holocene: single-barbed or multi-barbed, unilateral or bilateral, with or without a line fastening system, in bone, ivory or antler depending on the availability of these materials in the environment. Many authors have attempted to classify them according to the way they were used or their attributes, focusing sometimes on the line fastening system, sometimes on the barb shape, and often on a more or less complex mix of several features (Mason 1902; Mathiasen 1927; Leroi-Gourhan 1946; Julien 1982; Weniger 1992, 2000; Pétillon 2009, etc.). The diversity of these typologies (usually alphanumeric) highlights the complexity of the functional interpretations of these objects and the numerous interactions of their uses. This situation is particularly true in North America where almost all systems are represented up to the most complex, with regional and chronological variations.

Despite many counter-examples, some trends nonetheless appear. In Patagonia, for example, detachable harpoons are usually armed with single-barbed points (rarely with 2 or 3 barbs) and a base with a strong line attachment system, while "fixed" harpoons points are often multi-barbed (rarely single-barbed) and do not have a line attachment system (clamp hafting or lateral hafting, according Weniger 1992).

The proximal end, so often used in the typologies of the Arctic or Magdalenian harpoons, has revealed significant chronological change unrelated to function, as in Patagonia. Leroi-Gourhan (1946) thus spoke of the evolution from male bases (or pin-hafting) to female bases (or socketed heads) in the North Pacific, leading to the toggle harpoon of the Dorset and Inuit Indians. The principle of the toggling head was reinvented (or adapted from the Inuit?) by a blacksmith in New Bedford in 1850 and adopted by modern whalers as the most complete model of detachable harpoon heads.

But what is the specific nature of "harpoons" from Patagonia and what kind of information can they provide?

To Catch or to Kill, That is the Question…

The dichotomy between harpoons and barbed spears is particularly clear in Patagonia, enabling us to clearly distinguish between the two manners of functioning in all ethnographic records from both North and South America, regardless of the delivery mode:

- In the case of harpoons, the head, composed of a tip and sometimes a foreshaft,[9] is detachable and remains attached by a line to the hunter, his boat or a floater—the shaft in Patagonia or a bladder or seal skin floater in Arctic (Fig. 16.6a–d, "Type A" after Pétillon 2009:Fig. 5). In this case, the base must be equipped with a line attachment system such as the one or two shoulders or projections in Patagonia and the Arctic, and later, a hole in the latter

[9] If we consider certain Inuit harpoon or harpoons made of wood with a flint or a shell point, and a metallic bard in the northern Chile or at the Peruvian coast.

Fig. 16.5 Multibarbed spear heads. Early Period: (**a**, **b**, **c**) Bahía Colorada (col. Inst. Patagonia, cl. Christensen); (**d**, **e**) Bahía Buena (col. Inst. Patagonia, cl. Christensen); (**f**, **g**) Pizzulich 2 (col. Inst. Patagonia, cl. Christensen); and Intermediate period: (**h**) Túnel I (col. CADIC Ushuaia, cl. Legoupil et Christensen); (**i**) I. Isabel (Bird 1980:Fig. 3)

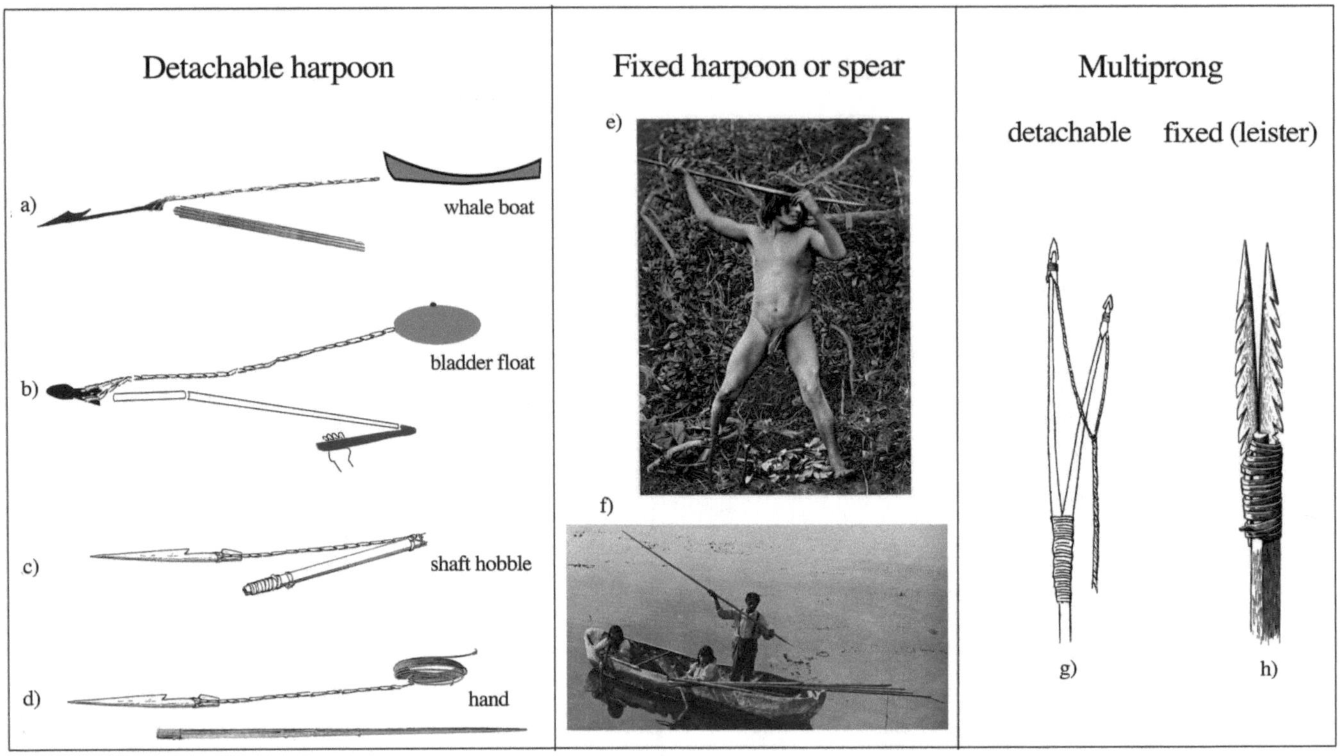

Fig. 16.6 Principal ways of functioning of harpoons *sensu lato*: (**a**) modern whaling; (**b**) Artic seal hunting; (**c** and **d**) Pinnipeds hunting in Patagonia; (**e** and **f**) Example of use of the barbed spear in Patagonia (after Hyades and Deniker 1891:pl. I and unknown author, *ca.* 1910, archivo Museo del Fin del Mundo); (**g**) multiprong/leister (Leroi-Gourhan 1946:Fig. 708); (**h**) mobile multiprong (Gusinde 1986:Fig. 36)

region. This classic harpoon type, defined by the presence of a line and a barbed and detachable head (Mason 1902; Leroi-Gourhan 1946[10]; Weniger 1992) is the most common, although exceptions exist.[11] The detachable harpoon is almost always a propelled weapon that can be thrown by hand, especially in Patagonia where this is the only delivery mode,[12] but it can also be launched with spearthrower, as is done by Arctic hunters. It is therefore both a capture device with a hook (with one or two barbs),[13] similar in some ways to a fishing hook, and a weapon. Consequently, unless lucky, the killing of an animal is separate from its harpooning, as is true with modern whalers who are systematically armed with a harpoon and spears, whose use was also associated with social factors, being reserved for different individuals.

- When the barbed head and shaft form an inseparable whole without a line (Fig. 16.6e, f), it is considered as the "fixed harpoon" of the Patagonian Indians according Hyades and Deniker (1891) and Gusinde (1986), or a spear by most other authors ("Type F" in Pétillon 2009:Fig. 5). This weapon is designed to kill. Its barbs, often multiple and small, are designed not only to keep the weapon inside the prey (Weniger 2000), but probably also to flesh it. The barbed spear can be hand-held or hurled. When hand-held it is generally called a spear, pike or "*épieu*", in accordance with codified military or hunting vocabulary. The efficiency of the weapon is enhanced by the weight and length of the shaft; in Patagonia it can reach up to 3 or 4 m. In the second case, when hurled, they are known as a spear, dart, assegai or arrow, since they are launched by hand or with a device: when hand launched, its weight enhances the efficiency as in the previous case. When launched with a spearthrower or a bow, the weapon is often shorter and lighter, lightness being compensated by the speed of propulsion.

To these two types, a third can be added, consisting of 2 or 3 multi-barbed points (leister prongs) mounted together. This weapon, usually called a leister, existed in the north, as well as in the south of southern America (Fig. 16.6g, type "F'" in Pétillon 2009:Fig. 5). This several pronged weapon

[10]"*ce qui distingue catégoriquement le harpon, c'est sa tête détachable, qui reste prise dans le corps de l'animal alors que la hampe de l'arme se libère. La tête est rattachée à une ligne de cuir ou de corde au moyen de laquelle on manoeuvre l'animal blessé*" (Leroi-Gourhan 1946:54).

[11]Barbless detachable harpoons exist as well as fixed harpoons with a line attachment system: "types B et C" in Pétillon (2009: Fig. 5).

[12]The bow, which appeared toward the beginning of modern times, less than 2000 years ago, was not used with osseous projectile points.

[13]The toggle harpoon is only a sophisticated sort of hook.

can be used as a true detachable harpoon with a detachable head, as it is by the Kwakiutl of British Columbia to hunt salmon, as mentioned by Mason (1902) (Fig. 16.6g). Depending on how it is used, this composite artefact is thus considered as a leister (or fork) or a detachable harpoon, double in the case of the Kwakiutl. It is therefore both a capture device and a weapon. It is usually reserved for smaller animals such as fish, birds and small mammals. Its barbs can be turned inward (enveloping the prey), which is most often the case in the Arctic and sub-Arctic, or outward (penetrating the animal), such as in Patagonia (Fig. 16.6h).

So What About the Magdalenian Harpoons...

Around 40,000 BP, the first anatomically Modern Humans arrived in Europe with a new culture called the Aurignacian, bringing with them the first osseous hunting weapons. From this time on, osseous projectile points persisted throughout the Upper Paleolithic in various types and forms. Around 16,000 cal BP, at the beginning of the Upper Magdalenian, a previously unknown type of osseous projectile point appeared in France, Spain and Germany; it was described as a "harpoon" by its first discoverers (Mérimée quoted by Woersaae 1875; Lartet 1861). Between 1500 and 2000 points of this kind are currently known. They played a significant role in the weapon kit for nearly 2000 years, from ca. 16,000 to 14,000 cal BP (see Langley et al. 2016). These uni- and bilaterally barbed points are generally made of antler, and rarely of bone. Their typology and morphology will not be described here at length since these artifacts have already been thoroughly analyzed in several detailed studies (e.g., Julien 1982; Weniger 1995; Langley 2013). Their function, however, has been repeatedly discussed with reference to ethnographic data, especially from North America. We are aware of only one paper (Estévez and Vila 2013) that uses data from southernmost South America to discuss the morpho-functional characteristics of Iberian Paleolithic barbed points. The authors stress that the ethnographic literature attests the specific use of barbed points to catch aquatic resources and that these resources were available along the Paleolithic coast of Iberia (this idea being derived from the presence of faunal remains of fish or marine mammals, and of indirect evidence such as depictions of marine animals). Therefore they conclude that the archaeological barbed points from the Mediterranean coastal sites must have been used against aquatic resources—probably for fishing given their size (too small for sea mammal hunting) and the associated zooarchaeological record (Estévez and Vila 2013:298 *sqq*.).

We would here like to widen the discussion adding other arguments such as barb morphology, hafting and line attachment systems and the possible prey.

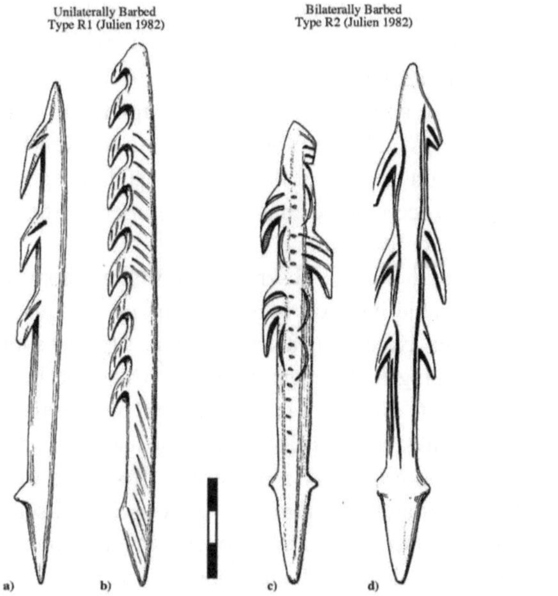

Fig. 16.7 Typical Magdalenian barbed points—arranged after the first degree of M. Julien's typology (Drawings from Julien 1982). (**a**) Unilaterally barbed harpoon—type A with less than 6 barbs (Mas d'Azil); (**b**) Unilaterally barbed harpoon—type B with more than 7 barbs (Bruniquel); (**c**) Bilaterally barbed harpoon with angled barbs (Lortet); (**d**) Bilaterally barbed harpoon with convex barbs (Sainte Eulalie)

When comparing Magdalenian barbed points to those of the *Canoeros*, it appears that the features of the Magdalenian points are evocative of both the functional patterns observed in Patagonia:

- The morphology of the barbs suggests detachable harpoons: these barbs are generally pronounced and able to cling tightly to the muscles of the prey. Furthermore, the male hafting system (pin-hafting) is accompanied by a line attachment system, either double for bilateral points or single for unilateral barbed points (Fig. 16.7). In some Cantabrian cases, this line attachment system is a perforation.
- At the same time, they are similar to the Patagonian "barbed spears" if you consider the number of barbs. While in Patagonia detachable harpoons almost always have a single barb, rarely 2, and 3 in exceptional cases only, unilateral Magdalenian points have an average of 2.8–9.7 barbs depending on the subtype, and bilateral points have between 4.5 and 7.5 (Julien 1982).

The "harpoons" from these two regions also differ in size. In Patagonia, the length of the points varies from 1 to 5, the smallest being around 10 cm while the largest can reach half a meter for both detachable single-barbed points and multi-barbed points. Such lengths are rarely reached in the Arctic, unless you include the wooden foreshaft of toggle harpoons. The Magdalenian harpoons rank among the smallest harpoons of Patagonia. The length of 82 unilateral Magdalenian harpoons ranges from 10.4 to 14.9 cm, and 90 bilateral harpoons vary from 12.3 to 14.5 cm (Julien 1982). Lefebvre (2011) gives an average 12.3 cm for 500 complete uni- and bilateral harpoons. It would therefore be useful to reassess the relationship between the size of the harpoons and the species hunted, taking into account the constraints imposed by the raw materials and the delivery mode.

Many authors have observed in the ethnographic record that the detachable harpoon is used primarily in aquatic environments—sea, rivers or lakes. Its purpose is to prevent the animal from escaping in an environment in which the hunter cannot not follow it (even with the eyes), or to prevent the loss of the prey if it sinks. Specifically, detachable harpoons are very often used to capture aquatic mammals, from the biggest ones (whales) through medium-sized species (seals, sea lions, and dolphins) down to the smallest ones (sea otters, otters, beavers). In rare cases, detachable harpoons are also used for fishing.

The Patagonian detachable harpoon conforms to this pattern: it is never used to hunt terrestrial mammals and is intended for hunting marine mammals, either medium (seals, dolphins) or large species (male sea lions, elephant seals), or more rarely to haul large cetaceans. Depending on the size of the animal, the harpoon head must pass through a more or less thick layer of fat to cling to the muscles. This could explain the variability in the length of the points.

The maritime context of the Arctic and sub-Arctic can be compared to that of the *Canoeros*, but this is not the case for the Magdalenian. If Magdalenian coastal sites existed, they have been submerged by the Flandrian rise in sea level. The sites currently known, including those that yielded the highest number of barbed points, are far from the Paleolithic shoreline—except sites on certain parts of the current coasts of the Iberian Peninsula where the shoreline has receded only a few kilometers since the Paleolithic. Regardless, evidence of the hunting of marine mammals remains very scarce in all contexts (Serangeli 2003; Corchón et al. 2008), including evidence of seal hunting, which could theoretically have been practiced from inland sites since seals sometimes swim up large rivers.

Is also seems unlikely that the Magdalenian hunters used harpoons to catch ungulates in aquatic environments, such as when they crossed a river. Most faunal assemblages from the Upper Magdalenian are dominated by the remains of large- (horse, bison) and medium-sized (reindeer, red deerr, ibex) ungulates. Although these species occasionally cross rivers or seek refuge in ponds, most of them are bad swimmers and all are unable to hide by diving. Among them, only reindeer are good swimmers, but ethnographic data show that barbed weapons were not often used to hunt reindeer in aquatic environments and the hunters managed to recover the animals killed anyway (e.g., Balikci 1970).

Among the animals hunted by the Upper Magdalenian groups, the only ones that could be disabled by a small barbed point and who were likely to disappear into a place where the hunter could not recover them are fish (likely to sink in water), birds (that can fall out of reach when hit) and mammals such as rabbit, hare and marmot (usually struck in their burrows where they can be impossible to retrieve).

At the beginning of the Upper Magdalenian, there was actually an increase in the exploitation of these small species (e.g., Cochard and Brugal 2004; Costamagno and Laroulandie 2004; Costamagno et al. 2008, 2009). In the preceding Lower and Middle Magdalenian, apart from local exceptions at some sites, these species were rarely sought. After the beginning of the Upper Magdalenian, however, fishing, fowling and small mammal hunting became regular practices and remains of these small species are often found in large quantities at Magdalenian sites, at least in the southern half of France (which yielded most barbed points). Whether this evolution is due only to a change in human prey selection choices, or is the result of a changing ecology leading to a greater abundance of these species after 16,000 cal BP, is still a matter of debate. In any case, it is likely that new hunting techniques adapted to this small game were developed (Boudadi-Maligne et al. 2012) and since this evolution is contemporaneous with the appearance of barbed points, it is tempting to correlate the two (Langlais et al. 2012).

A first attempt to test this hypothesis was made by comparing the faunal record with the percentage of barbed points in the total number of bone and antler points in all Upper Magdalenian sites on the northern side of the Pyrenees (Pétillon 2009). However, the number of sites where few and reliable and available data is few and no significant correlation was observed between the frequent use of barbed points and the abundance of a specific game type. A similar analysis made on the Upper Magdalenian sites of the Cantabrian Coast also yielded negative results (Fano et al. 2013). As stated in Pétillon 2009, enlarging the sample to other regions beyond the Pyrenees and the Cantabrian Coast might change the picture: for example, the Upper Magdalenian layer of Bois-Ragot (Vienne, west-central France) yielded both a very high percentage of barbed points and a faunal spectrum largely dominated by small game (birds, fish, and especially arctic hare: *Ibid*, 2009). The development of microscopic traceology on Magdalenian barbed points might also be a promising perspective (Fano et al. 2013).

We have seen that most of these small species were hunted in both Arctic and Patagonian contexts with barbed spears of all types and sizes and not with detachable harpoons. If the Magdalenian barbed points were indeed used for small game hunting, we must thus assume that these points were probably not detachable, despite their strong barbs (but actually not that strong considering the overall small size of the head) and their projections on the base (but these projections could also be used to secure the lashing of a fixed head on its shaft, or to prevent the shaft from splitting upon impact: Weniger 1992).

Theoretically, it is not possible to exclude that the Upper Magdalenian groups were among the very few hunter-gatherers who used detachable harpoons to catch terrestrial species. However, in the current state of the debate, fishing, fowling and small mammal hunting should probably be given specific attention as the most plausible hypothesis for the use of barbed elements in the Upper Magdalenian.

References

Balikci, A. (1970). *The Netsilik Eskimo*. Long Grove: Waveland Press.

Bird, J. (1980). Investigaciones arqueológicas en la isla Isabel, estrecho de Magallanes. *Anales del Instituto de la Patagonia, 11*, 75–85.

Borrero, L. A., & Borella, F. (2010). Harpoons and travellers: Fuegian ethnographic collections and the recent archaeological record. *Before Farming, 1*, 1–14.

Boudadi-Maligne, M., Mallye, J.-B., Langlais, M., & Barshay-Szmidt, C. (2012). Des restes de chien magdaléniens à l'abri du Morin (Gironde, France): implications socio-économiques d'une innovation zootechnique. *Paléo, 23*, 39–54.

Bridges, T. (1998). *Los indios del Último Confin sus escritos para la South American Missionary Society*. Ushuaïa: Zagier & Urruty Publications.

Cochard, D., & Brugal, J.-P. (2004). Importance des fonctions de site dans les accumulations paléolithiques de léporidés. In J.-P. Brugal & J. Desse (Eds.), *Petits Animaux et Sociétés Humaines. Du complément alimentaire aux ressources utilitaires, Actes des 24e Rencontres internationales d'Archéologie et d'Histoire, Antibes, 2003* (pp. 283–296). Antibes: APDCA.

Cooper, J. M. (1917). *Analytical and critical bibliography of the tribes of Tierra del Fuego and adjacent territory*. Washington: Government Printing Office.

Corchón, M. S., Mateos, A., Álvarez Fernández, E., Peñalver, E., Delclòs, X., & van der Made, J. (2008). Ressources complémentaires et mobilité dans le Magdalénien cantabrique. Nouvelles données sur les mammifères marins, les crustacés, les mollusques et les roches organogènes de la Grotte de Las Caldas (Asturies, Espagne). *L'Anthropologie, 112*, 284–327.

Costamagno, S., Cochard, D., Ferrié, J.-G., Laroulandie, V., Barbaza, M., Cazals, N., et al. (2008). Nouveaux milieux, nouveaux gibiers, nouveaux chasseurs ? Évolution des pratiques cynégétiques dans les Pyrénées du Tardiglaciaire au début du Postglaciaire. *Bulletin de la Société Préhistorique Française, 105*, 17–27.

Costamagno, S., & Laroulandie, V. (2004). L'exploitation des petits vertébrés dans les Pyrénées françaises du Paléolithique au Mésolithique: Un inventaire taphonomique et archéozoologique. In J.-P. Brugal & J. Desse (Eds.), *Petits animaux et sociétés humaines. Du complément alimentaire aux ressources utilitaires, Actes des 24e Rencontres internationales d'Archéologie et d'Histoire, Antibes, 2003* (pp. 403–416). Antibes: APDCA.

Costamagno, S., Laroulandie, V., Langlais, M., & Cochard, D. (2009). Exploitation du monde animal sur le versant nord des Pyrénées durant le Tardiglaciaire. In J. Fullola, N. Valdeyron, & M. Langlais (Eds.), *Els Pirineus i les àrees circumdants durant el Tardiglacial. Mutacions i filiacions tecnoculturals evolucion paleoambiental, Actes du XIVe Colloque international d'Archéologie de Puigcerdà, Hommage à G. Laplace, Novembre 2006* (pp. 185–209). Puigcerdà: Institut d'Estudis Ceretans.

Duplessis. (2003). *Périple de Beauchesne à la Terre de Feu (1698-1701)*. Paris: Transboréal.

Emperaire, J. (1955). *Les Nomades de la mer*. Paris: L'Espèce Humaine, NRF, Gallimard.

Estévez, J., & Vila, A. (2013). Analysis of Paleolithic barbed points from the Mediterranean Coast of the Iberian Peninsula: An ethnoarchaeological approach. In A. Pastoors & B. Aufferam Mettmann (Eds.), *Pleistocene Foragers on the Iberian Peninsula: Their culture and environment* (pp. 283–307). Wissenschaftliche Schriften des Neanderthal Museums 7.

Fano, M. A., Clemente, I., & Rivero, O. (2013). Apuntes en torno al uso de los arpones magdalenienses: primeras observaciones microscópicas a partir de los materiales de El Horno (Ramales de la Victoria, Cantabria). *Trabajos de Prehistoria, 70*, 332–345.

Gallardo, C. R. (1910). *Los Onas*. Buenos-Aires: Cabaut y Cia ed.

Gusinde, M. (1986). [1° ed. in German 1937]. *Los Indios des Tierra del Fuego. T. 2: Los Yamana*. Buenos Aires: Centro Argentino de Etnología Americana ed.

Hyades, P. D. J., & Deniker, J. (1891). *Anthropologie et Ethnographie. Mission Scientifique du Cap Horn, 1882-1883.* T. VII. Paris: Gauthiers-Villars.

Julien, M. (1982). *Les harpons Magdaléniens*. Paris: CNRS.

Laming-Emperaire, A. (1972). Pêcheurs des archipels et chasseurs des pampas. *Objets et Monde, XII*(2), 167–184.

Langlais, M., Costamagno, S., Laroulandie, V., Pétillon, J.-M., Discamps, E., Mallye, J.-B., et al. (2012). The evolution of Magdalenian societies in South-West France between 18,000 and 14,000 cal BP: Changing environments, changing tool kits. *Quaternary International, 272–273*, 138–149.

Langley, M. C. (2013). *Investigating maintenance and discard patterns for Middle to Late Magdalenian antler projectile points*. Ph.D. Dissertation, University of Oxford.

Langley, M. C., Pétillon, J.-M., & Christensen, M. (2016). Diversity and evolution of osseous hunting equipment during the Magdalenian (21,000–14,000 cal. BP). In M. C. Langley (Ed.), *Osseous projectile weaponry: Towards an understanding of Pleistocene cultural variability* (pp. 143–159). Dordrecht: Springer.

Lartet, E. (1861). Nouvelles recherches sur la coexistence de l'Homme et des grands mammifères fossiles réputés caractéristiques de la dernière période géologique. *Annales des Sciences naturelles, Zoologie, XV*, 177–253.

Lefèvre, A. (2011). Les pointes barbelées magdaléniennes. Étude typologique, géographique et chronologique. Unpublished Mémoire de Master 2, Université Toulouse Le Mirail.

Legoupil, D. (1981). La chasse aux cachalots aux Açores. *Le Chasse-Marée, 1*, 3–14.

Legoupil, D. (1989). *Ethno-archéologie dans les archipels de Patagonie: Les nomades marins de Punta Baja*. Mémoire n° 84. Paris: ADPF, Ed. Recherches sur les Civilisations.

Legoupil, D. (1993–1994). El archipíelago del Cabo de Hornos y la costa sur de la isla Navarino: poblamiento y modelos económicos. *Anales del Instituto de la Patagonia,* Ser. Cs. Hs., *22*, 101–121.

Legoupil, D. (1997). *Bahia Colorada: Les premiers chasseurs de mammifères marins de Patagonie australe*. Paris: ADPF, Ed. Recherches sur les Civilisations.

Legoupil, D. (2003). Les chasseurs-cueilleurs de Ponsonby (Patagonie australe) et leur environnement du VIème au IIIème mill. av. J.-C., *Magallania*, Tirada especial (Documentos), 31, bilingue (français/espagnol), Punta Arenas, Chili: Universidad de Magallanes.

Legoupil, D., Bearez, P., Lefèvre, C., San Román, M., & Torres, J. E. (2011). Estrategias de aprovisionamiento de cazadores recolectores de isla Dawson (estrecho de Magallanes) durante la segunda mitad del Holoceno: primeras aproximaciones. *Magallania, 39*(2), 153–165.

Leroi-Gourhan, A. (1946). *Archéologie du Pacifique nord. Matériaux pour l'étude des relations entre les peuples riverains d'Asie et d'Amérique*. Paris: Travaux et Mémoires de l'Institut d'Ethnologie.

Lothrop, S. K. (2002). [1928]. *The Indians of Tierra del Fuego*. New York: Museum of the American Indian, Heye Foundation.

Lovisato, D. (1883). *Di alcune armi e utensili dei Fueghini, e degli antichi patagoni. Reale Academia dei Linci, Serie 3.11, Memorie della Classe di scienze morali, storiche e filologiche, CCLXXX* (pp. 194–202). Roma: Academia dei Linci.

Mason, O. T. (1902). *Aboriginal American harpoons: A study in ethnic distribution and invention*. From United States National Museum Annual Report, 1900: 189–304. Washington, DC: Washington Government Printing Office.

Mathiasen, T. (1927). *Archaeology of the central Eskimos. Reports of the 5th Thule expedition 1921–24*. Copenhague: Gyldendal.

Molina, M. J. (1966–71). Arpones monodentados de la Patagonia méridional. *Acta Prehistórica, VIII/X*, 173–179.

Orquera, L. A., Legoupil, D., & Piana, E. L. (2011). La adaptación litoral en el extremo meridional de la Patagonia. *Quaternary International, 239*, 61–69.

Pétillon, J.-M. (2009). What are these barbs for? Preliminary reflections on the function of the Upper Magdalenian barbed weapon tips. In J.-M. Pétillon & M.-H. Dias-Meirinho (Eds.), *Recherches sur les armatures de projectiles du Paléolithique supérieur au Néolithique/ Research on projectile tips from the Upper Palaeolithic to the Neolithic, Palethnologie, 1* (pp. 69–102). Toulouse: Traces.

Piana, E. (1984). *Arriconamiento o adaptación en Tierra del Fuego*. Ensayos de Antropología Argentina. Buenos Aires. Editorial de Belgrano.

San Román, M. (2004). *L'exploitation des mammifères chez les chasseurs maritimes du site de Bahía Buena: économie des anciens canoeros de Patagonie* (Détroit de Magellan). Mémoire du D.E.A. Université de Paris I, Panthéon-Sorbonne.

San Román, M. (2010). La explotación de recursos faunísticos en el sitio Punta Santa Ana 1: Estrategias de subsistencia de grupos de cazadores marinos tempranos de Patagonia meridional. *Magallania, 38*(1), 183–198.

Scheinsohn, V. (1997). *Exploitación de materias primas oséas en la isla Grande de Tierra del Fuego*. Tesis de doctorado, Universidad Nacional de Buenos-Aires.

Scheinsohn, V. (2010). The good, the bad and the ugly: Prehispanic Harpoon heads from Beagle Channel, Isla Grande de Tierra del Fuego (Patagonia, Argentina). In A. Legrand-Pineau & I. Sidera (Eds.), *Ancient and Modern Bone Artefacts from America to Russia: Cultural, technological and functional signature, 6th meeting of the WBRG was held from the 27th –31st of August 2007. British Archaeological Reports International Series 2136* (pp. 295–301). Oxford: Archaeopress.

Serangeli, J. (2003). La zone côtière et son rôle dans les comportements alimentaires des chasseurscueilleurs du Paléolithique supérieur. In M. Patou-Mathis & H. Bocherens (Eds.), *Le Rôle de l'Environnement dans les Comportements des Chasseurs-cueilleurs Préhistoriques. British Archaeological Reports International Series 1105* (pp. 67–82). Oxford: Archaeopress.

Spears, J. R. (1895). *The gold digging of Cape Horn: A study of life in Tierra del Fuego and Patagonia*. New York: The Knickerbocker Press.

Torres, J. E. (2009). La pesca entre los cazadores recolectores terrestres de la isla grande de Tierra del Fuego, desde la prehistoria a tiempos etnográficos. *Magallania, 37*(2), 109–138.

Weniger, G.-C. (1992). Function and form: An ethnoarchaeological analysis of barbed points from northern; hunter-gatherers. In F. Audouze (Ed.), *Ethnoarchéologie: Justification, problèmes, limites, XIIe rencontres internationales d'archéologie et d'histoire d'Antibes, Juan-les-Pins* (pp. 257–268). Juan-les-Pins: APDCA.

Weniger, G.-C. (1995). *Widerhakenspitzen des Magdalénien Westeuropas. Ein Vergleich mit ethnohistorischen Jägergruppen Nordamerikas*. Mainz-am-Rhein: von Zabern.

Weniger, G.-C. (2000). Magdalenian barbed points: Harpoons, spears and arrowheads. In C. Bellier, P. Cattelain & M. Otte (Eds.), *La chasse dans la Préhistoire/Hunting in Prehistory, Actes du colloque international de Treignes, 3–7 octobre 1990* (pp. 79–87). Bruxelles: Société royale belge d'Anthropologie et de Préhistoire (Anthropologie et Préhistoire, 111)/Université de Liège - Service de Préhistoire (ERAUL, 51)/CEDARC (Artefacts, 8).

Woersaae, J. J. A. (1875). Communication relative à l'authenticité des os de renne présentant des dessins et trouvés en France dans les cavernes du Périgord. In V. Schmidt (Ed.), *Congrès international d'anthropologie et d'archéologie préhistoriques: Compte-rendu de la 4e session*, Copenhague, 1869 (pp. 127–134). Copenhague: Ed Thiele.

Index

A
Abris Blanchard, 63
Abris Castanet, 63
Abri Pataud, 98
Abri Poisson, 61, 62, 66
Abrigo do Lagar Velho, 131–135, 140
Ach Valley, 71, 74, 79, 81, 82, 84
Adhesive, 6, 16, 20, 25, 35, 78, 121, 148, 162, 210
Africa
 East Africa, 3
 North Africa, 4, 15, 18, 23, 24, 109, 127
 South Africa, 4–6, 24, 25, 34, 173
Agate Basin Site, 222
Ahrensburgian, 6, 152
Aire 2, 215
Aizpea, 127
Algarve, 3, 131, 138–140
Altamira, 144, 155
Ambrosio, 140
Anbangbang, 215
Andernach-Martinsberg, 5, 144, 148
Antler projectiles, 67, 127
Anzick, 222, 223, 226, 228, 230–232
Apollo 11, 24, 25
Aquitaine sites, 150
Archaic human, 16
Arcy-sur-Cure, 89, 91–93, 95, 98, 101
Art mobilier. *See* Portable Art
Aterian, 4, 15, 18, 22, 23
Atlatl. *See* Spearthrower
Aurensan, 144, 151, 152
Aurignacian, 5, 43, 45, 46, 49–52, 55–60, 63–67, 75–77, 82, 83, 89, 90, 93, 97–103, 137, 138, 151, 155, 163, 228, 249
 early Aurignacian, 55, 56, 58–60, 63, 65, 67, 78, 100
 evolved Aurignacian, 56, 64, 65, 100
 proto-Aurignacian, 46, 52, 55, 56, 100
Avdeevo, 103, 104
Awl, 15, 16, 34, 46, 50, 59, 66, 73, 111, 112, 123, 151, 161, 163, 186, 191, 203, 204, 209, 212, 213, 216, 221
Azilian, 4, 103

B
Backed bladelets, 50, 138
Badegoulian, 59, 110–112, 119, 122, 127, 151, 167
Baguette demi-ronde, 132, 133, 135, 137, 139, 145, 147, 162
Balangoda culture, 174
Barbed point, 3–5, 17, 18, 55, 100, 144–152, 155, 241, 243–246, 248–250
Barbed spear, 33, 211, 212, 215, 216, 241–244, 246–250
Bass point, 215
Batadomba-lena, 174–187
Bâtons percés, 90
Batu Ejaya, 202
Bayacian, 90
Bead, 24, 25, 58, 79, 174, 186, 190, 244
Bédeilhac, 144, 155
Bellan-bandi Palassa, 174, 186
Biconoical points, 93
Biface, 66, 222, 231–233
Bipoints, 4, 6, 94–95, 99, 110, 111, 114, 115, 122, 126, 127, 138, 193–199, 201, 203–206, 214, 215
Birds, 6, 20, 123, 126, 140, 148, 155, 192, 203, 204, 206, 212–214, 216, 241, 242, 249, 250
Blackwater Draw Locality No. 1, 221, 222, 224, 226–228, 231, 233
Bladelet, 32, 46, 50, 103, 138, 146–148
Blombos Cave, 5, 19–24, 32, 173
Blowgun/blowpipe, 3, 183, 203
Bockstein-complex, 71
Bola Batu, 202
Bone
 avian/bird, 212, 214, 216
 cave, 212
 mammal, 51, 58, 126, 132–136, 138–140, 238
 mammoth, 221, 222, 233
 marine/cetacean/whale, 4, 5, 146–148, 241
 point functional diversity, 31–37
 projectile points, 162, 217
Boomerang, 1, 6, 210, 212
Boomplaas, 22
Bora Gran d'en Carreras, 109, 110, 123, 127
Border Cave, 5, 22–24, 32, 35
Bourrouilla, 155
Bow, 1, 3, 5, 6, 20, 22, 31, 34, 37, 63, 84, 102, 104, 150, 152–154, 162, 237, 238, 243, 248
Brassempouy, 91, 103, 144
Brillenhöhle, 71, 72, 74, 77–83
Broken Hill, 17, 55
Bruniquel, 144, 155, 249
Bukovac pećina, 43, 48, 49, 52
Buraca Grande, 131–133, 135, 139, 140
Burial, 24, 136, 137, 216, 232

Michelle C. Langley (ed.), *Osseous Projectile Weaponry*, Vertebrate Paleobiology and Paleoanthropology,
DOI 10.1007/978-94-024-0899-7

Burin, 90, 91, 97–99, 110, 163
Burnisher, 80, 151
Bushmen, 20, 31, 34
Bushmen Rockshelter

C
Canoe, 214, 238, 241, 243, 244
Canoeros, 238, 240, 243, 244, 249, 250
Cantabria, 4, 50, 52, 60, 97, 109, 110, 123, 127, 155, 249, 250
Cap Blanc, 144, 150
Casa da Moura, 131, 133, 136, 137, 139
Cassowary Bone Knife, 202, 205
Cave Bay Cave, 213
Cave of Hearths, 25
Ceramics, 1, 31, 33, 37, 138, 204
Cetacean, 5, 147, 148, 238, 242, 243, 250
Chaîne opératoire concept, 166, 215
Châtelperronian, 56
Chattermarks, 36, 121
Chronological evolution, 98
Cierro, 57, 59
Clacton-on-Sea, 55
Cloggs Cave, 212
Clovis, 221–223, 226, 227, 229–233
  culture/period, 222
  point, 222, 223, 226, 227, 231–233
Cognition, 15, 16, 24, 25, 173
Combe-Buisson, 144, 155
Combe-Saunière I, 144, 153
Community of practice, 2, 5
Conde, 57, 59, 60
Cordage, 190, 200–202, 206
Core, 46, 110, 123, 143
Courbet, 144, 151
Cuzoul de Vers, 167
Cova de les Cendres, 109, 110, 123, 125, 127, 140
Cova del Parco, 109
Cova de Parpalló, 109, 110, 112, 113, 115–119, 121–123, 125, 127, 140
Cova dels Blaus, 109, 123
Cova Matutano, 109, 123
Covalejos, 59–61, 63
Crossbow, 63, 237
Cueto de la Mina, 144, 155
Cueva del Higuerón, 123
Cueva de Nerja, 109, 110, 123, 125–127, 140
Cueva Victoria, 123, 213, 214
Cultural variability, 1–7, 31
Cuzoul de Vers, 167

D
Daeo 2, 203
Dart. *See* Spear
Dawson, 242, 243
Decoration, 32, 37, 121, 122, 137, 139, 154, 155, 166, 244
Deer
  red deer, 4, 58, 59, 74, 81, 82, 111, 119, 121, 123, 125, 132, 135, 138, 140, 155, 250
  reindeer, 4, 58, 73, 74, 79, 81, 82, 90, 93, 96–98, 123, 144, 145, 147, 148, 153, 154, 164, 166, 167, 222, 250
Demography, 16, 25, 37, 102, 191, 205
Denticulate point, 110
Devil's Lair, 212, 213, 215
Diepkloof, 23, 24
Digging stick, 24, 210
Dog, 195, 197, 243
Dolphin, 250
Domestic animals, 138
Dorset culture, 231
Double-bevelled based point, 92, 96–98, 122, 145, 151, 152
Double longitudinal grooving (DLG), 59, 101–103
Drake, 232, 242
Dufour bladelets, 50, 138
Dugong, 214
Durras North, 214, 215

E
Early bone technology, 173–186
Early Paleoindian period, 221
Early Stone Age, 15–25
East Wenatchee/Richey–Roberts Clovis Cache, 223, 224, 226, 228, 229, 231, 232
Edge tool, 194, 202–203
El Castillo, 57–59, 61–63
Éléments bipointe, 148
El Espertín, 127
El Mnasara, 4, 5
Engravings, 16, 24, 46, 50, 79, 100, 104, 146, 151, 154, 155, 204, 205, 212, 215
Enlène, 144, 150, 155
Epigravettian, 46, 50, 52, 155
Epipaleolithic, 4
Espalungue, 155
Espélugues, 144, 155
Ethnography, 1–3, 5, 7, 16, 20, 21, 24, 32, 34, 103, 127, 155, 167, 185, 202, 205, 211, 213, 228, 230, 238, 240, 243–246, 249, 250
Etiolles, 144, 150
Expedient tools, 193–194

F
Fa Hien-lena, 174, 186
Federmessergroupen, 7
Female hafting system. *See* Hafting
Fish, 7, 19, 135, 139, 140, 155, 181, 185, 192, 204, 212–215, 221, 241–243, 249, 250
Fishhook, 1, 4, 6, 7, 99, 135, 140, 214, 215
Fishing, 4–7, 18, 140, 155, 183, 185, 209, 213–216, 241–243, 248–251
Float, 213, 214, 241
Folsom period, 222
Fontirobertian, 90
Font-Robert points, 90
Foradada d'Oliva, 123
Foreshaft, 1, 5, 97, 147, 148, 151, 162, 210, 221–228, 231, 233, 246, 249
Fork-Based Point, 123, 145, 149, 151, 154, 161
Fracturation, 102, 111, 113, 120, 151
Fromm's Landing, 215

G
Gandil, 144
Gazel, 144, 148
Geissenklösterle, 60, 61, 78
Golo Cave, 203–205
Gönnersdorf, 148
Gorge, 6, 43, 110, 125, 127, 185, 209, 214
Gourdan, 144, 150, 155

Grand Abri de Laussel, 96
Gravettian, 5, 6, 58, 60, 77, 83–84, 89–104, 132, 135–140, 153, 163
  early, 90, 92, 93, 95–98, 100, 101, 104
  final, 90, 92–98, 100, 101, 136
  middle, 92, 93, 96, 98
  recent, 90–98, 100, 101
Groove-and-snap technique, 190, 199, 205
Groove-and-Splinter, 37, 79, 101, 144, 147, 150, 180
Grosse Grotte, 74
Grotte de l'Abbe, 144, 148
Grotte du Bois-Ragot, 7, 144, 148, 150, 151
Gruta da Furninha, 131, 133, 135
Gruta das Salemas, 131, 133, 137–140
Gruta do Caldeirão, 131, 133, 135
Gua Mo'o hono, 191–195, 197, 202, 203
Gua Sambagowala, 192
Gua Talimbue, 191–193, 196, 202

H
Hafting
  contact system, 98
  female hafting system, 246
  male hafting system, 97, 249
  pin-hafting, 246, 249
  socket hafting, 94–95, 100, 144, 149, 226
Hardwood, 210, 214, 216
Harpons Cave, 144, 148
Harpoon, 3, 5, 18, 19, 22, 24, 96, 110, 123, 125–127, 133, 135, 136, 149, 161, 176, 205, 214, 237, 238, 240, 242–246, 248, 249, 251
Hauterive-Champréveyres, 167
Heat treatment, 16
Heinrich 4 Climatic Event, 66–67
Hohle Fels, 71, 72, 74, 76–82, 84
Hohlenstein-Stadel, 71, 72, 74, 75
Holmegaard, 6
Hominin, 2, 16, 24, 25, 44–46, 50, 189, 190
Homo ergaster, 16
Homo heidelbergensis, 17
Hornos de la Peña, 59
Horse, 79, 81, 84, 119, 154, 155, 250
Howiesons Poort (HP), 5, 15, 20, 21, 32, 173
  post-Howiesons Poort, 15
Hoyo de la Mina, 109, 123
Hunter-gatherers, 1, 2, 7, 20, 21, 24, 31, 32, 37, 67, 139, 140, 154, 186, 209, 251
Hunting kits, 5, 21, 144–149

I
Iberia, 58, 109, 110, 127, 140, 144, 249, 250
Iberomaurusien, 4
Ishango, 4
Isturitz, 4, 59, 61, 63, 64, 66, 90, 92–98, 100, 101, 103, 144, 147–151, 154, 155
Isturitz point, 5, 91, 92, 94–95, 98–100
Ivory, 1, 4–7, 16, 57–60, 71, 73, 74, 77–79, 81–84, 90, 91, 93, 97, 103, 104, 148, 161, 167, 189, 224, 237, 246
  mammoth, 4, 6, 7, 57, 73, 78, 82, 233

J
Javelin, 212, 222, 225, 237, 241–243
Jerimalai, 6
Jwalapuram, 176

K
Katanda, 4, 5, 17–19, 22, 24
Kebara, 4
Keilor, 212
Khok Phanom Di, 203
Kitulgala Beli-lena, 174, 186
Klasies River Mouth, 20, 21
Klein Kliphuis, 24
Klipdrift, 24
Kokorevo I, 155
Koongine Cave, 213
Kostienki-Avdeevo culture, 103, 104
Kria Cave, 204, 205
!Kung, 37
Kutikina, 213
Kwakiutl, 249

L
Labastide, 144, 148
Labattut shelter, 90, 92–95
Labeko Koba, 57–59, 61, 66
La Ferrassie, 55, 90, 95
La Garenne, 64, 144, 150, 162–167
La Gravette, 90, 92, 93, 96, 99, 100
Lake Mulurulu, 212
Lake Mungo, 216
La Madeleine, 144, 150–152
Lang Rongrien, 190
Lapa da Rainha, 131, 133, 135, 137, 139, 140
Lapa dos Coelhos, 131, 133, 135, 138, 140
La Paloma, 144, 149
La Quina-Aval, 59, 61–63, 66
La Roque Saint-Christophe, 96
Las Caldas, 144, 149
Lascaux, 144
Last Glacial Maximum (LGM), 82, 84, 89, 110, 127, 138, 140, 143, 176, 216, 217
Later Stone Age (LSA), 4, 18–21, 24, 33, 37
Laugerie-Basse, 144, 150, 151
Laugerie-Haute, 55, 90–98, 100, 101, 103, 148, 228, 229
La Vache, 127, 144, 147–152
Le Facteur, 92–95
Le Flageolet, 93, 99
Le Fourneau du Diable, 92, 97, 98
Leister, 213, 215, 238, 240, 242, 243, 248, 249
Lene Hara, 6
Le Placard, 167
Lespugue, 148, 155
Les Rideaux cave, 97
Les Vachons, 92, 93, 95, 96, 100
Levallois, 45, 51, 66
Levant, 4
Liang Lemdubu, 205
Liang Nabulei Lisa, 205
Ligature, 6, 96, 97, 166
Lind Coulee site, 222
Linkshaft, 19
Lone Valley, 71, 81, 82
Lussac-Angles point, 145, 148

M
M82/6, 212
Macropod, 212, 215, 216

Magdalenian, 3–5, 50, 52, 60, 64, 90, 93, 97, 98, 103, 124, 127, 132–140, 143–156, 163, 228, 238, 246, 249–251
early, 110
late, 4, 5, 7, 147, 148, 151, 166
middle, 5, 137, 143, 147–148, 150, 153, 154, 250
protomagdalenian, 90, 228
Malangangerr, 215
Male hafting system. *See* Hafting
Mammoth, 4, 6, 7, 57, 71, 73, 74, 77–84, 104, 221, 222, 227, 231, 233
Manis site, 229
Mannalargenna, 213
Mannheim, 144, 152
Maros point, 191, 206
Marsoulas, 148
Mas d'Azil, 137, 144, 148, 155, 249
Massat, 144, 155
Massive-Base point, 46, 49, 50, 52, 64, 65, 74, 79–83, 148
Mastic, 4, 190, 227
Mastodon, 229, 230
Matja Kuru, 186, 190
Mejillones, 123, 140
Mesolithic, 127, 138, 152, 155
Microlith, 33, 145, 149, 174, 186, 191, 203, 205
Middle Stone Age (MSA), 4, 15–25, 31, 55, 185
Minahasa, 191
Mladeč points, 46, 49, 51, 52
Modern human, 1, 3, 7, 15, 16, 24, 25, 46, 51, 52, 66, 67, 84, 173, 190, 191, 206, 249
Modern human behavior, 15, 16, 24, 189
Molluscs. *See* Shellfish
Monkey, 181, 183, 185, 204
Montconfort, 144, 148
Moravany-Zakovska, 96
Morín Cave, 57, 61, 65
Morphometric design, 60
Mousterian, 15, 43–46, 51, 56, 66, 74, 136, 138
Mudukian, 214
Mumbwa Cave, 17
Musang Cave, 203
Mutations, 100–104

N
Natufian, 4, 127, 163
Nawmoyn, 215
Neanderthal, 3, 16, 44, 45, 50–52, 56, 59, 66, 67, 74
Needle, 34, 112, 123, 126, 127, 162, 165, 198, 209, 215
Nelson Bay Cave, 21, 22
Neolithic, 136–138, 192
Niah Caves, 173, 174, 185, 186, 190, 202, 204–206
Noailles burins, 90, 97, 99
Noaillian, 90, 93–98, 100, 101
North America, 5, 6, 223–225, 229, 238, 242, 244, 246, 249

O
Oakhurst, 35, 37
Oblazowa, 6
Oenpelli, 185
Offing, 245
Olschewian, 49–51
Osseous hunting equipment, 143–156
Osseous points, 3, 4, 43–52, 55, 83, 84, 91, 92, 98–104, 110, 122, 139, 140, 149, 155, 203, 206, 237, 238, 246
Ostrich eggshells, 24, 25
Otter, 241–243, 250

P
Page-Ladson, 222
Panisi Ta'buttu, 202
Paso, 191, 203
Pavlov I, 102, 104
Pavlovian, 96, 102, 103
Peers Cave/Skildergatkop, 19–21
Pendant
Penguin, 242
Perak, 202
Perlis, 202
Personal ornaments, 16, 24, 50, 58, 78, 84, 174
Petit-Cloup-Barrat, 144, 167
Pigment, 16, 23, 24, 56, 151, 173
Pike, 237, 241, 248
Pincevent, 144, 149, 150
Pin-hafting. *See* Hafting
Pinniped. *See* Seal
Pins, 161, 227, 249
Pirulejo, 109, 140
Points *à Base Machonée*, 81
Points *à Base Raccourcie*, 113, 121, 123
Poison, 5, 19, 20, 24, 32, 35, 203
Ponsonby, 245
Portable art/art mobilier, 25, 58, 138
Potočka zijalka, 51, 52, 74
Poulton-le-Fylde, 155
Predmost, 103
Pressure flaker, 111, 221, 224, 231
Pulau Balambangan, 194
Pyrenean, 147–150

R
Raclage en Diabolo/Raccourcie, 81, 99, 120
Rainforest, 174, 183, 185, 186
Raysse burins, 90, 98
Rayssian, 90, 91, 93
Recycling, 51, 64, 121, 147, 149–153, 162, 166, 167
Reindeer cave, 98
Re-segmenting waste-products, 163, 166, 167
Reshaping, 60, 61, 63, 64, 206
Resharpening, 60, 61, 63, 64, 102, 113, 117, 120, 121, 126, 138, 151, 194, 202, 215
Re-splitting procedure, 102
Retoucher, 22, 46, 59, 66
Rhinoceros, 81
Rio Chico 3, 245
Robberg, 35, 37
Roc-aux-Sorciers, 148
Roc de Combe-Capelle, 96
Rock Art, 34, 152, 209–217
Rocky cape, 215
Roonka Flat, 216

S
Saint-Germain-La-Rivière, 144, 145, 147, 150
Saint Michel d'Arudy, 144, 150, 154, 155
Salmon, 7, 249
Salzgitter-Lebenstedt, 74
San, 19–24, 31
Šandalja II, 43, 48–52
Schöningen, 55
Scilles, 144, 145
Scraper, 33, 45, 46, 50, 51, 110, 214

Scraping marks, 198
Seal, 241, 242, 244, 246, 248, 250
Sea Lion, 241, 242, 244, 250
Self-barbed points, 144, 146, 148, 149
Shaving marks, 198
Sheaman site, 222, 226
Shell, 1, 6, 24, 25, 56, 58, 140, 174, 183, 185, 186, 191, 206, 210, 213–216, 242
Shellfish, 33, 140, 213, 215
Sheriden Cave, 222, 228–230, 232, 233
Sibudu Cave, 5, 19–25, 32, 173
Simple-based point, 92, 93, 95, 98, 99, 103, 111, 113, 114, 120, 121, 135–139
Single-bevelled based points, 77, 80–82, 92, 94–98, 115, 149, 151, 163
Sirgenstein, 71, 72, 74, 79
Siti Nafisah, 203
Sled-shoes, 231
Sloth Hole, 222
Social interaction, 155
Social signaling, 217
Socket hafting. *See* Hafting
Solutrean, 5, 93, 96–98, 110, 132–134, 136–140, 151, 153
  proto-Solutrean, 93, 97, 131, 133–134, 138
Song Terus, 190
Sorcery, 209, 216
South America, 209, 238, 244, 246, 249
Southeast Asia, 1, 4–6, 173, 174, 176, 186, 189–206
Spanish Aurignacian Projectile Points, 55–69
Spatula, 50, 151, 186, 190, 193, 194, 202, 205, 206, 215
Spear, 1, 3–5, 17–22, 33, 34, 55–60, 63, 64, 66, 78, 84, 98, 100, 122, 139, 153, 154, 161, 162, 189, 210–216, 221, 222, 227, 231, 237, 241–250
Spear-straightener, 1
Spearthrower (Atlatl, Woomera), 1, 3, 5, 6, 63, 153–155, 162, 209, 210, 212, 213, 216, 237, 248
Split-based points, 5, 55–57, 60–64, 66–67, 83, 100, 102, 161
Stellmoor, 6, 144, 152
Stencil, 210, 212, 216
Still Bay, 5, 15
  pre-Still Bay, 5, 15, 19
Still Bay (SB), 19–22, 173
Stillfried, 6
Stingray spines, 4, 204, 211, 216, 217
Stone plaquette, 139
Style, 2, 24, 123, 214, 217, 231
Suidae, 195, 196, 205
Sulawesi, 189–206
Swabian Jura, 60, 81–84
Symbolic behavior, 25, 50, 191
Szeletian, 50, 51

T
Taforalt, 4
Taillis-des-Coteaux, 144
Tierra del Fuego, 239
Tito Bustillo, 4, 144, 148
Toalean, 191, 202, 203, 205
Tongued piece, 63, 76, 78
Tossal de la Roca, 109, 123
Tower Hill Beach, 215
Trap/Trapping, 24, 185, 209, 215, 243
Trilobite cave, 89, 92, 95, 98
'True' Points versus Pointed Waste-Products, 161–167
Turtle, 214
Tusk tool, 193–195, 202, 205

U
Uattamdi, 203
Ulu Leang 1, 191, 202, 203
Umhlatuzana Rockshelter, 22
Unipoint, 193–199, 201, 204–206, 214
Use-trace study, 31, 34–37

V
Vale Boi, 131, 133, 134, 136–140
Velika pećina, 43, 46–49, 51, 52
Verberie, 144, 148, 150
Vindija, 43–47, 50–52
Vogelherd, 71, 72, 74, 76, 78, 79, 84
Volcán del Faro, 109, 123

W
Walandawe excavations, 189–206
Wareen, 213
Warfare, 1, 4–6
Wedges, 23, 63, 78, 80, 90, 147, 151, 155, 221, 224, 226, 227, 231–233
Weipa Shell Mounds, 214
Western Thule culture, 231
Whale, 5, 146–148, 241, 243, 244, 250
White Paintings Shelter, 4, 25, 32, 37
Willandra Lakes, 212
Wilton, 21, 35, 37
Winkle, 183
Wonderwerk, 25
Wood, 4–6, 24, 32, 35–37, 55, 59, 63, 78, 96, 97, 99, 101, 152, 166, 189, 194, 204, 210, 212–217, 224, 226–228, 231, 237, 238, 241, 249
Woomera. *See* Spearthrower
Wyrie Swamp, 210

MIX
Papier aus verantwortungsvollen Quellen
Paper from responsible sources
FSC® C105338

If you have any concerns about our products,
you can contact us on
ProductSafety@springernature.com

In case Publisher is established outside the EU,
the EU authorized representative is:
Springer Nature Customer Service Center GmbH
Europaplatz 3, 69115 Heidelberg, Germany

Printed by Libri Plureos GmbH
in Hamburg, Germany